Advanced Automotive Fault Diagnosis

DIAGNOSTICS: TEST DON'T GUESS

Learn all the skills you need to pass Level 3 and Level 4 Vehicle Diagnostics courses from IMI, City & Guilds, and BTEC, as well as ASE, AUR, and other higher-level qualifications.

Along with 25 new real-life case studies, this fifth edition of *Advanced Automotive Fault Diagnosis* includes new content on diagnostic tools and equipment: VCDS, decade boxes, scanners, pass-through, sensor simulators, breakout boxes, multimeter updates for HV use and more. It explains the fundamentals of vehicle systems and components, and it examines diagnostics principles and the latest techniques employed in effective vehicle maintenance and repair. Diagnostics, or faultfinding, is an essential part of an automotive technician's work, and as automotive systems become increasingly complex there is a greater need for good diagnostic skills.

Ideal for students, included throughout the text are useful definitions, key facts, and 'safety first' notes. This text will also assist experienced technicians to further improve their performance and keep up with recent industry developments.

Tom Denton has been researching and writing best-selling automotive textbooks for more than 25 years. His published work is endorsed by all leading professional organisations and used by automotive students and technicians across the world. His postgraduate education in all aspects of technology and education combined with many years of practical industry experience have given him a broad base from which to approach and teach this interesting, yet challenging, subject.

As a Fellow of the Institute of the Motor Industry and a Member of the Institute of Road Transport Engineers and the Society of Automotive Engineers, he keeps in contact with the latest technologies and innovations across all aspects of this fascinating industry. As well as publishing more than 30 textbooks, Tom has created amazing support materials and eLearning courses. Learn more about Tom's textbooks and his online Automotive Technology Academy at:

www.automotive-technology.org.

Advanced Automotive Fault Diagnosis

Automotive Technology: Vehicle Maintenance and Repair

Fifth Edition

Tom Denton

LONDON AND NEW YORK

Fifth edition published 2021
by Routledge
2 Park Square, Milton Park, Abingdon, Oxon, OX14 4RN

and by Routledge
52 Vanderbilt Avenue, New York, NY 10017

Routledge is an imprint of the Taylor & Francis Group, an informa business

First edition published by Elsevier 2000
Second edition published by Elsevier 2006
Third edition published by Routledge 2012
Fourth edition published by Routledge 2017

British Library Cataloguing-in-Publication Data
A catalogue record for this book is available from the British Library

Library of Congress Cataloging-in-Publication Data

Names: Denton, Tom, author.
Title: Advanced automotive fault diagnosis: automotive technology: vehicle maintenance and repair / Tom Denton.
Description: Fifth edition. | Abingdon, Oxon; New York, NY: Routledge, 2021. | Includes index.
Identifiers: LCCN 2020015717 (print) | LCCN 2020015718 (ebook) | ISBN 9780367330521 (paperback) | ISBN 9780367330545 (hardback) | ISBN 9780429317781 (ebook)
Subjects: LCSH: Automobiles–Maintenance and repair. | Fault location (Engineering)
Classification: LCC TL152 .D3934 2021 (print) | LCC TL152 (ebook) | DDC 629.28/7–dc23
LC record available at https://lccn.loc.gov/2020015717
LC ebook record available at https://lccn.loc.gov/2020015718

ISBN: 978-0-367-33054-5 (hbk)
ISBN: 978-0-367-33052-1 (pbk)
ISBN: 978-0-429-31778-1 (ebk)

Typeset in Sabon LT Std
by Cenveo® Publisher Services

Printed and bound in Great Britain by Bell & Bain Ltd, Glasgow

Contents

Preface

In this book, the fifth edition, you will find lots of useful information and methods relating to diagnosing faults on vehicles. It is the third in the series:

- *Automobile Mechanical and Electrical Systems*
- *Automobile Electrical and Electronic Systems*
- *Automobile Advanced Fault Diagnosis*
- *Electric and Hybrid Vehicles*
- *Alternative Fuel Vehicles*
- *Automated Driving and Driver Assistance Systems*

Ideally, you will have some experience before reading this book. But don't worry if you do not. All the basics are covered, and you will still learn a lot from it.

Lots of new content is included in this edition relating to tools and equipment as well as how to make the best use of this information. The final chapter is devoted almost entirely to case studies that I have chosen from those sent to me by technicians working in the real world. There are some great examples of the amazing work that is done, but more than anything they show the process of diagnostics to help students and new technicians learn the best methods.

DIAGNOSTICS: #TESTDONTGUESS

Comments, suggestions and feedback are always welcome at my website:

www.automotive-technology.org

On this site, you will also find lots of **free** online resources to help with your studies. Check out the final chapter for more information about the amazing resources to go with this and my other books. These resources work with the book and are ideal for self-study or for teachers helping others to learn.

We never stop learning, so I hope you find automotive technology as interesting as I still do.

Acknowledgements

Through the years many people have helped in the production of my books. I am therefore very grateful to the following companies who provided information and/or permission to reproduce photographs and/or diagrams:

AA
AC Delco
ACEA
Alpine Audio Systems
Audi
Autologic Data Systems
BMW UK
Bosch
Brembo brakes
CandK Components
Citroën UK
Clarion Car Audio
Continental
CuiCAR
Dana
Delphi Media
Eberspaecher
First Sensor AG
Fluke Instruments UK
Flybrid Systems
Ford Motor Company
FreeScale Electronics
General Motors
GenRad
Google (Waymo)
haloIPT (Qualcomm)
Hella
HEVT
Honda
Hyundai
Institute of the Motor Industry (IMI)
Jaguar Cars
Kavlico
Ledder
Loctite
Lucas UK
LucasVarity
Mahle
Matlab/Simulink
Mazda
McLaren Electronic Systems
Mennekes
Mercedes
MIT
Mitsubishi
Most Corporation
NASA
NGK Plugs
Nissan
Nvidia
Oak Ridge National Labs
Peugeot
Philips
PicoTech/PicoScope
Pierburg
Pioneer Radio
Pixabay
Porsche
Renesas
Rolec
Rover Cars
SAE
Saab Media
Scandmec
Shutterstock
SMSC
Snap-on Tools
Society of Motor Manufacturers and Traders (SMMT)
Sofanou
Sun Electric
TandM Auto-Electrical
Tesla Motors
Texas Instruments
Thrust SSC Land Speed Team
Toyota
Tracker
Tula
Unipart Group
Valeo
Vauxhall
VDO Instruments
Volkswagen
Volvo Cars
Volvo Trucks
Wikimedia
ZF Servomatic

If I have used any information, or mentioned a company name that is not listed here, please accept my apologies and let me know so it can be rectified as soon as possible.

Chapter 1

Introduction

1.1 DIAGNOSIS

1.1.1 What is needed to find faults?

Finding the problem when complex automotive systems go wrong is easy if you have the necessary knowledge. This knowledge consists of two parts:

- understanding of the system in which the problem exists;
- the ability to apply a logical diagnostic routine.

It is also important to be clear about these definitions:

- symptom(s) – what the user/operator/repairer of the system (vehicle or whatever) notices;
- fault(s) – the error(s) in the system that result in the symptom(s);
- root cause(s) – the cause(s) of the fault.

If a system is not operating to its optimum, then it should be repaired. This is where diagnostic and other skills come into play. It is necessary to recognise that something is not operating correctly by applying your knowledge of the system, and then by applying this knowledge further and combining it with the skills of diagnostics to be able to find out the reason.

DEFINITION

Diagnostics: The word 'diagnostics' centres around the principle of the four rights: Measuring the right thing, in the right way, at the right time, with the right tool. (Source: TechTopics, www.techtopics.co.uk)

The four main chapters of this book ('Engine systems', 'Chassis systems', 'Electrical systems' and 'Transmission systems') include a basic explanation of the vehicle systems followed by diagnostic techniques that are particularly appropriate for that area. Examples of faultfinding charts are also included. In the main text, references will be made to generic systems rather than to specific vehicles or marques. For specific details about a particular vehicle or system, the manufacturer's information is the main source.

DEFINITION

Diagnosis: The word 'diagnosis' comes from the ancient Greek word 'διάγνωσις', which means discernment. It is the identification of the nature and cause of anything. Diagnosis is used in many different disciplines, but all use logic, analysis and experience to determine cause and effect relationships. In automotive engineering, diagnosis is typically used to determine the causes of symptoms and solutions to issues.

KEY FACT

General diagnostic principles and techniques can be applied to any system, physical or otherwise.

Other chapters such as 'Sensors, actuators and oscilloscope diagnostics' and 'On-board diagnostics' are separated from the four previously mentioned chapters, because many operations are the same. For example, testing an inductive sensor is similar whether it is used on ABS or engine management.

An important note about diagnostics is that the general principles and techniques can be applied to any system, physical or otherwise. As far as passenger-carrying heavy or light vehicles are concerned, this is definitely the case. As discussed earlier, there is a need for knowledge of the particular system, but diagnostic skills are transferable (Figure 1.1).

Figure 1.1 Diagnostics in action.

1.2 SAFE WORKING PRACTICES

1.2.1 Introduction

Safe working practices in relation to diagnostic procedures, and indeed any work on a vehicle, are essential – for your safety as well as that of others. You only have to follow two rules to be safe:

Use your common sense – do not fool about.
If in doubt – seek help.

Further, always wear appropriate personal protective equipment (PPE) when working on vehicles.

The following section lists some particular risks when working with vehicle systems, together with suggestions for reducing them. This is known as risk assessment.

SAFETY FIRST

Aways wear appropriate personal protective equipment (PPE) when working on vehicles.

1.2.2 Risk assessment and reduction

Table 1.1 lists some identified risks involved with working on vehicles. The table is by no means exhaustive but serves as a good guide.

1.2.3 High voltage vehicles

This section covers some particular risks when working with electricity or electrical systems, together with suggestions for reducing them. This is known as risk assessment. The diagnostic process is no different but do not work on high voltage vehicles unless trained – stay safe.

DEFINITION

Risk assessment: A systematic process of evaluating the potential risks that may be involved in an activity or undertaking.

Table 1.1 Identifying and reducing risk

Identified risk	*Reducing the risk*
Battery acid	Sulphuric acid is corrosive, so always use good PPE – in this case overalls and if necessary rubber gloves. A rubber apron is ideal as are goggles if working with batteries a lot, particularly older types.
Electric shock	Ignition HT is the most likely place to suffer a shock – up to 25000V is quite normal. Use insulated tools if it is necessary to work on HT circuits with the engine running. Note that high voltages are also present on circuits containing windings due to back emf as they are switched off – a few hundred volts is common. Mains supplied power tools and their leads should be in good condition, and using an earth leakage trip is highly recommended.
Exhaust gases	Suitable extraction must be used if the engine is running indoors. Remember it is not just the CO that might make you ill or even kill you, other exhaust components could also cause asthma or even cancer.
Fire	Do not smoke when working on a vehicle. Fuel leaks must be attended to immediately. Remember the triangle of fire – (heat/fuel/oxygen) – do not let the three sides come together.
Moving loads	Only lift what is comfortable for you; ask for help if necessary and/or use lifting equipment. As a general guide, do not lift on your own if it feels too heavy.
Raising or lifting vehicles	Apply brakes and/or chock the wheels when raising a vehicle on a jack or drive on lift. Only jack under substantial chassis and suspension structures. Use axle stands in case the jack fails.
Running engines	Do not wear loose clothing – good overalls are ideal. Keep the keys in your possession when working on an engine to prevent others starting it. Take extra care if working near running drive belts.
Short circuits	Use a jump lead with an in-line fuse to prevent damage due to a short when testing. Disconnect the battery (earth lead off first and back on last) if any danger of a short exists. A very high current can flow from a vehicle battery – it will burn you as well as the vehicle.
Skin problems	Use a good barrier cream and/or latex gloves. Wash skin and clothes regularly.
Electric vehicle – high voltage electric shock	Do not work on these vehicles unless trained. Then carry out de-energisation before work and use appropriate high voltage protection such as electrical rubber gloves and insulated tools.

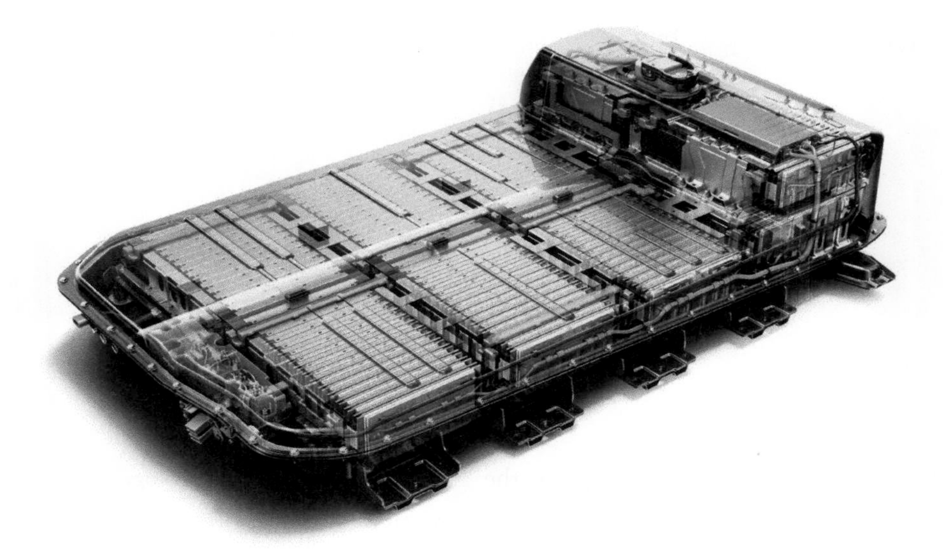

Figure 1.2 High voltage (HV) battery pack.

Source: Chevrolet Media.

Electric vehicles (pure or hybrid) use high voltage batteries (Figure 1.2) so that energy can be delivered to a drive motor or returned to a battery pack in a very short time. The Toyota Prius originally used a 273.6 V battery pack, but this was changed in 2004 to a 201.6 V pack. Voltages of 400 V are now common and some are now 800 V – so clearly, there are electrical safety issues when working with these vehicles.

EV batteries and motors have high electrical and magnetic potential that can severely injure or kill if not handled correctly. It is essential that you take note of all the warnings and recommended safety measures outlined by manufacturers. Any person with a heart pacemaker or any other electronic medical device should not work on an EV motor since the magnetic effects could be dangerous. In addition, other medical devices such as intravenous insulin injectors or meters can be affected.

SAFETY FIRST

EV batteries and motors have high electrical and magnetic potential that can severely injure or kill if not handled correctly.

The electrical energy is conducted to or from the motor via thick orange wires connected to the battery. If these wires have to be disconnected, SWITCH OFF or DE-ENERGISE the high voltage system. This will prevent the risk of electric shock or short circuit of the high voltage system.

The general advice about working on high voltage vehicles is split into four areas:

SAFETY FIRST

High voltage wires are always orange.

Note: Always follow manufacturers' instructions. It is not possible to outline all variations here.

Before work: You must:

- Turn OFF the ignition switch and remove the key.
- Switch OFF the Battery Module switch or de-energise the system.
- Wait for ten minutes before performing any maintenance procedures on the system. This allows any storage capacitors to be discharged.

During work: You should:

- Always wear insulating gloves when there is risk of touching a live component.
- Always use insulated tools when performing service procedures to the high voltage system. This precaution will prevent accidental short-circuits.

Interruptions: When maintenance procedures have to be interrupted while some high voltage components are uncovered or disassembled:

- Turn off the ignition and remove the key.
- Switch off the Battery Module switch.
- Prevent access to that area by untrained persons and prevent any unintended touching of the components.

Qualifiy:
Recognised award that includes pracitical competence at:
• Level 2
• Level 3
• Level 4

Join:
IMI Membership and gainTechSafe Professional Registration

Maintain:
Currency of competence
• General annual CPD
• TechSafe specific annual CPD
• Techsafe assessment every three years

Retain:
TechSafe™ professional registration

Figure 1.3 IMI TechSafe process.

After work: Before switching on or re-energising the battery module after repairs have been completed:

- Tighten a terminals to the specified torque.
- Make sure that no high voltage wires or terminals have been damaged or shorted to the body.
- Check the insulation resistance between each high voltage terminal of the part you disassembled and the vehicle's body.

For more information, please refer to Electric and Hybrid Vehicles – available from my website (www.tomdenton.org) or all good bookshops!

1.2.4 IMI TechSafe™

The IMI TechSafe™ professional registration scheme is designed to ensure complex automotive technologies are repaired safely and that technicians work safely. To be added to the register, a technician must successfully complete a specified qualification, join the IMI Professional Register and complete specified annual CPD to ensure current competency is maintained (Figure 1.3).

Safe working on modern technology covers all areas of the vehicle, but when working on high voltage systems, ADAS and connected vehicles, for example, current competence is essential. Technology safe means technician safe means customer safe. (tide.theimi.org.uk)

Figure 1.4 IMI TechSafe logo.

Table 1.2 Diagnostic terminology

Symptom	The effect of a fault noticed by the driver, user or technician
Fault	The cause of a symptom/problem
Root cause	This may be the same as the fault, but in some cases it can be the cause of it
Diagnostics	The process of tracing a fault by means of its symptoms, applying knowledge and analysing test results
Knowledge	The understanding of a system that is required to diagnose faults
Logical procedure	A step-by-step method used to ensure nothing is missed
Concern, cause, correction	A reminder of the process starting from what the driver reports, to the correction of the problem
Report	A standard format for the presentation of results

1.3 TERMINOLOGY

1.3.1 Introduction

The terminology included in Tables 1.2 and 1.3 is provided to ensure we are talking the same language. These tables are provided as a simple reference source.

1.4 REPORT WRITING

1.4.1 Introduction

As technicians you may be called on to produce a report for a customer. If you are involved in research of some kind, it is important to be able to present results in a professional way. The following sections describe the main headings that a report will often need to contain together with an example report based on the performance testing of a vehicle alternator.

Laying out results in a standard format is the best way to ensure all the important and required aspects of the test have been covered. Keep in mind that the

Table 1.3 General terminology

System	A collection of components that carry out a function.
Efficiency	This is a simple measure of any system. It can be scientific, for example, if the power out of a system is less than the power put in, its percentage efficiency can be determined (P-out/P-in × 100%). This could, for example, be given as say 80%. In a less scientific example, a vehicle using more fuel than normal is said to be inefficient.
Noise	Emanations of a sound from a system that is either simply unwanted or is not the normal sound that should be produced.
Active	Any system that is in operation all the time (steering for example).
Passive	A system that waits for an event before it is activated (an airbag is a good example).
Short circuit	An electrical conductor is touching something that it should not be touching (usually another conductor of the chassis).
Open circuit	A circuit that is broken (a switched off switch is an open circuit).
High resistance	In relation to electricity, this is part of a circuit that has become more difficult for the electricity to get through. In a mechanical system, a partially blocked pipe would have a resistance to the flow of fluid.
Worn	This word works better with further additions such as worn to excess, worn out of tolerance or even worn, but still within tolerance.
Quote	To make an estimate of or give exact information on the price of a part or service. A quotation may often be considered to be legally binding.
Estimate	A statement of the expected cost of a certain job (e.g. a service or repairs). An estimate is normally a best guess and is not legally binding.
Bad	Not good – and also not descriptive enough really.
Dodgy, knackered or @#%&*.	Words often used to describe a system or component, but they mean nothing. Get used to describing things so that misunderstandings are eliminated.

report should convey clearly to another person what has been done. Further, a 'qualified' person should be able to extract enough information to be able to repeat the test – and check your findings. Use clear simple language, remembering that in some cases the intended audience may not be as technically competent as you are.

KEY FACT

Setting out results of any test in a standard format is the best way to ensure all the important and required aspects of the test have been covered.

1.4.2 Example report

An example report is presented here relating to a simple alternator test where its actual output is to be compared to the rated output. Minimal details are included so as just to illustrate the main points.

1.4.2.1 Introduction

A 'Rotato' 12 V alternator was tested under different temperature conditions to check its maximum output. The manufacturer's specifications stated that the alternator, when hot, should produce 95 A at 6000 rpm.

1.4.2.2 Test criteria

Start at room temperature.

Run alternator at 3000 rpm, 30 A output for 10 minutes.

Run alternator at 6000 rpm, maximum output. Check reading every 30 seconds for 10 minutes.

Run alternator at 6000 rpm, maximum output for a further 20 minutes to ensure output reading is stable.

1.4.2.3 Facilities/Resources

A 'Krypton' test bench model R2D2 was used to drive the alternator. The test bench revcounter was used and a 'Flake' digital meter fitted with a 200 A shunt was used to measure the output. A variable resistance load was employed.

1.4.2.4 Test procedures

The alternator was run for 10 minutes at 3000 rpm and the load adjusted to cause an output of 30 A. This was to ensure it was at a nominal operating temperature. The normal fan was kept in place during the test.

Speed was then increased to 6000 rpm and the load adjusted to achieve the maximum possible output. The load was further adjusted as required to keep the maximum possible output in case the load resistance changed due to temperature. Measurements were taken every 30 seconds for a period of 10 minutes.

Table 1.4 Results

Time (±1 s)	0	30	60	90	120	150	180	210	240	270	300	330	360	390	420	450	480	510	540	570	600
Output (±0.2 A)	101	100	99	99	98	98	98	98	98	98	97	97	96	96	96	96	96	96	96	96	96

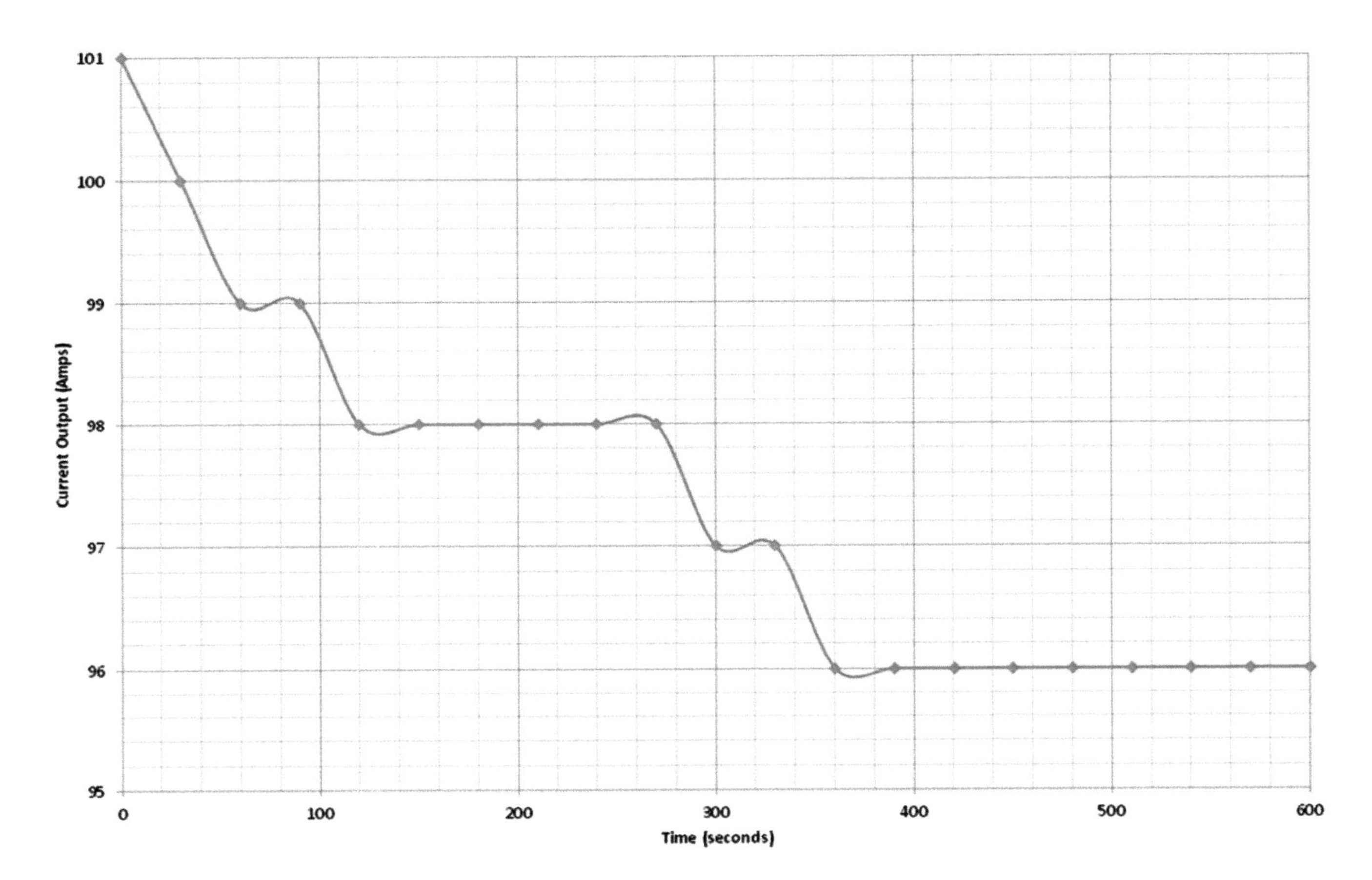

Figure 1.5 Alternator output current over time.

1.4.2.5 Measured results

Speed held constant at 6000 (±200) rpm.

Room temperature (18°C).

See Table 1.4.

To ensure the alternator output had stabilised it was kept running for a further 20 minutes at full output. It continued to hold at 96 A.

1.4.2.6 Analysis of results

Figure 1.5 shows the results in graphical format.

1.4.2.7 Conclusions

The manufacturer's claims were validated. The device exceeded the rated output by 6% at the start of the test and, under continuous operation at full load, continued to exceed the rated output by 1%.

The overall duration of this test was 40 minutes, it is possible, however, that the device would increase in temperature and the output may fall further after prolonged operation. Further tests are necessary to check this, for example, under more realistic vehicle operating conditions.

Overall the device performed in excess of its rated output in this test.

Tom Denton

(Always sign and date the report.)
Tom Denton, March 2020

Chapter 2

Diagnostic techniques

2.1 INTRODUCTION

2.1.1 Logic

Diagnostics or faultfinding is a fundamental part of an automotive technician's work. The subject of diagnostics does not relate to individual areas of the vehicle. If your knowledge of a vehicle system is at a suitable level, then you will use the same logical process for diagnosing the fault, whatever the system.

2.1.2 Data

Information and data relating to vehicles are available for carrying out many forms of diagnostic work. They used to come as a book or on CD/DVD. By far most are now online or part of a package. This information is essential to ensure that you find the fault – as long as you have developed the diagnostic skills to go with it, of course! The general type of information available is:

- engine testing and tuning;
- servicing, repair processes and times;
- fuel and ignition systems;
- circuit diagrams;
- component location;
- alignment data;
- diagnostic routines.

There are some excellent packages that you can buy on subscription, or they are included with a diagnostic tool. Some have a pay-as-you-go option. Some example sources are:

- Bosch;
- Snap-on;
- Hella;
- Delphi;
- Thatcham;
- AllData;
- Autodata;
- Haynes;
- Tech4Techs.

I am an advocate of using manufacturers' data (some of the above packages do this). My reason is simple: you know you have the correct and latest information. This can be bought as required from almost all manufacturers. For example, Volkswagen data is available from: https://erwin.volkswagen.de. It is necessary to register on each site, but then you can access the same information that is supplied to the main dealers. At the time of this writing, the Volkswagen site allowed a range of payment options, but an example was one hour of full access for 7EUR. This is more than enough time to find what you need for a specific repair or diagnostic job. Some workshops include this cost in their standard rates or add it as an extra on the customer's invoice.

On my website, www.automotive-technology.org, I have created a list of different manufacture data site links.

2.1.3 Where to stop?

This is one of the most difficult skills to learn. It is also one of the most important. The secret is twofold:

- know your own limitations – it is not possible to be good at everything;
- leave systems alone where you could cause more damage or even injury – for example, airbag circuits.

Often with the best of intentions, a person new to diagnostics will not only fail to find the fault but also introduce more faults into the system in the process. I would suggest you learn your own strengths and weaknesses; you may be confident and good at dealing with mechanical system problems but less so when electronics is involved. Of course you may be just the opposite of this.

KEY FACT

Know your own limitations.

Remember that diagnostic skill is in two parts – the knowledge of the system and the ability to apply diagnostics. If you do not yet fully understand a system, leave it alone until you do.

2.2 DIAGNOSTIC PROCESS

2.2.1 Six-stage process

A key checklist – the six stages of fault diagnosis – is given in Table 2.1 and Figure 2.1 shows this as a flow chart.

Here is a very simple example to illustrate the diagnostic process. The reported fault is excessive use of engine oil.

1. Question the customer to find out how much oil is being used (is it excessive?).
2. Examine the vehicle for oil leaks and blue smoke from the exhaust. Are there any service bulletins?
3. If leaks are found the engine could still be burning oil but leaks would be a likely cause.
4. A compression test, if the results were acceptable, would indicate a leak to be the most likely fault. Clean down the engine and run it for a while. The leak will show up better.
5. Change a gasket or seal, etc.

Table 2.1 Stages of diagnostics

1. Verify: Is there actually a problem, can you confirm the symptoms
2. Collect: Get further information about the problem, by observation and research
3. Evaluate: Stop and think about the evidence
4. Test: Carry out further tests in a logical sequence
5. Rectify: Fix the problem
6. Check: Make sure all systems now work correctly

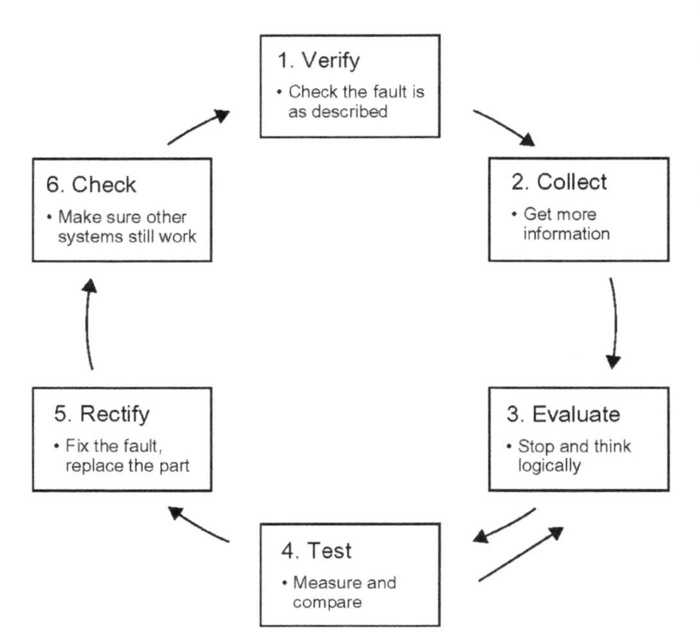

Figure 2.1 Six-stage diagnostic process.

6. Run through an inspection of the vehicle systems particularly associated with the engine. Double-check that the fault has been rectified and that you have not caused any further problems.

The six-stage diagnostic process will be used extensively to illustrate how a logical process can be applied to any situation.

KEY FACT

The six-stage diagnostic process is recommended but there are others that are similar – the important thing is to follow any 'process' logically:

1. Verify
2. Collect
3. Evaluate
4. Test
5. Rectify
6. Check

2.2.2 The art of diagnostics

The knowledge needed for accurate diagnostics is in two parts:

1. understanding of the system in which the problem exists;
2. having the ability to apply a logical diagnostic routine.

The knowledge requirement and use of diagnostic skills can be illustrated with a very simple example:

After connecting a hosepipe and turning on the tap, no water comes out of the end. Your knowledge of this system tells you that water should come out providing the tap is on, because the pressure from a tap pushes water through the pipe, and so on. This is where your diagnostic skills become essential. The following stages are now required:

1. Confirm that no water is coming out by looking down the end of the pipe.
2. Check if water comes out of the other taps, or did it come out of this tap before you connected the hose?
3. Consider what this information tells you; for example, if the answer is 'Yes' the hose must be blocked or kinked.
4. Walk the length of the pipe looking for a kink.
5. Straighten out the hose.
6. Check that water now comes out and that no other problems have been created.

SAFETY FIRST

Don't point any pipes at your eyes.

Much simplified I accept, but the procedure you have just followed made the hose work and it is also guaranteed to find a fault in any system. It is easy to see how it works in connection with a hosepipe and I'm sure anybody could have found that fault (well most people anyway).

The higher skill is to be able to apply the same logical routine to more complex situations. The routine (Table 2.1) is also represented by Figure 2.1. The loop will continue until the fault is located.

I will now explain each of these steps further in relation to a more realistic automotive workshop situation – not that getting the hose to work is not important! Often electrical faults are considered to be the most difficult to diagnose – but this is not true. I will use a vehicle cooling system fault as an example here, but electrical systems will be covered in detail in later chapters. Remember that the diagnostic procedure can be applied to any problem – mechanical, electrical or even medical.

However, let us assume that the reported fault with the vehicle is overheating. As is quite common in many workshop situations that's all the information we have to start with. Now work through the six stages:

- Stage 1 – Take a quick look to check for obvious problems such as leaks, broken drive belts or lack of coolant. Run the vehicle and confirm that the fault exists. It could be the temperature gauge, for example.
- Stage 2 – Is the driver available to give more information? For example, does the engine overheat all the time or just when working hard? Check records, if available, of previous work done to the vehicle.
- Stage 3 – Consider what you now know. Does this allow you to narrow down what the cause of the fault could be? For example, if the vehicle overheats all the time and it had recently had a new cylinder head gasket fitted, would you be suspicious about this? Do not let two and two make five, but do let it act as a pointer. Remember that in the science of logical diagnostics, two and two always makes four. However, until you know this for certain then play the best odds to narrow down the fault.
- Stage 4 – The further tests carried out would now be directed by your thinking at stage 3. You do not yet know if the fault is a leaking head gasket, the thermostat stuck closed or some other problem. Playing the odds, a cooling system pressure test would probably be the next test. If the pressure increases when the engine is running, then it is likely to be a head gasket or similar problem. If no pressure increase is noted, then move on to the next test and so on. After each test go back to stage 3 and evaluate what you know, not what you don't know.
- Stage 5 – Let us assume the problem was a thermostat stuck closed – replace it and top up the coolant, etc.
- Stage 6 – Check that the system is now working. Also check that you have not caused any further problems such as leaks or loose wires.

This example is simplified a little, but like the hosepipe problem it is the sequence that matters, particularly the 'stop and think' at stage 3. It is often possible to go directly to the cause of the fault at this stage, providing that you have an adequate knowledge of how the system works.

2.2.3 Concern, cause, correction

The three Cs, as concern, cause, correction are sometimes described, is another reminder that following a process for automotive repairs and diagnostics is essential.

It is in a way a simplified version of our six-stage process as shown in Table 2.2.

Table 2.3 is a further example where extra suggestions have been added as a reminder of how important it is to collect further information. It is also recommended that this information and process is included on the jobsheet so the customer is kept informed. Most customer complaints come about because of poor work or poor communication – this may be acceptable in some poor quality establishments but not in any that you and I are involved in – be professional and you will be treated like one (lecture over, sorry).

So, while the concern, cause, correction sequence is quite simple, it is very effective as a means of communication as well as a diagnosis and repair process. An example jobcard/jobsheet is available for download from www.automotive-technology.co.uk that includes the three Cs. It is ideal as a training aid as well as for real use.

Table 2.2 Repair and diagnostic processes

Six-stage process	*CCC*
Verify	Concern
Collect Evaluate Test	Cause
Rectify Check	Correction

Table 2.3 CCC process

Process outline	*Example situation*	*Notes*
Customer concern:	Battery seems to be discharged and will sometimes not start the car. It seems to be worse when the headlights are used	This should set you thinking that the cause is probably a faulty battery, a charging system fault, a parasitic discharge or a starter motor problem (the symptoms would suggest a charging fault is most likely but keep an open mind)
Vehicle service history information:	Car is five years old, has done 95 000 miles but has a good service history. A new battery was fitted one year ago and the cam belt was replaced two years ago	Battery probably OK and drive belt adjustment likely to be correct (still suspicious of a charging fault)
Related technical service bulletins:	New camshaft drive belt should be fitted every 50 000 miles	Not connected but it would be good to recommend that the belt was changed at this time
Diagnostic procedures performed:	Battery voltage and discharge test – OK Drive belt tension – OK (but a bit worn) Alternator charging voltage – 13 V Checked charging circuit for volt drop – OK	14 V is the expected charging voltage on most systems
Cause:	Alternator not producing correct voltage	An auto electrician may be able to repair the alternator but for warranty reasons a new or reconditioned one is often best (particularly at this mileage)
Correction:	Reconditioned alternator and new drive belt fitted and checked – charging now OK at 14 V	Note how by thinking about this process we had almost diagnosed the problem before doing any tests, also note that following this process will make us confident that we have carried out the correct repair, first time. The customer will appreciate this – and will come back again

2.2.4 Root cause analysis

The phrase 'root cause analysis' (RCA) is used to describe a range of problem-solving methods aimed at identifying the root causes of problems or events. I have included this short section because it helps to reinforce the importance of keeping an open mind when diagnosing faults, and again, stresses the need to work in a logical and structured way. The root cause of a problem is not always obvious; an example will help to illustrate this:

Let us assume the symptom was that one rear light on a car did not work. Using the six-stage process, a connector block was replaced as it had an open circuit fault. The light now works OK but what was missed was that a small leak from the rear screen washer pipe dripped on the connector when the washer was operated. This was the root cause.

The practice of RCA is based, quite rightly, on the belief that problems are best solved by attempting to address, correct or eliminate the root causes, as opposed to just addressing the faults causing observable symptoms. By dealing with root causes, it is more likely that problems will not reoccur. RCA is best considered to be an iterative process because complete prevention of recurrence by one corrective action is not always realistic.

Root causes of a problem can be in many different parts of a process. This is sometimes represented by a 'fishbone' diagram. Two examples are presented as Figures 2.2 and 2.3. These show how any one cause on any one branch (or rib) can result in a problem at the end of a more complex process.

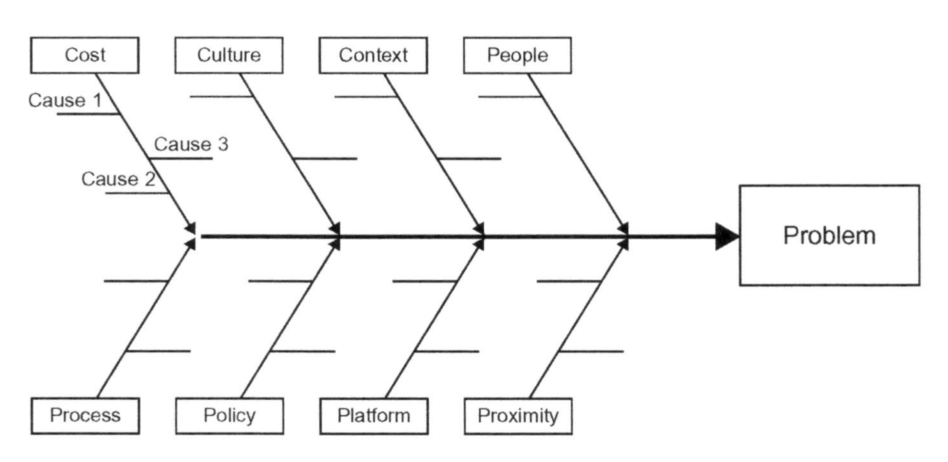

Figure 2.2 Fishbone diagram showing possible root causes of a problem in software development.

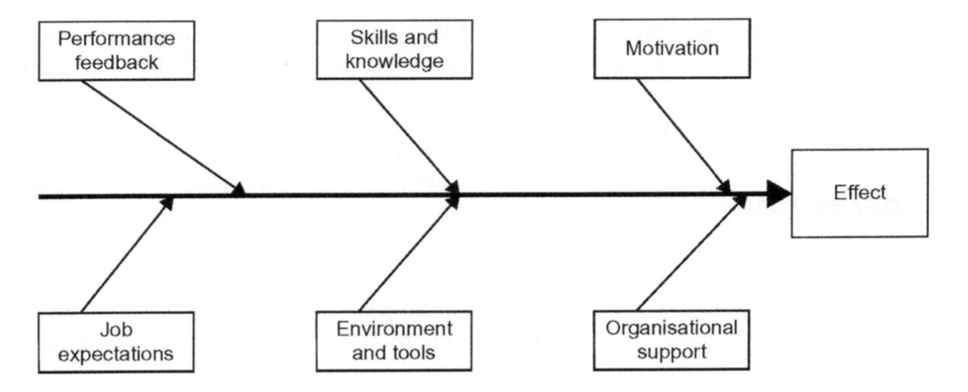

Figure 2.3 Fishbone diagram that could be used to look at diagnostic processes.

RCA is usually used as a reactive method of identifying causes, revealing problems and solving them and it is done after an event has occurred. However, RCA can be a useful proactive technique because, in some situations, it can be used to forecast or predict probable events.

DEFINITION

RCA: Root cause analysis.

RCA is not a single defined methodology. There are a number of different ways of doing the analysis. However, several very broadly defined methods can be identified:

- Safety-based RCA descends from the fields of accident analysis and occupational safety and health.
- Production-based RCA has its origins in the field of quality control for industrial manufacturing.
- Process-based RCA is similar to production-based RCA, but has been expanded to include business processes.
- Failure-based RCA comes from the practice of failure analysis used in engineering and maintenance.

KEY FACT

RCA directs the corrective action at the true root cause of the problem.

The following list is a much simplified representation of a failure-based RCA process. Note that the key steps are numbers 3 and 4. This is because they direct the corrective action at the true root cause of the problem.

1. Define the problem.
2. Gather data and evidence.
3. Identify the causes and root causes.
4. Identify corrective action(s).
5. Implement the root cause correction(s).
6. Ensure effectiveness (Figure 2.4).

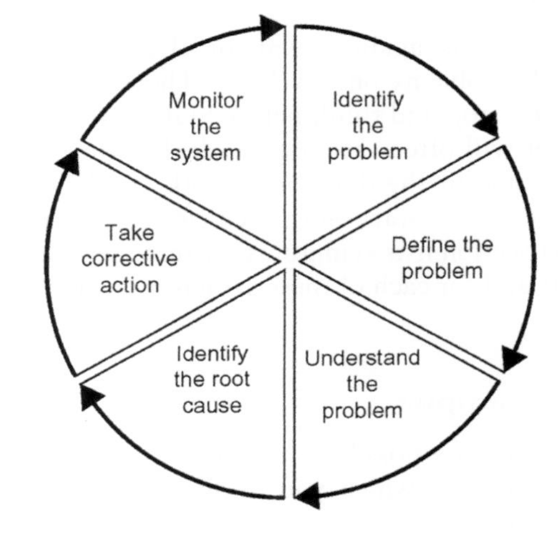

Figure 2.4 RCA process.

As an observant reader, you will also note that these steps are very similar to our six-stage faultfinding process.

KEY FACT

Six-stage process:

1. Verify
2. Collect
3. Evaluate
4. Test
5. Rectify
6. Check

2.2.5 Summary

I have introduced the six-stage process of diagnostics, not so that it should always be used as a checklist but to illustrate how important it is to follow a process. Much more detail will be given later, in particular about stages 3 and 4. The purpose of this set process is to ensure that 'we' work in a set, logical way.

DEFINITION

'Logic is the beginning of wisdom not the end'. (Spock to Valeris, *Star Trek II*)

2.3 DIAGNOSTICS ON PAPER

2.3.1 Introduction

This section is again a way of changing how you approach problems on a vehicle. The key message is that if you stop and think before 'pulling the vehicle to pieces', it will often save a great deal of time. In other words, some of the diagnostic work can be done 'on paper' before we start on the vehicle. To illustrate this, the next section lists symptoms for three separate faults on a car and for each of these symptoms, three possible faults.

2.3.2 Examples

All the faults are possible in the following example, but in each case see which you think is the 'most likely' option (Table 2.4).

The most likely fault for example A is number 3. It is possible that all the lights have blown but unlikely. It could not be the auxiliary relay because this would affect other systems.

For example B, the best answer would be number 2. It is possible that the pump pressure is low but this would be more likely to affect operation under other conditions. A loose wire on the engine speed sensor could cause the engine to stall but it would almost certainly cause misfire under other conditions.

The symptoms in example C would suggest answer 1. The short circuit suggested as answer 3 would be more likely to cause lights and others to stay on rather than not work, equally the chance of a short between these two circuits is remote if not impossible. If the lighting fusible link were blown then none of the lights would operate.

The technique suggested here relates to stages 1–3 of 'the six stages of fault diagnosis' process. By applying a little thought before even taking a screwdriver to the car, a lot of time can be saved. If the problems suggested in the previous table were real we would at least now be able to start looking in the right area for the fault.

KEY FACT

Stop and think before pulling the vehicle to pieces.

2.3.3 How long is a piece of string?

Yes, I know, twice the distance from the middle to one end. What I am really getting at here though is the issue about what is a valid reading or measurement and what is not – when compared to data. For example, if the 'data source' says the resistance of the component should be between 60 and 90 Ω, what do you do when the measured value is 55 Ω? If the measured value was 0 Ω or 1000 Ω then the answer is easy – the component is faulty. However, when the value is very close you have to make a decision. In this case (55 Ω) it is very likely that the component is serviceable.

The decision over this type of issue is difficult and must, in many cases, be based on experience. As a general guide, however, I would suggest that if the reading is in the right 'order of magnitude', then the component has a good chance of being OK. By this I mean that if the value falls within the correct range of 1s, 10s, 100s or 1000s, etc., then it is probably good.

Do notice that I have ensured that words or phrases such as 'probably', 'good chance' and 'very likely' have been used here. This is not just to make sure I have a get out clause; it is also to illustrate that diagnostic work can involve 'playing the best odds' – as long as this is within a logical process.

Table 2.4 Example faults

Symptoms	*Possible faults*
A: The brake/stop lights are reported as not operating. On checking it is confirmed that neither of the two bulbs or the row of high-mounted LEDs are operating as the pedal is pressed. All other systems work correctly	1. Two bulbs and 12 LEDs blown 2. Auxiliary systems relay open circuit 3. Brake light switch not closing
B: An engine fitted with full management system tends to stall when running slowly. It runs well under all other conditions and the reported symptom is found to be intermittent	1. Fuel pump output pressure low 2. Idle control valve sticking 3. Engine speed sensor wire loose
C: The off side dip beam headlight not operating. This is confirmed on examination and also noted that the off side tail lights do not work	1. Two bulbs blown 2. Main lighting fusible link blown 3. Short circuit between off side tail and dip beam lights

DEFINITION

Order of magnitude:

- A degree in a continuum of size or quantity;
- A number assigned to the ratio of two quantities;
- Two quantities are of the same order of magnitude if one is less than 10 times as large as the other;
- The number of magnitudes that the quantities differ is specified to within a power of 10.

2.4 REAL-WORLD DIAGNOSTICS

2.4.1 Introduction

This section will look at carrying out diagnostics in the real world. It is **not** intended to be the definitive answer; rather I hope it will make you think about the subject and how you go about it, as well as the customer's experience and expectations.

2.4.2 What do you know?

When faced with a fault on a vehicle, we have to make decisions about what to do first, and then what to do next! Let's take a simple example where the brake lights are not working. It is very easy to start thinking about all the things it could be, but at this stage we need to find out some facts. In other words, build up the information from what we know, NOT from what we don't know. To do this, observe or measure something, consider what the observation means or compare the measurement to data. Then, based on your conclusions so far, take the next step (Figure 2.5).

In our simple example, after having observed than none of the brake lights work, we would probably measure the fuse. I used the word 'measure' on purpose because observing a fuse is often not good enough. Switch on the ignition and check both sides of the fuse with a test lamp, for example. Assuming that this is OK, we now have some facts to build upon. We would probably decide to check the supply into and out of the brake light switch next. Based on this result we could then decide the next step and so on – each time building up what we know about the fault.

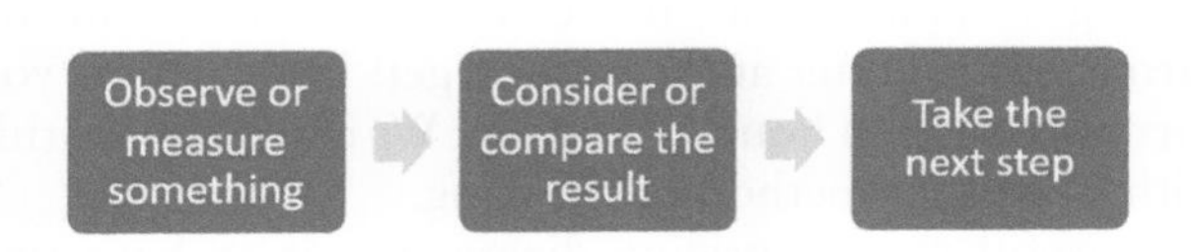

Figure 2.5 This process can be thought of as: Sense, think, act.

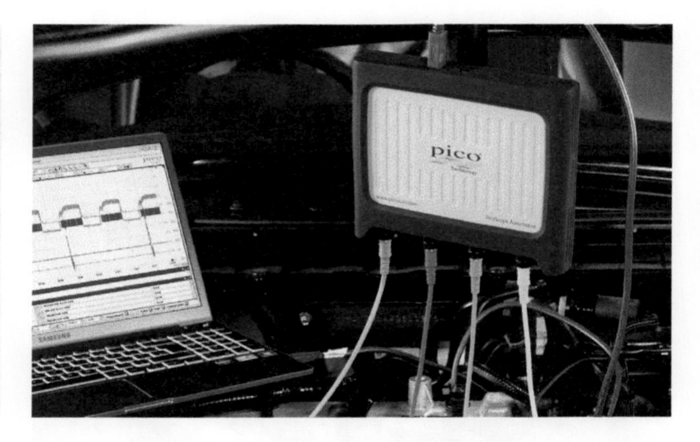

Figure 2.6 PicoScope connected to view waveforms.

Source: PicoScope Media.

2.4.3 Training

There are many excellent training and CPD courses available to help you practice and develop your diagnostic skills. You can learn a lot from videos, magazine articles and textbooks, but nothing beats hands-on training from an experienced teacher. I am pleased to recommend: Simply Diagnostics Network (www.simplydiag.net), TechTopics (www.techtopics.co.uk) and Pro-Moto (www.pro-moto.co.uk).

2.4.4 The customer

Let's start with the most important point about our customers: without them we would not have a job. That said, some can be challenging to work with! The answer to this is to be very clear about what you will be doing and not doing as part of a diagnostic process. Also, be willing to explain things to help them understand. One key skill to develop is how to get the appropriate information from a customer. For example, if a driver says: *'My car won't start,'* this could mean several things, but three in particular:

- The starter is not operating;
- The starter is operating but not properly;
- The starter is operating properly but the engine will not run.

In this example we would need to narrow it down by talking to them. Another really important customer-skill is to be very clear about what you will spend and what you will cost, and at what point you will stop and consult them further.

A third customer relationship skill is to understand that their lack of knowledge means they can come to the wrong conclusion over something, or worse, there are some who try and take advantage of a situation. How many times have you heard something like this? *'Ever since you did X to my car, Y has stopped working'.*

Figure 2.7 ArtiPad scanner in use.

My favourite from some years ago, was: *'Ever since you changed the starter motor, one of the rear indicators has stopped working'*. In this case it was a genuine lack of understanding. They assumed because I was working on 'electrical things,' it must have been caused by me. To complete the story, I fitted the new bulb for free – but did explain the situation to them! To help counter some of these issues, a set diagnostic routine or sequence can be very useful.

2.4.5 How much does diagnostics cost?

Multimeters, oscilloscopes, scanners and lots of other equipment is needed in our armoury to diagnose faults. This equipment costs serious money and it is important to remember this when pricing diagnostic work.

Training is important and courses cost money. We also need to generate our wages and a suitable profit margin, whether employed by an organisation or self-employed.

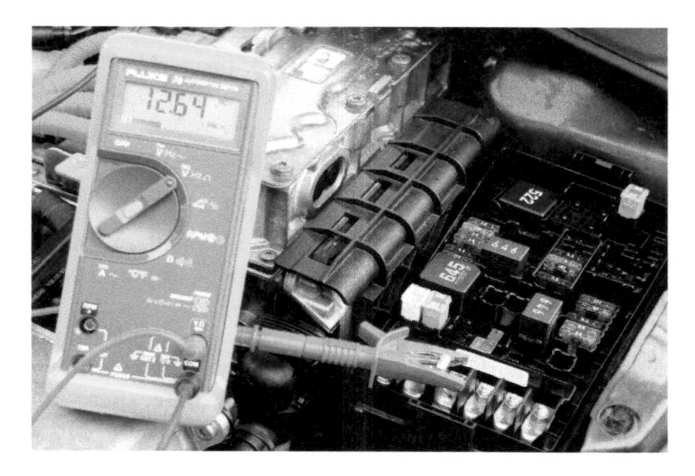

Figure 2.8 Voltage reading at the fuse/relay distribution box.

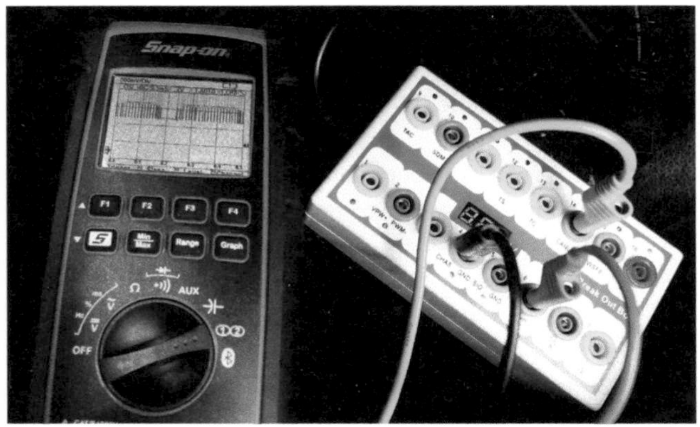

Figure 2.9 CAN high and low on a scope using a breakout box connected to the diagnostic socket.

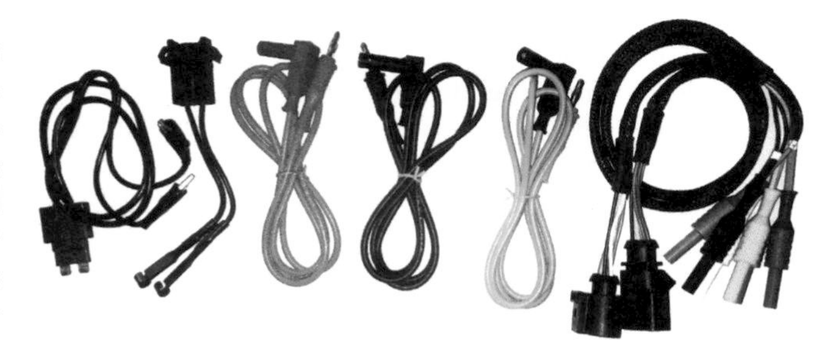

Figure 2.10 Test leads and adapters: at the time of writing, even this simple set of leads cost over £50.

Equipment and training should be thought of as an investment, not a cost. To get the return on this investment, it is essential to charge appropriately for diagnostic work.

2.4.6 How much should you charge?

Prices change and vary depending on geographic area, reputation, volume of demand and over time. The figures stated here are just an outline guide, but make sure you charge what you are worth. If a customer is not willing to pay for proper diagnostic work, then they may not be the type of customer you want.

One thing is easy to state as a given, and that is diagnostic charges should be fixed, NOT time related (don't diagnose at an hourly rate). This is because the more money you invest in equipment, the more experience and skill you develop through attending and paying for courses, for example, the quicker you will become at finding the fault. A time related charge would mean you earn less, the better at the job you get! Let's assume you currently have an hourly rate of X. We can compare this with a two-tier method of charging.

A useful way of pricing diagnostics is to have two tiers:

- **Tier 1: 1.25 times X:** For this you will carry out non-invasive diagnostic work that does not involve removing any components other than maybe engine covers or similar. It also does not involve complex testing using specialist equipment. At the end of this process you will be able to state the fault OR will be able to explain to the customer why the next tier (level) is needed. Also explain that this may involve removing some components that have a risk of breakage. The diagnostic sequence in Table 2.5 below is an example of a tier 1 process. Most faults will be traced at tier 1, but repairs will cost extra.

Table 2.5 Diagnostic sequence (tier 1)

Steps	*Comments*	*Time (minutes)*
Prepare Check oil and coolant levels	This is to make sure it is safe to run the engine. In some cases, it may be necessary to run the engine at high speed, so better to be safe than sorry. It may also start to give you some information. For example if the oil is low and very dirty it may indicate lack of servicing.	2
Carry out battery, charging and starting test	Not essential but if the tests you carry out require the engine to be cranked over several times, a starter motor on its last legs could fail altogether – better to know before you start. Once again, before we even begin looking in any detail, this stage does give us more information. I use the device shown in Figure 2.13 for this test as this equipment allows you to print out the results.	5
Connect battery saver	Any test routine will involve using up the battery to some extent. A suitable battery saver, such as the GYS shown in Figure 2.14, will ensure the it remains at a suitable level.	1
Verify the symptoms	In this case it is easy, the warning light is on. It is still worth running and listening to the engine at idle, low and high speeds because there may be other symptoms that the customer did not notice like excess smoke or a misfire.	2
Collect more information	Is there a service history or has any other work done? Access manufacturer's or other form of data.	5
Carry out a full scan for DTCs	Most scanners will allow you to save and/or print this file. Doing this regardless of the symptoms gives you more detailed information about the fault but it also acts as a record in case the: 'Ever since you… ' comment is used! Let's assume an EGR diagnostic trouble code (DTC) is shown, which caused the MIL to be illuminated. Also present is a braking system fault (Figure 2.12).	5
Hand and eye (and ear!) checks	Look and listen for anything obvious, loose wires, loose connections, oil leaks, noises, etc., particularly in areas related to the DTC.	2
Evaluate	Stop and think to decide what specific tests should now be done.	3
Test Check live data	In this case we would probably look at the signal or voltages to and from the EGR sensor. Compare any readings to manufacturer's date and look for anomalies: Does something look like an apple when it should be an orange?!	5
Carry out cylinder balance check	This can be done if appropriate to the symptoms, but also as a generic process to learn more and allow you to provide detailed information about the condition of the engine/vehicle to the customer. For example, if there is a serious loss of compression on one cylinder, will they really want you to chase down a smaller EGR fault? Some diagnostic technicians choose to carry out this test in the preparation stage. It can even be done using a DLC breakout box and by measuring voltage drop on a scope from inside the vehicle. However, the ideal is to measure starter current using an amp clamp and a scope. Compare the current on each cylinder for balance. If the average current is about three times the battery Ah capacity, then the compression is probably OK. (Note: There is an excellent free phone app from Exide that gives useful data about what battery should be on a vehicle.)	5
Guided by any DTCs and the live data, carry out appropriate tests on sensors, actuators and wiring	This can be done with a multimeter set to measure voltage. If very detailed checks with a multichannel oscilloscope are necessary, it may be appropriate to stop at this stage and report to the customer that you need to go to tier 2.	5
Rectify	A sensor may need to be fitted or an actuator removed and cleaned. If it is a simple wiring fault, you may be able to fix the fault but remember the diagnostic routine is just that, it does not cover the cost of repairs. If appropriate, clear the DTC.	0
Check	Physically check all areas of the vehicle where you have worked (and more), but also carry out another full scan for DTCs. This then shows either that you have rectified the fault or not caused any more compared to the first scan.	5
Report	Let the customer know what has been done, what has or has not been fixed and what your recommendations are, for example the brake system fault may need to be followed up depending on its priority.	5

- **Tier 2: 2 times X:** At this level, and with the customer's permission to proceed, we can start digging much deeper. This will involve carrying out more invasive measurements that may involve some dismantling work. Even at this stage, the cost of the actual repair may not be included, unless it is relatively simple to carry out. At the end of this process you would guarantee to either have fixed the fault or be able to state exactly what is needed.

Figure 2.11 Malfunction indicator lamp (MIL) or check engine warning light.

2.4.7 Diagnostic sequence

Every set of symptoms you come across will be different and so the diagnostic routine will need to adapt to find the fault. However, it is useful to have a generic sequence in mind. Doing this means you cover all the steps but also cover your back if needed! Further, there could be opportunities for generating more work.

Let's take an example where all we know at this stage is that the malfunction indicator lamp (MIL) is on (Figure 2.11). Remember, this is a guide to make you consider different sequences and methods, it is not a definitive routine. It follows the six stages of diagnosis we examined earlier, with an extra one at the start (prepare) and at the end (report). I have estimated the time each step will take, but again this is a rough guide only. The routine here (Table 2.5) totals to about 50 minutes, but in many cases, it will be less. Your experience will tell you when certain tests, like the battery charging and starting check, could be avoided, for example, if you know of and normally service the vehicle. It is good practice to keep a note of what you do. Sometimes it

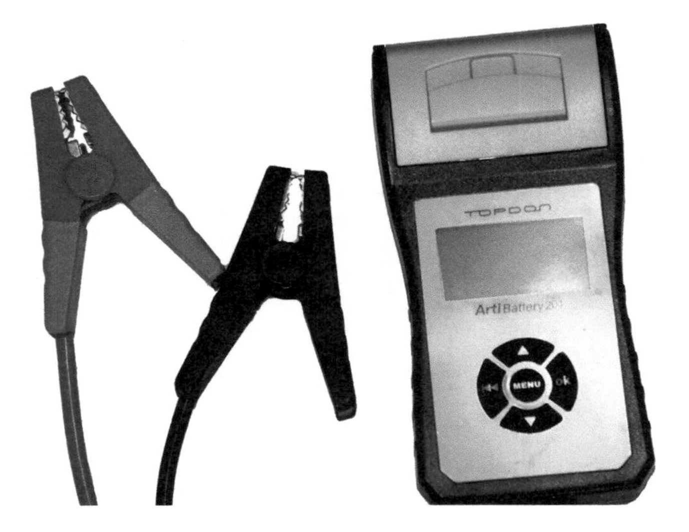

Figure 2.13 Battery, charging and starting tester.

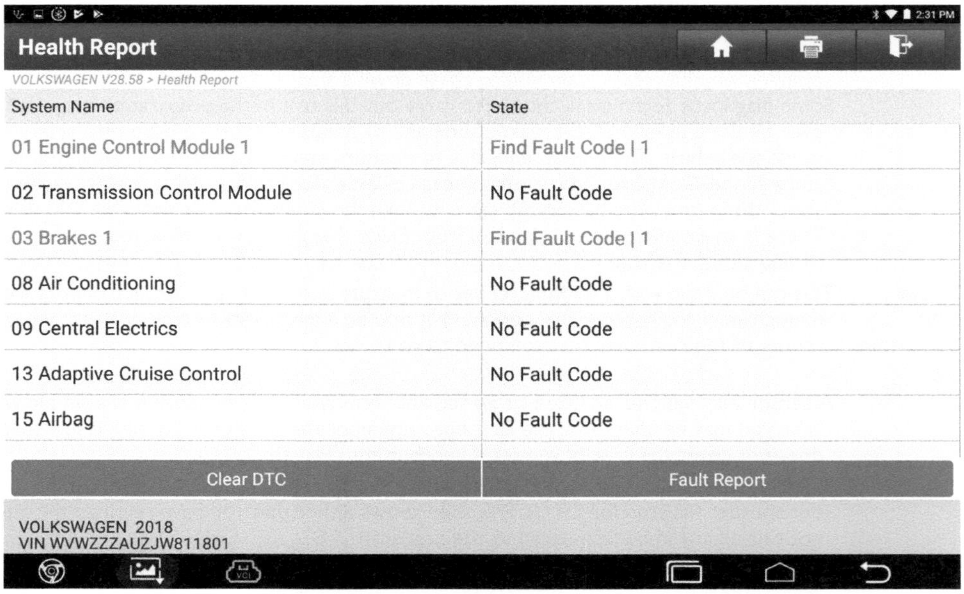

Health Report

VOLKSWAGEN V28.58 > Health Report

System Name	State
01 Engine Control Module 1	Find Fault Code | 1
02 Transmission Control Module	No Fault Code
03 Brakes 1	Find Fault Code | 1
08 Air Conditioning	No Fault Code
09 Central Electrics	No Fault Code
13 Adaptive Cruise Control	No Fault Code
15 Airbag	No Fault Code

Clear DTC | Fault Report

VOLKSWAGEN 2018
VIN WVWZZZAUZJW811801

Figure 2.12 Scanner health report showing two DTCs.

Figure 2.14 GYS Battery Charger/Saver connected to a vehicle.

may take longer, but on average you should spend less than an hour on tier 1.

2.4.8 Efficiency

The following was written by my colleague James Dillon. See www.techtopics.co.uk for more information about the excellent training in diagnostics and much more.

An efficiency figure is often calculated to see how well time is being utilised in a workshop. This efficiency measure usually stems from some form of comparison between job time sold versus job time taken. With a standard 'pure' mechanical labour job, the specification of 'time allowed' is fairly straightforward, and the efficiency measurement resulting from the time taken may work.

For diagnostics, where there are both known unknowns and unknown unknowns, this measure is very difficult. The time allowed cannot be known in advance, as the second step of the diagnostic process cannot be defined until the first steps results are in. There is no 'book time' or 'standard time' for diagnostics because the process is dynamic and results based and unfolds as the data is gathered and analysis performed. Often, the 'time allowed' is defined at the end of the successful diagnosis, based on the 'time taken'. This should give 100% efficiency because you took as long as the diagnostic process took.

Many businesses then look back at the job from the 'finish line' and consider what the job should have taken, based on the fix. This paradigm shift causes them to write down the 'time taken' using their gut feel and their own perception of how long it should have taken. Really, the job should be priced with the perspective view from the start line (when the challenge and the value of the diagnosis was high), and 100% time charged.

Arguably, if the technician made mistakes in their data gathering, analysis or conclusions, the time taken should not be charged at 100%, training needs should be identified and skills/knowledge gaps filled.

The other factor often overlooked is the labour rate applied to diagnostics. It cannot and should not be charged at the same rate as mechanical, as the knowledge, skills and resources required to deliver it are much higher. This should also have a positive impact on labour sales if managed correctly within the business structure, as essentially the technician time should be charged out at ×1.5, ×2 or ×3 times the standard rate. This won't help efficiency (which can be argued is the wrong tool to measure the effectiveness of diagnosis), but it will help profitability.

Because of the unknown unknowns, one method to size the task accurately (and to make a profit in diagnostics) is to offer a standard rate diagnostic assessment. Once preliminaries are completed, use an agreed budget to move to the next diagnostic way-marker. This technique pre-sells the time taken (as time that likely/going to be taken). In this case, the efficiency will be 100%.

Efficiency deficiency in diagnostics is equally likely to show up front counter pricing problems as it is to show technical deficiency.

2.4.9 Summary

Diagnostics is not the easiest part of our job, but it can be the most satisfying. It can also generate a good income if carried out correctly and charged for accordingly. It is also an opportunity to generate more work if communication with the customer is good.

2.5 MECHANICAL DIAGNOSTIC TECHNIQUES

2.5.1 Check the obvious first

Start all hands-on diagnostic routines with 'hand and eye checks'. In other words, look over the vehicle for obvious faults. For example, if automatic transmission fluid is leaking on to the floor then put this right before carrying out complicated stall tests. Here are some further suggestions that will at some point save you a lot of time.

- If the engine is blowing blue smoke out of the exhaust – consider the worth of tracing the cause of a tapping noise in the engine.
- When an engine will not start – check that there is fuel in the tank (Figure 2.15).

KEY FACT

All diagnostic routines should include 'hand and eye checks'.

Figure 2.15 Honda Civic 1.0 VTEC Turbo Engine.

Source: Honda Media.

2.5.2 Noise, vibration and harshness

Noise, vibration and harshness (NVH) concerns have become more important as drivers have become more sensitive to these issues. Drivers have higher expectations of comfort levels. NVH issues are more noticeable due to reduced engine noise and better insulation in general. The main areas of the vehicle that produce NVH are:

- tyres;
- engine accessories;
- suspension;
- driveline.

DEFINITION

NVH: Noise, vibration and harshness.

It is necessary to isolate the NVH into its specific area(s) to allow more detailed diagnosis. A road test, as outlined later, is often the best method.

The five most common sources of non-axle noise are exhaust, tyres, roof racks, trim and mouldings, and transmission. Ensure that none of the following conditions is the cause of the noise before proceeding with a driveline strip down and diagnosis.

1. In certain conditions, the pitch of the exhaust may sound like gear noise or under other conditions like a wheel bearing rumble.
2. Tyres can produce a high-pitched tread whine or roar, similar to gear noise. This is particularly the case for non-standard tyres.
3. Trim and mouldings can cause whistling or whining noises.
4. Clunk may occur when the throttle is applied or released due to backlash somewhere in the driveline.
5. Bearing rumble sounds like marbles being tumbled.

KEY FACT

The five most common sources of non-axle noise are exhaust, tyres, roof racks, trim and mouldings, and transmission.

2.5.3 Noise conditions

Noise is very difficult to describe. However, the following are useful terms and are accompanied by suggestions as to when they are most likely to occur.

- Gear noise is typically a howling or whining due to gear damage or incorrect bearing preload. It can occur at various speeds and driving conditions or it can be continuous.
- 'Chuckle' is a rattling noise that sounds like a stick held against the spokes of a spinning bicycle wheel. It usually occurs while decelerating.
- Knock is very similar to chuckle though it may be louder and occurs on acceleration or deceleration.

Check and rule out tyres, exhaust and trim items before any disassembly to diagnose and correct gear noise.

2.5.4 Vibration conditions

Clicking, popping or grinding noises may be noticeable at low speeds and be caused by the following:

- inner or outer CV joints worn (often due to lack of lubrication, so check for split gaiters);
- loose drive shaft;
- another component contacting a drive shaft;
- damaged or incorrectly installed wheel bearing, brake or suspension component.

The following may cause vibration at normal road speeds:

- out-of-balance wheels;
- out-of-round tyres.

The following may cause shudder or vibration during acceleration:

- damaged powertrain/drivetrain mounts;
- excessively worn or damaged out-board or in-board CV joints.

The cause of noise can often be traced by first looking for leaks. A dry bearing or joint will produce significant noise.

1. Inspect the CV joint gaiters (boots) for cracks, tears or splits.
2. Inspect the underbody for any indication of grease splatter near the front wheel half shaft joint boots.
3. Inspect the in-board CV joint stub shaft bearing housing seal for leakage at the bearing housing.
4. Check the torque on the front axle wheel hub retainer.

2.5.5 Road test

A vehicle will produce a certain amount of noise. Some noise is acceptable and may be audible at certain speeds or under various driving conditions such as on a new road.

Carry out a thorough visual inspection of the vehicle before carrying out the road test. Keep in mind anything that is unusual. A key point is to not repair or adjust anything until the road test is carried out. Of course this does not apply if the condition could be dangerous or the vehicle will not start.

Establish a route that will be used for all diagnostic road tests. This allows you to get to know what is normal and what is not. The roads selected should have sections that are reasonably smooth, level and free of undulations as well as lesser quality sections needed to diagnose faults that only occur under particular conditions. A road that allows driving over a range of speeds is best. Gravel, dirt or bumpy roads are unsuitable because of the additional noise they produce.

KEY FACT

Establish a standard route that will be used for all diagnostic road tests so you know what to expect.

If a customer's concern is a noise or vibration on a particular road and only on a particular road, the source of the concern may be the road surface. Test the vehicle on the same type of road. Make a visual inspection as part of the preliminary diagnosis routine prior to the road test; note anything that does not look right. For example,

1. tyre pressures, but do not adjust them yet;
2. leaking fluids;
3. loose nuts and bolts;
4. bright spots where components may be rubbing against each other;
5. check the luggage compartment for unusual loads.

Road test the vehicle and define the condition by reproducing it several times during the road test. During the road test recreate the following conditions:

1. Normal driving speeds of 20–80 km/h (15–50 mph) with light acceleration – a moaning noise may be heard and possibly a vibration is felt in the front floor pan. It may get worse at a certain engine speed or load.
2. Acceleration/deceleration with slow acceleration and deceleration – a shake is sometimes noticed through the steering wheel seats, front floor pan, front door trim panels, etc.
3. High speed – a vibration may be felt in the front floor pan or seats with no visible shake, but with an accompanying sound or rumble, buzz, hum, drone or booming noise. Coast with the clutch pedal down or gear lever in neutral and engine idling. If vibration is still evident, it may be related to wheels, tyres, front brake discs, wheel hubs or wheel bearings.
4. Engine rpm sensitive – a vibration may be felt whenever the engine reaches a particular speed. It may disappear in neutral coasts. Operating the engine at the problem speed while the vehicle is stationary can duplicate the vibration. It can be caused by any component, from the accessory drive belt to the clutch or torque converter, which turns at engine speed when the vehicle is stopped.
5. Noise and vibration while turning – clicking, popping or grinding noises may be due to the following: damaged CV joint; loose front wheel half shaft joint boot clamps; another component contacting the half shaft; worn, damaged or incorrectly installed wheel bearing; damaged powertrain/drivetrain mounts.

After a road test, it is often useful to do a similar test on a hoist or lift. When carrying out a 'shake and vibration' diagnosis or 'engine accessory vibration' diagnosis on a lift, observe the following precautions:

- If only one drive wheel is allowed to rotate, speed must be limited to 55 km/h (35 mph) indicated on the speedometer. This is because the actual wheel speed will be twice that indicated on the speedometer.
- The suspension should not be allowed to hang free. If a CV joint were run at a high angle, extra vibration as well as damage to the seals and joints could occur.

Support the front suspension lower arm as far outboard as possible. This will ensure that the vehicle is at its correct ride height. The procedure is outlined by the following steps:

1. Raise and support the vehicle.
2. Explore the speed range of interest using the road test checks as previously discussed.

3. Carry out a coast down (overrun) in neutral. If the vehicle is free of vibration when operating at a steady indicated speed and behaves very differently in drive and coast, a transmission concern is likely.

A test on the lift may produce different vibrations and noises than a road test because of the effect of the lift. It is not unusual to find a vibration on the lift that was not noticed during the road test. If the condition found on the road can be duplicated on the lift, carrying out experiments on the lift may save a great deal of time.

2.5.6 Engine noises

How do you tell a constant tapping from a rattle? Worse still, how do you describe a noise in a book? I'll do my best. Table 2.6 is a non-definitive guide to the source or cause of engine or engine ancillary noises.

2.5.7 Sources of engine noise

Table 2.7 is a further guide to engine noise. Possible causes are listed together with the necessary repair or further diagnosis action as appropriate.

Table 2.6 Noise diagnostics

Noise description	*Possible source*
Tap	Valve clearances out of adjustment, cam followers or cam lobes worn
Rattle	A loose component, broken piston ring or component
Light knock	Small-end bearings worn, earn or cam follower
Deep knock or thud	Big-end bearings worn
Rumble	Main bearings worn
Slap	Worn pistons or bores
Vibration	Loose or out-of-balance components
Clatter	Broken rocker shaft or broken piston rings
Hiss	Leak from inlet or exhaust manifolds or connections
Roar	Air intake noise, air filter missing, exhaust blowing or a seized viscous fan drive
Clunk	Loose flywheel, worn thrust bearings or a loose front pulley/ damper
Whine	Power steering pump or alternator bearing
Shriek	Dry bearing in an ancillary component
Squeal	Slipping drive belt

Table 2.7 Engine noises

Sources of engine noise	*Possible cause*	*Required action*
Misfiring/backfiring	Fuel in tank has wrong octane/cetane number, or is wrong type of fuel Ignition system faulty Engine temperature too high Carbon deposits in the combustion chamber start to glow and cause misfiring Timing incorrect, which causes misfiring in the intake/exhaust system	Determine which type of fuel was last put in the tank Check the ignition system Check the engine cooling system Remove the carbon deposits by using fuel additives and driving the vehicle carefully Check the timing
Valve train faulty	Valve clearance too large due to faulty bucket tappets or incorrect adjustment of valve clearance Valve timing incorrectly adjusted valves and pistons are touching Timing belt broken or damaged	Adjust valve clearance if possible and renew faulty bucket tappets – check cam condition Check the valve timing and adjust if necessary Check timing belt and check pistons and valves for damage – renew any faulty parts
Engine components faulty	Pistons Piston rings Cylinder head gasket Big-end and/or main bearing journals	Disassemble the engine and check components
Ancillary components	Engine components or ancillary components loose or broken	Check that all components are secure, tighten/adjust as required. Renew if broken

2.6 ELECTRICAL DIAGNOSTIC TECHNIQUES

2.6.1 Check the obvious first

Start all hands-on diagnostic routines with 'hand and eye checks'. In other words, look over the vehicle for obvious faults. For example, if the battery terminals are loose or corroded then put this right before carrying out complicated voltage readings. Here are some further suggestions that will at some point save you a lot of time.

- A misfire may be caused by a loose plug lead – it is easier to look for this than interpret the ignition waveforms on a scope.
- If the ABS warning light stays on – look to see if the wheel speed sensor(s) are covered in mud or oil (Figure 2.16).

KEY FACT

Start all hands-on diagnostic routines with 'hand and eye checks'.

2.6.2 Test lights and analogue meters – warning

A test lamp is ideal for tracing faults in say a lighting circuit because it will cause a current to flow, which tests out high-resistance connections. However, it is this same property that will damage delicate electronic circuits – so don't use it for any circuit that contains an electronic control unit (ECU).

SAFETY FIRST

A test lamp will cause a current to flow, which can damage delicate electronic circuits.

Figure 2.16 Bentley wiring harness.

Source: Bentley Media.

Even an analogue voltmeter can cause enough current to flow to at best give you a false reading and at worst damage an ECU – so do not use it.

A digital multimeter is ideal for all forms of testing, most have an internal resistance in excess of 10 MΩ, which means that the current they draw is almost insignificant. An LED test lamp or a logic probe is also acceptable.

KEY FACT

A digital multimeter is ideal for all forms of electrical testing.

2.6.3 Generic electrical testing procedure

The following procedure is very generic but with little adaptation can be applied to any electrical system. Refer to manufacturer's recommendations if in any doubt. The process of checking any system circuit is represented by Figure 2.17.

2.6.4 Volt drop testing

Volt drop is a term used to describe the difference between two points in a circuit. In this way we can talk about a voltage drop across a battery (normally about 12.6 V) or the voltage drop across a closed switch (ideally 0 V but may be 0.1 or 0.2 V).

The first secret to volt drop testing is to remember a basic rule about a series electrical circuit:

'The sum of all volt drops around a circuit always add up to the supply'.

The second secret is to ensure the circuit is switched on and operating – or at least the circuit should be 'trying to operate'.

In Figure 2.18 this means that, if the circuit is operating correctly, $V_1 + V_2 + V_3 = V_s$. When electrical testing therefore, and if the battery voltage is measured as say 12 V, a reading of less than 12 V at V_2 would indicate a volt drop between the terminals of V_1 and/or V_3. Likewise the correct operation of the switch, that is, it closes and makes a good connection, would be confirmed by a very low reading on V_1.

What is often described as a 'bad earth' (when what is meant is a high resistance to earth) could equally be determined by the reading on V_3. To further narrow the cause of a volt drop down, simply measure across a smaller area. The voltmeter V_4, for example, would only assess the condition of the switch contacts.

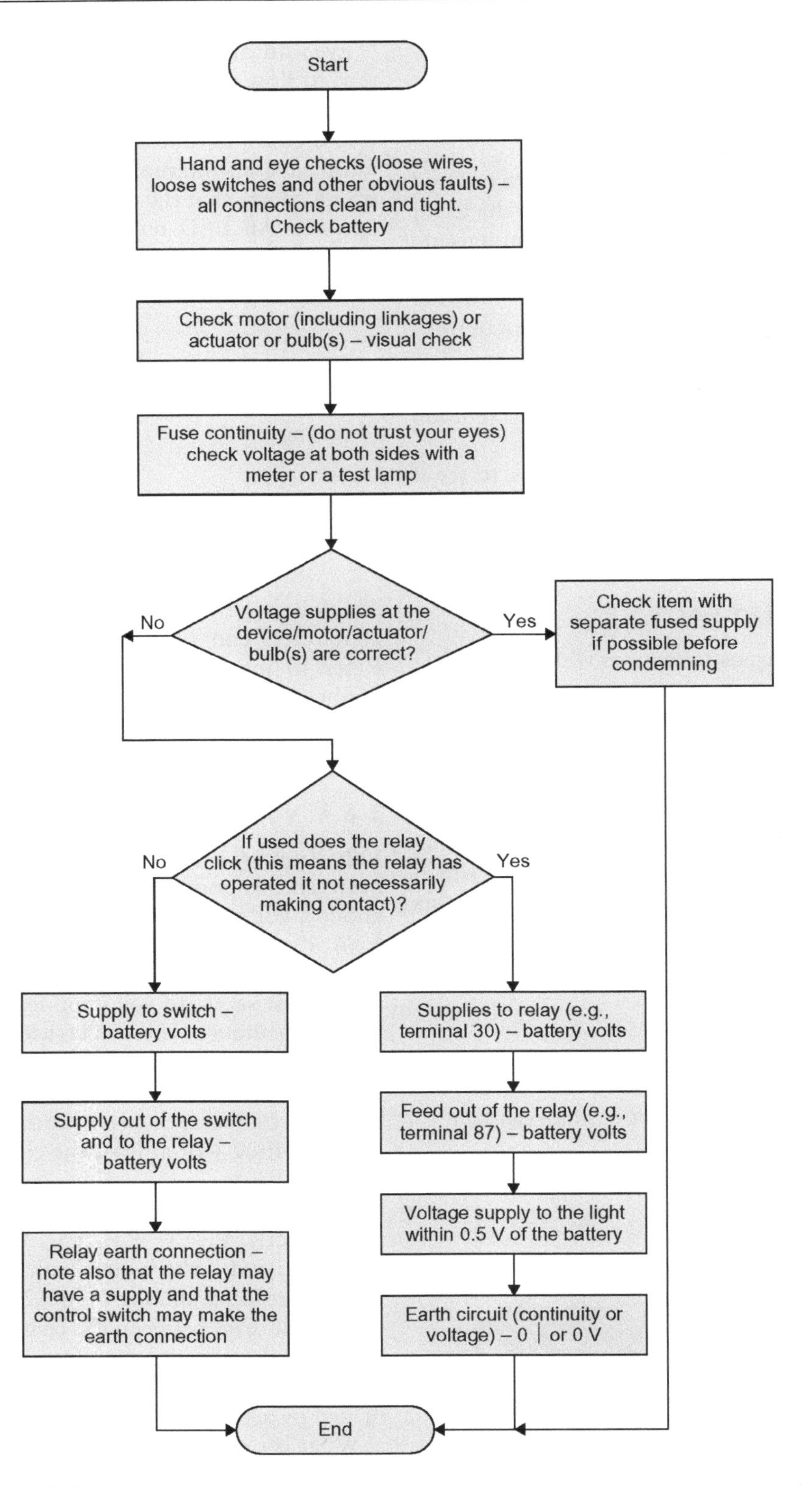

Figure 2.17 Generic electrical diagnostics chart.

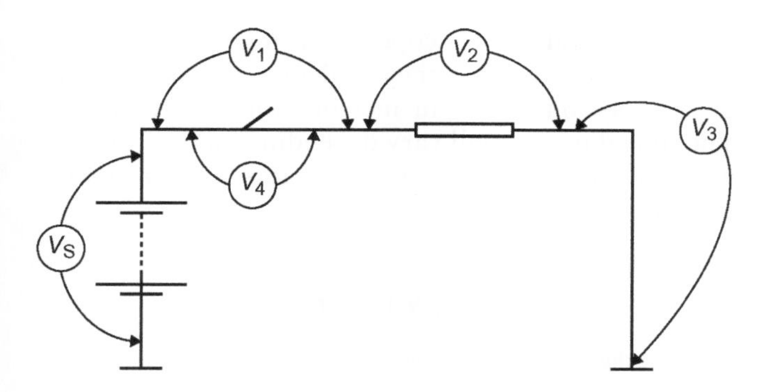

Figure 2.18 Volt drop testing.

2.6.5 Testing for short circuits to earth

This fault will normally blow a fuse – or burn out the wiring completely. To trace a short circuit is very different to looking for a high-resistance connection or an open circuit. The volt drop testing above will trace an open circuit or a high-resistance connection.

My preferred method of tracing a short, after looking for the obvious signs of trapped wires, is to connect a bulb or test lamp across the blown fuse and switch on the circuit. The bulb will light because on one side it is connected to the supply for the fuse and on the other side it is connected to earth via the short circuit fault.

Now disconnect small sections of the circuit one at a time until the test lamp goes out. This will indicate the particular circuit section that has shorted out.

KEY FACT

The sum of all volt drops around a circuit always add up to the supply.

2.6.6 On and off load tests

On load means that a circuit is drawing a current; off load means it is not. One example where this may be an issue is when testing a starter circuit. Battery voltage may be 12 V (well, 12.6 V) off load, but may be as low as 9 V when on load (cranking a cold engine perhaps).

A second example is the supply voltage to the positive terminal of an ignition coil via a high-resistance connection (corroded switch terminal for example). With the ignition on and the vehicle not running, the reading will almost certainly be battery voltage because the ignition ECU switches off the primary circuit and no volt drop will show up. However, if the circuit were switched on (with a fused jumper lead if necessary) a lower reading would result showing up the fault.

2.6.7 Black box technique

The technique outlined here is known as 'black box faultfinding'. This is an excellent technique and can be applied to many vehicle systems from engine management and ABS to cruise control and instrumentation.

As most systems now revolve around an ECU, the ECU is considered to be a 'black box'; in other words, we know what it should do but the exact details of how it does it are less important.

KEY FACT

Most vehicle systems involve an ECU.

Figure 2.19 shows a block diagram that could be used to represent any number of automobile electrical or electronic systems. In reality the arrows from the 'inputs' to the ECU and from the ECU to the 'outputs' are wires. Treating the ECU as a 'black box' allows us to ignore its complexity. The theory is that if all the sensors and associated wiring to the 'black box' are OK, all the output actuators and their wiring are OK and the supply/earth (ground) connections are OK, then the fault must be the 'black box'. Most ECUs are very reliable however, and it is far more likely that the fault will be found in the inputs or outputs.

Normal faultfinding or testing techniques can be applied to the sensors and actuators. For example, if an ABS system uses four inductive-type wheel speed sensors, then an easy test is to measure their resistance. Even if the correct value were not known, it would be very unlikely for all four to be wrong at the same time so a comparison can be made. If the same resistance reading is obtained on the end of the sensor wires at the ECU then almost all of the 'inputs' have been tested with just a few ohmmeter readings.

The same technique will often work with 'outputs'. If the resistance of all the operating windings in say a

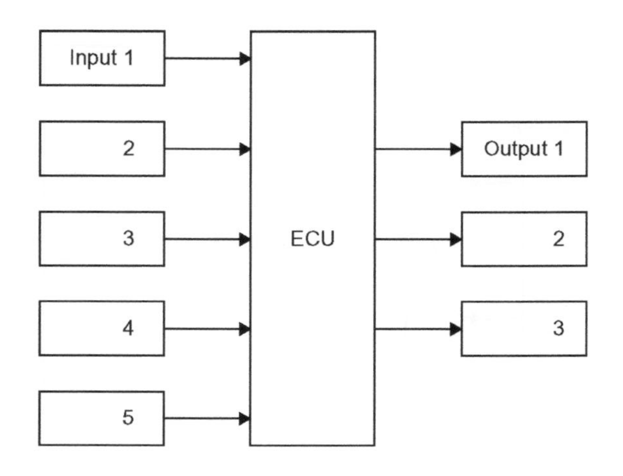

Figure 2.19 System block diagram.

hydraulic modulator were the same, then it would be reasonable to assume the figure was correct.

Sometimes, however, it is almost an advantage not to know the manufacturer's recommended readings. If the 'book' says the value should be between 800 and 900 Ω, what do you do when your ohmmeter reads 905 Ω? Answers on a postcard please... (or see Section 2.3.3).

Finally, don't forget that no matter how complex the electronics in an ECU, they will not work without a good power supply and an earth.

KEY FACT

If the resistance of all similar items connected to an ECU is the same, then it is reasonable to assume the figure is almost certainly correct.

2.6.8 Sensor to ECU method

This technique is simple but very useful. Figure 2.20 shows a resistance test being carried out on a component. Ω_1 is a direct measure of its resistance, whereas Ω_2 includes the condition of the circuit. If the second reading is the same as the first then the circuit must be in good order.

Warning: The circuit supply must always be off when carrying out ohmmeter tests.

2.6.9 Flight recorder tests

It is said that the best place to sit in an aeroplane is on the black box flight recorder. Personally, I would prefer to be in 'first class'! Also – apart from the black box usually being painted bright orange so it can be found after a crash – my reason for mentioning it is to consider how the flight recorder principle can be applied to automotive diagnostics.

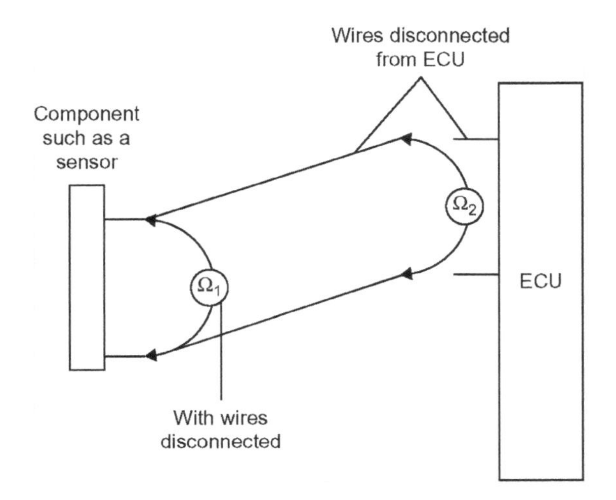

Figure 2.20 Ohmmeter testing.

Most digital oscilloscopes have flight record facilities. This means that they will save the signal from any probe connection in memory for later playback. The time duration will vary depending on the available memory and the sample speed but this is a very useful feature.

KEY FACT

Most digital oscilloscopes have flight record facilities.

As an example, consider an engine with an intermittent misfire that only occurs under load. If a connection is made to the suspected component (coil HT output for example), and the vehicle road tested, the waveforms produced can be examined afterwards.

Many engine (and other system) ECUs have built-in flight recorders in the form of self-diagnostic circuits. If a wire breaks loose causing a misfire but then reconnects, the faulty circuit will be 'remembered' by the ECU.

2.6.10 Faultfinding by luck – or is it logic?

Actually, what this section considers is the benefit of playing the odds which, while sometimes you get lucky, is still a logical process.

If four electric windows stopped working at the same time, it would be very unlikely that all four motors had burnt out. On the other hand if just one electric window stopped working, then it may be reasonable to suspect the motor. It is this type of reasoning that is necessary during faultfinding. However, be warned that it is theoretically possible for four motors to apparently burn out all at the same time.

Using this 'playing the odds' technique can save time when tracing a fault in a vehicle system. For example, if both stop lights do not work and everything else on the vehicle was OK, I would suspect the switch (stages 1–3 of the six-stage process). At this stage though, the fault could be anywhere – even two or three blown bulbs. Nonetheless a quick test at the switch with a voltmeter would prove the point. Now, let's assume the switch is OK and it produces an output when the brake pedal is pushed down. Testing the length of wire from the front to the back of the vehicle further illustrates how 'luck' comes into play.

Figure 2.21 represents the main supply wire from the brake switch to the point where the wire 'divides' to each individual stop light (the odds say the fault must be in this wire). For the purpose of this illustration we will assume the open circuit is just before point 'I'. The procedure continues in one of the two following ways:

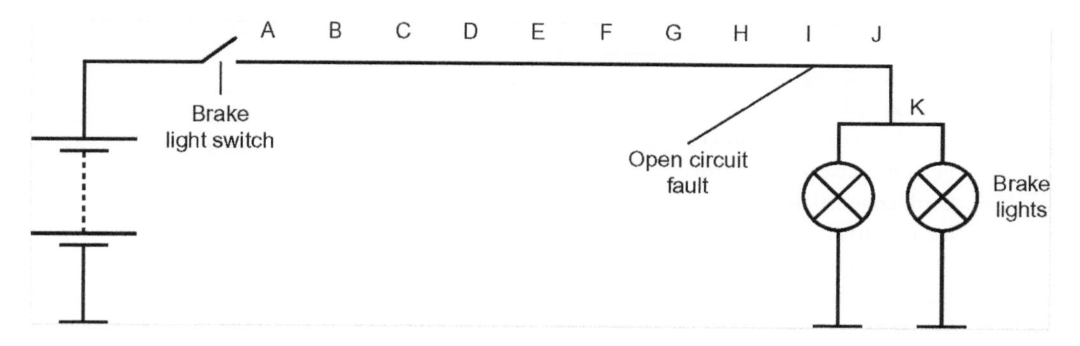

Figure 2.21 Faultfinding by playing the odds – sometimes you get lucky.

One

- Guess that the fault is in the first half and test at point F.
- We were wrong. Guess that the fault is in the first half of the second half and test at point I.
- We were right. Check at H and we have the fault ... In only 3 tests

Two

- Test from A to K in a logical sequence of tests.
- We would find the fault ... In 9 tests

You may choose which method you prefer.

2.6.11 Colour codes and terminal numbers

It is useful to become familiar with a few key wire colours and terminal numbers when diagnosing electrical faults. As seems to be the case for any standardisation a number of colour code systems are in operation.

A system used by a number of manufacturers is based broadly on the information in Table 2.8. After some practice with the use of colour codes the job of the technician is made a lot easier when faultfinding an electrical circuit.

KEY FACT

Further reference should always be made to manufacturer's information for specific details.

A system now in use almost universally is the terminal designation system in accordance with DIN 72 552. This system is to enable easy and correct connections to be made on the vehicle, particularly in after-sales repairs. Note that the designations are not to identify individual wires but are to define the terminals of a device. Listed in Table 2.9 are some of the most popular numbers.

Ford motor company, and many others, now uses a circuit numbering and wire identification system. This is in use worldwide and is known as Function-System-Connection (FSC). The system was developed to assist in vehicle development and production processes. However, it is also very useful to help the technician with faultfinding. Many of the function codes are based on the DIN system. Note that earth wires are now black.

The system works as follows: 31S-AC3A || 1.5 BK/RD

Function:
31 = ground/earth
S = additionally switched circuit

System:
AC = headlamp levelling

Table 2.8 Colour codes in use in Europe and elsewhere

Colour	*Symbol*	*Destination/Use*
Red	Rt	Main battery feed
White/Black	Ws/Sw	Headlight switch to dip switch
White	Ws	Headlight main beam
Yellow	Ge	Headlight dip beam
Grey	Gr	Sidelight main feed
Grey/Black	Gr/Sw	Left-hand sidelights
Grey/Red	Gr/Rt	Right-hand sidelights
Black/Yellow	Sw/Ge	Fuel injection
Black/Green	Sw/Gn	Ignition controlled supply
Black/White/Green	Sw/Ws/Gn	Indicator switch
Black/White	Sw/Ws	Left-side indicators
Black/Green	Sw/Gn	Right-side indicators
Light Green	LGn	Coil negative
Brown	Br	Earth
Brown/White	Br/Ws	Earth connections
Pink/White	KW	Ballast resistor wire
Black	Sw	Reverse
Black/Red	Sw/Rt	Stop lights
Green/Black	Gn/Sw	Rear fog light

Table 2.9 DIN terminal numbers (examples)

1	Ignition coil negative
4	Ignition coil high tension
15	Switched positive (ignition switch output)
30	Input from battery positive
31	Earth connection
49	Input to flasher unit
49a	Output from flasher unit
50	Starter control (solenoid terminal)
53	Wiper motor input
54	Stop lamps
55	Fog lamps
56	Headlamps
56a	Main beam
56b	Dip beam
58L	Left-hand sidelights
58R	Right-hand sidelights
61	Charge warning light
85	Relay winding out
86	Relay winding input
87	Relay contact input (change over relay)
87a	Relay contact output (break)
87b	Relay contact output (make)
L	Left side indicators
R	Right side indicators
C	Indicator warning light (vehicle)

Connection:
3 = switch connection
A = branch

Size:
1.5 = 1.5 mm^2

Colour:
BK = Black (determined by function 31)
RD = Red stripe (Tables 2.10 and 2.11)

Table 2.10 Colour codes table

Code	*Colour*
BK	Black
BN	Brown
BU	Blue
GN	Green
GY	Grey
LG	Light Green
OG	Orange
PK	Pink
RD	Red
SR	Silver
VT	Violet
WH	White
YE	Yellow

Table 2.11 Ford system codes

Letter	*Main system*	*Examples*
D	Distribution systems	DE = earth
A	Actuated systems	AK = wiper/washer
B	Basic systems	BA = charging BB = starting
C	Control systems	CE = power steering
G	Gauge systems	GA = level/pressure/ temperature
H	Heated systems	HC = heated seats
L	Lighting systems	LE = headlights
M	Miscellaneous systems	MA = air bags
P	Powertrain control systems	PA = engine control
W	Indicator systems ('indications' not turn signals)	WC = bulb failure
X	Temporary for future features	XS = could mean too much?

It should be noted that the colour codes and terminal designations given in this section are for illustration only.

2.6.12 Back probing connectors

If you are testing for a supply (for example, at a sensor, actuator or ECU), then use the probes of your digital meter or scope with care. Connect to the back of the terminals, as this will not damage the connecting surfaces as long as you do not apply excessive force (Figure 2.22). Sometimes a pin clamped in the test lead's crocodile/alligator clip is ideal for connecting 'through' the insulation of a wire without having to disconnect it. However, piercing the insulation of a

Figure 2.22 Test the voltage by back probing a connector with care.

Figure 2.23 Terminal adapters.

wire should be the last resort and it should be sealed afterwards.

Using a special adapter kit or leads is a better way and also ensures a definite connection (Figure 2.23).

2.7 SYSTEMS

2.7.1 What is a system?

System is a word used to describe a collection of related components, which interact as a whole. A motorway system, the education system or computer systems are three varied examples. A large system is often made up of many smaller systems which in turn can each be made up of smaller systems and so on. Figure 2.24 shows how this can be represented in a visual form. One further definition: 'A group of devices serving a common purpose'.

DEFINITION

System: From the Latin *systēma*, in turn from Greek *σύστημα systēma*, system is a set of interacting or interdependent system components forming an integrated whole.

Using the systems approach helps to split extremely complex technical entities into more manageable parts. It is important to note, however, that the links between the smaller parts and the boundaries around them are also very important. System boundaries will overlap in many cases.

The modern motor vehicle is a complex system and in itself forms just a small part of a larger transport system.

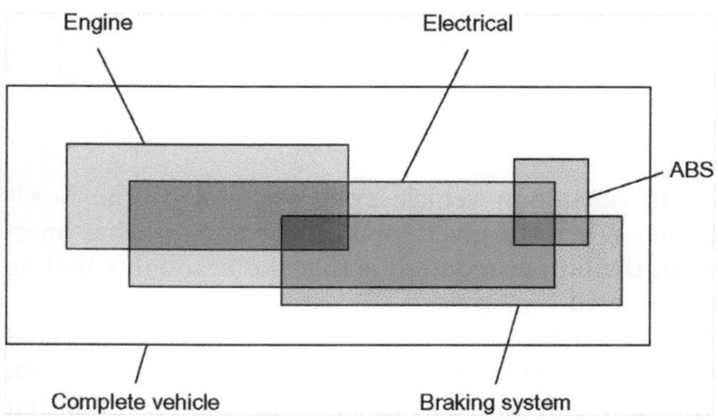

Figure 2.24 Systems in systems representation.

It is the ability for the motor vehicle to be split into systems on many levels which aids both in its design and construction. The systems approach helps in particular with understanding of how something works and further how to go about repairing it when it doesn't.

2.7.2 Vehicle systems

Splitting the vehicle into systems is not an easy task because it can be done in many different ways. A split between mechanical systems and electrical systems would seem a good start. However, this division can cause as many problems as it solves. For example, in which half do we put antilock brakes, mechanical or electrical? The answer is of course both. Nonetheless, it still makes it easier to be able to just consider one area of the vehicle and not have to try to comprehend the whole.

Once a complex set of interacting parts such as a motor vehicle has been 'systemised', the function or performance of each part can be examined in more detail. In other words, what each part of the system should do in turn helps to determine how each part actually works. It is again important to stress that the links and interactions between various sub-systems are a very important consideration. Examples of this would be how the power demands of the vehicle lighting system will have an effect on the charging system operation, or in the case of a fault, how an air leak from a brake servo could cause a weak air/fuel ratio.

To further analyse a system whatever way it has been sub-divided from the whole, consideration should be given to the inputs and the outputs of the system. Many of the complex electronic systems on a vehicle lend themselves to this form of analysis. Considering the ECU of the system as the control element and looking at its inputs and outputs is the recommended approach.

2.7.3 Open-loop systems

An open-loop system is designed to give the required output whenever a given input is applied. A good example

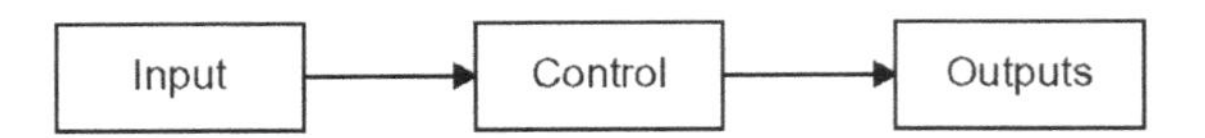

Figure 2.25 Open-loop system.

of an open-loop vehicle system would be the headlights. When the given input is the switch being operated, the output required is that the headlights will be illuminated.

This can be taken further by saying that an input is also required from the battery and a further input from, say, the dip switch. The feature, which determines that a system is open loop, is that no feedback is required for it to operate. Figure 2.25 shows this example in block diagram form.

2.7.4 Closed-loop systems

A closed-loop system is identified by a feedback loop. It can be described as a system where there is a possibility of applying corrective measures if the output is not quite what is wanted. A good example of this in a vehicle is an automatic temperature control system. The interior temperature of the vehicle is determined by the output from the heater which is switched on or off in response to a signal from a temperature sensor inside the cabin. The feedback loop is the fact that the output from the system, temperature, is also an input to the system. This is represented by Figure 2.26.

KEY FACT

A closed-loop system always has a feedback loop that may be negative or positive.

The feedback loop in any closed-loop system can be in many forms. The driver of a car with a conventional heating system can form a feedback loop by turning the heater down when he or she is too hot and turning it back up when cold. The feedback on an ABS system is a signal that the wheel is locking, where the system reacts by reducing the braking force – until it stops locking, when braking force can be increased again – and so on to maintain a steady state.

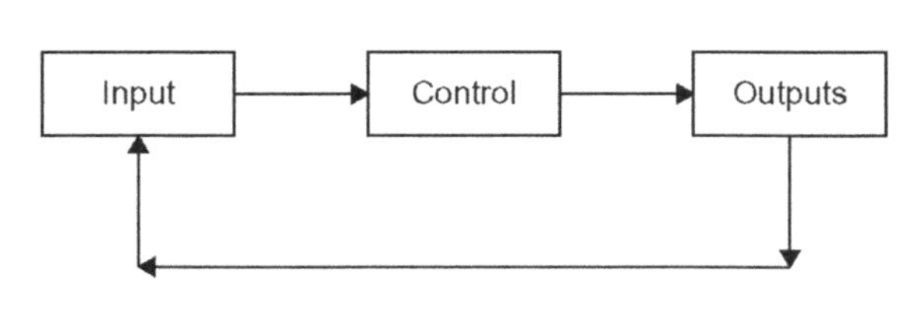

Figure 2.26 Closed-loop system.

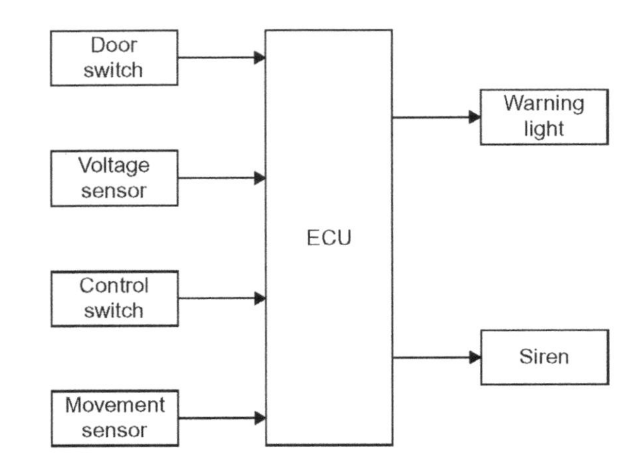

Figure 2.27 Block diagram.

2.7.5 Block diagrams

Another secret to good diagnostics is the 'block diagram' approach. Most systems can be considered as consisting of 'inputs to a control which has outputs'. This technique means that complex systems can be considered in manageable 'chunks'. It is similar to the black box method but just a different approach.

Many complex vehicle electronic systems can be represented as block diagrams. In this way several inputs can be shown supplying information to an ECU that in turn controls the system outputs. As an example of this, consider the operation of a vehicle alarm system (Figure 2.27). In its simplest form the inputs would be the 'sensors' (such as door switches) and the 'outputs' the actuators (such as the siren). The 'control' section is the alarm ECU.

The diagnostic approach is that if all the sensors are providing the correct information to the control and the actuators respond when tested, then the fault must be the control unit. If a sensor does not produce the required information then the fault is equally evident.

2.8 PASS-THROUGH

2.8.1 Overview

A pass-through device (Figures 2.28 to 2.31) is used in conjunction with a computer to reprogram vehicle control modules through the OBD-II/CANbus port.

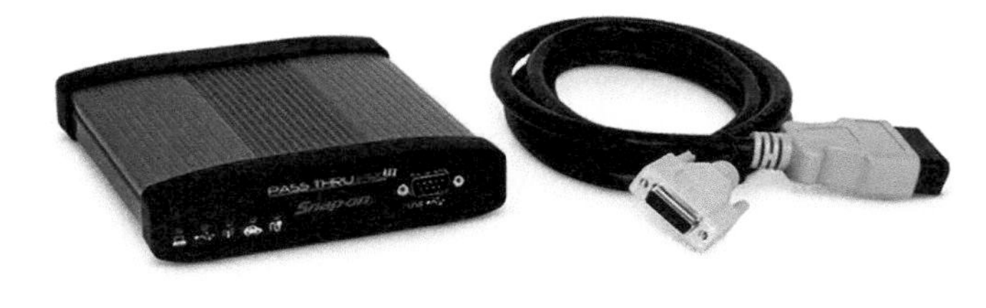

Figure 2.28 Pass-through II device.

Source: Snap-on.

Figure 2.29 Pass-through wireless device.

Source: Autologic.

It is sometimes necessary to reprogram ECUs to regulate and repair vehicles equipped with OBD systems which do not conform with pollution emission values. Alternatively, these ECUs, or others, may require updating to improve other functions or to recognise new components that have been fitted.

Figure 2.30 Bosch pass-through device.

Source: Bosch Media.

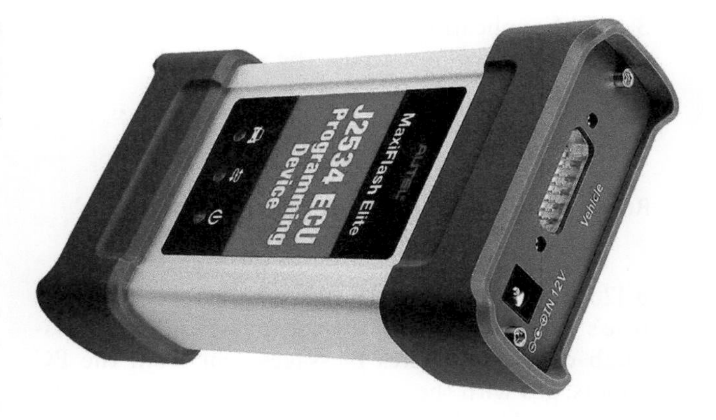

Figure 2.31 Autel Maxiflash Elite.

Source: Autel.

Each manufacturer has their own methods, but SAE International standardised the J-2534 universal requirements in 2004. This required all manufacturers of vehicles sold in the USA and Europe to accept powertrain reprogramming through specific universal parameters. In the USA and Europe, vehicle manufacturers must therefore provide ECU reprogramming functionality to all workshops, whether independent or franchised.

KEY FACT

SAE International standardized the J-2534 universal requirements in 2004.

DEFINITION

SAE: SAE International, initially established as the Society of Automotive Engineers (as in the SAE 10/40 on the oil can!), is a USA-based, globally active professional association and standards developing organization for automotive, aerospace and commercial vehicles.

Because of the existence of pass-through, vehicle manufacturers have had to ensure that the reprogramming software applications (APIs) are compatible with standardised J2534 vehicle communication interfaces (VCIs). Because of this, independent workshops can access OEM applications by subscribing to the appropriate websites. By downloading the software to a PC and connecting it to the vehicle with a J2534 VCI, you have the same level of access to the vehicle as a main dealer.

In summary, to reprogram a vehicle ECU you need:

1. Computer equipped with a Windows operating system;
2. J2534 vehicle communication interface (VCI);

3. OEM application programming interface (API);
4. Knowledge of how to use the software!

KEY FACT

Reprogramming is sometimes described as 'flashing'.

The J2534 hardware works like a bridge between the vehicle's ECU and the PC (Figure 2.32). This pass-through device translates messages sent from the PC into messages with the same protocol being used in the vehicle ECU. J2534 supports a range of protocols.

The connection between the PC and the J2534 hardware can be chosen by the manufacturer of the device, for example, RS-232, USB or a wireless interface. The vehicle manufacturer's programming application is not dependent on the hardware connection. Therefore, any device can be used for programming any vehicle regardless of the manufacturer.

The connection between the J2534 hardware and the vehicle should be the SAE J1962 connector, more commonly called the OBD connector or the data link connector (DLC) (Figure 2.33). The maximum recommended length of the cable between the J2534 device and the vehicle is 5 metres. If the vehicle manufacturer doesn't use DLC, necessary information for connection must be provided.

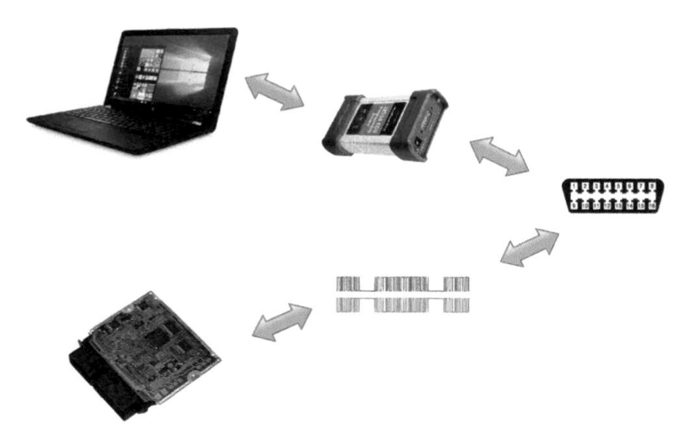

Figure 2.32 Pass-through principle and stages.

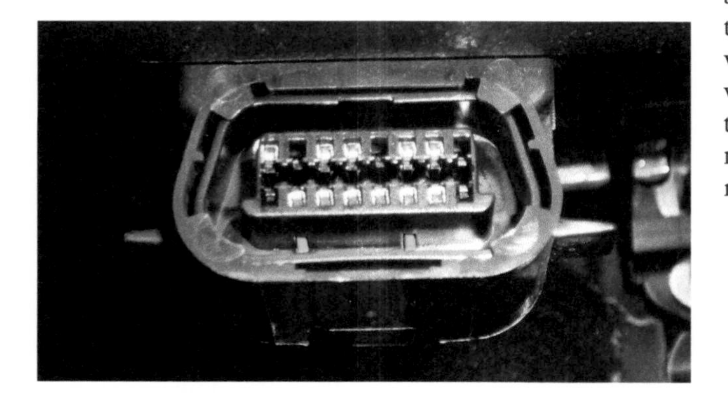

Figure 2.33 J1962 data link connector (DLC).

2.8.2 Software

Reprogramming an ECU using J2534 is done from a PC, preferably a laptop computer, with a Windows operating system. Each vehicle manufacturer has their own software application (API) used for analysing and programming their vehicles. The application will have complete information on the ECUs that are supported by it. The application also includes a user interface where choices can be made, depending on the ECU, and what action to perform.

The APIs can be downloaded from the internet or installed from a DVD. How this API is provided depends on the manufacturer, but they do make a charge. The price differs a lot between manufacturers, a one-year subscription, for example, currently costs between £75 and £2500.

Each manufacturer of the J2534 tool (hardware device) must have a DLL file that includes functions and routines for communicating with the PC. This DLL is then loaded into the vehicle manufacturer's application.

DEFINITION

DLL: Dynamic link library. DLLs provide a mechanism for shared code and data, allowing a developer of shared code/data to upgrade functionality without requiring applications to be re-linked or re-compiled.

The intention is that every J2534 tool should be capable of communicating with all protocols supported by the standard. The connection and initialisation process starts by information being sent to the hardware tool about which protocol is being used. Thereafter it is up to the hardware tool to manage the connection to the vehicle with the desired protocol. The PC application will send messages in the previously determined protocol format to the hardware tool, which buffers the messages and transmits them in the order they were received.

Chapter 3

Tools and equipment

3.1 RANGE OF EQUIPMENT

3.1.1 Introduction

Diagnostic techniques are very much linked to the use of test equipment. In other words, you must be able to interpret the results of tests. In most cases this involves comparing the result of a test to the reading given in a data source. By way of an introduction, Table 3.1 lists some of the basic words and descriptions relating to tools and equipment.

3.1.2 Basic hand tools

You will not learn how to use tools by reading a book; it is clearly a very practical skill. However, you can follow the recommendations made here and by the manufacturers. Even the range of basic hand tools is now quite daunting and very expensive. It is worth repeating the general advice given by Snap-on for the use of hand tools:

- Only use a tool for its intended purpose.
- Always use the correct size tool for the job you are doing.
- Pull a wrench rather than pushing whenever possible.
- Do not use a file or similar without a handle.
- Keep all tools clean and replace them in a suitable box or cabinet.
- Do not use a screwdriver as a pry bar.
- Always follow manufacturer's recommendations (you cannot remember everything).
- Look after your tools and they will look after you!

3.1.3 Multimeter

A digital multimeter (Figure 3.1) is an essential tool for working on vehicle electrical and electronic systems. A way of determining the quality of a digital multimeter as well as the range of facilities provided, is to consider the following:

- accuracy;
- loading effect (resistance) of the meter;
- protection circuits (fused when measuring current).

The loading effect relates to the internal resistance of the meter. It is recommended that the internal resistance of a meter should be a minimum of 10 MΩ. This not only ensures greater accuracy but also prevents the

Table 3.1 Tools and equipment

Hand tools	Spanners and hammers and screwdrivers and all the other basic bits!
Special tools	A collective term for items not held as part of a normal tool kit. Or items required for just one specific job.
Test equipment	In general, this means measuring equipment. Most tests involve measuring something and comparing the result of that measurement to data. The devices can range from a simple ruler to an engine analyser.
Dedicated test equipment	Some equipment will only test one specific type of system. The large manufacturers supply equipment dedicated to their vehicles, for example, a diagnostic device that plugs into a certain type of fuel injection ECU.
Accuracy	Careful and exact, free from mistakes or errors and adhering closely to a standard.
Calibration	Checking the accuracy of a measuring instrument.
Serial port	A connection to an electronic control unit, diagnostic tester or computer, for example. Serial means the information is passed in a 'digital' string like pushing black and white balls through a pipe in a certain order.
Code reader or scanner	This device reads the 'black and white balls' mentioned above or the on-off electrical signals and converts them into language we can understand.
Combined diagnostic and information system	These PC-based systems can be used to carry out tests on vehicle systems and they also contain electronic data. Test sequences guided by the computer can also be carried out.
Oscilloscope	The main part of a 'scope' is the display, which is a TV or computer screen. A scope is a voltmeter, but instead of readings in numbers it shows the voltage levels by a trace or mark on the screen. The marks on the screen can move and change very fast allowing us to see the way voltages change.

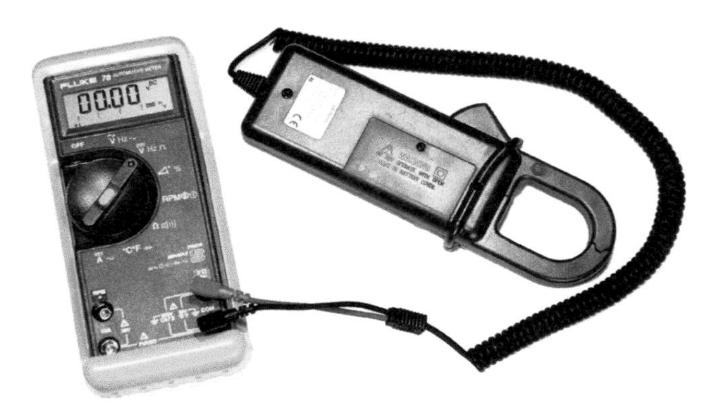

Figure 3.1 Multimeter with an amp clamp connected.

meter from damaging sensitive circuits. Meters and the leads also have category ratings that give the voltage levels to which they are safe to use. CATIII or CATIV is recommended for working on high voltage vehicles.

DEFINITION

DMM: Digital multimeter

Figure 3.2 shows two equal resistors connected in series across a 12 V supply. The voltage across each resistor should be 6 V. However, the internal resistance of the meter will affect the circuit conditions and change the voltage reading. If the resistor values were 100 kΩ, the effect of meter internal resistance would be as follows:

Meter resistance 1 MΩ

The parallel combined value of 1 MΩ and 100 kΩ = 91 kΩ. The voltage drop in the circuit across this would be:

$$91/(100 + 91) \times 12 = 5.71 \text{ V}$$

This is an error of about 5%.

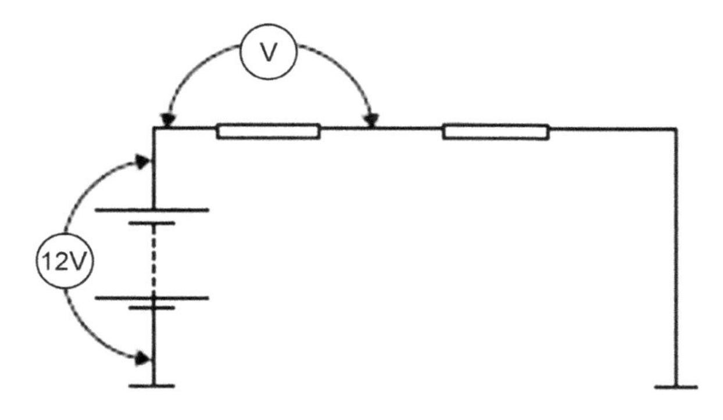

Figure 3.2 Loading effect of a meter.

Meter resistance 10 MΩ

The parallel combined value of 10 MΩ and 100 KΩ = 99 KΩ. The voltage drop in the circuit across this would be:

$$99/(100 + 99) \times 12 = 5.97 \text{ V}$$

This is an error of about 0.5%.

KEY FACT

An 'invasive measurement' error is in addition to the basic accuracy of the meter.

Understanding accuracy is important, but there are two further skills that are important when using a multimeter: where to put the probes and what the reading you get actually means!

KEY FACT

A voltmeter connects in parallel across a circuit.
An ammeter connects in series.
An ohmmeter connects across a component – but the circuit must be isolated.

3.1.4 Diagnostic scanners overview

As with most things, you get what you pay for, and the more you pay the greater the coverage of vehicles. However, it is accepted in the trade at large that to cover 99% of the different vehicles out there, and be able to carry out all the required functions, you may need up to three different devices.

3.1.4.1 VCDS

This system (Figure 3.3) is designed specifically for VW, Audi, Seat and Skoda passenger cars and remains current due to regular software updates. Like all scanners, it has a vehicle communication interface that plugs into the diagnostic socket on the car. It can then be connected by USB or wireless using a Windows computer running the VCDS software. The facilities available are almost as comprehensive as the dealership tools. They include scanning for faults as well as reading live data, resetting service interval warnings and much more. It costs a fraction of the price of a top range system but, of course, the coverage is restricted.

Figure 3.3 VCDS vehicle connection interface (VCI).

3.1.4.2 Snap-on Verdict

The Verdict (Figure 3.4) diagnostic and information system includes a wireless display, scanner and meter/scope. The advantage of this is that the information is mirrored to the main unit, which in turn can be connected to a large display screen or used on its own. It runs the software in Windows. You can view the readings and control the scanner and scope remotely from the main display unit. This is very convenient, for example, if you want to compare a scope waveform with a known good example. A repair information system is built into the software package. The Verdict is a few years old now and newer models are available, but it is still a good example – and I like it!

The wireless scanner module connects to the display tablet, which is operated using a stylus. One keyless adapter covers OBD2 compliant vehicles. It will connect to dozens of systems on most vehicle makes, including European, Asian and USA vehicles.

The wireless meter has two channels for component testing. The high-speed oscilloscope and digital graphing meter have CATIII/CATIV certification, so they are safe for use on hybrid and electric vehicles. They can be used with the Verdict display or as a standalone tool.

There are newer systems available from Snap-on, but this tool is a good example of an integrated system that combines a scanner, meter, scope and information system.

3.1.4.3 TopDon ArtipadI

I have included this device (Figure 3.5) because it has received excellent feedback from technicians working in the trade as being a high-quality device at a very competitive price. It also allows some pass-through operations. It allows quick and complete diagnoses, provided the user operates it correctly, of course, and it can interpret the information presented.

With the associated Bluetooth VCI device, this diagnostic tool is capable of ECU programming for Benz, BMW, VW, Audi and more. This diagnostic scanner is suitable for over 2000 European-based and newer OBD2 vehicles.

The device has rubberised outer protection and a rugged internal housing, making it sturdy and durable. The 30 cm touchscreen enables several functions to be run at the same time to allow comparison between test results.

Figure 3.4 Snap-on Verdict display unit, scanner and multimeter/oscilloscope.

Figure 3.5 TopDon AtripadI main unit and VCI.

I recently used this device on a PHEV and, as an example of its capabilities, I could log the hybrid battery voltage and current while driving the car. This could then be replayed and graphed. The interface is easy to use and very responsive. Adapters are available, but in its basic form it does not work as a scope or test meter.

3.1.5 Oscilloscope

An oscilloscope draws a graph of voltage (the vertical scale or Y-axis) against time (the horizontal scale or X-axis). The trace is made to move across the screen from left to right and then to 'fly back' and start again. The frequency at which the trace moves across the screen is known as the time base, which can be adjusted either automatically or manually. The voltage from the item under test can either be amplified or attenuated (reduced), much like changing the scale on a voltmeter.

This oscilloscope (a PicoScope is shown in Figure 3.6) connects to a USB port and can take up to 32 million samples per trace, making it possible to capture complex automotive waveforms, and then zoom in on areas of interest. Since they are PC-based, these waveforms can then be saved for future reference, printed or emailed.

3.1.6 Breakout box

A breakout box (Figure 3.7) is a simple piece of electrical test equipment used to support diagnostics by providing easy access to test signals. The type shown here connects to a vehicle diagnostic link connector (DLC) and because it has a plug and corresponding socket, other test equipment can be used at the same time. The terminals on the box can then be used, for example, to connect an oscilloscope to monitor CAN signals.

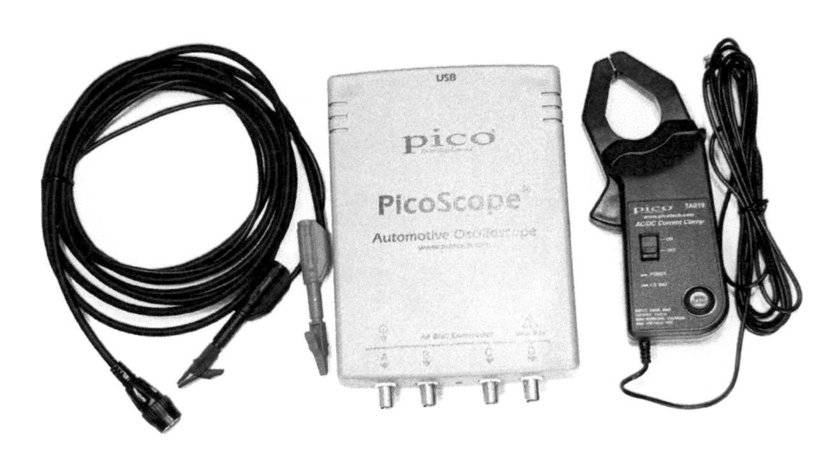

Figure 3.6 This PicoScope is a few years old now but still working well and a great piece of equipment.

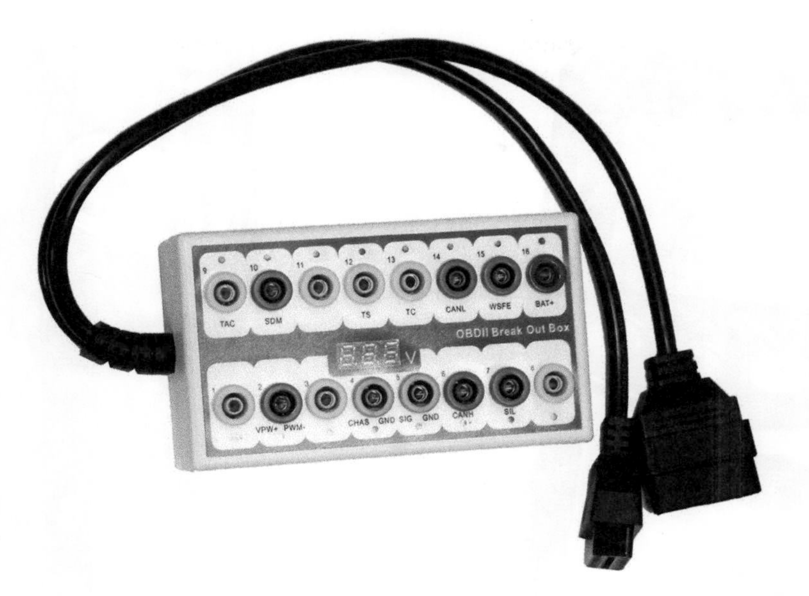

Figure 3.7 OBDII breakout box.

3.1.7 Battery tester

Traditionally batteries were tested with a voltmeter and a high rate discharge tester. This monitored the battery voltage while causing a current discharge of 100s of amps. Some batteries exploded during the process! A safer and more reliable way to check batteries is conductance testing.

The device shown here will test 12 V batteries rated from 100 to 2000 CCA (Figure 3.8). Using advanced conductance measurement technology, it shows battery health status including voltage, cranking power, state of charge and more. It displays one of five easy-to-understand results that can be printed out for the customer:

- good battery;
- good battery, need recharge;
- replace, life cycle;
- replace, bad cell;
- charge, retest.

Built-in reverse-connection protection prevents damage to the battery or tester. It can be used for batteries on or off the vehicle.

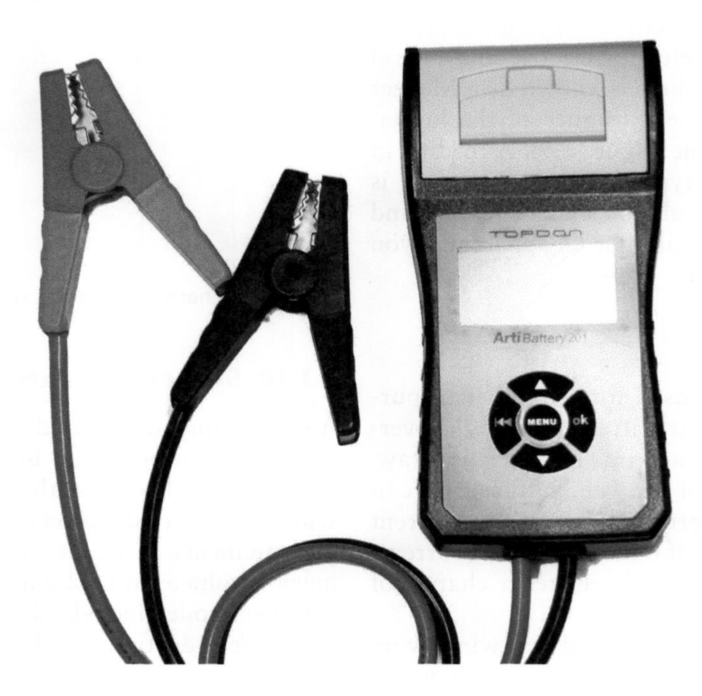

Figure 3.8 TopDon battery tester.

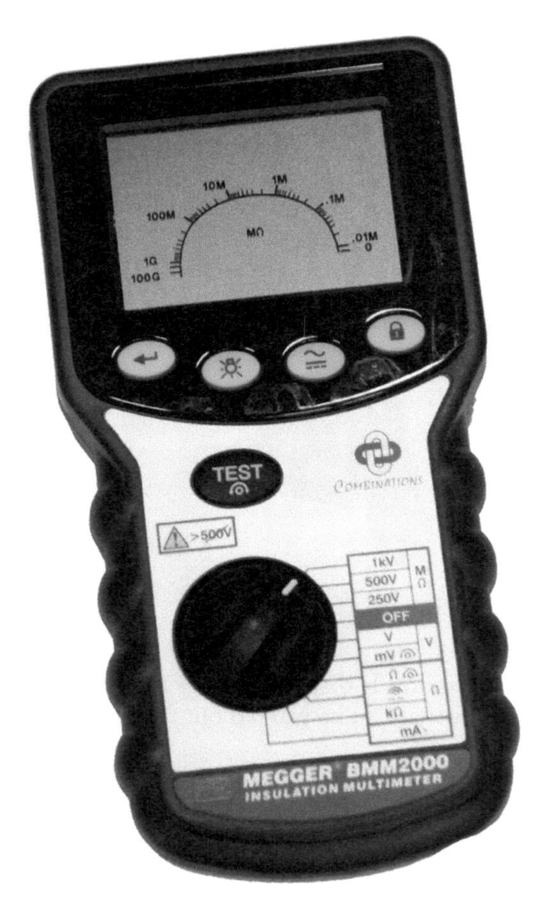

Figure 3.9 Megger multimeter and insulation tester.

Figure 3.10 Flir thermal camera fitted to an iPhone.

3.1.8 Insulation tester

An insulation tester does exactly as its name suggests. This test is particularly useful for electric and hybrid vehicles. The device shown here is known as a Megger and uses 1000 V to test the resistance of the insulation on a wire or component (Figure 3.9). A reading in excess of 10 megaohms is typical. The high voltage is used because it puts the insulation under pressure and will show up faults that would not be apparent if you used an ordinary ohmmeter.

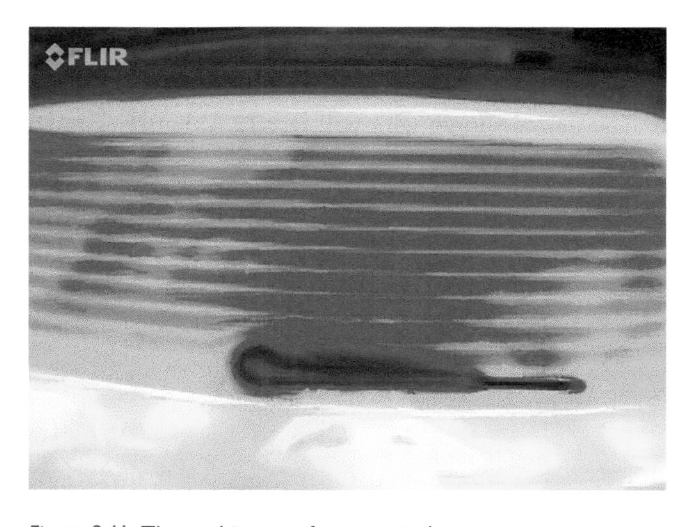

Figure 3.11 Thermal image of a rear window.

3.1.9 Thermal camera

A thermal camera can be used for a number of purposes. If, for example, a car battery discharges overnight, it may be due to a parasitic current draw. Taking a thermal image of the vehicle fuse box can help to show where the problem is because current flow causes heat, not much if it is a very small current, but a sensitive camera can still register a change of temperature (Figure 3.10).

The image shown here is of a heated rear window on a car as it was first switched on. In this case it shows all the elements to be intact (Figure 3.11).

3.1.10 Sensor simulator

A sensor simulator is used to replace a suspected faulty sensor for test purposes. By simulating the sensor output, you can check that the signals generated are being sent and received correctly by an ECU. Simulation measurements available on this device include frequency, voltage and O2 sensor signals.

Other modes include simulation of crankshaft and camshaft inductive and Hall effect sensors. In these modes it is possible to adjust the number of active teeth and the number of missing teeth.

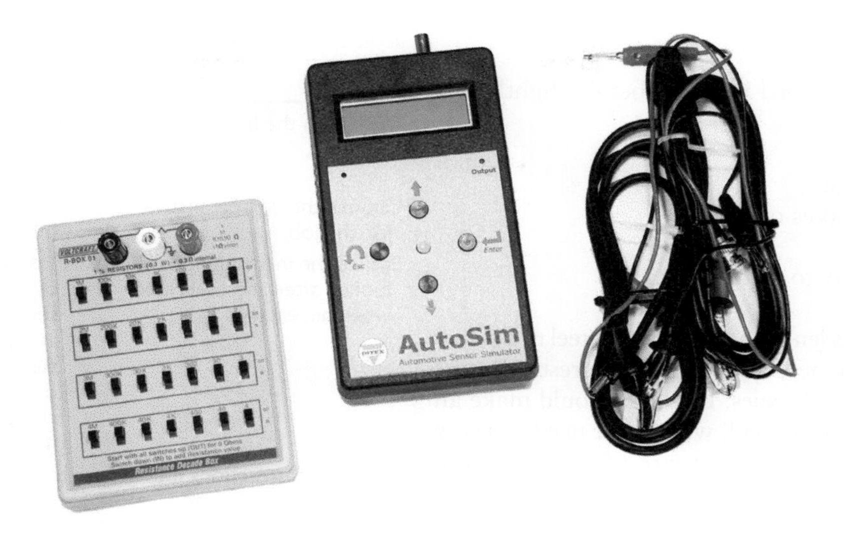

Figure 3.12 A decade box that allows resistance values to be selected and an AutoSim sensor simulator.

The simulator is useful for circuit tests because it can simulate the voltages and you can check the live data readings on a scan tool to confirm the wiring. Also, simulating an O2 sensor signal, for example, can be used to check ECU response and operation.

3.1.11 Actuator driver

Automotive actuators work by being supplied appropriate electrical signals. There is the simple on or off signal as supplied to most petrol fuel injectors, or a more complex set of signals to operate a stepper motor.

The device shown here can supply a pulse width modified (PWM) signal at a current of a few amps if necessary (Figure 3.13). PWM signals are used for many different actuators, but a typical example would be a throttle control valve. When connected in place of the normal supply from the ECU, it will prove if the actuator is working or not.

3.1.12 Logic probe

This device is a useful way of testing logic circuits, but it is also useful for testing some types of sensor. Most types consist of two power supply wires and a metal 'probe'. The display consists of three LEDs labelled 'high', 'low' and 'pulse'. These LEDs light up together with an audible signal in some cases, when the probe touches either a high, low or pulsing voltage. Above or below 2.5 V is often used to determine high or low on a 5 V circuit.

Figure 3.13 The unit will produce a pulse width modified (PWM) output to drive an actuator such as a throttle controller.

3.1.13 Accuracy of test equipment

Accuracy can be described in a number of slightly different ways:

- careful and exact;
- free from mistakes or errors;
- precise;
- adhering closely to a standard.

Consider measuring a length of wire with a steel rule. How accurately could you measure it? To the nearest 0.5 mm? This raises a number of issues. First, you could make an error reading the ruler. Second, why do we need to know the length of a bit of wire to the nearest 0.5 mm? Third, the ruler may stretch and not give the correct reading!

The first and second issues can be dispensed with by knowing how to read the test equipment correctly and also knowing the appropriate level of accuracy required. A micrometer for a plug gap? A ruler for valve clearances? I think you get the idea. The accuracy of the equipment itself is another issue.

Accuracy is a term meaning how close the measured value of something is to its actual value. For example, if a length of approximately 30 cm is measured with an ordinary wooden ruler, then the error may be up to 1 mm too high or too low. This is quoted as an accuracy of ±1 mm. This may also be given as a percentage, which in this case would be 0.33%.

DEFINITION

Accuracy: How close the measured value of something is to the actual value.

Resolution or, in other words, the 'fineness' with which a measurement can be made is related to accuracy. If a steel ruler was made to a very high standard but only had markings of 1/cm, it would have a very low resolution even though the graduations were very accurate. In other words, the equipment is accurate but your reading will not be!

DEFINITION

Resolution: The 'fineness' with which a measurement can be made.

To ensure instruments are, and remain accurate, there are just two simple guidelines:

1. Look after the equipment; a micrometer thrown on the floor will not be accurate.
2. Ensure instruments are calibrated regularly – this means being checked against known good equipment.

Table 3.2 Accurate measurement process

Step	*Example*
Decide on the level of accuracy required.	Do we need to know that the battery voltage is 12.6 V or 12.635 V?
Choose the correct instrument for the job.	A micrometer to measure the thickness of a shim.
Ensure the instrument has been looked after and calibrated when necessary.	Most instruments will go out of adjustment after a time. You should arrange for adjustment at regular intervals. Most tool suppliers will offer the service or in some cases you can compare older equipment to new stock.
Study the instructions for the instrument in use and take the reading with care. Ask yourself if the reading is about what you expected.	Is the piston diameter 70.75 or 170.75 mm?
Make a note if you are taking several readings.	Don't take a chance, write it down.

Table 3.3 Multimeter functions

Function	*Range*	*Accuracy*
DC voltage	500 V	0.3%
DC current	10 A	1.0%
Resistance	0–10 MΩ	0.5%
AC voltage	500 V	2.5%
AC current	10 A	2.5%
Dwell	3, 4, 5, 6, 8 cylinders	2.0%
RPM	10 000 rpm	0.2%
Duty cycle	% on/off	0.2%/kHz
Frequency	Over 100 kHz	0.01%
Temperature	>9000°C	0.3% + 30°C
High current clamp	1000 A (DC)	Depends on conditions
Pressure	3 bar	10.0% of standard scale

Table 3.2 provides a summary of the steps to ensure a measurement is accurate.

The list of functions presented in Table 3.3, broadly in order starting from essential to desirable, should be considered.

3.2 MULTIMETERS

3.2.1 Introduction

There is a range of different multimeters available and choosing the right one is essential. Indeed, like me, you may choose to own more than one, and, of course, you get what you pay for. Figure 3.14 shows my current selection (pun intended). Each of these meters has advantages and disadvantages, and some examples are listed in Table 3.4:

Table 3.4 Meter advantages and disadvantages

Multimeter	*Advantages*	*Disadvantages*
Megger	1000 V insulation tester Standard range of other functions	No CAT rating No additional features
Snap-on Verdict	CAT III 1000 V, Cat IV 600 V so ideal for H/EV work Comprehensive automotive options Accurate Oscilloscope	Expensive compared to some others, but the number of features is much larger
Fluke 78 Automotive	Comprehensive range of functions related to automotive testing such as duty cycle and RPM Accurate	This version is CAT II 300 V so not to be used on H/EVs, but this meter is over 10 years old; more up-to-date Fluke versions have higher CAT ratings
Uni-T amp clamp	Quick and easy to measure current up to 100 A – even on H/EVs as it is non-contact	Cheaper end of the market so less accurate and fewer features
Sealey pocket	Great for a first look and quick measurements as it is easily portable	Cheap, not very strong and only CAT II

My Fluke 78 automotive meter, shown as Figure 3.15, is several years old now but perfectly functional and ideal for low voltage applications. It has the following features and specifications:

- volts, amps, continuity and resistance;
- frequency for pulsed-DC and AC frequency tests;
- Duty cycle to verify operation of sensors and actuator supply signals;
- direct reading of dwell for 3-, 4-, 5-, 6- and 8-cylinder engines;
- temperature readings up to 999°C (F or C) using the thermocouple bead probe and adapter plug;

Figure 3.14 Multimeters (left to right): Megger insulation tester and meter, Snap-on Verdict meter, Fluke 78 Automotive meter, Uni-T amp clamp meter and Sealey pocket meter.

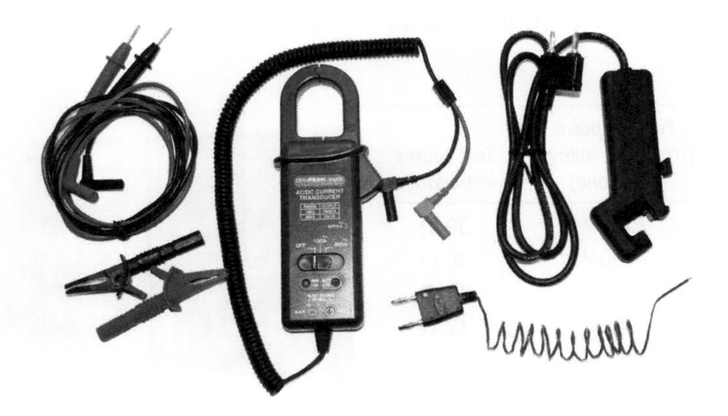

Figure 3.15 Fluke 78 Automotive meter and accessories (test leads, 600 A clamp, plug lead RPM, temperature thermocouple).

- min/max recording that works with all meter functions;
- precision analog bar graph;
- RPM inductive pickup for both conventional and distributorless (DIS) ignitions;
- 10 MΩ input impedance;
- Cat II 300 V (ideal for low voltage automotive and even mains voltages, but not recommended for H/EV high voltage use).

3.2.2 Categories

Meters and their leads have category ratings that give the voltage levels up to which they are safe to use. CAT ratings can be a little confusing but there is one simple rule of thumb: select a multimeter rated to the highest category in which it could possibly be used. In other words, err on the side of safety. Table 3.5 lists some of the different ratings.

The voltages listed in Table 3.5 are those that the meter will withstand without damage or risk to the user. A test procedure (known as IEC 1010) is used and takes three main criteria into account:

- steady-state working voltage;
- peak impulse transient voltage;
- source impedance [Impedance is the total opposition to current flow in an AC circuit (in a DC circuit it is described as resistance)].

These three criteria together will tell you a multimeter's true ***'voltage withstand'*** values. However, this is confusing because it can look as if some 600 V meters offer more protection than 1000 V ones.

Within a category, a higher working voltage is always associated with a higher transient voltage. For example, a CAT III 600 V meter is tested with 6000 V transients while a CAT III 1000 V meter is tested with 8000 V transients. This indicates that they are different, and that the second meter clearly has a higher rating. However, the 6000 V transient CAT III 600 V meter and the 6000 V transient CAT II 1000 V meter are not the same even though the transient voltages are. This is because the source impedance has to be considered.

Ohm's Law (I = V/R) shows that the 2 Ω test source for CAT III will have six times the current of the 12 Ω test source for CAT II. The CAT III 600 V meter, therefore, offers better transient protection, compared to the CAT II 1000 V meter, even though in this case the voltage rating appears to be lower.

The combination of working voltage and category determines the total 'voltage withstand' rating of a multimeter (or any other test instrument), including the very important 'transient voltage withstand' rating. Remember, for working on vehicle high voltage systems, you should **choose CAT III or CAT IV meter** (Figure 3.16) **AND leads** (Figure 3.17).

Figure 3.16 Cat III 1000 V and CAT IV 600 V meter.

Table 3.5 CAT ratings

Category	*Working voltage (voltage withstand)*	*Peak impulse (transient voltage withstand)*	*Test source impedance*
CAT I	600 V	2500 V	30 Ω
CAT I	1000 V	4000 V	30 Ω
CAT II	600 V	4000 V	12 Ω
CAT II	1000 V	6000 V	12 Ω
CAT III	600 V	6000 V	2 Ω
CAT III	1000 V	8000 V	2 Ω
CAT IV	600 V	8000 V	2 Ω

Figure 3.17 Cat III 1000 V and CAT IV 600 V leads.

3.2.3 Taking measurements

There are lots of different options or settings available when using a multimeter, but the three most common measurements are: voltage (volts), resistance (ohms) and current (amps).

3.2.3.1 Voltage

To measure voltage the meter (Figure 3.18) is connected in parallel with the circuit. Most voltage measurements on a vehicle are DC. Remember to set the range of the meter (some are auto-ranging) and, if in doubt, start with a higher range and work downwards.

As an example of how a multimeter can tell you much more than basic voltage information, it is possible, for instance, to check alternator diodes on some vehicles. Use the AC voltage function to measure ripple voltage at the rear of the alternator. With the engine running, a good alternator will show less than 0.5 V AC. A higher reading could indicate damaged alternator diodes.

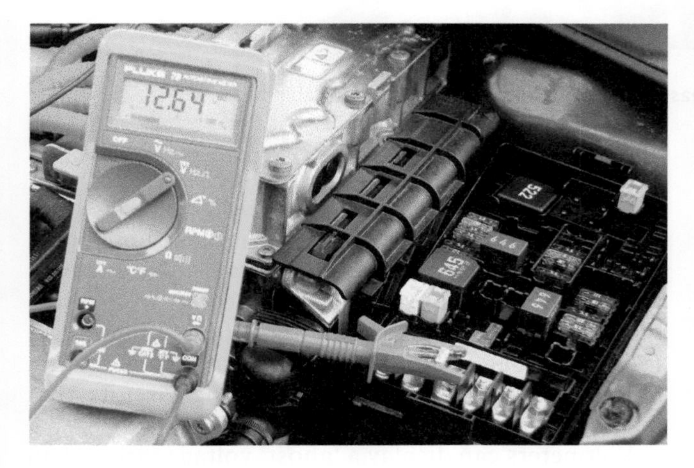

Figure 3.18 Voltage supply to a fuse box.

3.2.3.2 Resistance

To measure resistance the meter (Figure 3.19) must be connected across (in parallel with) the component or circuit under test. However, the circuit must be switched off or isolated. If not, the meter will be damaged. Likewise, because an ohmmeter causes a current to flow, there are some circuits, such as Hall effect sensors, that can be damaged by the meter.

3.2.3.3 Current

Current can be measured in two ways:

1. connecting the meter in series with the circuit (in other words break the circuit and reconnect it through the meter);
2. using an inductive amp clamp around the wire (Figure 3.20), which is a safer way to measure, but less accurate at low values.

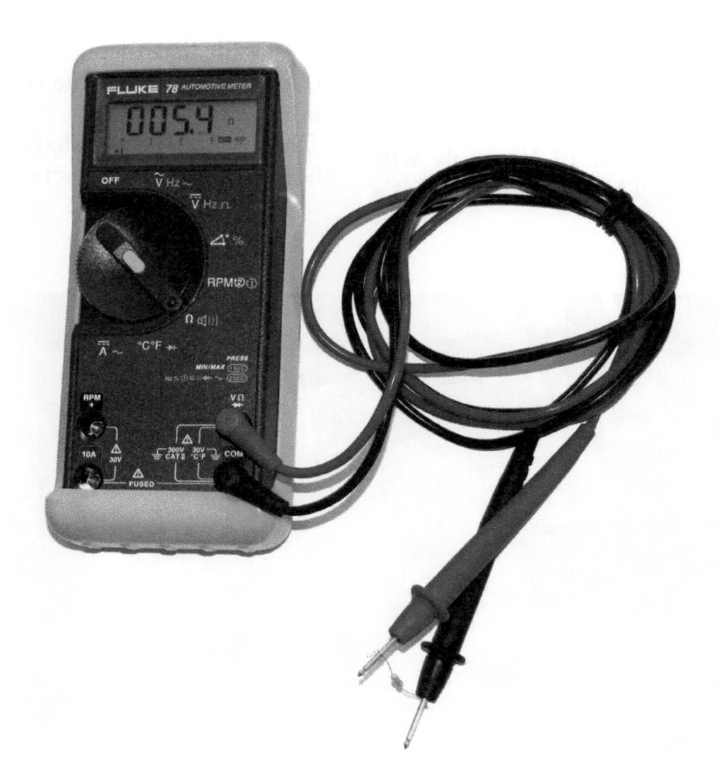

Figure 3.19 Checking a simple resistor.

Figure 3.20 Inductive ammeter clamp on a high voltage cable (measuring the current drawn by the EV cabin heater).

3.2.4 What is an open circuit voltage and how do you know zero is zero?

The internal resistance of a meter can affect the reading it gives on some circuits. A minimum of 10 MΩ is recommended, which ensures accuracy because the meter only draws a very tiny (almost insignificant) current. This stops the meter loading the circuit and giving an inaccurate reading, and it prevents damage to sensitive circuits (in an ECU, for example).

However, the very tiny current draw of a good multimeter can also be a problem. A supply voltage of 12 V, for example, can be shown on a meter when testing a circuit but does not prove the integrity of the supply. This is because a meter with a 10 MΩ internal resistance connected to a 12 V supply will only cause a current of 1.2 μA ($I = V/R$) – that's 1.2 millionths of an amp, which will not cause any noticeable voltage loss even if there is an unwanted resistance of several thousand ohms in the supply circuit. A test lamp can be connected in parallel with the meter to load the circuit (make more current flow), but should be used carefully so you don't damage sensitive electronic switching circuits that may be present.

Voltmeters can display a 'ghost' voltage rather than zero when the leads are open circuit. In other words, if checking the voltage at an earth/chassis connection we would expect a 0 V reading. However, the meter will also display zero before it is connected, so how do we know the reading is correct when it is connected?

The answer is to shake the multimeter leads. A 'ghost' voltage will fluctuate (Figure 3.21), while a

Figure 3.21 Ghost voltage caused by shaking the red lead.

real voltage will not! Thanks to my colleague Paul Danner for this tip; he has some amazing resources at: www.scannerdanner.com.

3.2.5 Voltage drop testing example

Figure 3.22 is a slightly simplified lighting circuit. The vehicle symptoms show that the left-hand side high beam headlight is dimmer than the right-hand one. The voltmeter negative lead is connected to a good earth, on the battery negative terminal (position marked 0). The ignition and the lights are switched on and the following voltage readings are obtained when the meter positive lead is connected to the numbered locations:

1. 12.62 V
2. 12.59 V
3. 12.58 V
4. 12.52 V
5. 4.33 V
6. 0.06 V (while shaking the leads!)

What is the fault? What were the reasons for testing voltage at points 1 and 2?

Answer: The fault is a high resistance in the earth wire between 5 and 6. Testing battery voltage (1) is a reference for all other tests, testing the voltage at 2 means you can compare it with the voltage at 3.

3.2.6 Insulation testing

An insulation tester does exactly as its name suggests. On automotive systems, this test is mostly used on electric and hybrid vehicles. Refer to manufacturers' information before carrying out any tests on the high voltage system – and be TechSafe™. (Find out more about TechSafe™ at tide.theimi.org.uk.)

The device in Figure 3.23 is known as a Megger. It is a multimeter but also able to use up to 1000 V to test the resistance of insulation on a wire or component. A reading well in excess of 10 MΩ (as shown in Figure 3.23) is what we would normally expect if the insulation is in good order. The high voltage is used because it puts the insulation under pressure and will show up faults that would not be apparent if you used an ordinary ohmmeter.

Take care when using insulation testers, the high voltage used for the test will not kill you because it cannot sustain a significant current flow. But it still hurts!

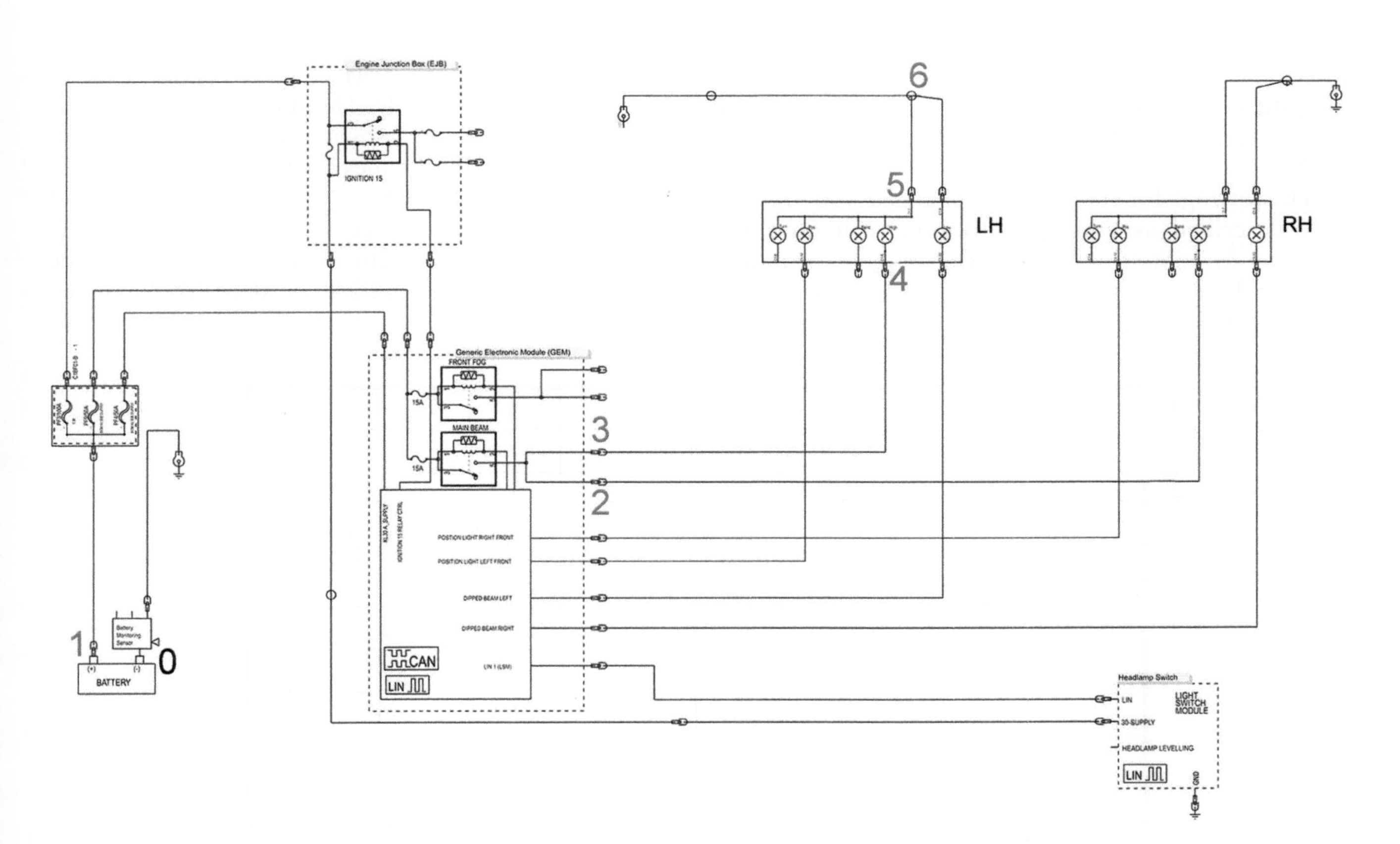

Figure 3.22 Simplified lighting circuit from a Ford vehicle.

Source: Ford Motor Company.

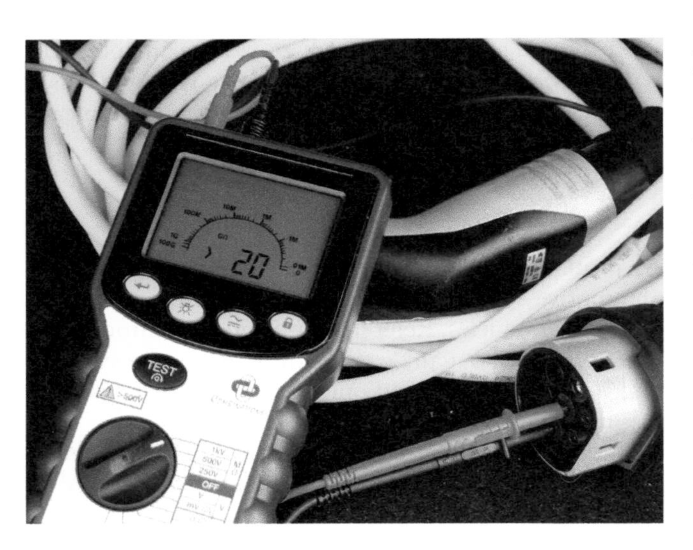

Figure 3.23 Checking the insulation resistance between conductors in an EV charging lead (in this case the reading is greater than 20 GΩ).

3.3 OSCILLOSCOPES

3.3.1 Operation

An oscilloscope (often shortened to 'scope') is an instrument used to display amplitude and period of a signal, as well as its shape. Amplitude is its height, and period is the time over which it repeats. A scope draws a two-dimensional graph. Normally, voltage is displayed on the vertical, or Y-axis, and time on the horizontal, or X-axis.

The settings for these can be changed and are usually described as volts per division (volts/div) and time per division (time/div). The graphs can be moved up and down on the screen with a control known as Y-shift. This means that the zero voltage position can, for example, be set to the middle of the screen. This means that the blue line in Figure 3.24 shows a DC voltage of 0 V. The red line represents a DC voltage of about 12 V because the volts per division (each small vertical square) in this case, is set at 5 V. The green trace shows an AC signal that, assuming the volts/div is still set at 5 V, has a peak-to-peak voltage of 20 V.

Much more information is shown compared to other instruments such as multimeters or frequency meters. For example, when using a scope, you can determine the amplitude and shape, how much noise is present and the frequency of a signal.

The frequency of a signal can be worked out from the time per division settings. In Figure 3.24 the time/div (each small horizontal square), is set to 25 ms. This means that the sinewave repeats every 100 ms (four divisions). Because frequency is measured in cycles per second (hertz, or Hz), we simply need to divide 1 s (1000 ms) by the time period, 100 ms in this case, which means that this signal has a frequency of 10 Hz.

Besides directly displaying electric signals, scopes can measure non-electrical values if appropriate transducers are used. Transducers change one kind of variable into another. For example, a pressure transducer (often described as a sensor), produces a voltage that is proportional to pressure.

The speed at which the trace moves across the screen is known as the time base, which can be adjusted either automatically or manually. The start of the trace moving across the screen is known as the trigger. This can be internal, such that it flies back and re-starts every two seconds or whatever, or it can be external, so it starts every time a fuel injector operates, for example. The voltage from the item under test can either be amplified or attenuated (reduced), much like changing the scale on a voltmeter.

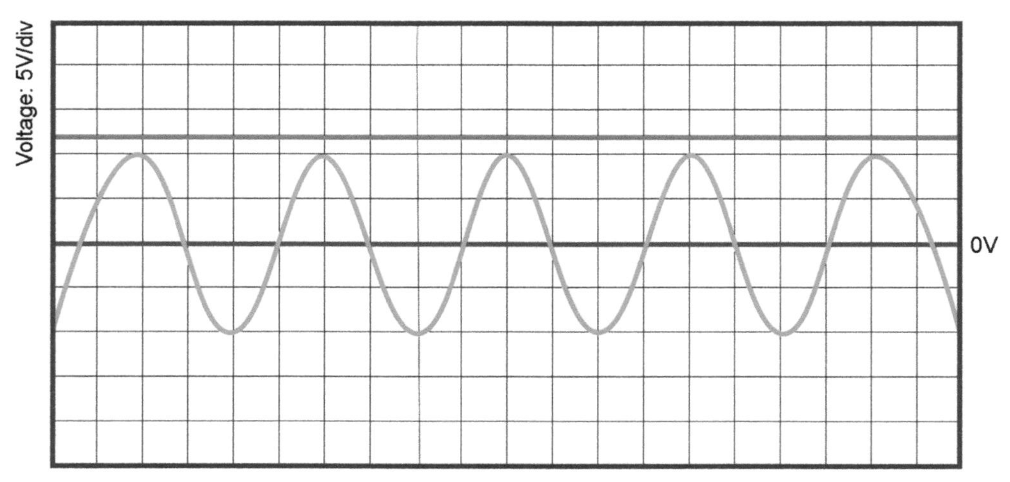

Figure 3.24 Oscilloscope screen.

Years ago, oscilloscopes were analogue and used cathode ray tubes. Now almost all automotive oscilloscopes are digital and use a computer screen to display signals. This also allows values such as the voltage and time base scales and frequency to be shown on the screen. It also means the waveforms can be saved and shared. Waveforms gathered over a period of time can also be saved so that they can be replayed after a road test, for example.

3.3.2 High voltages and safety

Some scopes, such as the Snap-on Verdict shown below, have a CAT rating (see the section on multimeters for more information), so it can be used directly with higher voltages. However, in most cases when high voltages are measured, special attenuating leads are used. These reduce the voltage by, say, a factor of ten so a 300 V signal would be attenuated to 30 V before entering the instrument. The scale is then set and adjusted accordingly.

3.3.3 AC–DC

In most cases a scope is set to measure DC voltage, even though it may appear to be showing an AC signal! This is because at any point in time it is actually measuring DC – it is just that this DC voltage varies quickly.

However, for some measurements it is useful to display and measure just the AC component of a signal. This is known as AC coupling. The voltage across the battery on a car is normally about 14 V when the engine is running, and the alternator is charging the battery. However, because the alternator rectifies AC into DC and the voltage regulator controls voltage, the voltage across the battery is 14 V but with very small variations. Figure 3.25 shows this variation (ripple) to be about 0.2 V. If the display was set to show the DC components, then the scale would not be sensitive enough to show details of the 0.2 V ripple.

3.3.4 Channels

The number of traces that can be displayed on an oscilloscope at the same time is known as the number of channels. Most scopes designed for automotive use have a minimum of two channels and more often four. This is so that different waveforms (traces, patterns, signals, etc.) can be displayed at the same time, using different voltage scales if necessary, so that comparisons can be made. Figure 3.28 shows a fuel injector signal compared to the fuel pressure. In this figure, three

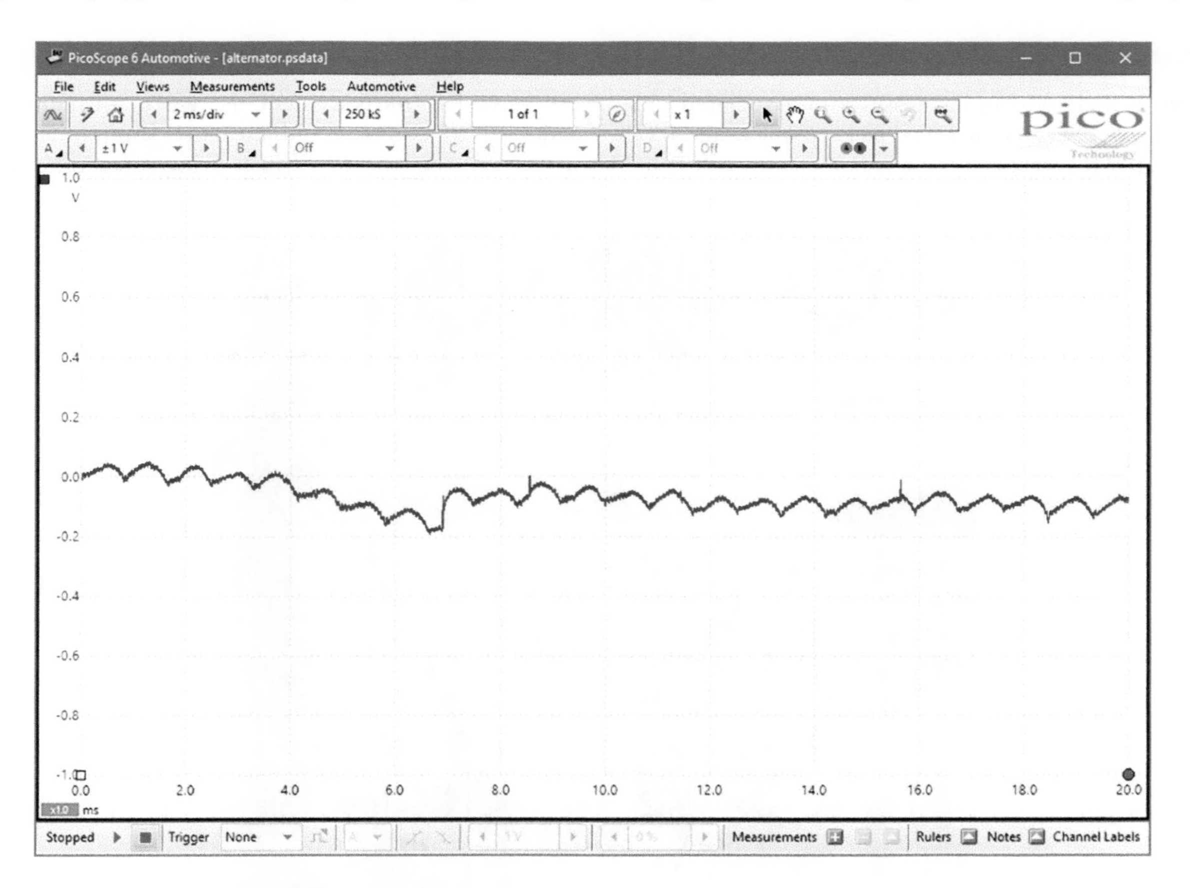

Figure 3.25 Voltage ripple show with AC coupling.

Source: PicoScope.

Figure 3.26 Pico 8-channel scope.

Source: PicoScope.

signals are being compared: injector voltage, secondary waveform and intake manifold pressure.

Pico recently released an 8-channel scope (Figure 3.26), which has already been given the appropriate nickname, the Octoscope.

3.3.5 Making connections

Modern vehicle wiring harnesses and associated plugs and sockets are very well sealed to keep out water. (Well, most are!) This means that it is sometimes difficult to make a connection with scope leads. It is very important not to damage connections when taking a reading, and unless absolutely essential, the insulation on a wire should not be pierced. Figure 3.27 shows a connection to a lambda sensor using a set of six small adapter leads.

3.3.6 Non-invasive measurement

An invasive measurement is when dismantling is required to connect and measure something. It can also mean that the act of measuring affects the reading

Figure 3.27 Adapter leads to make a secure connection to a lambda sensor.

you get. Non-invasive measurement is a great way to get a first impression of a vehicle without removing any major items, which can introduce a risk of damage.

A good example of this is to carry out a relative compression test. This is done by measuring either starter current or battery voltage while the engine is cranking. The voltage and time scales are set so that the current flow or voltage drop during compression of each cylinder can be captured and compared.

3.3.7 Wiggly lines

The phrase 'wiggly lines' is an affectionate term for waveforms or scope traces. On multichannel scopes, comparing one waveform with another is a very useful diagnostic technique. In Figure 3.28, for example, the injector signal is being compared with a fuel pressure signal. In this case a small drop in pressure is shown as the injector fires.

In Figure 3.29 the injector waveform, secondary waveform (the high voltage that creates the spark) and intake manifold pressure are being compared. This is a great way to check for correct operation of the electrical components, but also by comparing each cylinder, it can also tell you a lot about engine condition generally and valve condition in particular.

3.3.8 Example equipment

3.3.8.1 *Snap-on Verdict*

The Verdict diagnostic and information system (Figure 3.30) includes a wireless display, scanner and meter/scope. The advantage of this is that the information is mirrored to the main unit, which in turn can be connected to a large display screen, or used on its own. It runs the software in a locked down version of Windows 7. You can view the readings and control the scanner and scope remotely from the main display unit. This is very convenient, for example, if you want to compare a scope waveform with a known good example. A repair information system is built into the software package.

The two-channel, high-speed oscilloscope is CATIII/CATIV certified, so it is safe for use on hybrid and electric vehicles. It can be used with the Verdict display, or as a standalone tool.

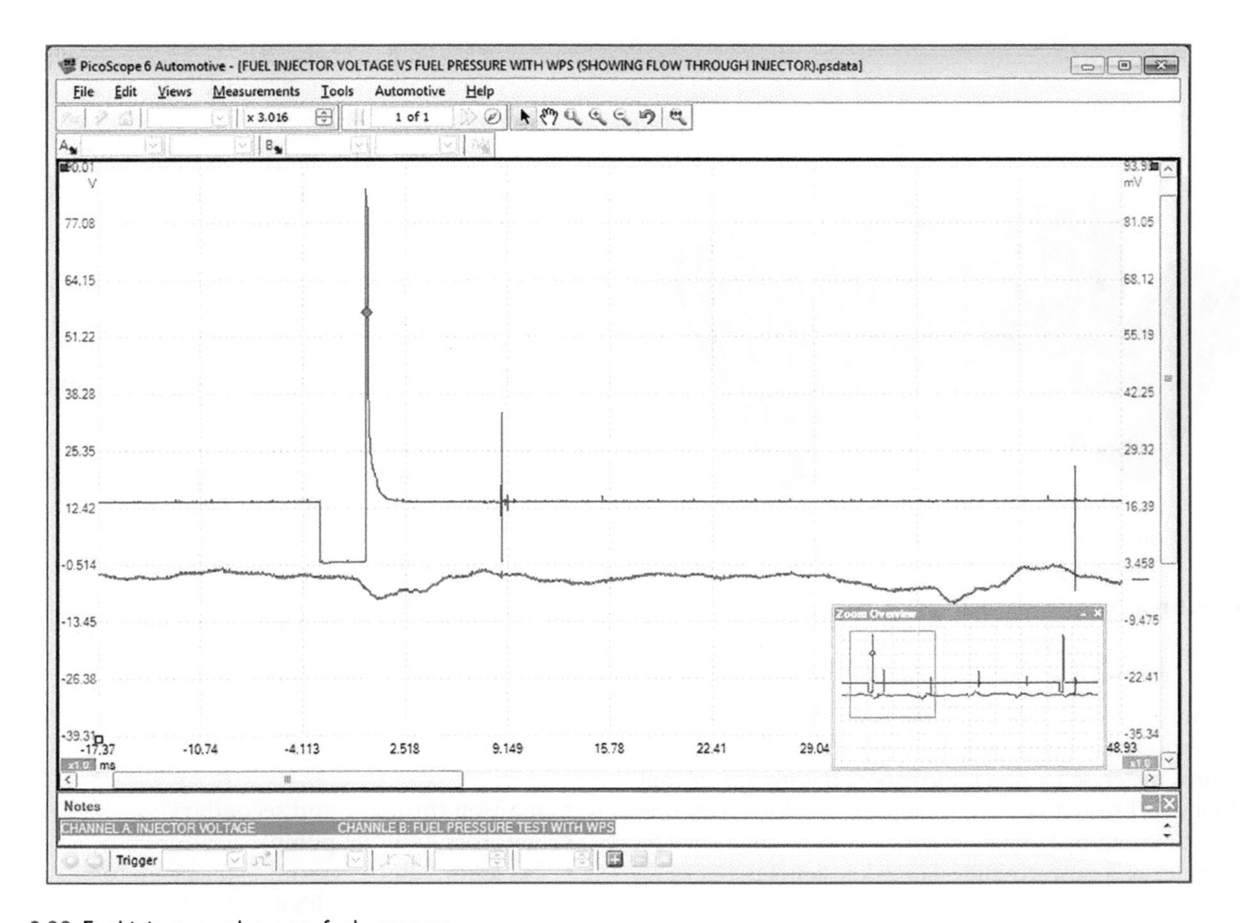

Figure 3.28 Fuel injector voltage vs. fuel pressure.

Source: PicoScope.

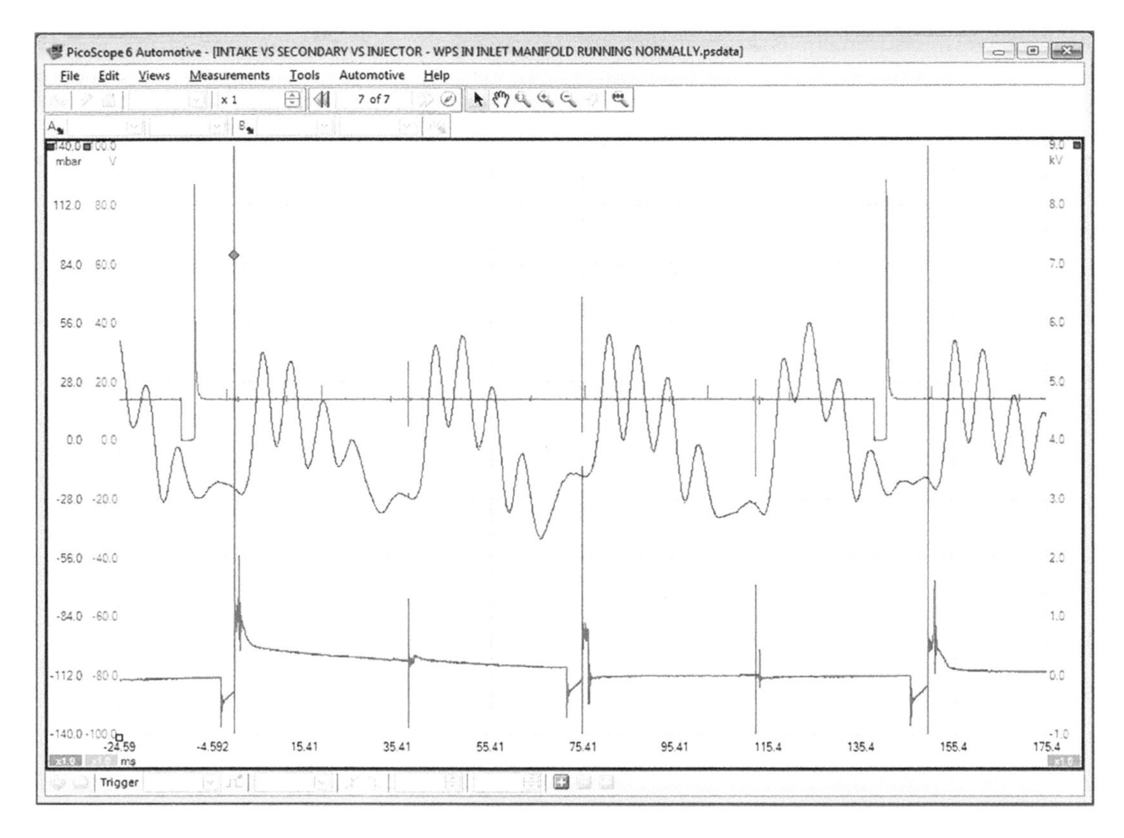

Figure 3.29 Intake pressure vs. injector voltage vs. secondary waveform.

Source: PicoScope.

Figure 3.30 Snap-on Verdict display unit, scanner and multimeter/ oscilloscope.

Source: Snap-on.

3.3.8.1.1 What I like about this equipment

There are newer systems available from Snap-on, but this tool is a good example of an integrated system that combines a scanner, meter, scope and information. It is my 'go-to' tool for first checks on a system because the multimeter/scope unit is self-contained and does not need to be connected to a computer or a power supply. Some key benefits:

- Standalone device includes all multimeter functions and a two-channel scope.
- Bluetooth connection to the main unit is very reliable.
- Testing environment is integrated.

The main downside for me is that the software will not be updated beyond a certain point; I guess this is done as a way to 'encourage' users to upgrade to newer equipment. Nonetheless, this is still a great piece of kit that I will continue to use.

For more information (and newer devices): https://www.snapon.com/diagnostics/UK/diagnostic-tools

3.3.8.2 PicoScope

There is a range of Pico Automotive Diagnostics Kits. They combine with software on a PC to measure and test virtually all of the electrical and electronic components and circuits in any modern vehicle. Common measurements include:

- ignition (primary and secondary);
- injectors and fuel pumps;
- batteries, alternators and starter motors;
- lambda, airflow, ABS and MAP sensors;
- electronic throttle control;
- CAN bus, LIN bus and FlexRay.

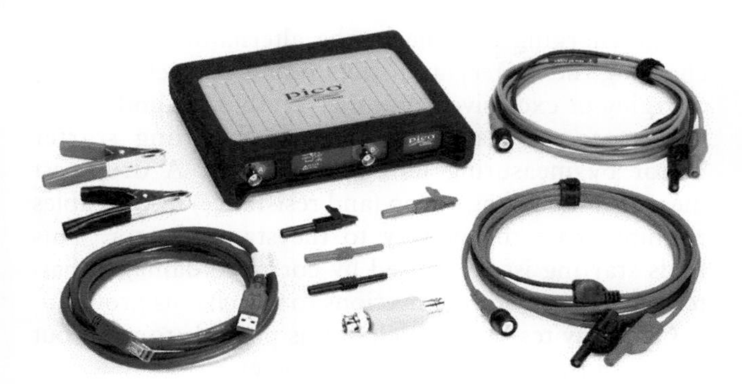

Figure 3.31 PicoScope two-channel starter kit.

Source: PicoScope.

A two- or four-channel PicoScope is recommended for general workshop use (Figures 3.31 and 3.32). On these devices, a separate ground connection is used for each channel and the instrument is protected up to 200 V.

For experienced oscilloscope users involved in training and complex diagnostics, there is now an eight-channel PicoScope. This instrument uses common grounds that are protected by self-resetting fuses. This means that you do not need one ground per channel, which can save time when connecting up lots of channels. The common-ground design limits the maximum input range to 50 V (protected to 100 V) so attenuators must be used when measuring voltage signals from inductive components such as injectors and primary ignition.

The software version at the time of this writing is version 6. Features such as auto setup, tutorials and guided tests are available to get new users up to speed. Advanced features such as math channels, waveform buffers, advanced triggers and reference waveforms ensure the experienced user will not run out of power. The regular software updates include new features and new tests, and are free for the life of the product.

With sampling rates of up to 400 million samples per second, it is possible to capture complex automotive

Figure 3.32 PicoScope four-channel master kit.

Source: PicoScope.

waveforms, and then zoom in on areas of interest. These PC-based waveforms can then be saved for future reference, printed or emailed.

3.3.8.2.1 *What I like about this equipment*

The PicoScope is my favourite device for all complex work because of the advanced features previously mentioned. Key benefits include:

- The software is intuitive and regularly updated.
- Guided testing and auto-setup mean new users learn quickly.
- A database of known good waveforms is available for making comparisons.
- Automotive options range from basic kits (at a price point to get everyone started) to master kits that contain everything needed for even the most complex diagnostic work.
- Maths channel options are available for advanced users.

For more information: www.picoauto.com

3.3.9 Waveform library

The library enables users to share and back up your waveforms online and also search a global database of waveforms uploaded by PicoScope users from around the world.

The powerful search options let you search for known good waveforms from vehicles identical to the one you are working on. For example, when checking for a slipped timing belt, there are hundreds of cam vs crank waveforms to compare against (Figure 3.33). The total number of waveforms at the end of the year 2015 was well over 2000!

3.3.10 PicoDiagnostics

With a simple connection to the battery, PicoDiagnostics software quickly performs the following tests:

- compression test;
- cylinder balance/misfire detect;
- battery test;
- starter motor test;
- alternator test.

For the compression test, simply clip the test leads to the battery and crank the engine. The relative compression of each cylinder is shown in an easy-to-understand bar graph format (Figure 3.34). It also checks for misfires during this test. With the addition of an optional pressure transducer to one cylinder the software will calculate the pressure for all cylinders in bar or PSI.

The battery test checks the state of charge, the voltage drop during starting and compares the cold cranking amps (CCA) with the rating for the battery.

The charging test checks the alternator is correctly charging the battery and spots problems such as overcharging or excessive ripple due to a blown diode.

The starter test helps identify a failing starter motor by measuring its coil resistance. A drop test measures the voltage drop (and resistance) of the cables running from the battery to the starter motor. This finds starting issues caused by corroded/damaged battery cables or poor connections to the chassis ground.

Once any test is completed, it is possible to print out a copy of the report for your customer.

3.3.11 Pressure sensor

Adding a pressure sensor kit to a PicoScope opens up a new world of diagnostics by turning pressure readings into waveforms (Figure 3.35). In-cylinder waveforms show valve and timing issues, mechanical failures, blown gaskets and blocked catalytic converters.

The Pico sensor has three ranges so it is flexible enough to work right around the vehicle: fuel pressure, transmission pressure, intake vacuum and even pulses from the exhaust.

The first range gives high resolution and accuracy for high-pressure tests such as cranking and running cylinder compression or fuel pressure testing. This test is not only a great way to find compression issues but also an excellent way to identify cam timing issues such as jumped timing belts and stretched timing chains.

The second range measures from –15 to 50 psi (about –1 to 3.45 bar). This range is ideal for vacuum tests and fuel system tests. When testing these systems, the zoom function is especially useful as it makes it easy to analyse the valves operating with the vacuum waveform, or the injectors through the fuel waveform.

With the third range it is possible to measure –5 to 5 psi (about –0.34 to 0.34 bar). This setting is sensitive enough to analyse small pressures or pulses such as those from the exhaust.

3.3.12 Noise and vibration

Improvements in damping and sound insulation in modern vehicles mean that vibrations and noises that would previously be ignored (by turning up the radio) are now diagnostic headaches.

The NVH kit works with a PicoScope to quickly identify the source of the problem (Figure 3.36). If the customer complaint is noise, the best option is to use the microphone; if it's a vibration, the accelerometer. Sophisticated software then quickly pinpoints the source of the problem. The kit can also be enhanced with accessories to also allow on-vehicle driveline balancing.

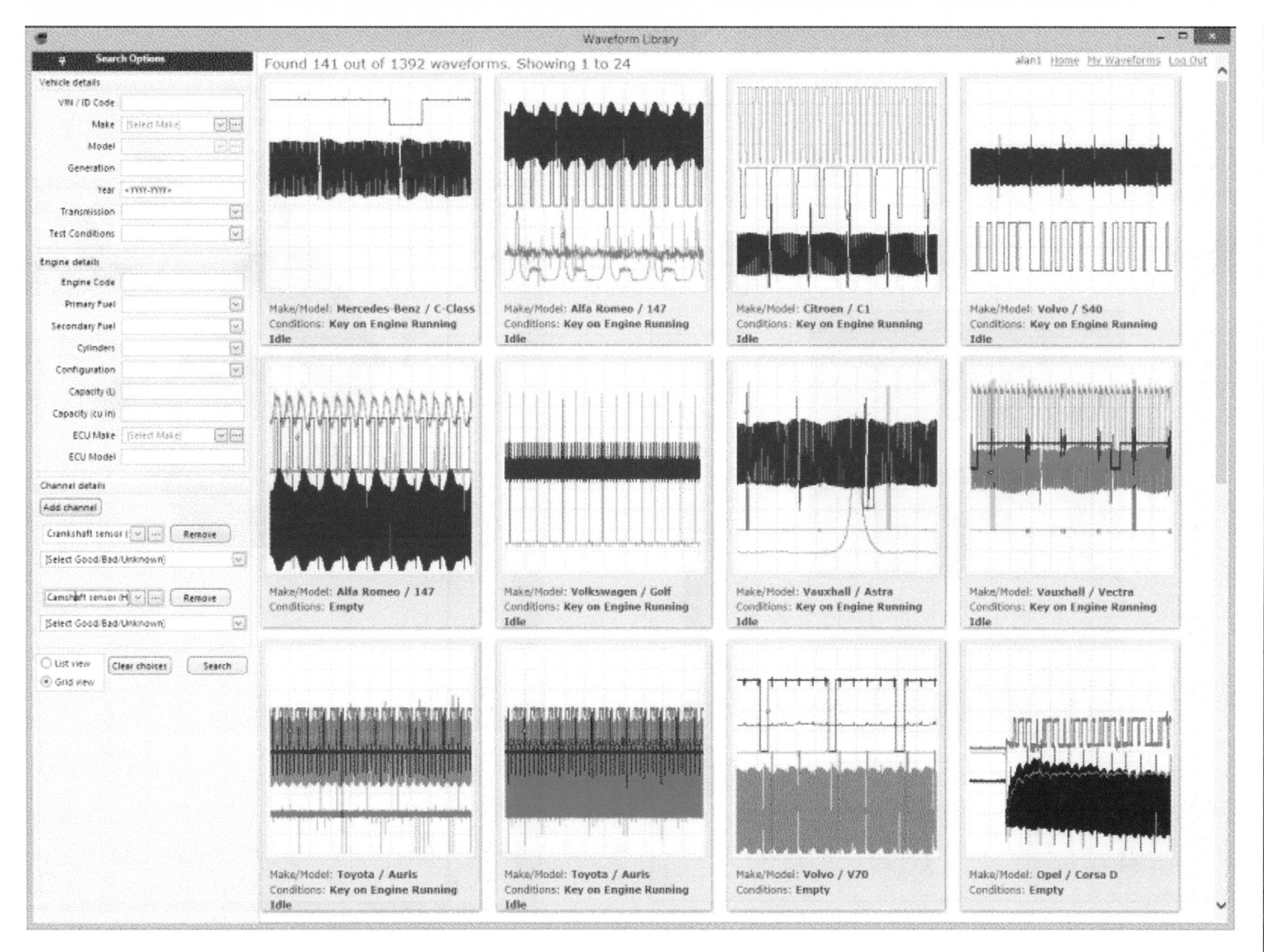

Figure 3.33 Cam vs crank waveforms from the library.

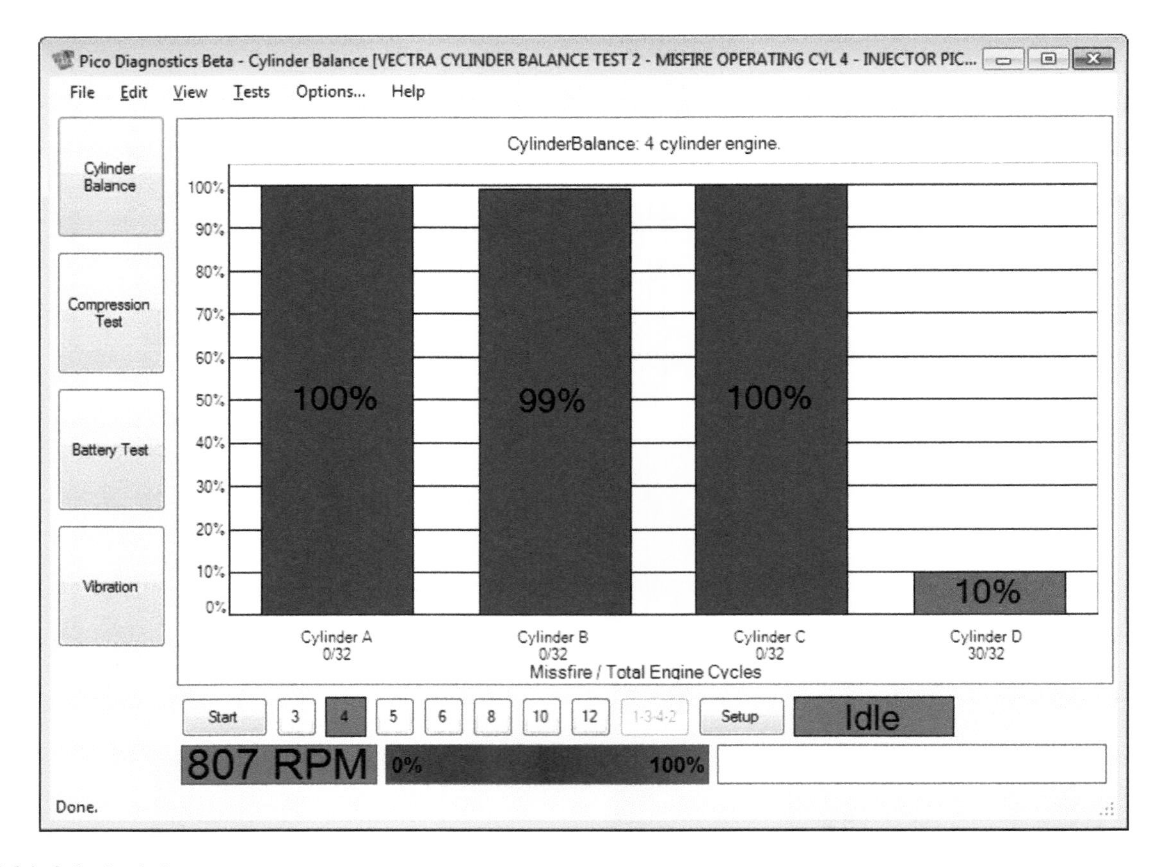

Figure 3.34 Cylinder balance.

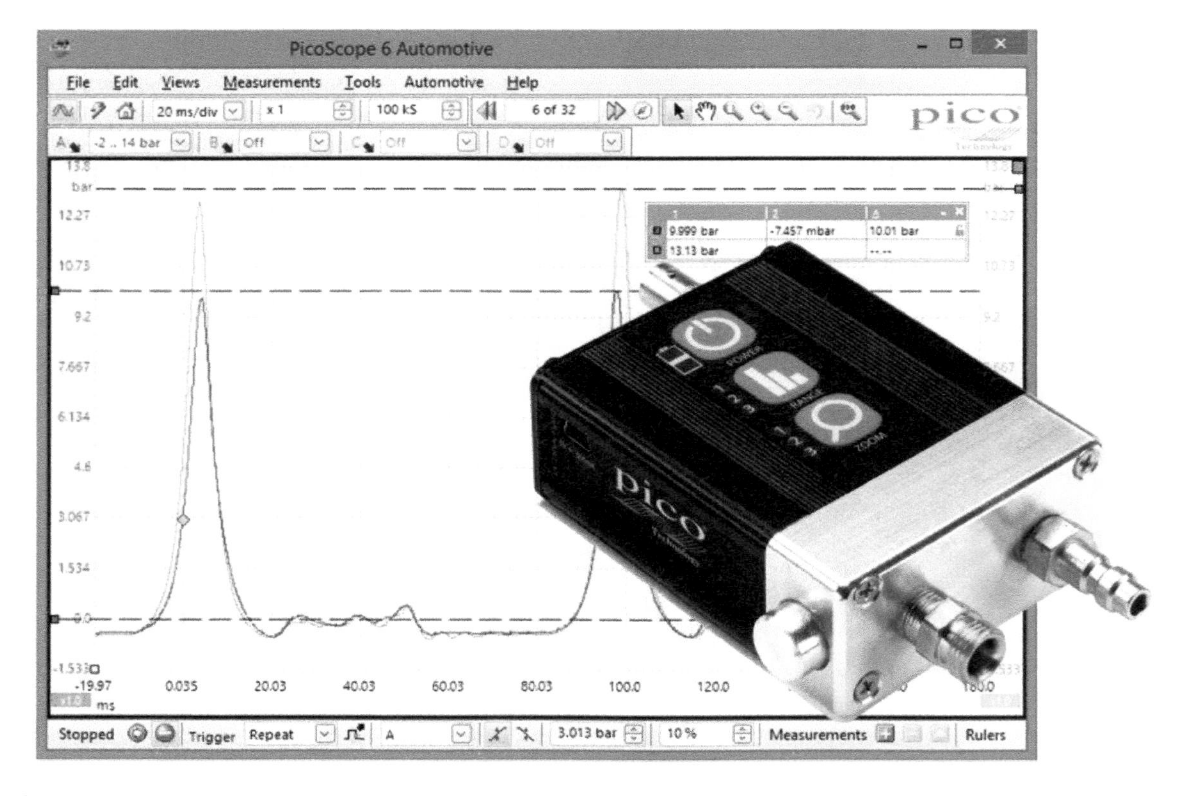

Figure 3.35 Pressure sensor and waveform.

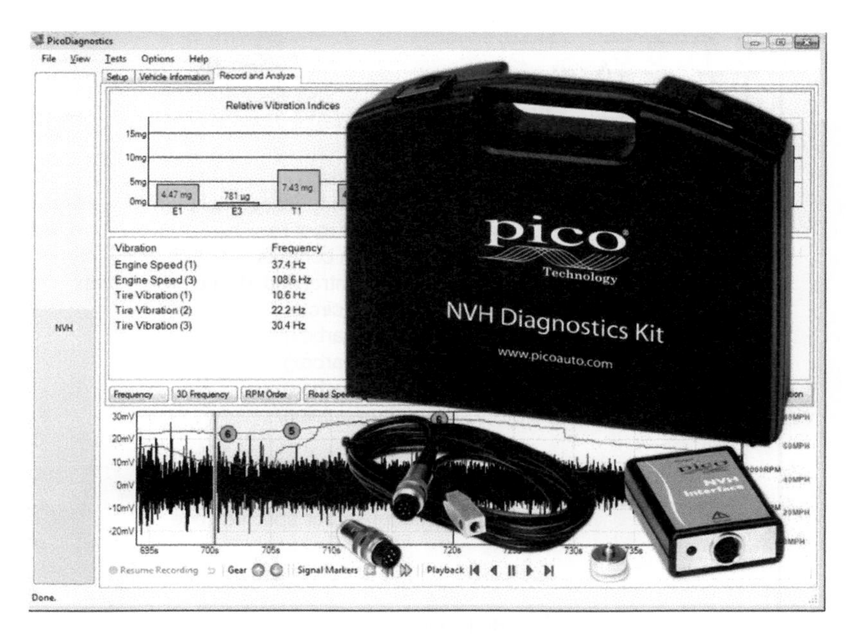

Figure 3.36 NVH diagnostics kit.

3.4 DIAGNOSTIC TOOLS

"Just plug it in and the computer will tell you what's wrong!"

3.4.1 On-board diagnostics

'On-board diagnostics' (OBD) is a generic term referring to a vehicle's self-diagnostic and reporting system. The amount of diagnostic information available via OBD has varied considerably since its introduction in the early 1980s. Early versions of OBD would simply illuminate a malfunction indicator light (MIL) if a problem was detected, but did not provide any information about the problem. The current versions are OBD2 and in Europe EOBD2. These two standards are quite similar. Modern OBD systems use a standardized digital communications port to provide real-time data in addition to a standardized series of diagnostic trouble codes (DTCs).

All 1996 and newer vehicles are OBD2 compatible. However, the amount of OBD2 parameters obtained will depend on the vehicle's specific OBD2 protocol. Under the original OBD2 specification, up to 36 parameters were available. Newer OBD2 vehicles that support the CAN-BUS protocol can have up to 100 generic parameters. This includes trouble codes in systems such as ABS, transmission and air bags. Most scanners now also access information via CAN as well as the OBD information.

3.4.2 Diagnostic trouble codes

Diagnostic trouble codes (DTCs), or fault codes, are stored by an on-board computer diagnostic system. These codes are stored when, for example, a sensor in the car produces a reading that is outside its expected range.

DTCs identify a specific problem area and are a guide as to where a fault might be occurring within the vehicle. Parts or components should not be replaced with reference only to a DTC. No matter what some customers may think, the computer does not tell us exactly what is wrong! For example, if a DTC reports a sensor fault, replacement of the sensor is unlikely to resolve the underlying problem. The fault is more likely to be caused by the systems that the sensor is monitoring. But it can also be caused by the wiring to the sensor.

DTCs may also be triggered by faults earlier in the operating process. For example, a dirty MAF sensor might cause the car to overcompensate its fuel-trim adjustments. As a result, an oxygen sensor fault may be set as well as a MAF sensor code. Figure 3.37 shows how to interpret OBD2 codes.

The following are example DTCs:

- P0106 Manifold Absolute Pressure/Barometric Pressure Circuit Range/Performance Problem;
- P0991 Transmission Fluid Pressure Sensor/Switch E Circuit Intermittent;
- C1231 Speed Wheel Sensor Rear Center Circuit Open (DTCs use USA spellings);
- B1317 Battery Voltage High;
- U2004 Audio Steering Wheel Control Unit is Not Responding.

All but the most basic DIY scanners will translate the codes and present the text so no need to refer to a list.

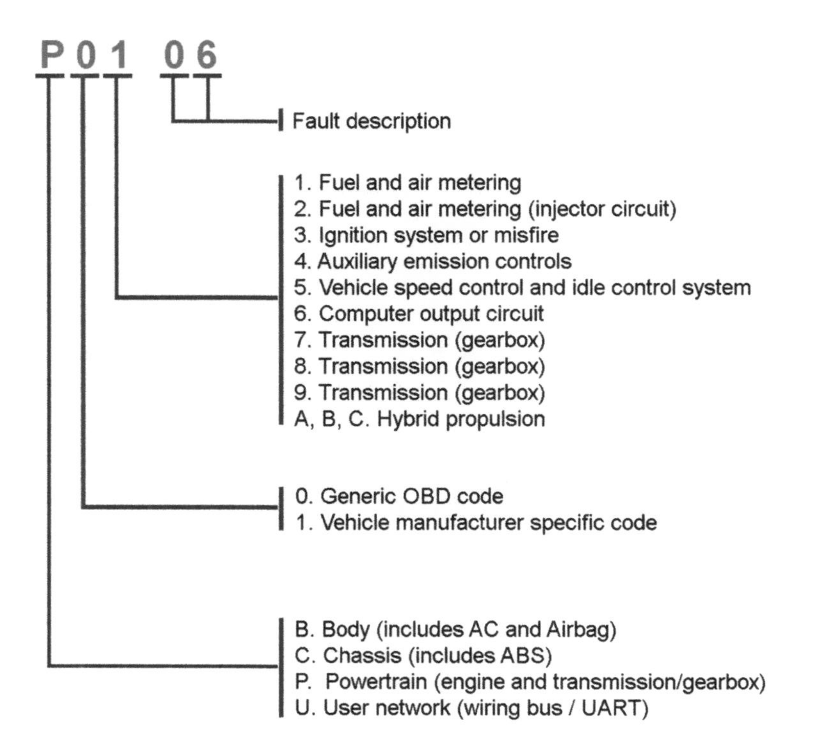

Figure 3.37 On-board diagnostic trouble codes.

3.4.3 Example scanners

3.4.3.1 Basic DIY

There is a huge range of scanners available and the prices range from a few pounds to several thousand pounds. This section will look at some options that vary from DIY to professional.

As an example of a very basic DIY scanner, I have chosen the MaxiScan device as it was one of the cheapest I could find. At the time of this writing it was less than £20 (Figure 3.38). It reads fault codes and will reset the warning light for OBD and EOBD vehicles. The codes are given numerically so they need to be looked up in a list (supplied). It is quite an impressive tool for the money. MaxiScan is part of Autel.

A very useful development in OBD tools was being able to use Bluetooth or Wi-Fi devices to connect to a computer, smartphone or tablet (with suitable software). USB-connected devices are still in use (and may be recommended for reprogramming), but a wireless connection is very convenient. The actual connection method doesn't affect the results, but a wireless vehicle communication interface (VCI) can save time and allows the 'computer' to be situated in a convenient place. A generic device is shown in Figure 3.39.

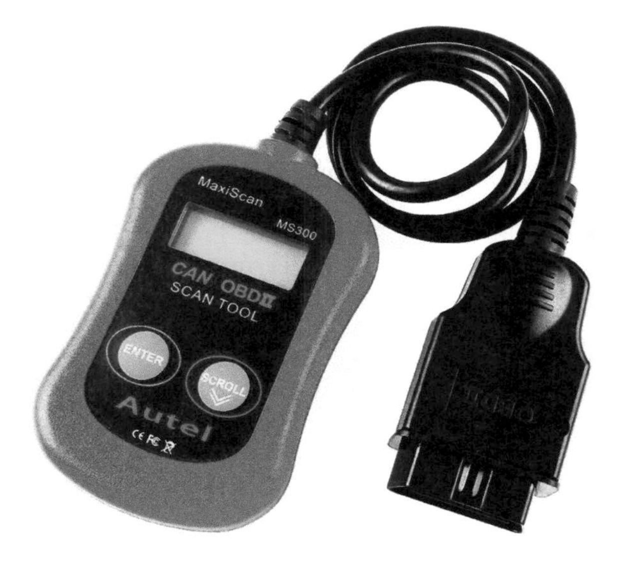

Figure 3.38 MaxiScan fault code reader.

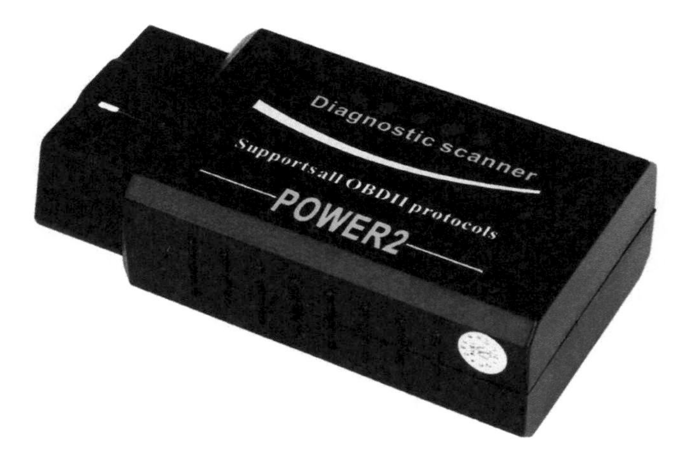

Figure 3.39 OBD Bluetooth adapter.

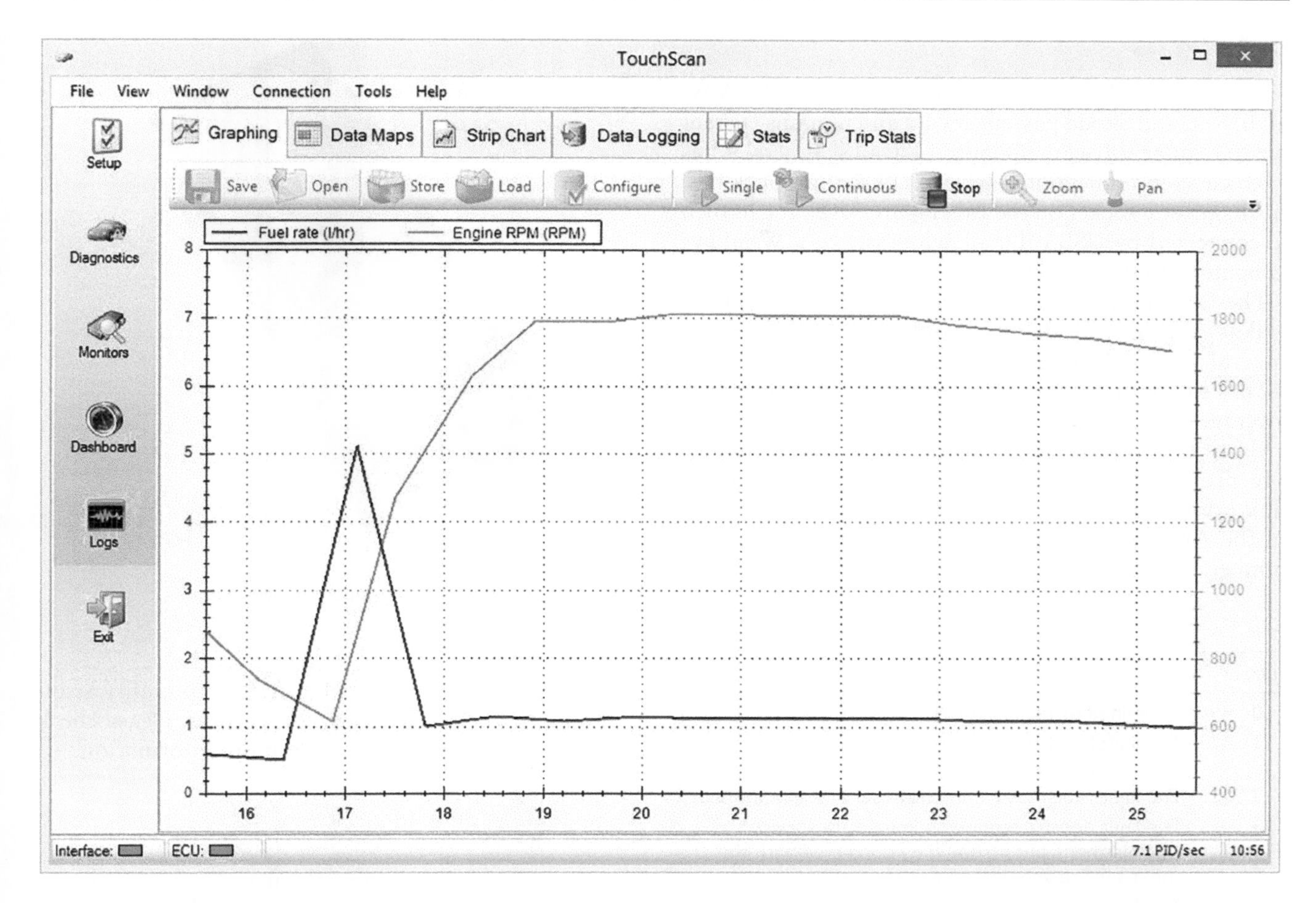

Figure 3.40 Logging of data allows variables to be examined over a set time, in this case comparing fuel rate to engine speed.

One software application I use for connecting a laptop to a vehicle using the wireless VLI is known as TouchScan. It allows fault code scans, monitor tests, logging of data and more. It is generally limited to OBD2 functions but only costs about £35 so is very good value. Figure 3.40 shows an example of live data plotted as a graph using this system.

I also use an iPhone to run an app called OBD Fusion, which costs about £5. It supports ELM327, which is a common protocol for connecting wireless devices. Figure 3.41 is an example of data that can be shown for a quick overview of a vehicle. A very wide range of phone apps are now available.

A simple interface and a smartphone app for quick diagnosis on the side of the road can be very useful. Some systems are more expensive than the ones I examined here and require dedicated adapters. However, because of this low cost, more customers will buy the equipment and apps and scan their vehicle for faults before bringing it to us for repair. We, therefore, need to fully understand how these systems work in even more detail so we can explain that reading a fault code is just the first step in the diagnostic process.

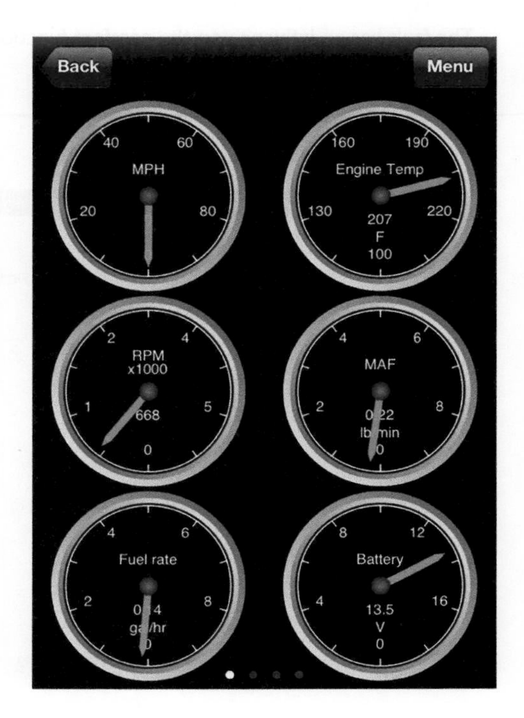

Figure 3.41 OBD Fusion customisable display.

3.4.3.2 VCDS

This professional system is designed specifically for VW, Audi, Seat and Skoda passenger cars and remains current due to regular software updates. Like all scanners, it has a vehicle communication interface (VCI) that plugs into the diagnostic socket on the car (Figure 3.42). It can then be connected by USB cable or wirelessly to a Windows computer running the VCDS software (Figure 3.43).

The facilities available include scanning for faults as well as reading and logging live data, resetting service interval warnings and much more. It costs a fraction of the price of a top range system, but of course the vehicle coverage is restricted.

Using the VCDS I scanned for faults and found the tyre pressure warning as shown in Figure 3.44. The fault priority is 3 (note that this shows in a different way on the ArtiPad live data examples). The DTC priority numbers are shown in Table 3.6.

Table 3.6 DTC priority numbers

Number	*Meaning*
0	Undefined by manufacturer.
1	The fault has a strong influence on drivability, immediate stop is required.
2	The fault requires an immediate service appointment.
3	The fault doesn't require an immediate service appointment, but it should be corrected with the next service appointment.
4	The fault recommends an action to be taken, otherwise drivability might be affected.
5	The fault has no influence on drivability.
6	The fault has a long-term influence on drivability.
7	The fault has an influence on the comfort functions, but doesn't influence the car's drivability.
8	General Note

Figure 3.42 VCDS vehicle connection interface.

The VCDS is a very popular and high-quality scanner and it has very wide range of features. Please check www.ross-tech.com/vag-com for more information.

3.4.3.3 ArtiPad

The ArtiPad outlined here (Figure 3.45) is just one example of a comprehensive, professional-level scanner. However, my experience, and that of others, shows it to be a very powerful tool for the price. Check out www.diagnosticconnections.co.uk for more information.

This diagnostic tool is based on the Android operating system. It will show fault codes from a very wide range of vehicles and associated live data. However, it is

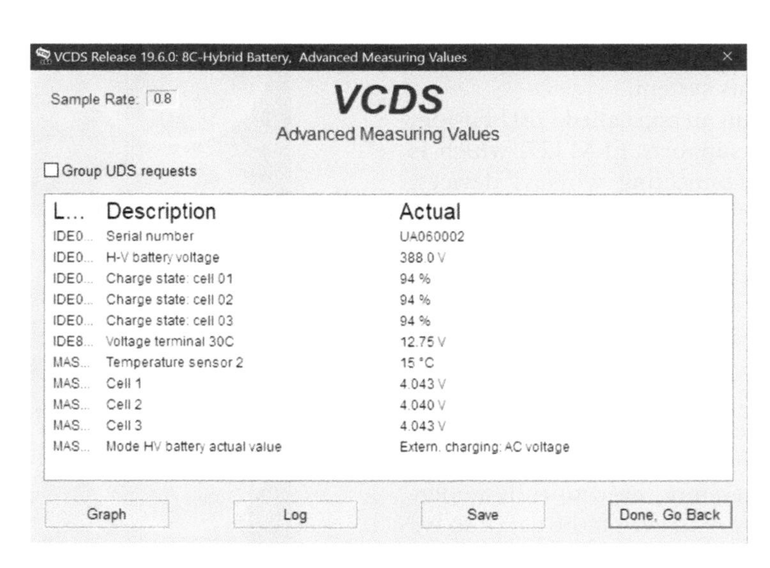

Figure 3.43 Screenshot from the VCDS software showing live data from the high voltage battery on a PHEV.

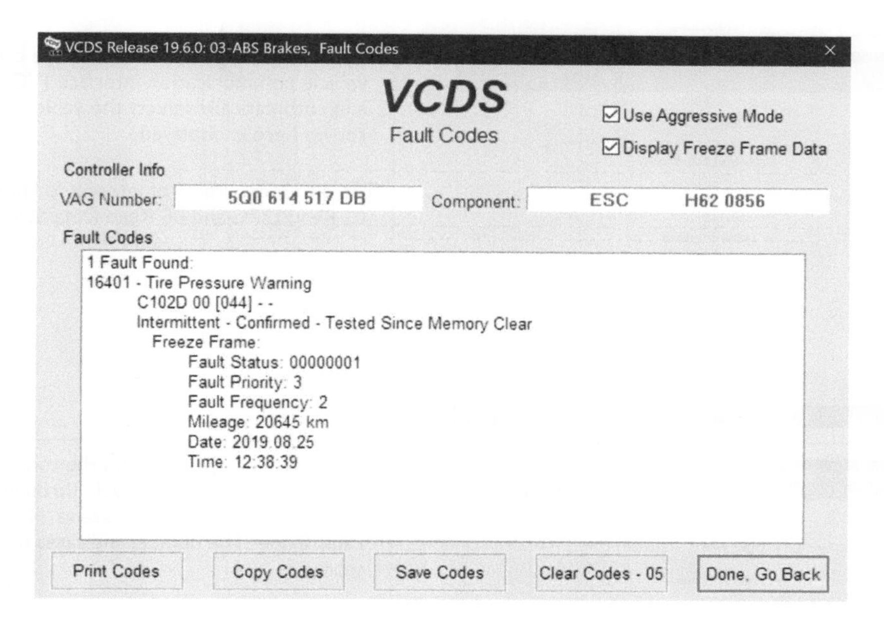

Figure 3.44 VCDS showing a tyre pressure warning fault code.

Figure 3.45 TopDon ArtiPadI main unit, the VCI is plugged into the diagnostic socket.

also capable of ECU programming for Mercedes-Benz, BMW, VW, Audi and Ford. The device itself is very well packaged and robust, as is the software, which is updated regularly. The battery has an excellent life and the screen is clear and easy to read.

3.4.3.4 How many scanners do you need?

Notwithstanding claims by some manufacturers, it is not possible to buy one scanner that will cover the entire range of vehicles. All will read E/OBD2 data but when it comes to digging deeper, every device has some gaps in its capabilities (for example, when coding a new component that has been fitted, or resetting some service intervals).

Most of the professional-level scanners cover a significant range but the real-world consensus is that you need three different devices to be certain of covering everything. Some excellent scanners (and many include lots of other features such as pass-through, data access, scopes, multimeters, etc.) are available from companies such as:

- Autel www.autel.uk
- Autologic www.autologic.com
- Bosch www.boschaftermarket.com/gb/en
- Delphi www.delphiautoparts.com
- Hella www.hella.com
- Launch www.launchtech.co.uk
- Snap-on www.snapon.co.uk

Finally, which is the best scanner? It is hard to say, but probably the one that covers most of the range of vehicles you work on. In addition, and perhaps more importantly, the best one is one you are familiar with and comfortable using.

"The computer does not tell you exactly what is wrong when you plug it in, it is just the first step in a process!"

3.4.4 Live data

In addition to OBD codes, most scanners now communicate with all the electronic modules on the data bus and display appropriate DTCs and live data. The following images, taken using my ArtiPad, show some examples of the scanning process and the live data that can be accessed using this professional-level scanner.

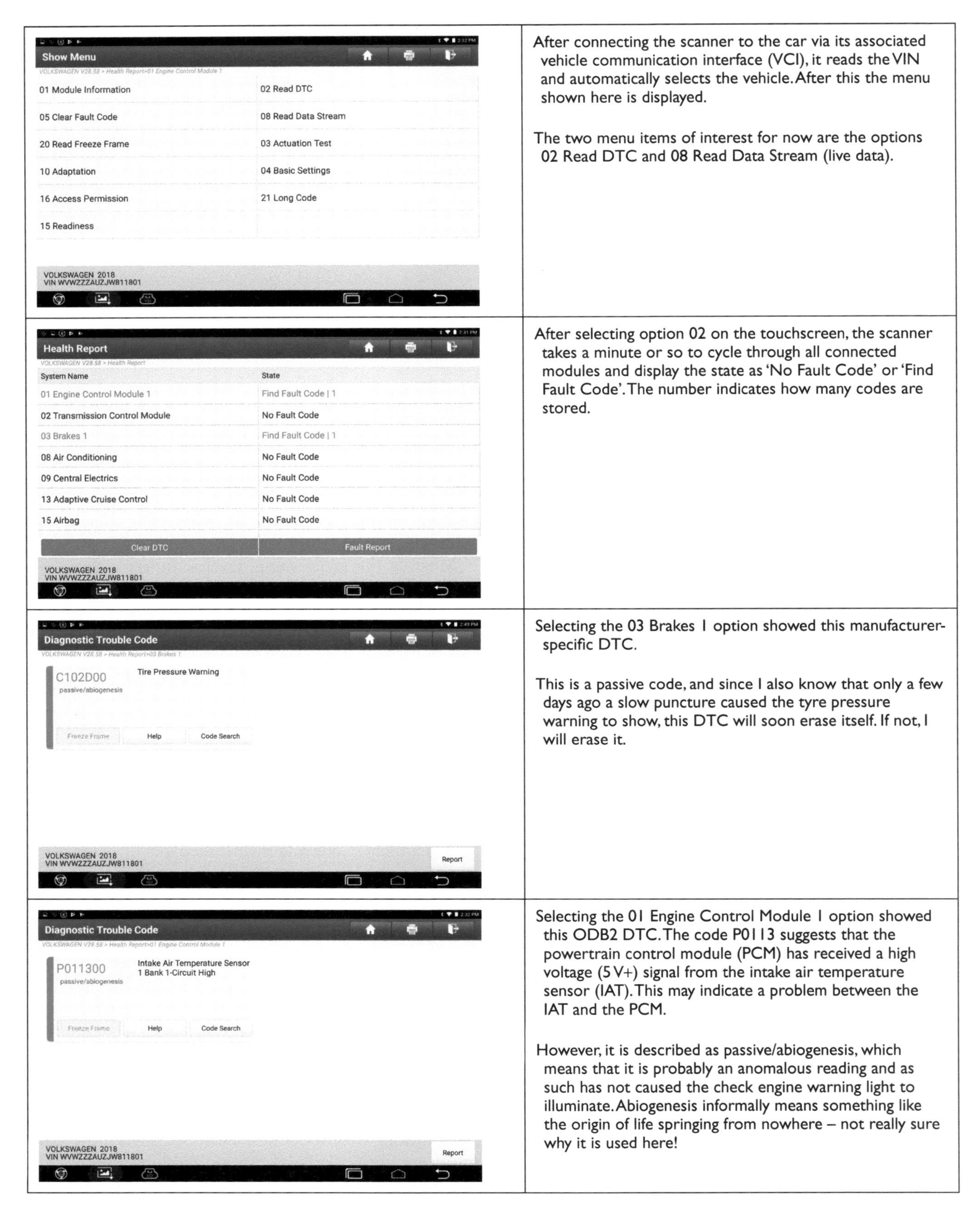

After connecting the scanner to the car via its associated vehicle communication interface (VCI), it reads the VIN and automatically selects the vehicle. After this the menu shown here is displayed.

The two menu items of interest for now are the options 02 Read DTC and 08 Read Data Stream (live data).

After selecting option 02 on the touchscreen, the scanner takes a minute or so to cycle through all connected modules and display the state as 'No Fault Code' or 'Find Fault Code'. The number indicates how many codes are stored.

Selecting the 03 Brakes 1 option showed this manufacturer-specific DTC.

This is a passive code, and since I also know that only a few days ago a slow puncture caused the tyre pressure warning to show, this DTC will soon erase itself. If not, I will erase it.

Selecting the 01 Engine Control Module 1 option showed this ODB2 DTC. The code P0113 suggests that the powertrain control module (PCM) has received a high voltage (5 V+) signal from the intake air temperature sensor (IAT). This may indicate a problem between the IAT and the PCM.

However, it is described as passive/abiogenesis, which means that it is probably an anomalous reading and as such has not caused the check engine warning light to illuminate. Abiogenesis informally means something like the origin of life springing from nowhere – not really sure why it is used here!

Figure 3.46 Scanning process and live data displays.

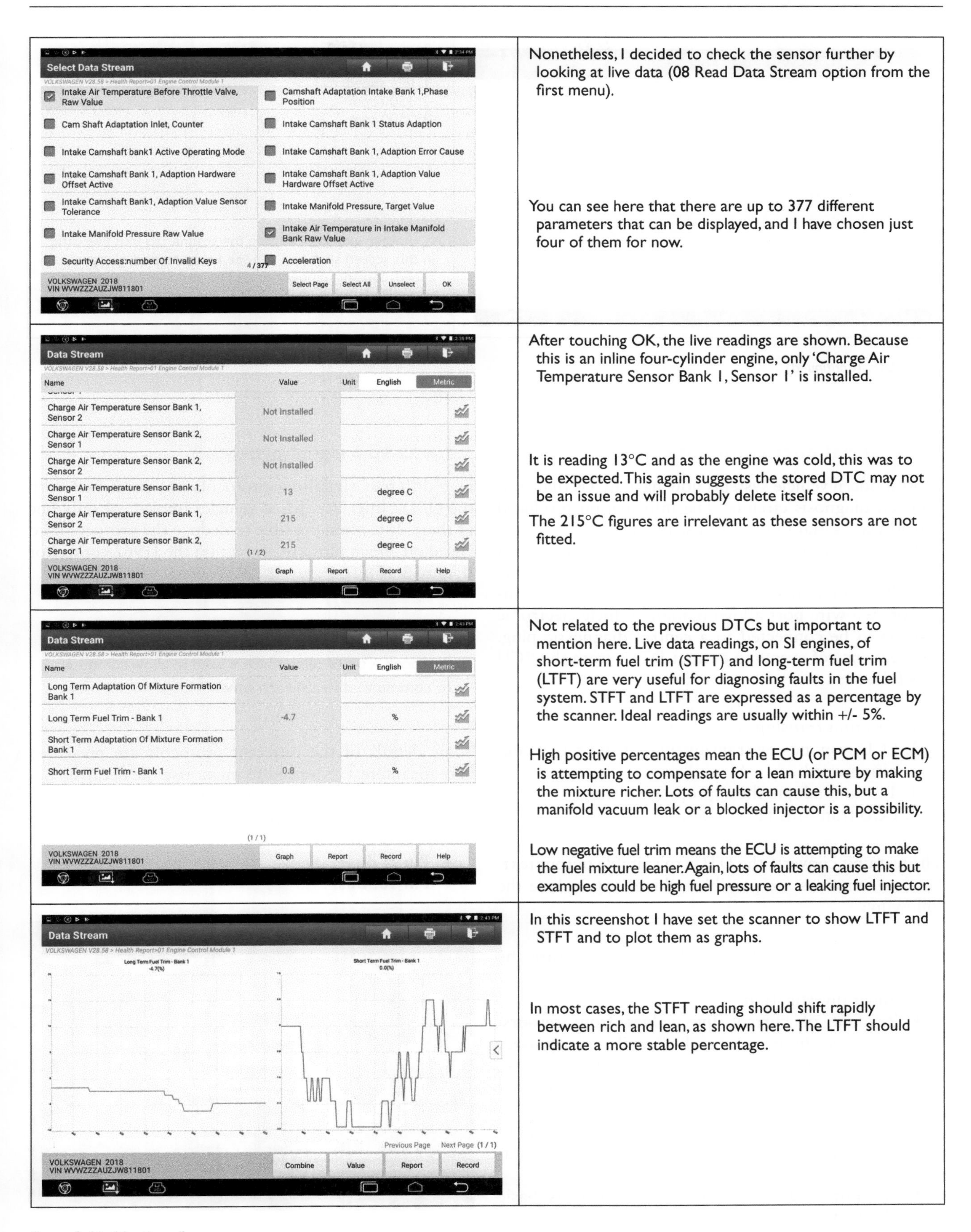

Nonetheless, I decided to check the sensor further by looking at live data (08 Read Data Stream option from the first menu).

You can see here that there are up to 377 different parameters that can be displayed, and I have chosen just four of them for now.

After touching OK, the live readings are shown. Because this is an inline four-cylinder engine, only 'Charge Air Temperature Sensor Bank 1, Sensor 1' is installed.

It is reading 13°C and as the engine was cold, this was to be expected. This again suggests the stored DTC may not be an issue and will probably delete itself soon.

The 215°C figures are irrelevant as these sensors are not fitted.

Not related to the previous DTCs but important to mention here. Live data readings, on SI engines, of short-term fuel trim (STFT) and long-term fuel trim (LTFT) are very useful for diagnosing faults in the fuel system. STFT and LTFT are expressed as a percentage by the scanner. Ideal readings are usually within +/- 5%.

High positive percentages mean the ECU (or PCM or ECM) is attempting to compensate for a lean mixture by making the mixture richer. Lots of faults can cause this, but a manifold vacuum leak or a blocked injector is a possibility.

Low negative fuel trim means the ECU is attempting to make the fuel mixture leaner. Again, lots of faults can cause this but examples could be high fuel pressure or a leaking fuel injector.

In this screenshot I have set the scanner to show LTFT and STFT and to plot them as graphs.

In most cases, the STFT reading should shift rapidly between rich and lean, as shown here. The LTFT should indicate a more stable percentage.

Figure 3.46 (Continued)

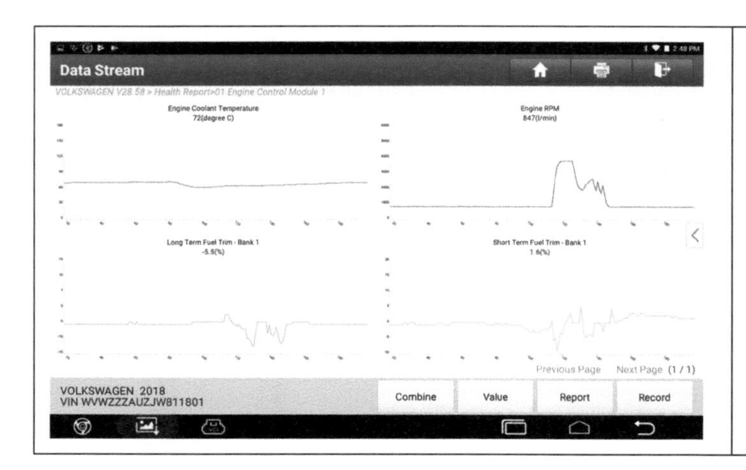

Fuel trim values should ideally be checked in three stages:

1. Idle speed
2. 1500 rpm
3. 2500 rpm

As another example of using the scanner to plot live data, in this screen shot you will see I am comparing engine speed, coolant temperature, LTFT and STFT. The engine was still warming up, but it is all looking good so far.

Figure 3.46 *(Continued)*

3.4.5 Serial port communications

Most modern vehicle systems now have ECUs that contain self-diagnosis circuits. The information produced is read via a serial link using a scanner.

A special interface, stipulated by one of a number of standards (see Section 3.3.3), is required to read the data. The standards are designed to work with a single- or two-wire port allowing many vehicle electronic systems to be connected to a central diagnostic plug. The sequence of events to extract DTCs from the ECU is as follows:

1. Test unit transmits a code word.
2. ECU responds by transmitting a baud rate recognition word.
3. Test unit adopts the appropriate setting.
4. ECU transmits fault codes.

The test unit converts the DTCs to suitable output text. Further functions are possible, which may include the following:

- Identification of ECU and system to ensure the test data is appropriate to the system currently under investigation.
- Readout of current live values from sensors. Spurious figures can be easily recognised. Information such as engine speed, temperature, airflow and so on can be displayed and checked against the test data.
- System function stimulation to allow actuators to be tested by moving them and watching for suitable response.
- Programming of system changes such as basic idle CO or changes in basic timing can be programmed into the system.

3.4.6 OBD2 signal protocols

Five different signalling protocols are permitted with the OBD2 interface. Most vehicles implement only one of them. It is often possible to deduce the protocol used based on which pins are present on the J1962 connector (Figure 3.47).

DEFINITION

Protocol: A set of rules which is used to allow computers to communicate with each other.

Some details of the different protocols are presented here for interest. No need to memorise the details!

SAE J1850 PWM (pulse-width modulation): A standard of Ford Motor Company

- Pin 2: Bus+.
- Pin 10: Bus–.

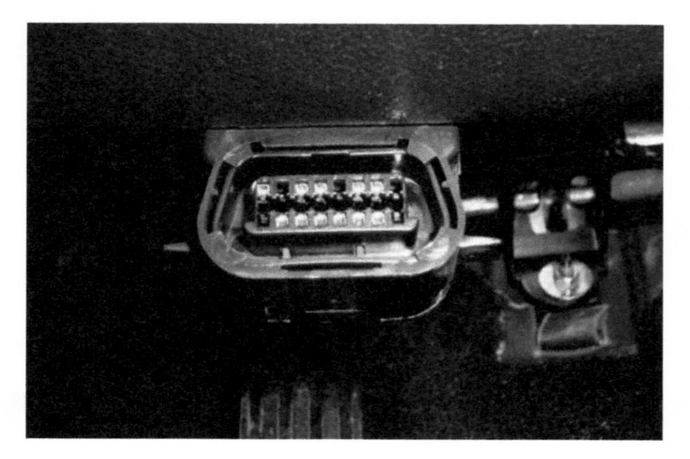

Figure 3.47 Diagnostic data link connector (DLC).

- High voltage is +5 V.
- Message length is restricted to 12 bytes, including cyclic redundancy check (CRC).
- Employs a multi-master arbitration scheme called 'Carrier Sense Multiple Access with Non-Destructive Arbitration' (CSMA/NDA).

SAE J1850 VPW (variable pulse width): A standard of General Motors

- Pin 2: Bus+.
- Bus idles low.
- High voltage is +7 V.
- Decision point is +3.5 V.
- Message length is restricted to 12 bytes, including CRC.
- Employs CSMA/NDA.

ISO 9141–2: Primarily used by Chrysler, European and Asian vehicles

- Pin 7: K-line.
- Pin 15: L-line (optional).
- UART signalling.
- K-line idles high, with a 510Ω resistor to Vbatt.
- The active/dominant state is driven low with an open-collector driver.
- Message length is restricted to 12 bytes, including CRC.

ISO 14230 KWP2000 (Keyword Protocol 2000)

- Pin 7: K-line.
- Pin 15: L-line (optional).
- Physical layer identical to ISO 9141–2.
- Message may contain up to 255 bytes in the data field.

ISO 15765 CAN: The CAN protocol was developed by Bosch for automotive and industrial control. Since 2008, all vehicles sold in the United States (and most others) are required to implement CAN as one of their signalling protocols.

- Pin 6: CAN high.
- Pin 14: CAN low.

All OBD2 pin-outs use the same connector but different pins, with the exception of pin 4 (battery ground) and pin 16 (battery positive) (Figure 3.48).

3.4.7 Engine analysers

Some form of engine analyser was an essential tool for fault finding modern vehicle engine systems. The latest machines are now based around a personal computer.

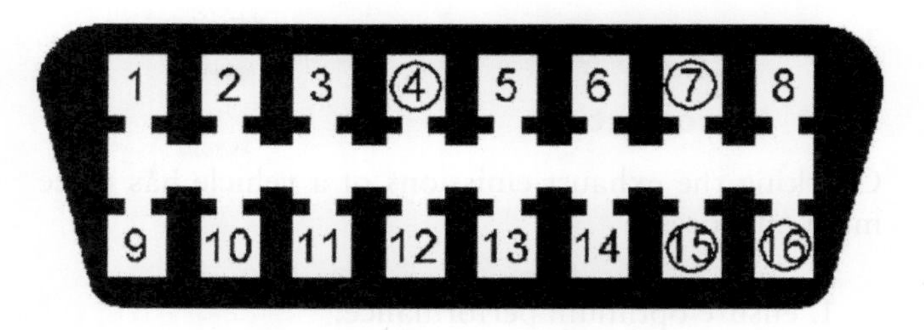

Figure 3.48 Connector pin-out: 4 – battery ground/earth, 7 – K-line, 15 – L-line, 16 – battery positive.

This allows more facilities that can be added to by simply changing the software.

While engine analysers (Figure 3.49) are designed to work specifically with the motor vehicle, it is worth remembering that the machine consists basically of three parts:

- multimeter;
- gas analyser;
- oscilloscope.

However, separate systems such as the Pico Automotive kit will now do as many tests as the engine analyser, currently with the exception of exhaust emissions.

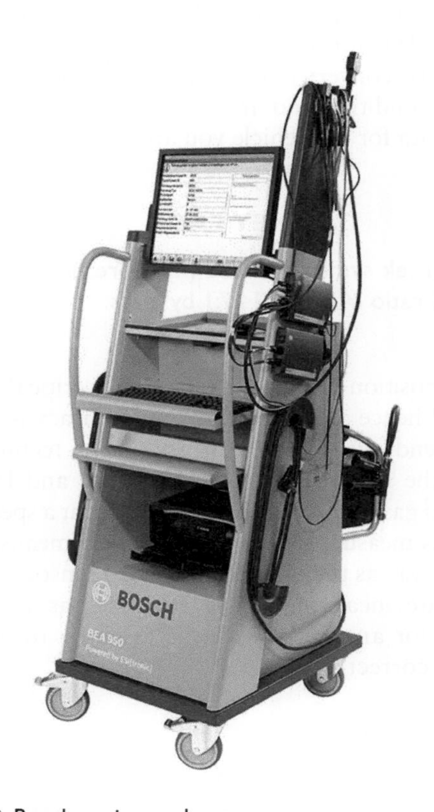

Figure 3.49 Bosch engine analysers.

Source: Bosch Media.

3.5 EMISSION TESTING

3.5.1 Introduction

Checking the exhaust emissions of a vehicle has three main purposes:

1. ensure optimum performance;
2. comply with regulations and limits;
3. provide diagnostic information.

There are many different exhaust testing systems available.

3.5.2 Exhaust gas measurement

It has now become standard to measure four of the main exhaust gases namely:

- carbon monoxide (CO);
- carbon dioxide (CO_2);
- hydrocarbons (HC);
- oxygen (O_2).

On many analysers, lambda value and the air fuel ratio are calculated and displayed in addition to the four gases. The Greek symbol lambda (λ) is used to represent the ideal air fuel ratio (AFR) of 14.7:1 by mass. In other words, just the right amount of air to burn up all the fuel. Table 3.7 lists gas, lambda and AFR readings for a closed loop lambda control system, before (or without) and after the catalytic converter. These are for a modern engine in excellent condition and are a guide only – always check current data for the vehicle you are working on.

KEY FACT

The Greek symbol lambda (λ) represents the ideal air fuel ratio (AFR) of 14.7:1 by mass.

The composition of exhaust gas is now a critical measurement and hence a certain degree of accuracy is required. To this end the infrared measurement technique has become the most suitable for CO, CO_2 and HC. Each individual gas absorbs infrared radiation at a specific rate. Oxygen is measured by electro-chemical means in much the same way as the on-vehicle lambda sensor.

Accurate measurement of exhaust gas is not only required for annual tests but is essential to ensure an engine is correctly tuned.

3.5.3 Exhaust analyser

There are lots of different pieces of equipment that can be used to measure exhaust emissions. Some are fixed (Figure 3.50) and some are mobile and battery operated (Figure 3.51).

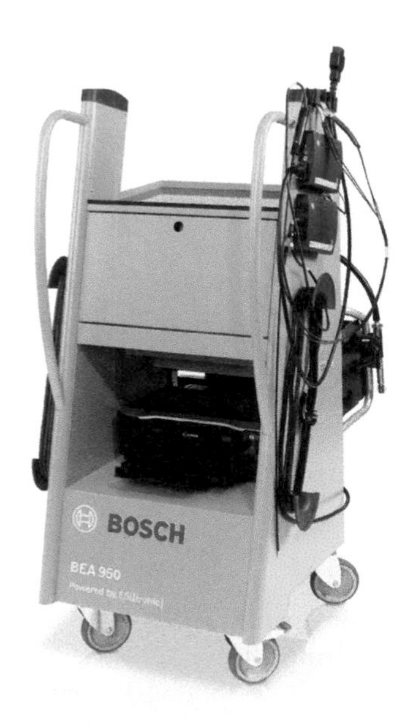

Figure 3.50 Fixed (workshop based) exhaust gas measuring equipment.

Source: Bosch media.

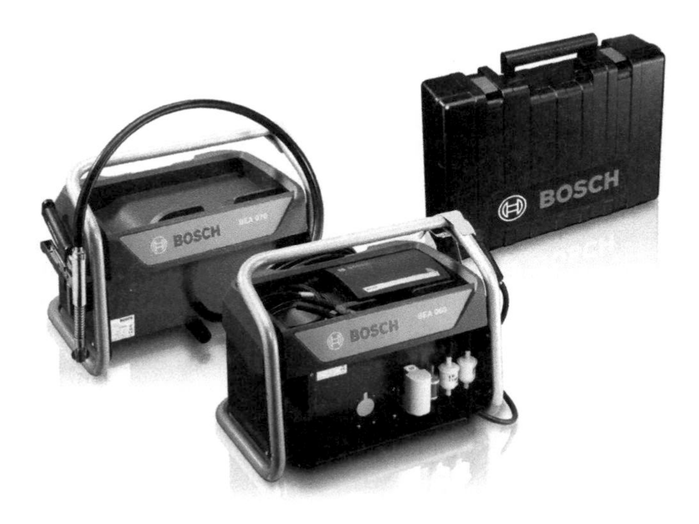

Figure 3.51 Mobile exhaust gas measuring equipment.

Source: Bosch media.

Table 3.7 Exhaust examples

Reading:	*CO%*	*HC ppm*	*CO_2%*	*O_2%*	*Lambda(λ)*	*AFR*
Before catalyst	0.5	100	14.7	0.7	1.0	14.7
After catalyst	0.1	12	15.3	0.1	1.0	14.7

The mobile equipment shown here as Figure 3.51 uses Bluetooth connectivity and lithium-ion batteries. It will test petrol/gasoline emissions and diesel.

DEFINITION

Bluetooth: A proprietary open wireless protocol for exchanging data over short distances from fixed and mobile devices, creating personal area networks (PANs).

3.5.4 Emission limits

Limits and regulations relating to exhaust emissions vary in different countries and in different situations. For example, in the UK certain limits have to be met during the annual test. The current test default limits (for passenger vehicles since September 2002 fitted with a catalytic converter) are:

Minimum oil temperature 60°C
Fast idle (2500 to 3000 rpm)

- CO <= 0.2%
- HC <= 200 ppm
- Lambda 0.97 to 1.03

Idle (450 to 1500 rpm)

- CO <= 0.3%

Passenger cars and light commercial vehicles with a diesel engine (first used on or after July 2008):

- 1.5 m^{-1} (a measure of smoke opacity)

Manufacturers, however, have to meet stringent regulations when producing new vehicles. In Europe the emission standards are defined in a series of EU directives staging the progressive introduction of increasingly stringent standards (see Table 3.8). Euro 7 legislation, expected in 2020, requires a reduction of CO_2 levels from 130 g/km to 95 g/km.

Table 3.8 European emission standards for passenger cars (Category M)[a], g/km

Tier	*Date (type approval)*	*Date (first registration)*	*CO*	*THC*	*NMHC*	*NOx*	*HC + NOx*	*PM*	*PN [#/km]*
Diesel									
Euro 1[d]	July 1992	January 1993	2.72 (3.16)	–	–	–	0.97 (1.13)	0.14 (0.18)	–
Euro 2	January 1996	January 1997	1.0	–	–	–	0.7	0.08	–
Euro 3	January 2000	January 2001	0.66	–	–	0.50	0.56	0.05	–
Euro 4	January 2005	January 2006	0.50	–	–	0.25	0.30	0.025	–
Euro 5a	September 2009	January 2011	0.50	–	–	0.180	0.230	0.005	–
Euro 5b	September 2011	January 2013	0.50	–	–	0.180	0.230	0.0045	6 × 1011
Euro 6b	September 2014	September 2015	0.50	–	–	0.080	0.170	0.0045	6 × 1011
Euro 6c	–	September 2018	0.50	–	–	0.080	0.170	0.0045	6 × 1011
Euro 6d-Temp	September 2017	September 2019	0.50	–	–	0.080	0.170	0.0045	6 × 1011
Euro 6d	January 2020	January 2021	0.50	–	–	0.080	0.170	0.0045	6 × 1011
Petrol/gasoline									
Euro 1[d]	July 1992	January 1993	2.72 (3.16)	–	–	–	0.97 (1.13)	–	–
Euro 2	January 1996	January 1997	2.2	–	–	–	0.5	–	–
Euro 3	January 2000	January 2001	2.3	0.20	–	0.15	–	–	–
Euro 4	January 2005	January 2006	1.0	0.10	–	0.08	–	–	–
Euro 5a	September 2009	January 2011	1.0	0.10	0.068	0.060	–	0.005[b]	–
Euro 5b	September 2011	January 2013	1.0	0.10	0.068	0.060	–	0.0045[b]	–
Euro 6b	September 2014	September 2015	1.0	0.10	0.068	0.060	–	0.0045[b]	6 × 1011[c]
Euro 6c	–	September 2018	1.0	0.10	0.068	0.060	–	0.0045[b]	6 × 1011
Euro 6d-Temp	September 2017	September 2019	1.0	0.10	0.068	0.060	–	0.0045[b]	6 × 1011
Euro 6d	January 2020	January 2021	1.0	0.10	0.068	0.060	–	0.0045[b]	6 × 1011

Note: See: https://en.wikipedia.org/wiki/European_emission_standards for updates.

[a] Before Euro 5, passenger vehicles > 2500 kg were type approved as light commercial vehicles N1 Class I.

[b] Applies only to vehicles with direct injection engines.

[c] 6 × 1012/km within first three years from Euro 6b effective dates.

[d] Values in parentheses are conformity of production (COP) limits.

3.6 PRESSURE TESTING

3.6.1 Introduction

Measuring the fuel pressure on a fuel injection engine is of great value when faultfinding. Many types of pressure testers are available and they often come as part of a kit consisting of various adapters and connections (Figure 3.52). The principle of mechanical gauges is that they contain a very small tube wound in a spiral.

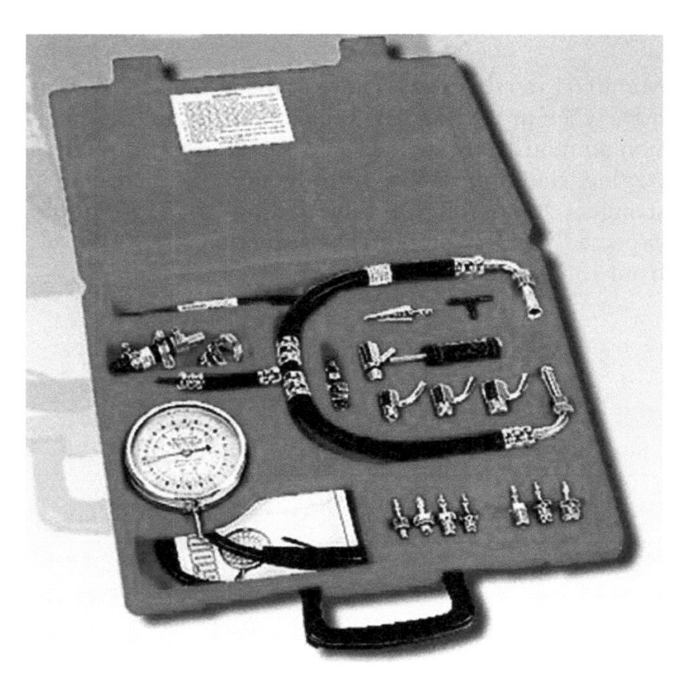

Figure 3.52 Fuel pressure gauge kit.

As fluid or gas under pressure is forced into the spiral tube, it unwinds causing a needle to move over a graduated scale.

Measuring engine cylinder compression or leakage is a useful test. Figure 3.53 shows an engine compression tester. This device is used to compare cylinder compressions as well as to measure actual values.

3.6.2 Oscilloscope transducer

PicoTech has developed an accurate pressure transducer (Figure 3.54) that can be used for pressure analysis of many automotive systems.

DEFINITION

Transducer: A device that converts a physical quantity (e.g. force, torque, pressure, rotation) to an electrical signal.

Some of the key features are as follows:

- range accurate from 0.07 psi (5 mbar) to 500 psi (34.5 bar);
- 100 µs response time;
- zoom function for enhanced analysis;
- temperature compensation.

These result in an accurate representation of rapidly changing signals that span a broad pressure range.

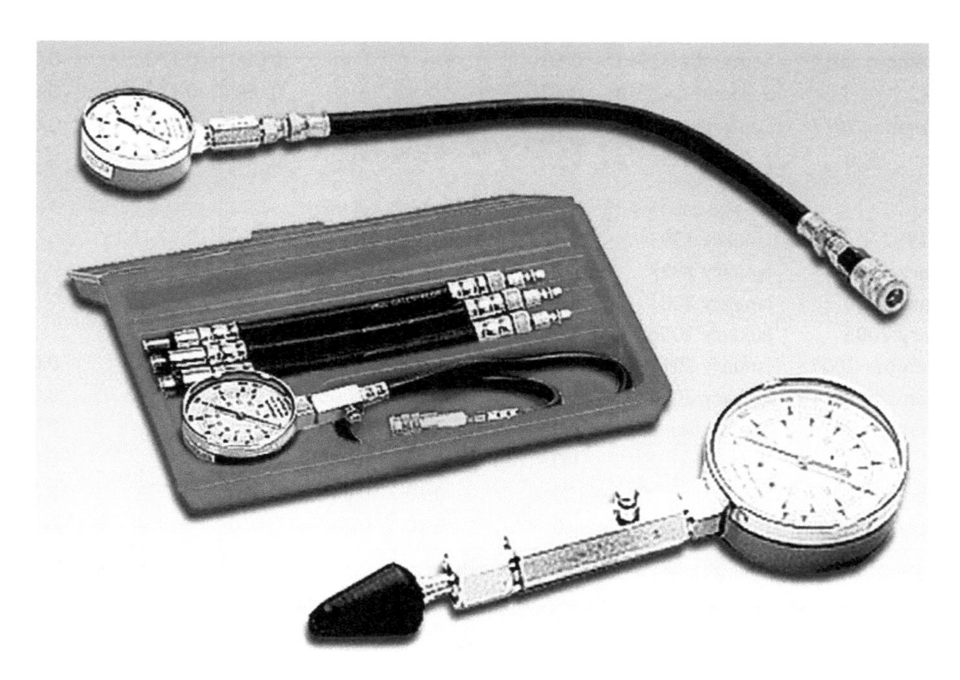

Figure 3.53 Compression tester.

Figure 3.54 Automotive pressure transducer.

Source: PicoTech.

The three pressure ranges of the device allow for accurate measurement and analysis of many automotive pressures, including

- cylinder compression;
- fuel pressure;
- intake manifold vacuum;
- pulses from the exhaust.

The first range gives high resolution and accuracy for high-pressure tests such as cranking and running cylinder compression or fuel pressure testing.

The second range measures from –15 to 50 psi (approximately –1 to 3.45 bar). This range is ideal for vacuum tests and fuel system tests. The zoom function is especially useful on these tests as it makes it easy to analyse the valves operating with the vacuum waveform, or the injectors through the fuel waveform (Figure 3.55).

With the third range you can measure –5 to 5 psi (approximately –0.34 to 0.34 bar). This setting is sensitive enough to allow analysis of small pressures or pulses such as from the exhaust. This is an excellent way of checking even for running cylinders.

3.6.3 Pressure analysis

PicoScope automotive scopes are top of the range for waveform analysis, and with the addition of a pressure transducer, now allow detailed examination of pressures at all levels. The Pico Technology pressure transducer simply converts pressure values to voltage, which is then relayed to the scope, allowing pressure to be displayed against time.

This device will actually replace vacuum, coolant, compression (petrol and diesel), oil pressure, turbo boost and fuel pressure gauges and more.

When used with suitable adaptors, the WPS500X Pressure Transducer, shown in Figure 3.56, not only replaces these gauges, but far exceeds their accuracy and reveals infinite detail about the transitions in pressure that were never visible with most mechanical gauges.

The compression test is an example of how using a transducer coupled to a PicoScope will reveal far more than the maximum compression pressure.

Connecting the device to a compression hose will allow for a compression measurement to be taken identical to a typical compression tester, but with the results displayed on the scope screen rather than reading from a mechanical gauge. Setting up PicoScope

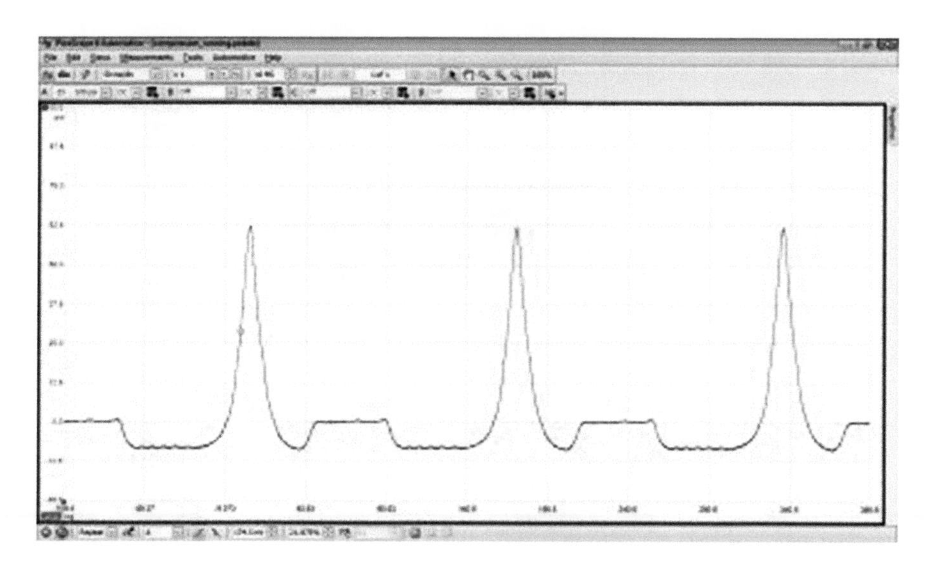

Figure 3.55 Running compression waveform.

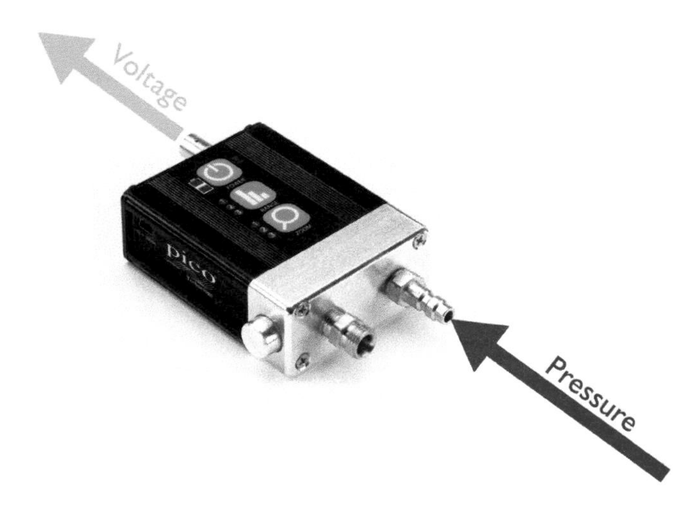

Figure 3.56 WPS500X Pressure Transducer.

Source: PicoScope.

to measure compression is straightforward using the Guided Tests built into the PicoScope software (see Figure 3.57).

To appreciate the advantages of this testing method, we need to understand the waveform acquired during a typical compression test. Using a suitable adaptor, the transducer is connected after removing a spark plug or glow plug. The injectors should also be disconnected to prevent fuel being delivered.

The waveform in Figure 3.58 reveals the compression peaks at 170 psi as would a typical compression tester. However, we can now see repeated even compression peaks (towers) as the crankshaft rotates, and more importantly, events taking place between compressions that would not be visible with a standard compression tester.

Using PicoScope we can equally divide the distance between compression events to reveal the position of the crankshaft (degrees of rotation) using our rotation markers. If we know the position of the crankshaft, we can identify each of the four stroke cycles between compressions.

Looking closely at the base of each compression tower, you can see the expansion pocket dropping below the zero psi rule (line) indicating the cylinder pressure momentarily dropped to negative (below atmospheric pressure). This indicates both intake and exhaust valves remain closed with adequate sealing as the piston descends down the cylinder towards the end of the power stroke. The power stroke is referred to here as the expansion stroke as there is no combustion (the integrity of the piston compression rings and cylinder face can also be confirmed via the expansion pocket).

Using the time-rulers, we can also measure the time it takes (frequency) for the crankshaft to rotate 360 degrees and multiply this value by 60 to reveal the cranking speed (278 RPM).

As with a conventional gauge, we can confirm peak pressure to be correct for the engine under test. However, by using the pressure transducer, we can also

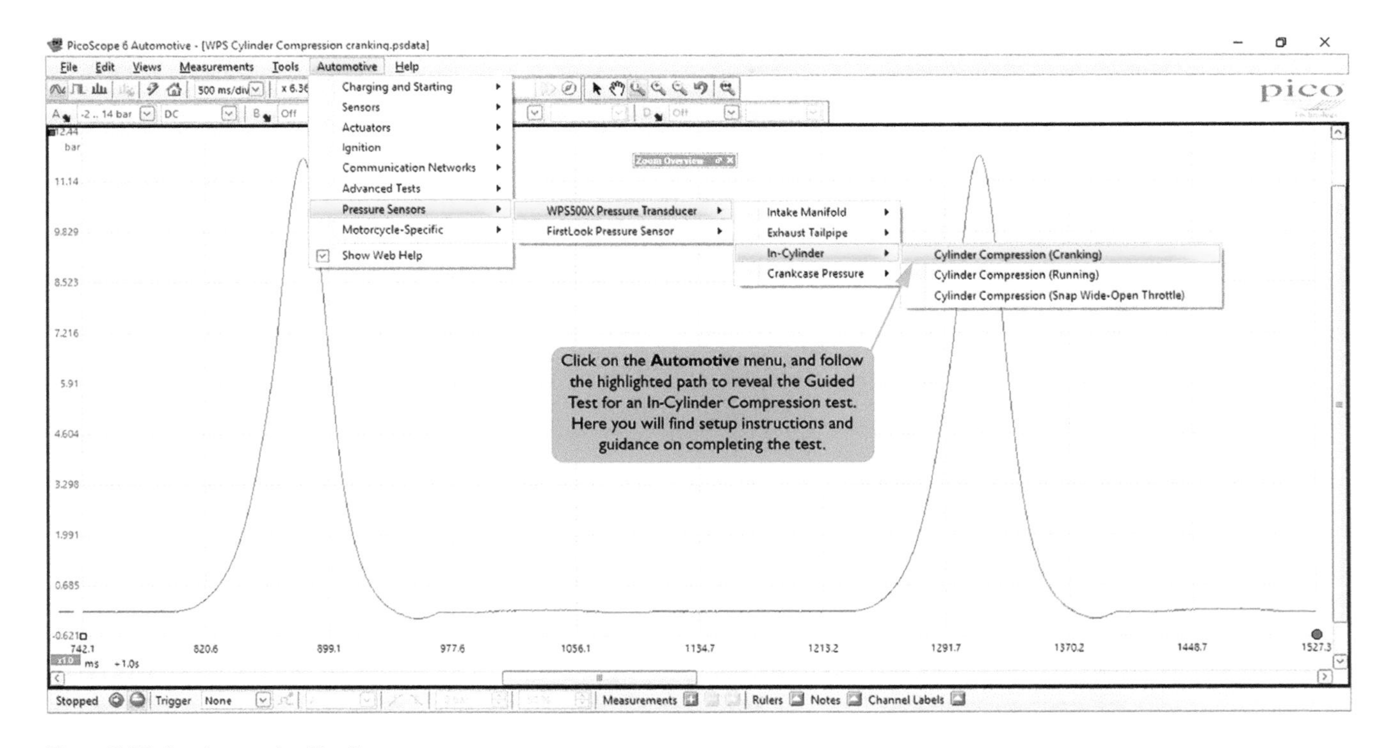

Figure 3.57 Setting up the PicoScope.

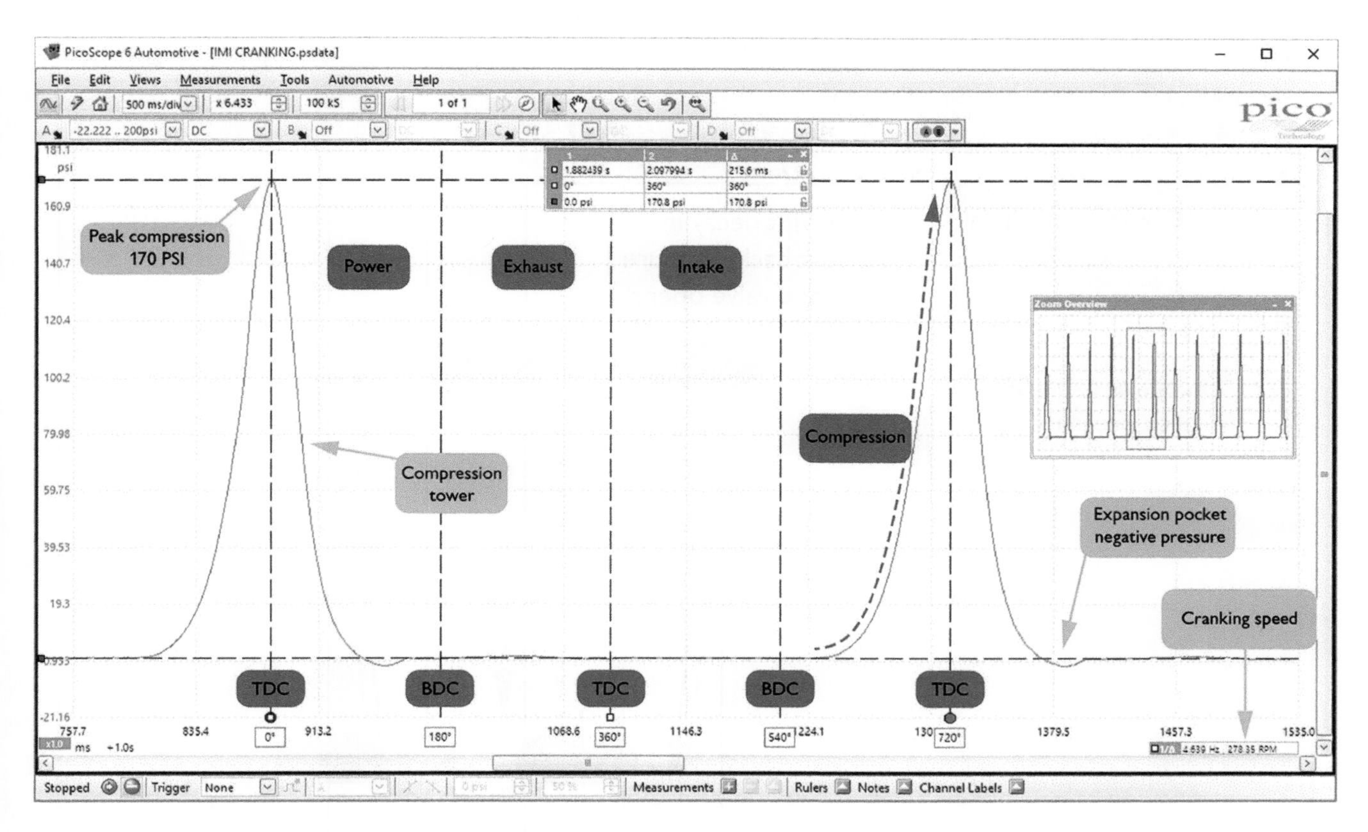

Figure 3.58 Good compression waveform – (Cranking WOT).

see uniform compression towers confirming cylinder efficiency, not only as the pressure builds but also as it naturally decays during the expansion stroke. For the compression towers to be symmetrical, the mechanical integrity of the piston/cylinder and valve gear must be efficient.

The presence of the expansion pocket confirms our cylinder can hold a vacuum and must, therefore, be airtight (valve seat and piston ring integrity are OK). We can also note that there is sufficient cranking speed, adequate intake and exhaust flow so achieving the correct peak pressure (no restrictions), and the repeatability of peak compression for every completion of the four-stroke cycle is good.

In Figure 3.59, which shows an engine with a fault, we can see an additional pressure peak during the exhaust event that should not be present. As the piston rises from BDC (bottom dead centre) of the expansion stroke, the exhaust valve will open to release the cylinder pressure out to the atmosphere (atmospheric pressure) via the exhaust system. The waveform indicates a pressure increase to 125 psi approximately during the exhaust stroke because the exhaust valve is not opening. In effect what we have here is another compression stroke during the four-stroke cycle: compression-expansion-compression-intake. A typical compression tester cannot detect this condition and would read a normal compression value.

Looking a little deeper at the exhaust stroke, we can see what looks to be another compression tower as a result of the exhaust valve remaining closed, only this time the tower is no longer symmetrical.

In this scenario we have approximately 125 psi present inside our cylinder when the intake valve opens, abruptly releasing this pressure into the intake manifold, hence the rapid drop in pressure and asymmetric tower. Such an event would manifest itself as a popping sound via the intake manifold.

The previous example highlighted just a small number of the advantages when viewing pressure against time given the accuracy and responsiveness of the transducer. The possibilities for accurate diagnosis increase further with a running engine.

It is extremely important, in any in-cylinder pressure analysis, to disconnect the relevant fuel injector to prevent bore wash, oil contamination and catalyst damage. In the case of a diesel engine, combustion would take place, resulting in potential injury, as well as damage to the compression hose and transducer.

With fuel removed from the cylinder in question, we are left with a simple air pump system where the air is drawn in via the intake, compressed, decompressed and then released via the exhaust (the four-stroke cycle). Each stage of the four-stroke cycle reveals information about the efficiency of the cylinder and the timings of each stroke relevant to the degrees of crankshaft rotation.

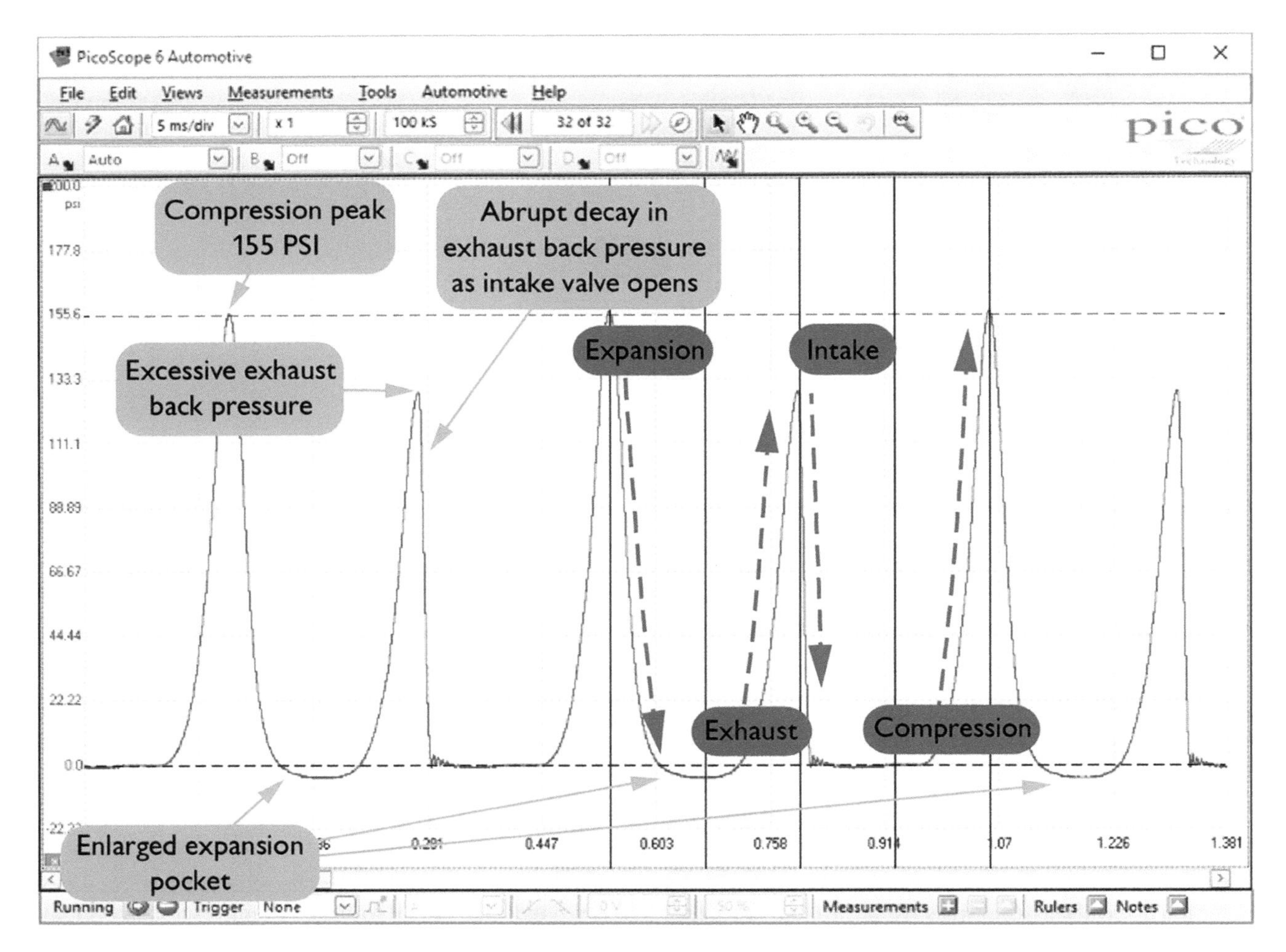

Figure 3.59 Faulty compression waveform.

Dynamic in-depth analysis of the four-stroke cycle for each cylinder is now possible. Specific errors, such as timing chain elongation, worn camshaft lobes, poor alignment (or installation) of camshafts, broken and insecure rockers, compressed hydraulic lifters or camshaft lobes that have spun independently of the camshaft can all be determined. All of these conditions will affect the independent valve timing of each cylinder.

Figure 3.60 is a known good cylinder waveform from a petrol engine, running at idle speed. Like all diagnostic techniques, we need to know what the waveform from a good cylinder looks like before we begin to formulate theories about the capture we have taken from the engine under diagnosis.

The waveforms below have additional features in comparison to the in-cylinder pressure waveform from a cranking engine. It is now possible to detect and measure the intake and exhaust events based on the formation of the expansion and intake pockets. As a general rule of thumb, the expansion and intake pockets should be equal in depth, and measure the typical negative pressure found in the intake manifold at idle speed (approximately –650 to –750 mbar).

In addition to the depth of the expansion pockets (Figure 3.61), their duration relates directly to the opening and closing of both the exhaust and intake valves. This time can be measured against crankshaft rotation by using the rotation rulers found in the PicoScope software. The compression towers should be symmetrical and uniform in structure, with an equal build and decay in pressure (remember there is no combustion taking place inside the cylinder).

Once familiar with the features of a good waveform, it is now possible to consider what effects certain faults will have.

For example, valve overlap activity can be monitored and measured for correct timing and duration in relation to crankshaft rotation. This is done during the transition of the piston at top dead centre (TDC) on the exhaust stroke, through to the commencement of the intake stroke. Valve timing and lift errors can be identified if the valve overlap events do not take place at the correct rotational position of the crankshaft.

The waveform in Figure 3.62 confirms a retarded exhaust valve open event based on the oversized expansion pocket, as well as a momentarily retarded intake

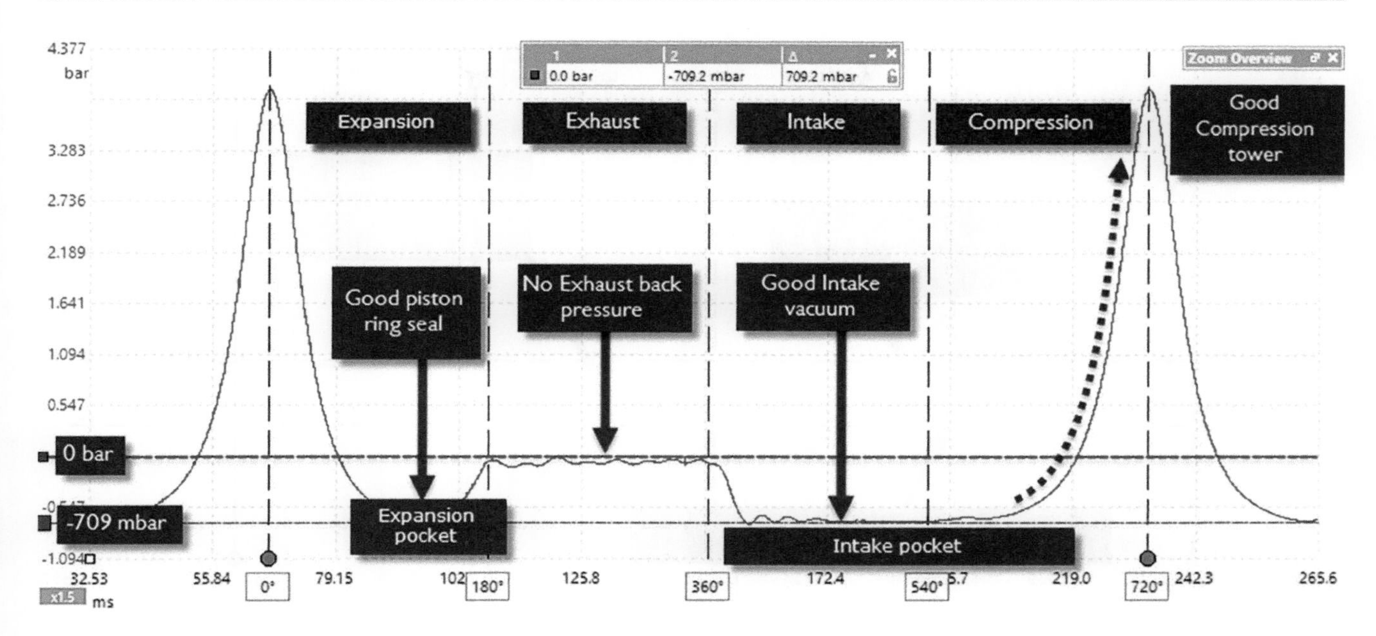

Figure 3.60 Good example of the in-cylinder waveform revealing the four stroke cycle at idle (Petrol).

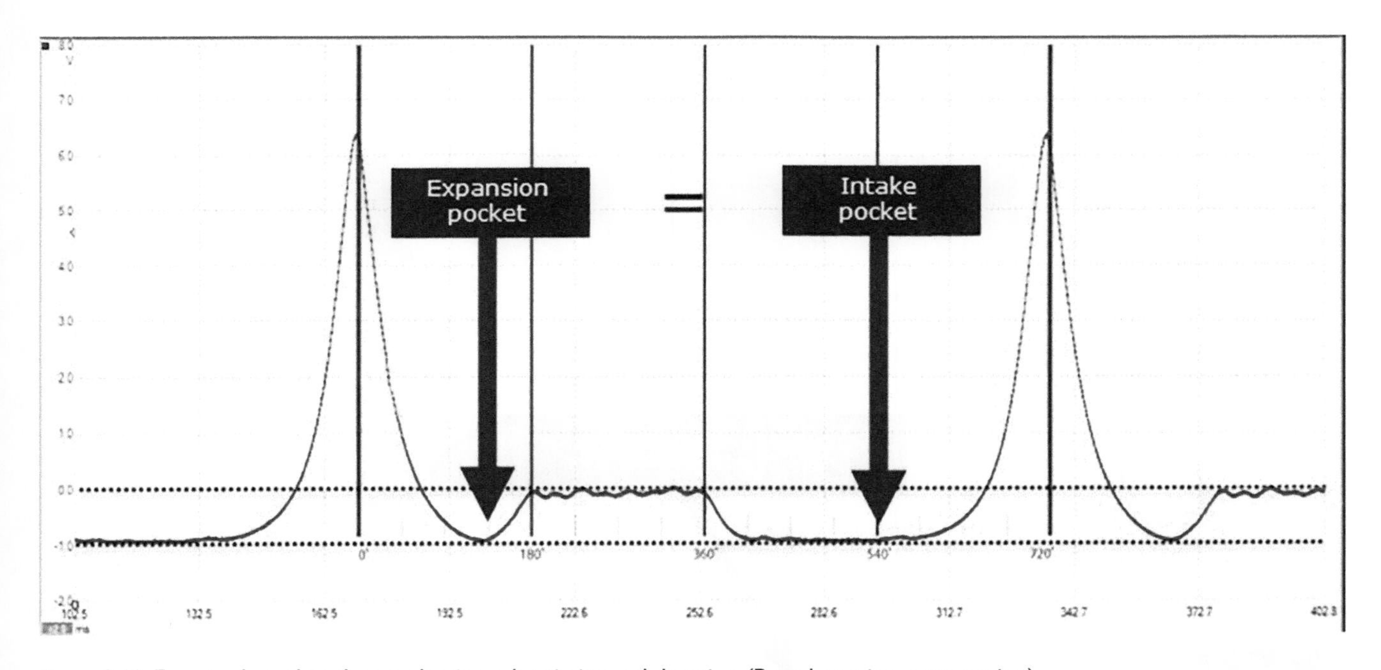

Figure 3.61 Expected results when evaluating valve timing and duration (Petrol running compression).

valve open event with minimal duration. Here we have a classic valve timing error on both camshafts, along with insufficient valve lift and duration of the intake valve.

Restrictions present in the exhaust system can also be rapidly confirmed, with the presence of backpressure measured inside the cylinder, as the piston moves upwards during the exhaust stroke. Here we can identify diesel particulate filter (DPF) and catalyst restrictions without intrusion to the exhaust system (exhaust valve open should result in cylinder pressure equalising to atmospheric pressure). The waveform in Figure 3.63 indicates over 600 mbar of backpressure at idle speed. This would only increase with engine speed and will result in power loss and, as a consequence, damage to relevant components.

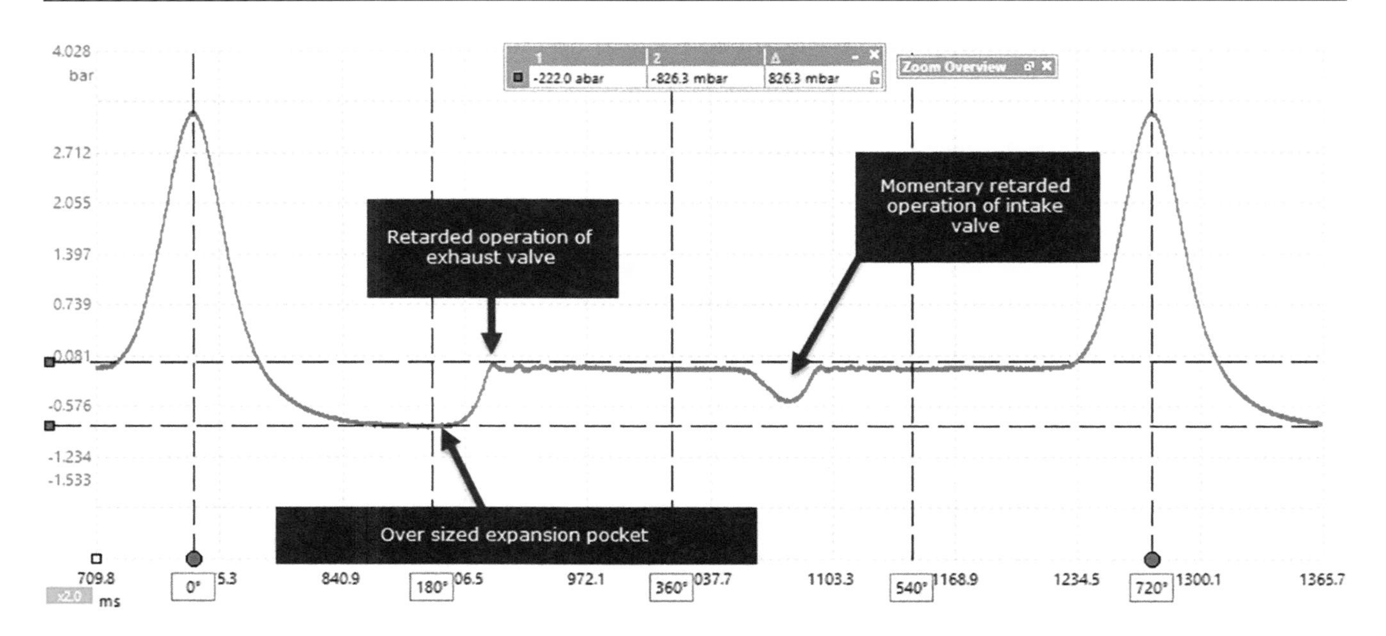

Figure 3.62 Incorrect valve timing and duration.

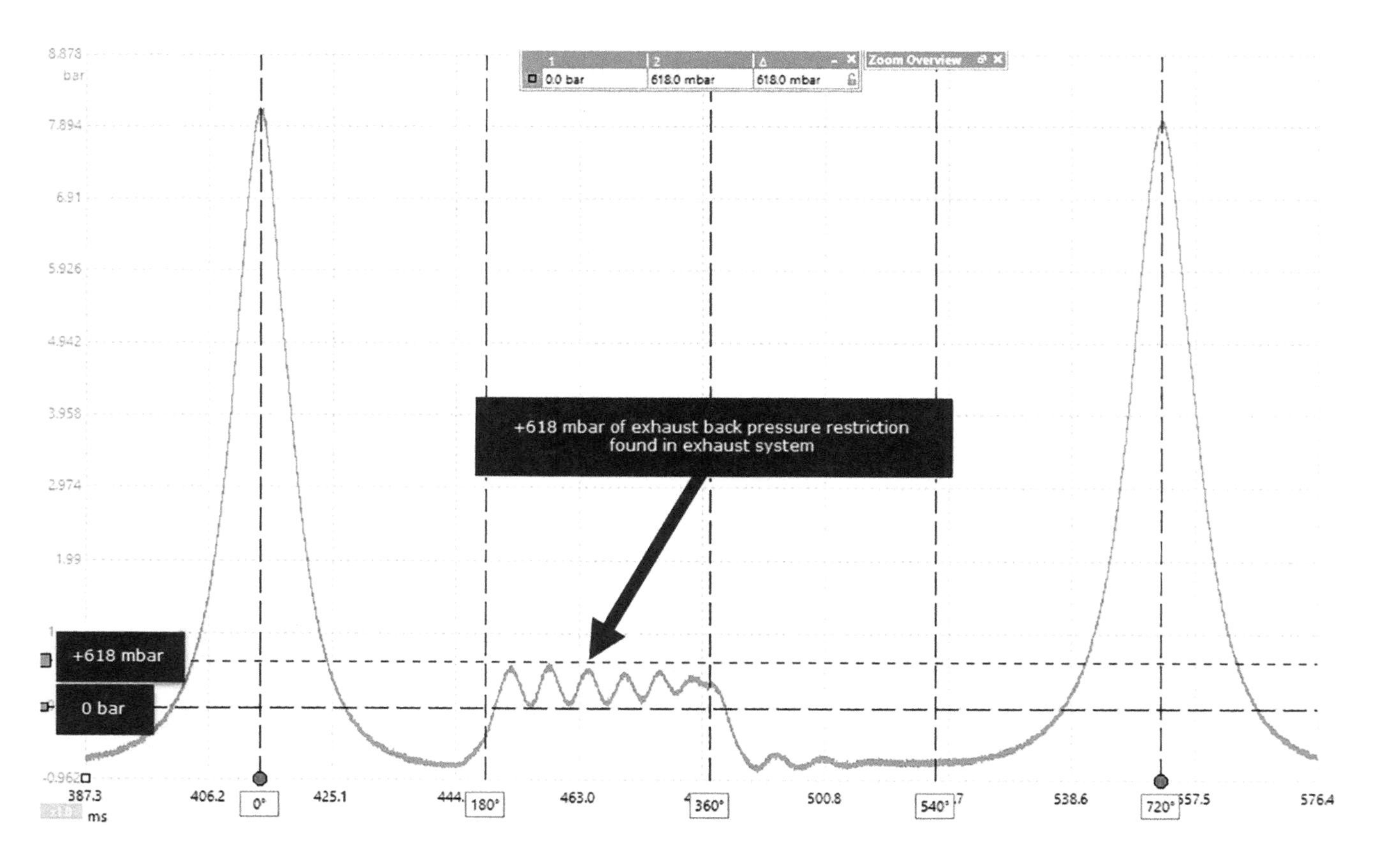

Figure 3.63 Restrictions within the exhaust system.

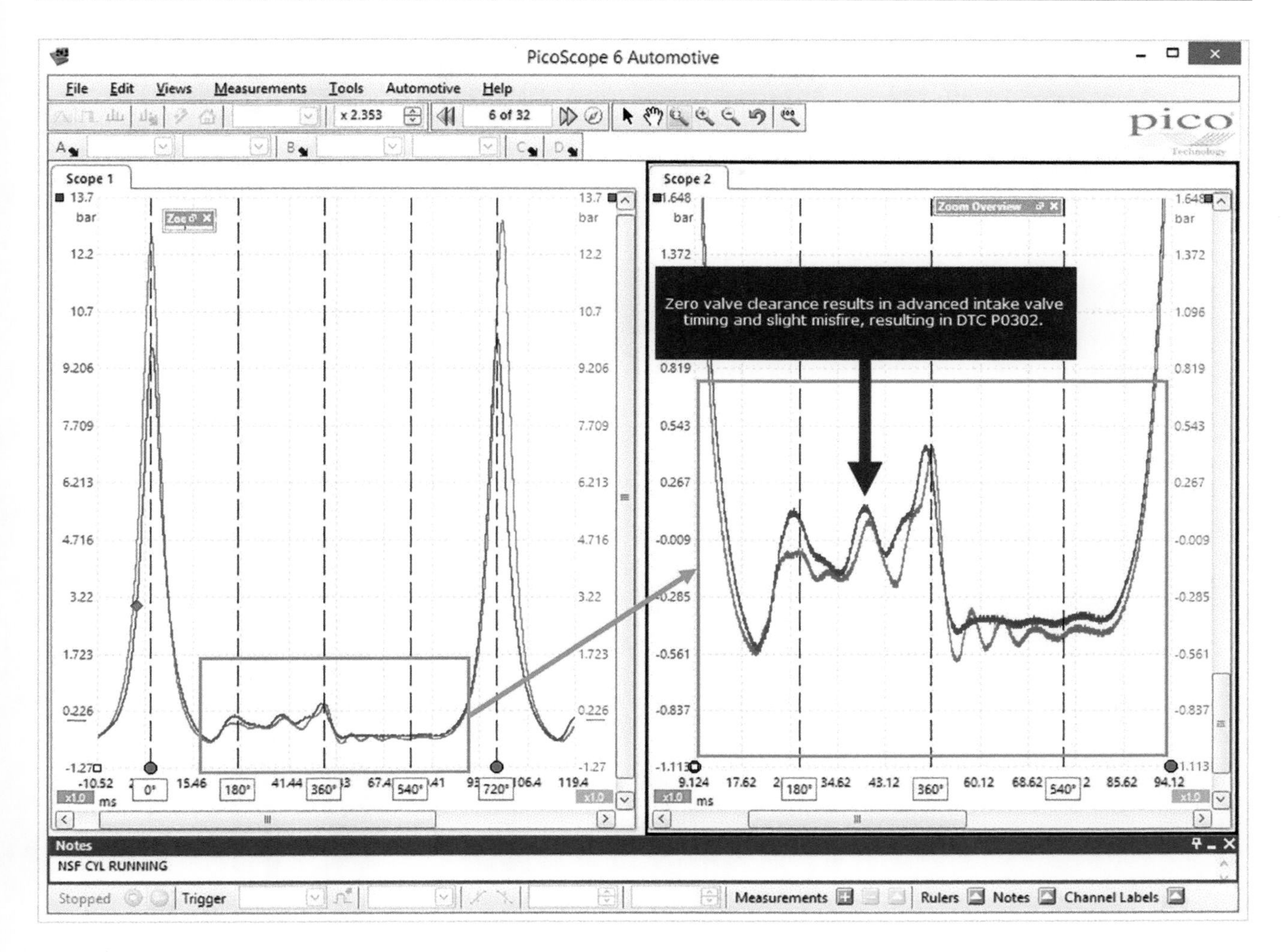

Figure 3.64 Valve clearance error (Zero valve clearance-intake valve).

Often, in-cylinder waveforms require close analysis where errors are not immediately apparent. The Pico pressure transducer has an ultra-fast response time of 100 ms, and when combined with the zoom and scaling features of the scope, even the smallest anomalies can be revealed (as in Figure 3.64).

These features allow us to measure variations in individual cylinder valve timings, as a result of valve clearance errors, camshaft lobe wear or valve lifter and rocker failure.

Connecting the pressure transducer to a PicoScope opens lots of new diagnostic possibilities, for example visualising in-cylinder events under various engine operating conditions. However, the pressure transducer can be connected to the intake manifold, exhaust tail-pipe, lambda sensor aperture or crankcase via the dip-stick tube, to name just a few accessible locations.

We are just scratching the surface of this tool's potential, and it is now as essential for many engine (and other system) diagnostic procedures as using a current clamp to determine cylinder balance.

The source of this section and more details about the highly-recommended PicoScope and the associated pressure transducer can be found at: www.picoauto.com.

Chapter 4

Sensors, actuators and oscilloscope diagnostics

4.1 INTRODUCTION

The issues and diagnostic techniques used for sensors and actuators are common to many systems. For example, the testing procedure for an inductive engine speed sensor on a fuel injection is the same as for an inductive speed sensor on an antilock brake system (ABS). Testing sensors to diagnose faults is usually a matter of measuring their output signal. In some cases, the sensor will produce this on its own (e.g. an inductive sensor). In other cases, it will be necessary to supply the correct voltage to the device to make it work (e.g. Hall sensor). It is normal to check that the vehicle circuit is supplying the voltage before proceeding to test the sensor.

KEY FACT

Testing sensors to diagnose faults is usually a matter of measuring their output signal.

At the beginning of the sections on sensors and actuators, a table is included listing the device, equipment, test method(s), results of the tests and, in most cases, a reference to a scope waveform. A waveform is often the recommended method of testing. The waveform shown will either be the output of a sensor or the signal supplied to an actuator.

Note: Any figures given are average or typical values. Refer to a good reference source such as a workshop manual or 'data book' for specific values.

Author's note: The waveforms in this chapter were captured using the PicoScope® automotive oscilloscope. I am most grateful to the PicoTech team for supplying information and equipment to assist in the production of this chapter (http://www.picoscope.com).

4.2 SENSORS

4.2.1 Introduction and sensor diagnostics

A sensor is a device that measures a physical quantity and converts it into a signal which can be read by an electronic control unit (ECU), an observer or an instrument. For accuracy, most sensors are calibrated against known standards. Most vehicle sensors produce an electrical signal, so checking their output on an oscilloscope is often the recommended method. However, many can also be checked using a multimeter (Table 4.1).

KEY FACT

Most vehicle sensors produce an electrical signal.

4.2.2 Inductive sensors

Inductive-type sensors are used mostly for measuring the speed and position of a rotating component. They work on the very basic principle of electrical induction (a changing magnetic flux will induce an electromotive force in a winding). The output voltage of most inductive-type sensors approximates to a sine wave. The amplitude of this signal depends on the rate of change of flux. This is determined mostly by the original design as in the number of turns, magnet strength and gap between the sensor and the rotating component. Once in use, though, the output voltage increases with the speed of rotation. In the majority of applications, it is the frequency of the signal that is used.

KEY FACT

The amplitude of an inductive sensor signal depends on the rate of change of flux.

Table 4.1 Sensor diagnostic methods

Sensor	*Equipment*	*Method(s)*	*Results*	*Scope waveform*
Thermistor Coolant sensor Air intake temperature sensor Ambient temperature sensor Etc.	Ohmmeter	Connect across the two terminals, or if only one, from this to earth	Most thermistors have a negative temperature coefficient (NTC). This means the resistance falls as temperature rises. A resistance check should give readings broadly as follows: 0°C = 4500 Ω 20°C = 1200 Ω 100°C = 200 Ω	Figure 4.14
Inductive Crankshaft speed and position ABS wheel speed Camshaft position	Ohmmeter AC voltmeter	A resistance test with the sensor disconnected AC voltage output with the engine cranking	Values vary from approx. 200–400 Ω on some vehicles to 800–1200 Ω on others. The 'sine wave' output should be approx. 5 V (less depending on engine speed)	Figure 4.2 Figure 4.4 Figure 4.7
Hall effect Ignition distributor Engine speed Transmission speed Wheel speed Current flow in a wire (ammeter amp clamp)	DC voltmeter Logic probe Do NOT use an ohmmeter as this will damage the Hall chip	The voltage output measured as the engine or component is rotated slowly. The sensor is normally supplied with a 5 V or a 10–12 V	This distributor switches between 0 and approx. 8 V as the Hall chip is magnetised or not. Others switch between 0 and approx. 4 V A logic probe will read high and low as the sensor output switches	Figure 4.3 Figure 4.17 Figure 4.19
Optical Ignition distributor Rotational speed	DC voltmeter	The device will normally be supplied with a stabilised voltage. Check the output wire signal as the device is rotated slowly	Clear switching between low and high voltage	N/A
Variable resistance Throttle potentiometer Flap-type airflow sensor Position sensor	DC voltmeter Ohmmeter	This sensor is a variable resistor. If the supply is left connected then check the output on a DC voltmeter With the supply disconnected, check the resistance	The voltage should change *smoothly* from approx. 0 V to the supply voltage (often 5 V) Resistance should change smoothly	Figure 4.8 Figure 4.10
Strain gauges MAP sensor Torque stress	DC voltmeter	The normal supply to an externally mounted manifold absolute pressure (MAP) sensor is 5 V. Check the output as manifold pressure changes either by snapping the throttle open, road testing or by using a vacuum pump on the sensor pipe	The output should change between approx. 0 and 5 V as the manifold pressure changes. As a general guide 2.5 V at idle speed	N/A
Variable capacitance	DC voltmeter	Measure the voltage at the sensor	Small changes as the input to the sensor is varied – this is not difficult to assess because of very low capacitance values	N/A
Accelerometer Knock sensors	Scope	Tap the engine block lightly (13 mm spanner) near the sensor	Oscillating output that drops back to zero If the *whole* system is operating, the engine will slow down if at idle speed	Figure 4.21
Hot wire Airflow	DC voltmeter or duty cycle meter	This sensor includes electronic circuits to condition the signal from the hot wire. The normal supply is either 5 or 12 V. Measure the output voltage as engine speed/load is varied	The output should change between approx. 0 and 5 V as the airflow changes. 0.4–1 V at idle is typical. Or depending on the system in use the output may be digital	Figure 4.12

(*Continued*)

Table 4.1 (Continued)

Sensor	*Equipment*	*Method(s)*	*Results*	*Scope waveform*
Oxygen Lambda sensor EGO sensor HEGO sensor	DC voltmeter	The lambda sensor produces its own voltage a bit like a battery. This can be measured with the sensor connected to the system	A voltage of approx. 450 mV (0.45 V) is the normal figure produced at lambda value of one The voltage output, however, should vary smoothly between 0.2 and 0.8 V as the mixture is controlled by the ECU	Figure 4.24 Figure 4.26 Figure 4.27
Acceleration switch Dynamic position	DC voltmeter	Measure the supply and output as the sensor is subjected to the required acceleration	A clear switching between say 0 and 12 V	N/A
Rain and other unknown types	DC voltmeter	Locate output wire – by trial and error if necessary and measure dry/wet output (splash water on the screen with the sensor correctly fitted in position)	A clear switching between distinct voltage levels	N/A

4.2.2.1 Crankshaft and camshaft sensors

Inductive-type crank and cam sensors work in the same way. A single tooth, or toothed wheel, induces a voltage into a winding in the sensor. The cam sensor provides engine position information as well as which cylinder is on which stroke. The crank sensor provides engine speed. It also provides engine position in many cases by use of a 'missing' tooth (Figure 4.1).

In this particular waveform, we can evaluate the output voltage from the crank sensor. The voltage will differ between manufacturers, and it also increases with engine speed. The waveform will be an alternating voltage signal.

If there is a gap in the trace, it is due to a 'missing tooth' on the flywheel or reluctor and is used as a reference for the ECU to determine the engine's position. Some systems use two reference points per revolution (Figure 4.2).

The camshaft sensor is sometimes referred to as the cylinder identification (CID) sensor or a 'phase' sensor and is used as a reference to time sequential fuel injection.

Figure 4.1 Crank sensor in position near the engine flywheel.

DEFINITION

CID: Cylinder identification.

The voltage produced by the camshaft sensor will be determined by several factors, these being the engine's speed, the proximity of the metal rotor to the pick-up and the strength of the magnetic field offered by the sensor. The ECU needs to see the signal when the engine is started for its reference; if absent, it can alter the point at which the fuel is injected. The driver of the vehicle may not be aware that the vehicle has a problem if the CID sensor fails, as the driveability may not be affected. However, the MIL should illuminate.

The characteristics of a good inductive camshaft sensor waveform is a sine wave that increases in magnitude as the engine speed is increased, and usually provides one signal per 720° of crankshaft rotation (360° of camshaft rotation). The voltage will be approximately 0.5 V peak to peak while the engine is cranking, rising to around 2.5 V peak to peak at idle.

Some crankshaft sensors (CAS) are now Hall effect types and will therefore show a broadly square wave signal (Figure 4.3).

4.2.2.2 ABS speed sensor

The ABS wheel speed sensors have become smaller and more efficient in the course of time. Recent models not only measure the speed and direction of wheel rotation but can be integrated into the wheel bearing as well (Figure 4.4).

ABS relies upon information coming in from the sensors to determine what action should be taken. If, under heavy braking, the ABS ECU loses a signal from one of the road wheels, it assumes that the wheel has locked

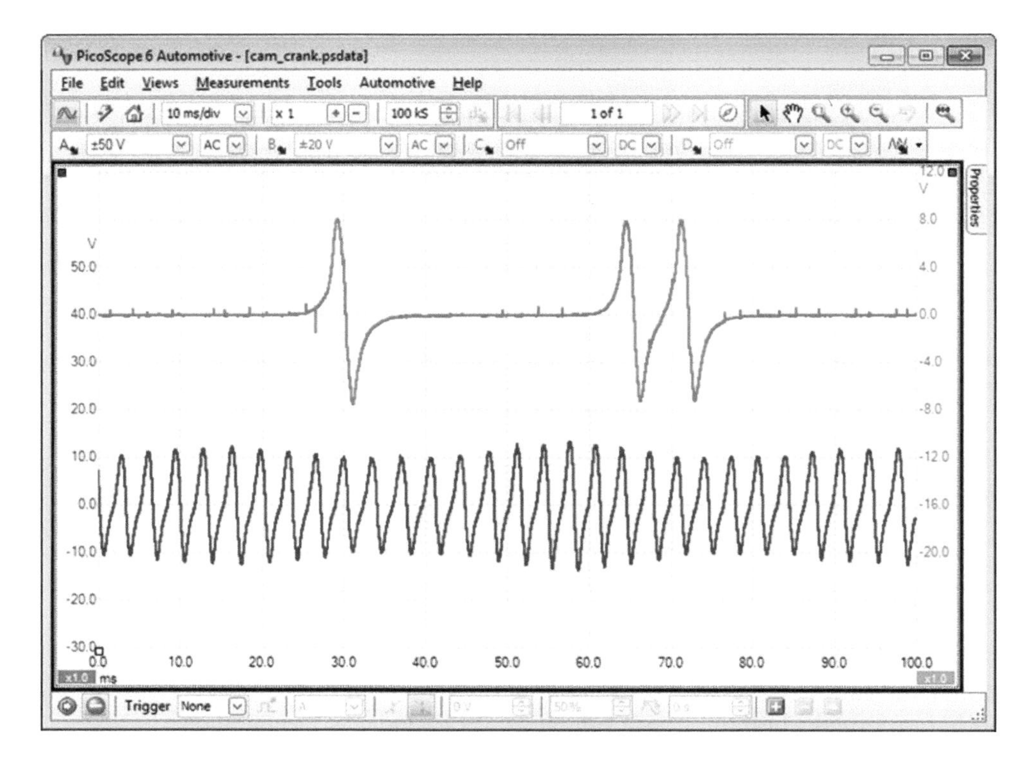

Figure 4.2 Crank and cam sensor output signals.

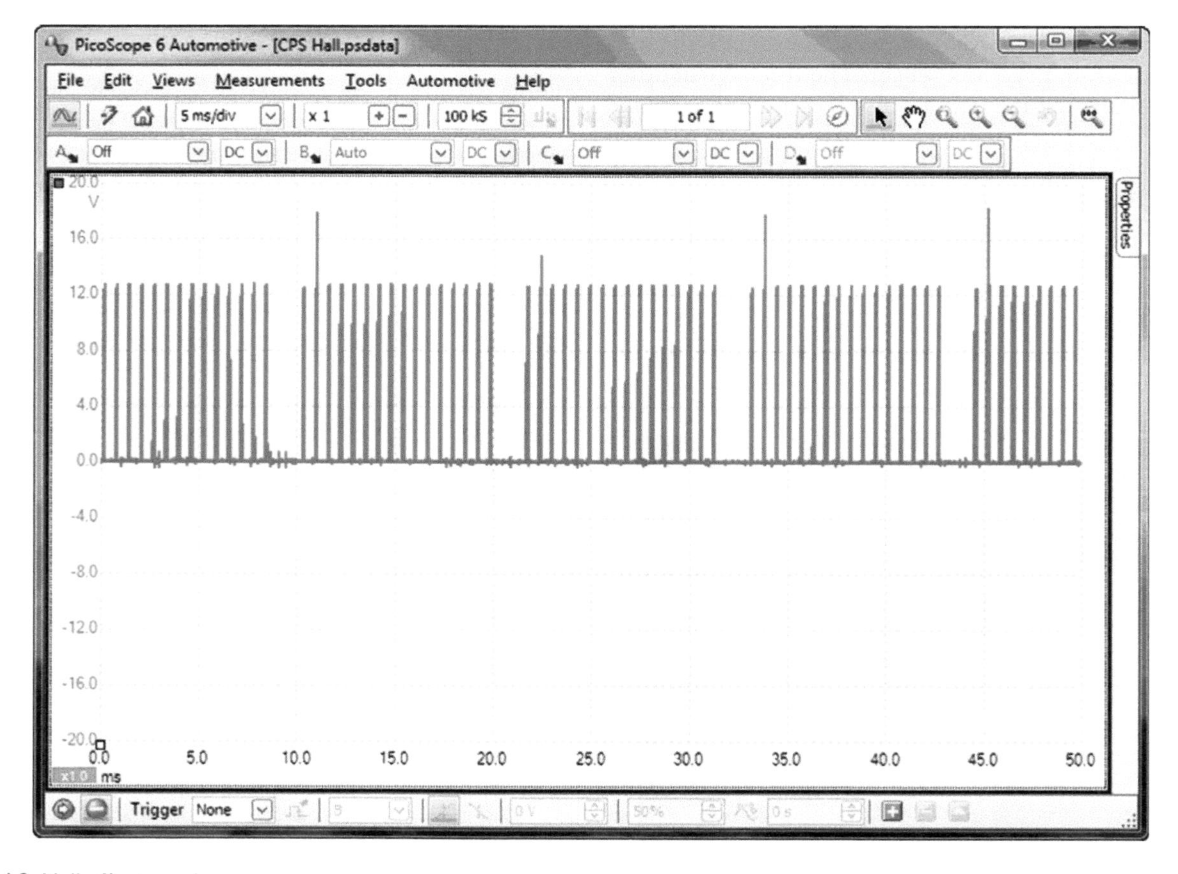

Figure 4.3 Hall effect crank sensor.

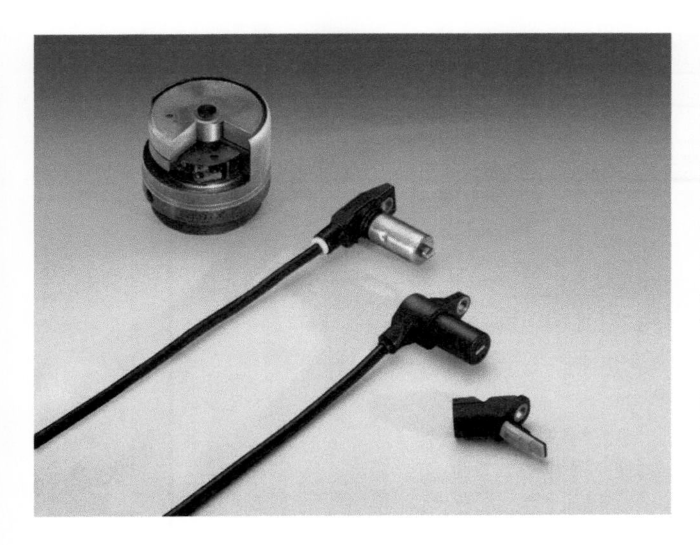

Figure 4.4 ABS wheel speed sensors.

Source: Bosch Press.

and releases that brake momentarily until it sees the signal return. It is therefore imperative that the sensors are capable of providing a signal to the ABS ECU. If the signal produced from one wheel sensor is at a lower frequency than the others, the ECU may also react (Figures 4.5 and 4.6).

The operation of most ABS sensors is similar to that of a crank angle sensor (CAS). A small inductive pick-up is affected by the movement of a toothed wheel, which moves in close proximity. The movement of the wheel next to the sensor results in a 'sine wave'. The sensor, recognisable by its two electrical connections (some may have a coaxial braided outer shield), will produce an output that can be monitored and measured on the oscilloscope. Some are now Hall effect types so expect to see a square wave output.

DEFINITION

CAS: Crank angle sensor.

4.2.2.3 Inductive distributor pick-up

Not used on modern cars, but there are still plenty out there! The pick-up is used as a signal to trigger the ignition amplifier or an ECU. The sensor normally has two connections. If a third connection is used, it is normally a screen to reduce interference.

As a metal rotor spins, a magnetic field is altered, which induces an AC voltage from the pick-up. This

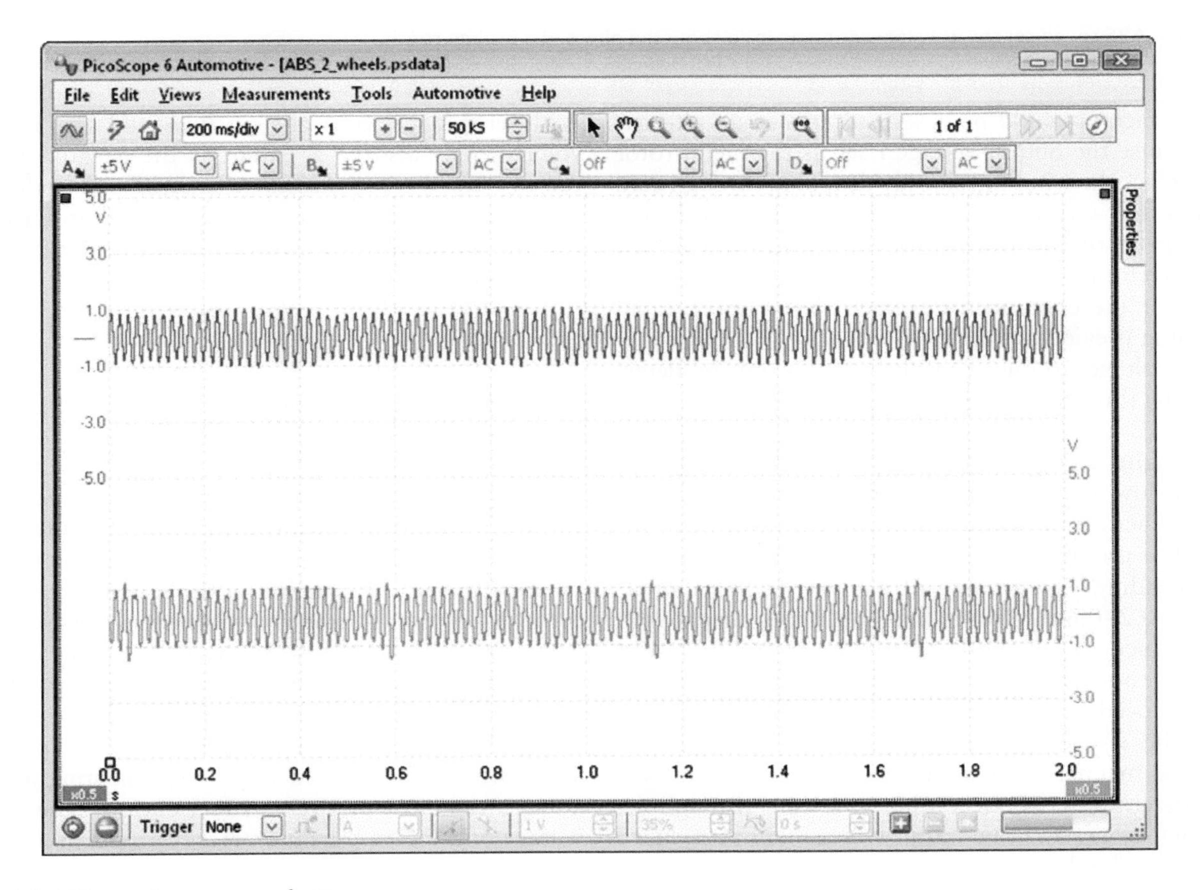

Figure 4.5 ABS speed sensor waveform.

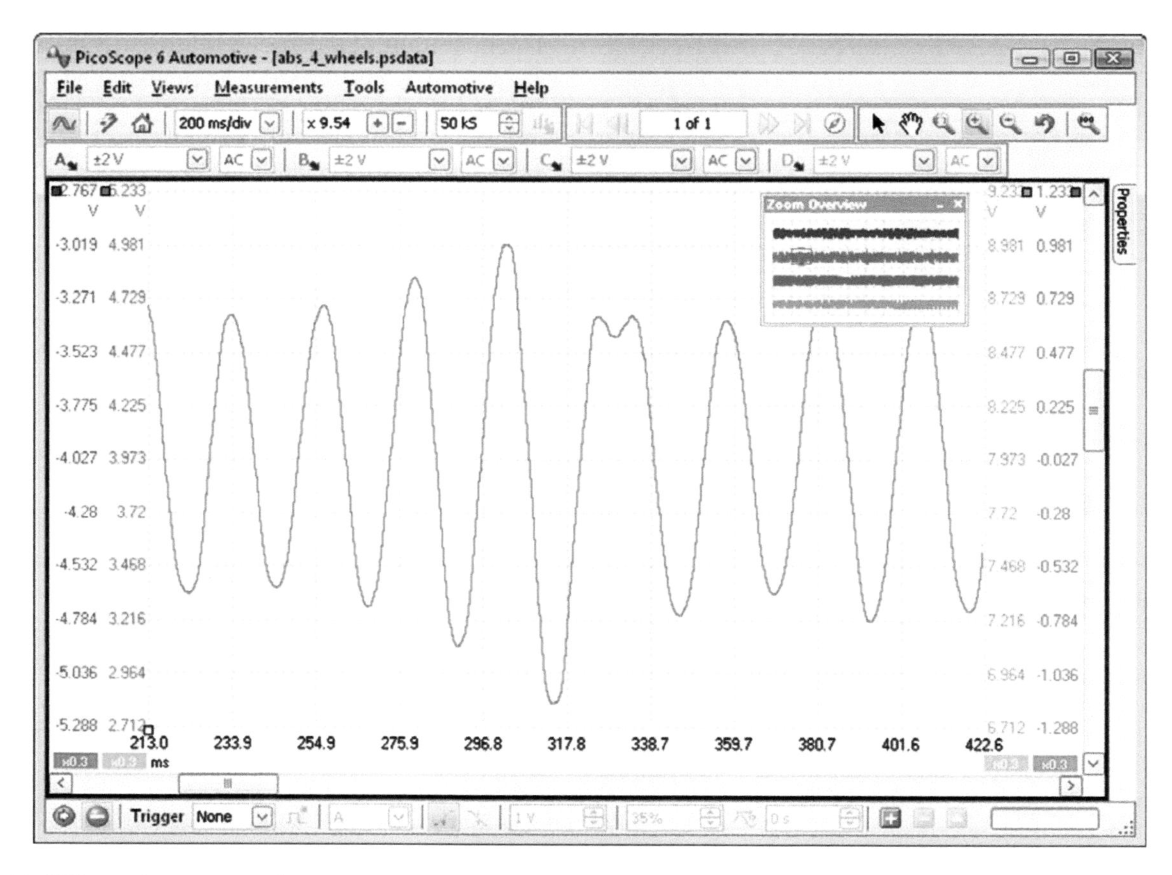

Figure 4.6 ABS speed sensor waveform zoomed in to show the effect of a broken tooth.

type of pick-up could be described as a small alternator because the output voltage rises as the metal rotor approaches the winding, sharply dropping through zero volts as the two components are aligned and producing a voltage in the opposite direction as the rotor passes. The waveform is similar to a sine wave; however, the design of the components is such that a more rapid switching is evident (Figure 4.7).

The voltage produced by the pick-up will be determined by three main factors:

- Engine speed – the voltage produced will rise from as low as 2–3 V when cranking, to over 50 V at higher engine speeds.
- The proximity of the metal rotor to the pick-up winding – an average air gap will be in the order of 0.2–0.6 mm (8–14 thou), a larger air gap will reduce the strength of the magnetic field seen by the winding and the output voltage will be reduced.
- The strength of the magnetic field offered by the magnet – the strength of this magnetic field determines the effect it has as it 'cuts' through the windings, and the output voltage will be reduced accordingly.

A difference between the positive and the negative voltages may also be apparent as the negative side of the sine wave is sometimes attenuated (reduced) when connected to the amplifier circuit, but will produce perfect AC when disconnected and tested under cranking conditions.

4.2.3 Variable resistance

The two best examples of vehicle applications for variable resistance sensors are the throttle position sensor and the flap-type airflow sensor. Although variable capacitance sensors are used to measure small changes, variable resistance sensors generally measure larger changes in position. This is due to lack of sensitivity inherent in the construction of the resistive track. The throttle position sensor is a potentiometer in which, when supplied with a stable voltage, often 5 V, the voltage from the wiper contact will be proportional to throttle position. The throttle potentiometer is mostly used to indicate rate of change of throttle position. This information is used when implementing acceleration enrichment or overrun fuel cut-off. The output voltage of a rotary potentiometer is proportional to its position.

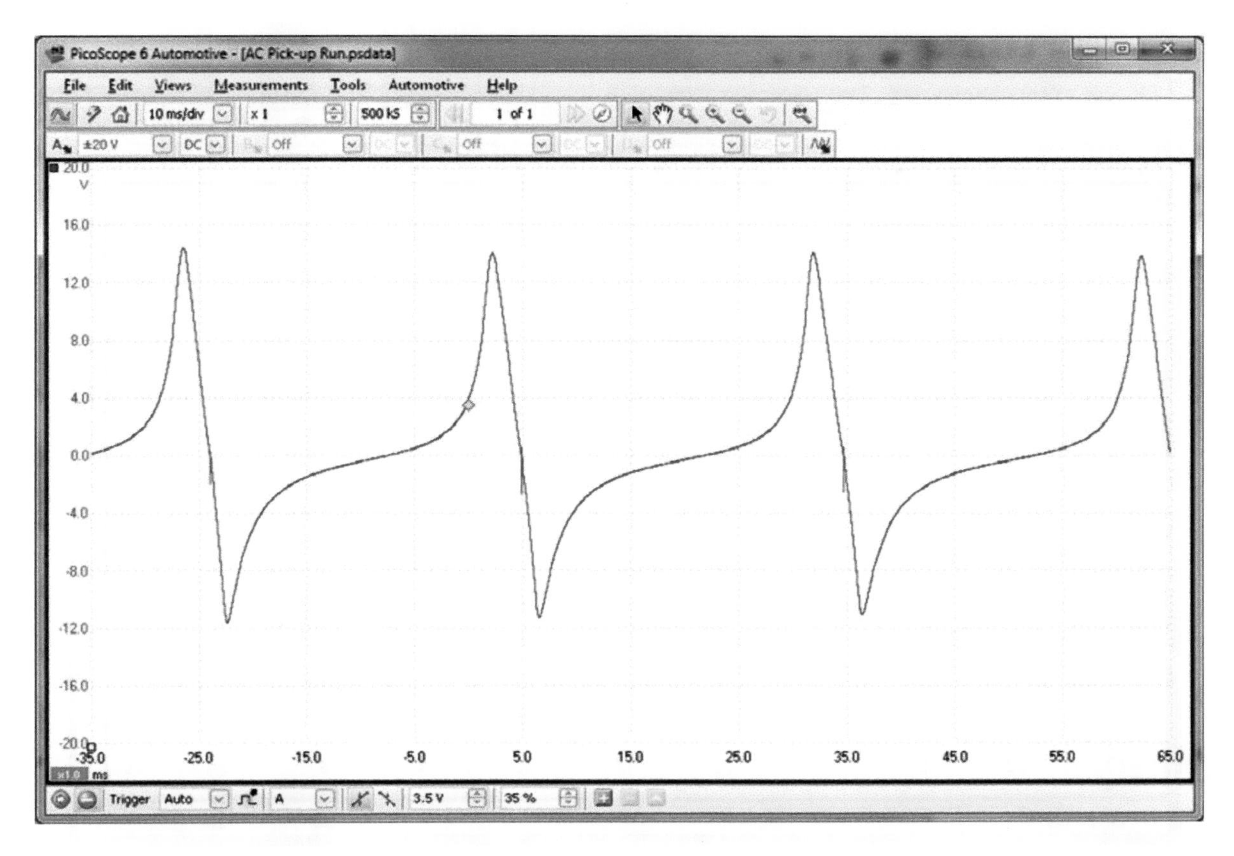

Figure 4.7 Inductive pick-up output signal (engine running).

KEY FACT

Variable capacitance sensors are used to measure small changes; variable resistance sensors generally measure larger changes in position.

The airflow sensor shown in Figure 4.8 works on the principle of measuring the force exerted on the flap by the air passing through it. A calibrated coil spring exerts a counter force on the flap such that the movement of the flap is proportional to the volume of air passing through the sensor. To reduce the fluctuations caused by individual induction strokes, a compensation flap is connected to the sensor flap.

The fluctuations therefore affect both flaps and are cancelled out. Any damage due to backfiring is also minimised due to this design. The resistive material used for the track is a ceramic metal mixture, which is burnt into a ceramic plate at very high temperature. The slider potentiometer is calibrated such that the output voltage is proportional to the quantity of inducted air. This sensor type is not used on modern vehicles as more accurate measurement is possible using other techniques.

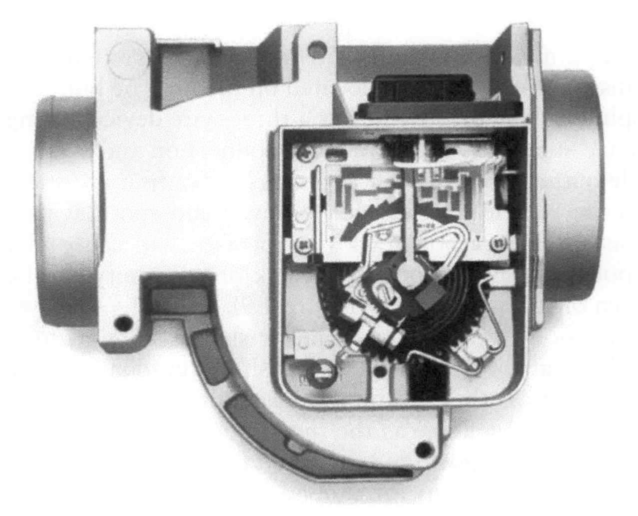

Figure 4.8 Vane- or flap-type airflow sensor.

Source: Bosch.

4.2.3.1 Throttle position potentiometer

This sensor or potentiometer is able to indicate to the ECU the exact amount of throttle opening due to its linear output.

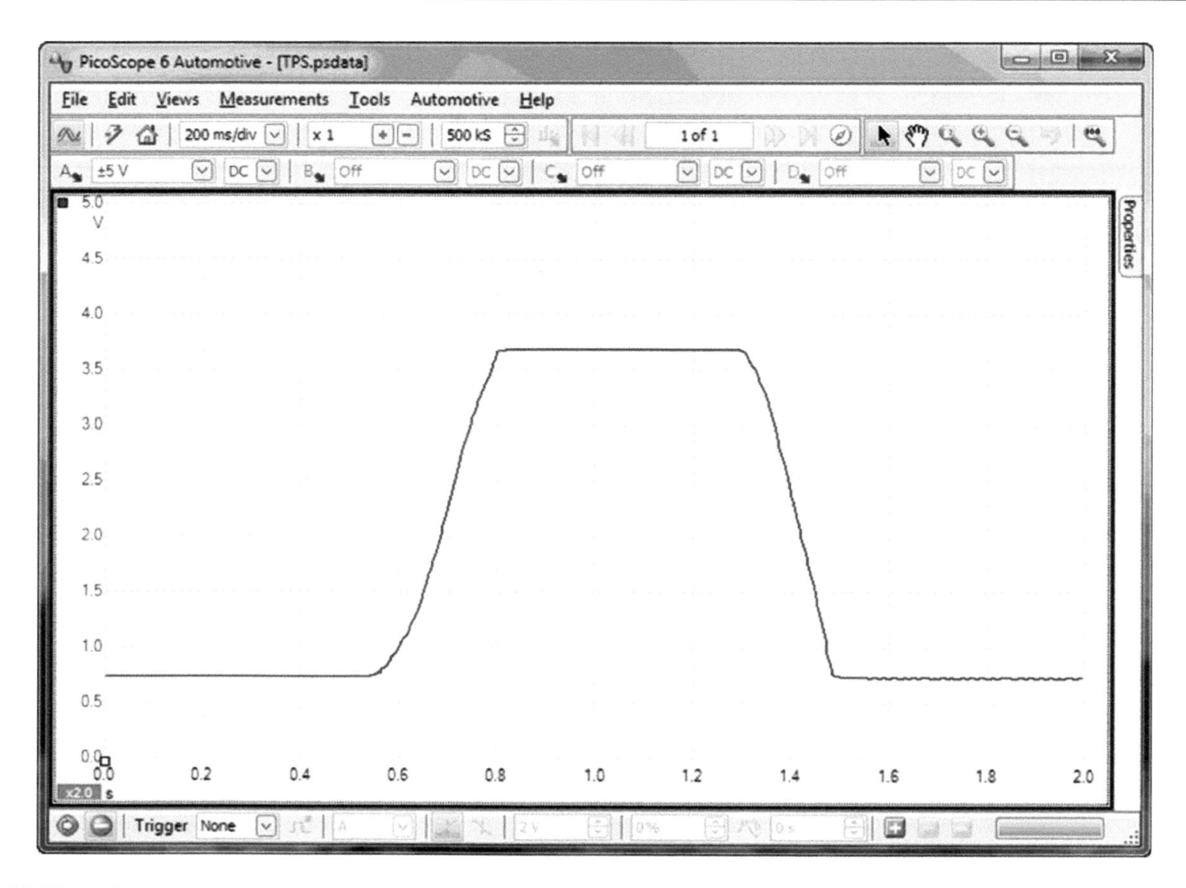

Figure 4.9 Throttle pot output voltage.

The majority of modern management systems use this type of sensor. It is located on the throttle butterfly spindle. The 'throttle pot' is a three-wire device having a 5 V supply (usually), an earth connection and a variable output from the centre pin. As the output is critical to the vehicle's performance, any 'blind spots' within the internal carbon track's swept area will cause 'flat spots' and 'hesitations'. This lack of continuity can be seen on an oscilloscope (Figure 4.9).

A good throttle potentiometer should show a small voltage at the throttle closed position, gradually rising in voltage as the throttle is opened and returning back to its initial voltage as the throttle is closed. Although many throttle position sensor voltages will be manufacturer specific, many are non-adjustable and the voltage will be in the region of 0.5–1.0 V at idle, rising to 4.0 V (or more) with a fully opened throttle. For the full operational range, an oscilloscope time scale around two seconds is used.

4.2.3.2 Airflow meter – air vane

The vane-type airflow meter is a simple potentiometer that produces a voltage output that is proportional to the position of a vane. The vane in turn positions itself proportional to the amount of air flowing (Figure 4.8).

The voltage output from the internal track of the airflow meter should be linear to flap movement; this can be measured on an oscilloscope and should look similar to the example shown in Figure 4.10.

The waveform should show approximately 1.0 V when the engine is at idle; this voltage will rise as the engine is accelerated and will produce an initial peak. This peak is due to the natural inertia of the air vane and drops momentarily before the voltage is seen to rise again to a peak of approximately 4.0–4.5 V. This voltage will, however, depend on how hard the engine is accelerated, so a lower voltage is not necessarily a fault within the airflow meter. On deceleration, the voltage will drop sharply as the wiper arm, in contact with the carbon track, returns back to the idle position. This voltage may in some cases 'dip' below the initial voltage before returning to idle voltage. A gradual drop will be seen on an engine fitted with an idle speed control valve (ISCV) as this will slowly return the engine back to base idle as an anti-stall characteristic.

A time base of approximately two seconds plus is used; this enables the movement to be shown on one screen, from idle through acceleration and back to idle again. The waveform should be clean with no 'drop-out' in the voltage, as this indicates a lack of electrical continuity. This is common on an airflow meter with a

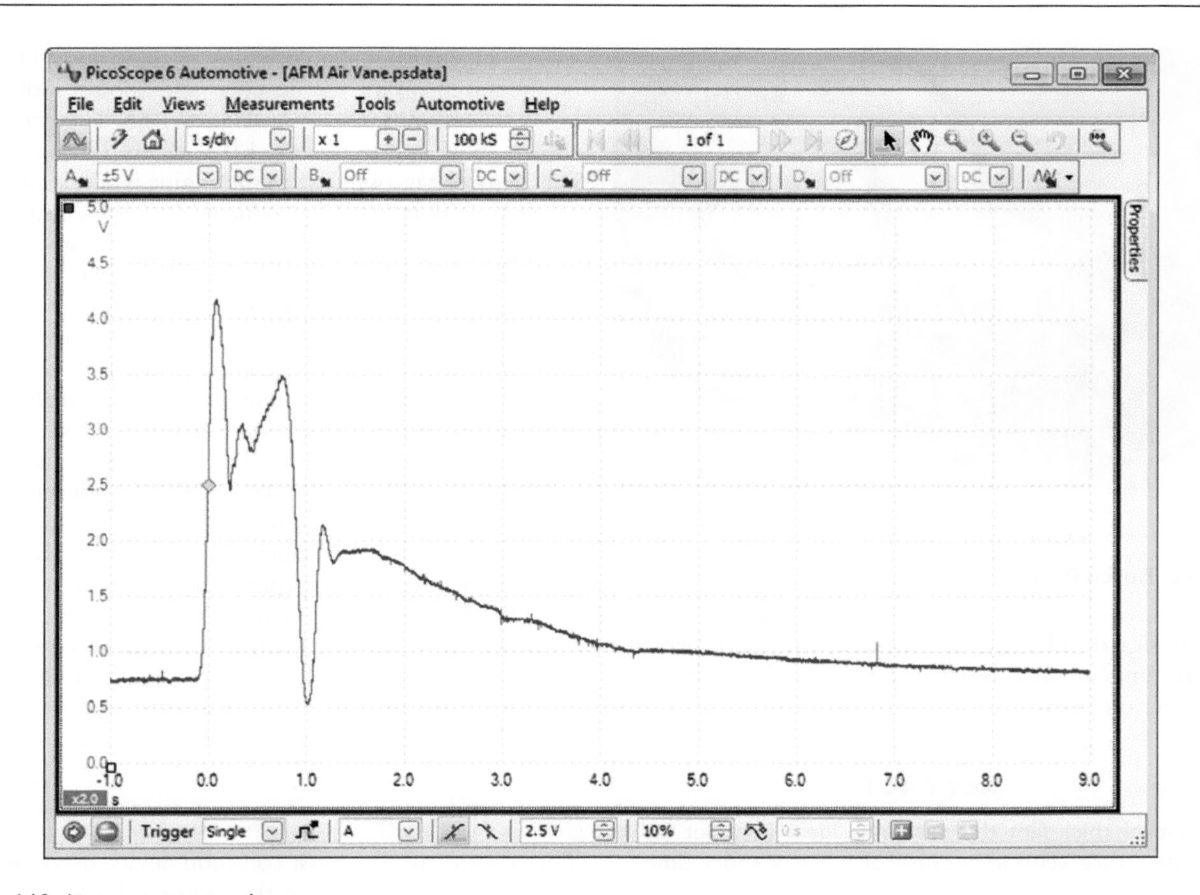

Figure 4.10 Air vane output voltage.

dirty or faulty carbon track. The problem will appear as a 'flat spot' or hesitation when the vehicle is driven, this is a typical problem on vehicles with high mileage that have spent the majority of their working life with the throttle in one position.

4.2.4 Hot wire airflow sensor

The advantage of this sensor is that it measures air mass flow. The basic principle is that as air passes over a hot wire, it tries to cool the wire down. If a circuit is created such as to increase the current through the wire, then this current will be proportional to the airflow. A resistor is also incorporated to compensate for temperature variations. The 'hot wire' is made of platinum and is only a few millimetres long and approximately 70 μm thick. Because of its small size, the time constant of the sensor is very short, in fact in the order of a few milliseconds. This is a great advantage as any pulsations of the airflow will be detected and reacted to in a control unit accordingly.

The output of the circuit involved with the hot wire sensor is a voltage across a precision resistor. The resistance of the hot wire and the precision resistor are such that the current to heat the wire varies between 0.5 and 1.2 A with different air mass flow rates. High-resistance resistors are used in the other arm of the bridge and so current flow is very small. The temperature-compensating resistor has a resistance of approximately 500 Ω, which must remain constant other than by way of temperature change. A platinum film resistor is used for these reasons. The compensation resistor can cause the system to react to temperature changes within about three seconds.

The output of this device can change if the hot wire becomes dirty. Heating the wire to a very high temperature for one second every time the engine is switched off prevents this, by burning off any contamination. In some air mass sensors, a variable resistor is provided to set idle mixture. The nickel film airflow sensor is similar to the hot wire system. Instead of a hot platinum wire, a thin film of nickel is used. The response time of this system is even shorter than the hot wire. The advantage which makes a nickel thick-film thermistor ideal for inlet air temperature sensing is its very short time constant. In other words, its resistance varies very quickly with a change in air temperature.

4.2.4.1 Airflow meter – hot wire

Figure 4.11 shows a mass airflow sensor from Bosch. This type has been in use since 1996. As air flows over the hot wire, it cools it down, and this produces the

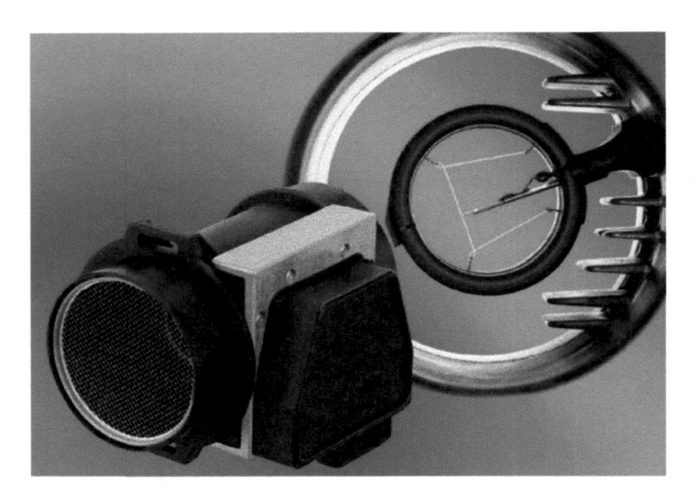

Figure 4.11 Hot wire air mass meter.

Source: Bosch Press.

output signal. The sensor measures air mass because the air temperature is taken into account due to its cooling effect on the wire.

KEY FACT

A nickel thick-film thermistor is ideal for inlet air temperature sensing because of its very short time constant.

The voltage output should be linear to airflow. This can be measured on an oscilloscope and should look similar to the example shown in Figure 4.12. The waveform should show approximately 1.0 V when the engine is at idle. This voltage will rise as the engine is accelerated and air volume is increased producing an initial peak. This peak is due to the initial influx of air and drops momentarily before the voltage is seen to rise again to another peak of approximately 4.0–4.5 V. This voltage will, however, depend on how hard the engine is accelerated; a lower voltage is not necessarily a fault within the meter.

On deceleration, the voltage will drop sharply as the throttle butterfly closes, reducing the airflow, and the engine returns back to idle. The final voltage will drop gradually on an engine fitted with ISCV as this will slowly return the engine back to base idle as an anti-stall characteristic. This function normally only affects the engine speed from around 1200 rpm back to the idle setting.

A time base of approximately two seconds plus is used because this allows the output voltage on one screen, from idle through acceleration and back to idle again. The 'hash' on the waveform is due to airflow changes caused by the induction pulses as the engine is running.

4.2.5 Thermistors

Thermistors are the most common device used for temperature measurement on the motor vehicle. The principle of measurement is that a change in temperature

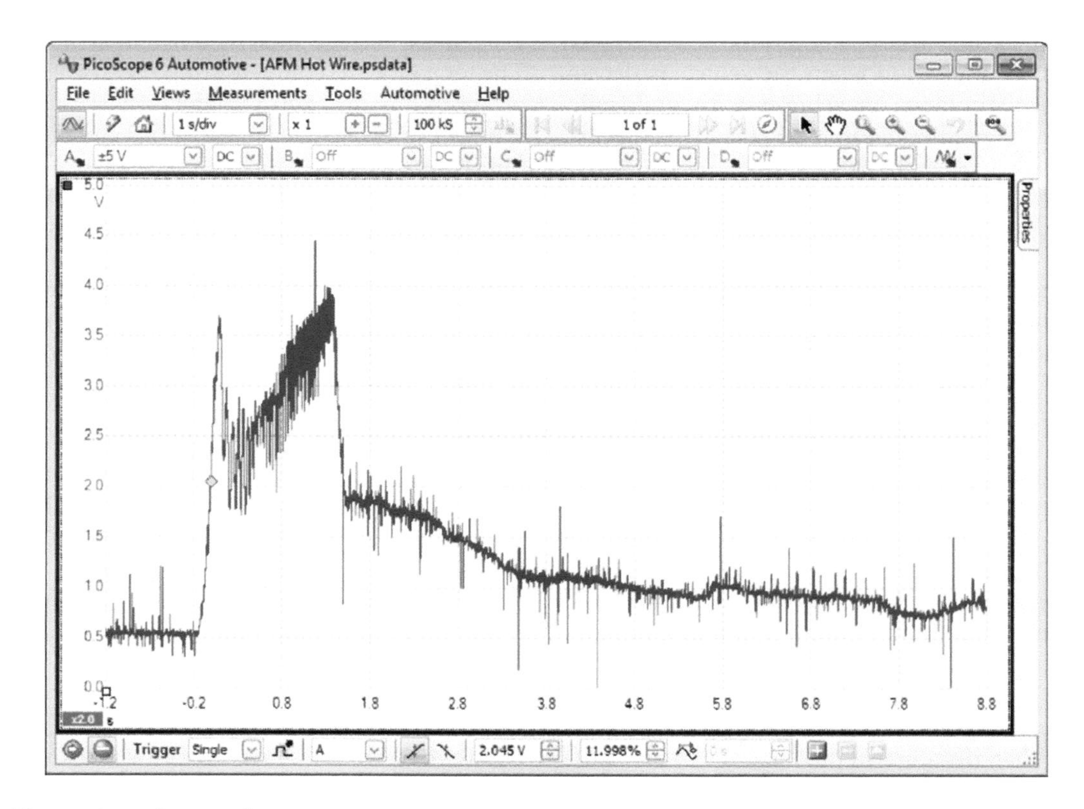

Figure 4.12 Air mass hot wire waveform.

will cause a change in resistance of the thermistor and hence an electrical signal proportional to the temperature being measured. Most thermistors in common use are of the negative temperature coefficient (NTC) type. The actual response of the thermistors can vary but typical values for those used on motor vehicles will vary from several kilo-ohms at 0°C to few hundred ohms at 100°C. The large change in resistance for a small change in temperature makes the thermistor ideal for most vehicles uses. It can also be easily tested with simple equipment. Thermistors are constructed of semiconductor materials. The change in resistance with a change in temperature is due to the electrons being able to break free more easily at higher temperatures.

4.2.5.1 Coolant temperature sensor

Most coolant temperature sensors (CTSs) are NTC thermistors; their resistance decreases as temperature increases. This can be measured on most systems as a reducing voltage signal.

DEFINITION

NTC: Negative temperature coefficient.

Figure 4.13 Temperature sensor.

The CTS is usually a two-wire device with a voltage supply of approximately 5 V (Figure 4.13).

The resistance change will therefore alter the voltage seen at the sensor and can be monitored for any discrepancies across its operational range. By selecting a time scale of 500 seconds and connecting the oscilloscope to the sensor, the output voltage can be monitored. Start the engine and in the majority of cases the voltage will start in the region of 3–4 V and fall gradually. The voltage will depend on the temperature of the engine (Figure 4.14).

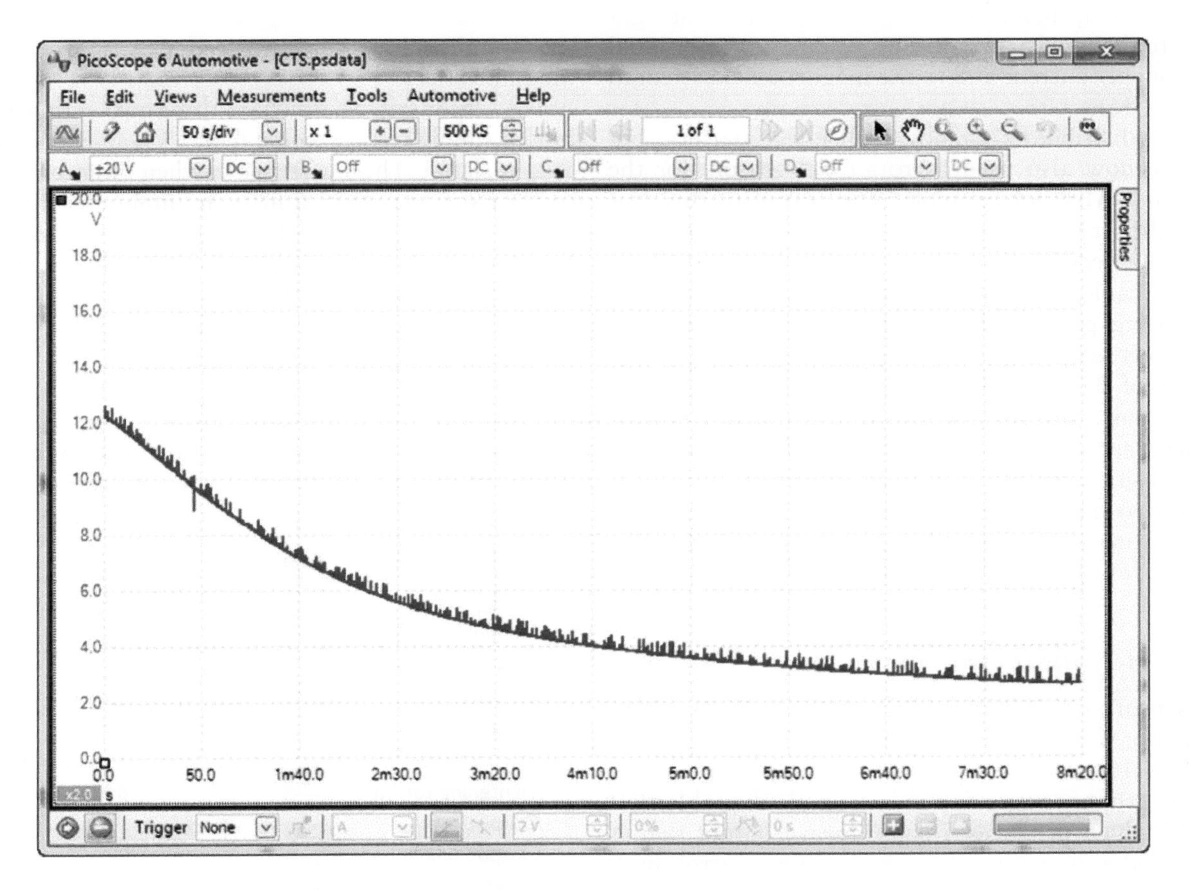

Figure 4.14 Decreasing voltage from the temperature sensor.

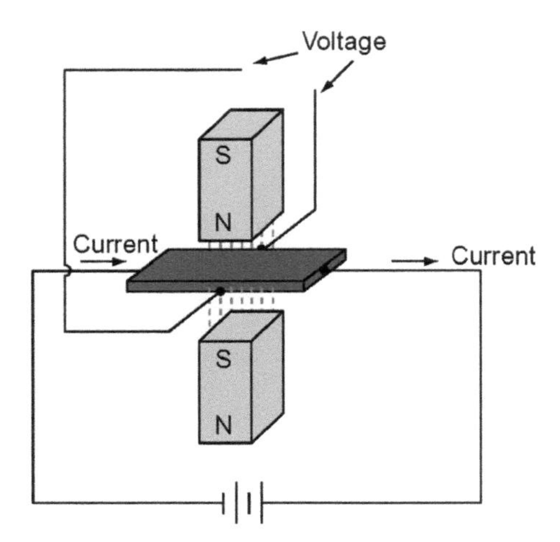

Figure 4.15 Hall effect.

Figure 4.16 Distributors usually contain a Hall effect or inductive pulse generator.

Source: Bosch Press.

The rate of voltage change is usually linear with no sudden changes to the voltage, if the sensor displays a fault at a certain temperature, it will show up in this test.

4.2.6 Hall effect sensors

If a conductor is carrying a current in a transverse magnetic field, then a voltage will be produced at right angles to the supply current. This voltage is proportional to the supply current and to the magnetic field strength (Figure 4.15).

Many distributors employ Hall effect sensors, but they are now also used as rotational sensors for the crank and ABS, for example. The output of this sensor is almost a square wave with constant amplitude. The Hall effect can also be used to detect current flowing in a cable, the magnetic field produced round the cable being proportional to the current flowing. The Hall effect sensors are becoming increasingly popular because of their reliability and also because they produce a constant amplitude square wave in speed measurement applications and a varying DC voltage for either position sensing or current sensing.

The two main advantages are that measurement of lower (or even zero) speed is possible and that the voltage output of the sensors is independent of speed.

4.2.6.1 Hall effect distributor pick-up

Hall sensors are used in a number of ways. The ignition distributor was very common but they are not used now (Figure 4.16).

This form of trigger device is a simple digital 'on/off switch' which produces a square wave output that is recognised and processed by the ignition control module or engine management ECU (Figure 4.17).

The trigger has a rotating metal disc with openings that pass between an electromagnet and the semiconductor (Hall chip). This action produces a square wave that is used by the ECU or amplifier.

4.2.6.2 ABS Hall sensor

The sensor when used by ABS for monitoring wheel speed and as transmission speed sensors works using the same effect.

The sensor will usually have three connections: a stabilised supply voltage (often 4 or 5 V), an earth and the output signal. The square wave when monitored on an oscilloscope may vary a little in amplitude; this is not usually a problem as it is the frequency that is important. However, in most cases, the amplitude/voltage will remain constant. Table 4.2 provides a list of technical data for the sensor shown in Figure 4.18.

4.2.6.3 Road speed sensor (Hall effect)

To measure the output of this sensor, jack up the driven wheels of the vehicle and place on axle stands on firm

Table 4.2 Hall sensor data

Supply voltage	4.5 V_{DC}
Nominal sensing distance	1.5 mm
Current (typical)	10 mA
Current (max)	20 mA
Weight	30 g
Temperature range	–30 to +130°C
Tightening torque	6 Nm
Output	PNP
Output sink voltage	0.4 V_{max}
Trigger type	Ferrous

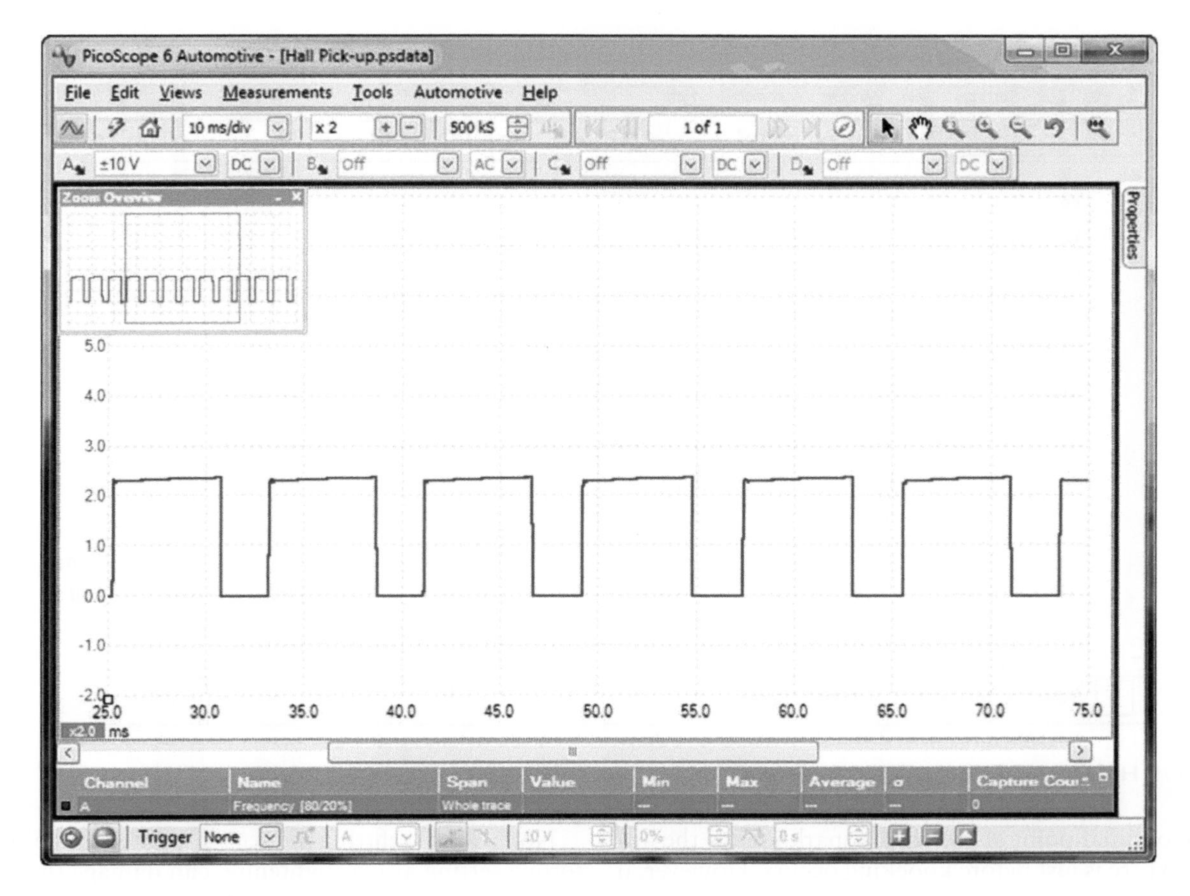

Figure 4.17 Hall output waveform.

level ground. Run the engine in gear and then probe each of the three connections (+, – and signal) (Figure 4.19).

As the road speed is increased, the frequency of the switching should be seen to increase. This change can also be measured on a multimeter with frequency capabilities. The sensor will be located on either the speedometer drive output from the gearbox or to the rear of the speedometer head if a speedo cable is used. The signal is used by the engine ECU and, if appropriate, the transmission ECU. The actual voltage will vary with sensor design.

Figure 4.18 Hall effect ABS or CAS sensor.

4.2.7 Piezo accelerometer

A piezoelectric accelerometer is a seismic mass accelerometer using a piezoelectric crystal to convert the force on the mass due to acceleration into an electrical output signal. The crystal not only acts as the transducer but as the suspension spring for the mass. The crystal is sandwiched between the body of the sensor and the seismic mass and is kept under compression. Acceleration forces acting on the seismic mass cause variations in the amount of crystal compression and hence generate the piezoelectric voltage.

DEFINITION

Piezoelectric effect: The production of electrical potential in a substance as the pressure on it changes.

4.2.7.1 Knock sensor

The sensor, when used as an engine knock sensor, will also detect other engine vibrations. These are kept to a minimum by only looking for 'knock', a few degrees before and after top dead centre (TDC). Unwanted signals are also filtered out electrically.

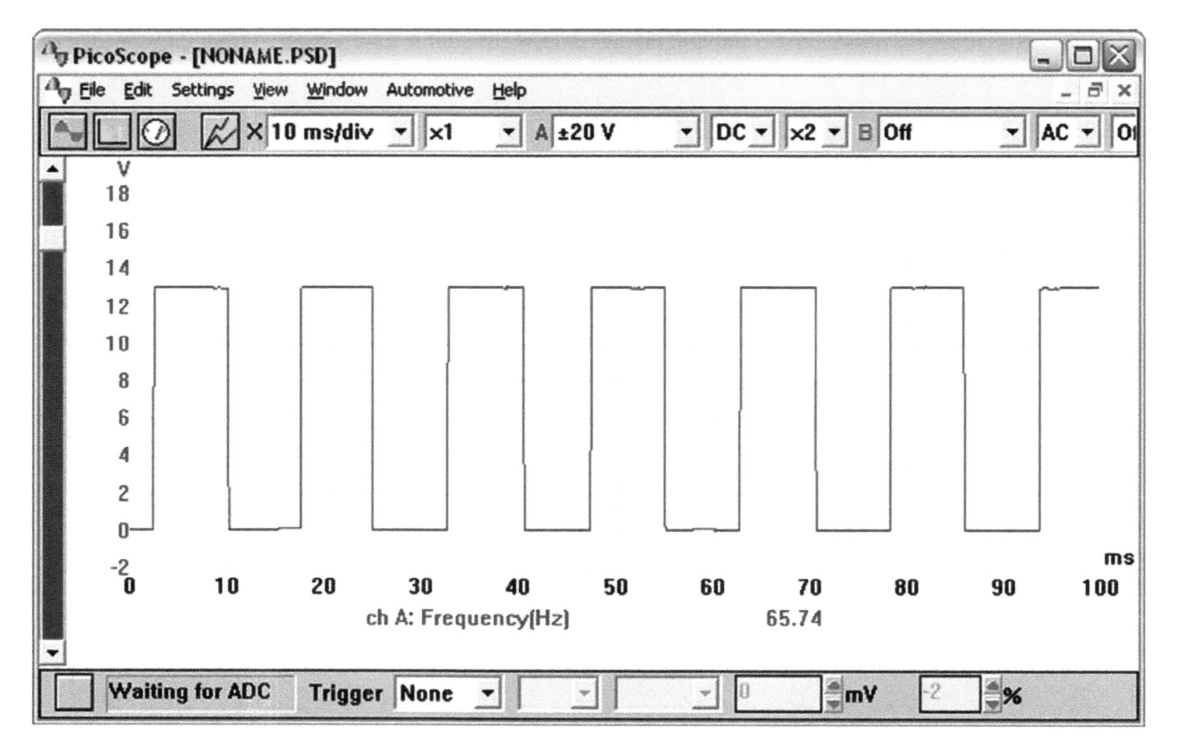

Figure 4.19 Hall effect road speed sensor waveform.

The optimal point at which the spark ignites the air/fuel mixture is just before knocking occurs. However, if the timing is set to this value, under certain conditions knock (detonation) will occur. This can cause serious engine damage as well as increase emissions and reduce efficiency.

A knock sensor is used by some engine management systems (Figure 4.20). When coupled with the ECU, it can identify when knock occurs and retard the ignition timing accordingly.

The frequency of knocking is approximately 15 kHz. As the response of the sensor is very fast, an appropriate time scale must be set, in the case of the example waveform a 0 to 500 ms timebase and a –5 to +5 V voltage scale. The best way to test a knock sensor is to remove the knock sensor from the engine and to tap it with a small spanner – the resultant waveform should be similar to the example shown in Figure 4.21.

Figure 4.20 Knock sensor.

Note: When refitting the sensor, tighten to the correct torque setting as overtightening can damage the sensor and/or cause it to produce incorrect signals.

4.2.8 Oxygen sensors

The vehicle application for an oxygen sensor is to provide a closed loop feedback system for engine management control of the air/fuel ratio. The amount of oxygen sensed in the exhaust is directly related to the mixture strength or air/fuel ratio. The ideal air/fuel ratio of 14.7:1 by mass is known as a lambda (λ) value of 1.

Exhaust gas oxygen (EGO) sensors are placed in the exhaust pipe near to the manifold to ensure adequate heating (Figure 4.22). The sensors operate best at temperatures over 300°C. In some cases, a heating element is incorporated to ensure that this temperature is reached quickly. This type of sensor is known as a heated exhaust gas oxygen sensor (HEGO). The heating element (which consumes approximately 10 W) does not operate all the time to ensure that the sensor does not exceed 850°C, at which temperature damage may occur to the sensor. This is why the sensors are not fitted directly in the exhaust manifold. The main active component of most types of oxygen sensors is zirconium dioxide (ZrO_2). This ceramic is housed in gas permeable electrodes of platinum. A further ceramic coating is applied to the side of the sensor exposed to the exhaust gas as a protection against residue from the combustion process. The principle of operation is that

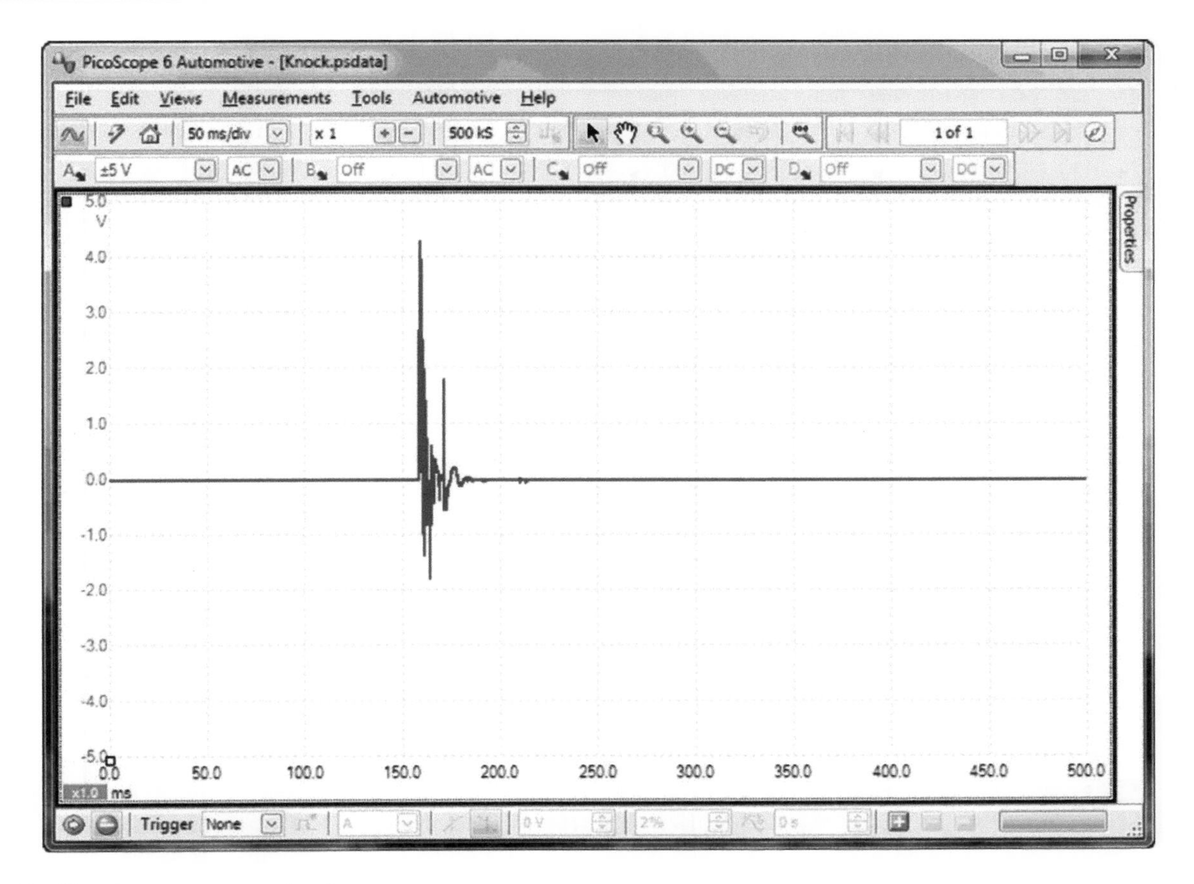

Figure 4.21 Knock sensor output signal.

at temperatures in excess of 300°C, the ZrO_2 will conduct the negative oxygen ions. The sensor is designed to be responsive very close to a lambda value of 1. As one electrode of the sensor is open to a reference value of atmospheric air, a greater quantity of oxygen ions will be present on this side. Because of electrolytic action, these ions permeate the electrode and migrate through the electrolyte (ZrO_2). This builds up a charge rather like a battery.

Figure 4.22 Zirconia lambda sensor in the exhaust downpipe.

The size of the charge is dependent on the oxygen percentage in the exhaust. The closely monitored closed loop feedback of a system using lambda sensing allows very accurate control of engine fuelling. Close control of emissions is therefore possible.

4.2.8.1 Oxygen sensor (Titania)

The lambda sensor, also referred to as the oxygen sensor, plays a very important role in the control of exhaust emissions on a catalyst equipped vehicle (Figure 4.23).

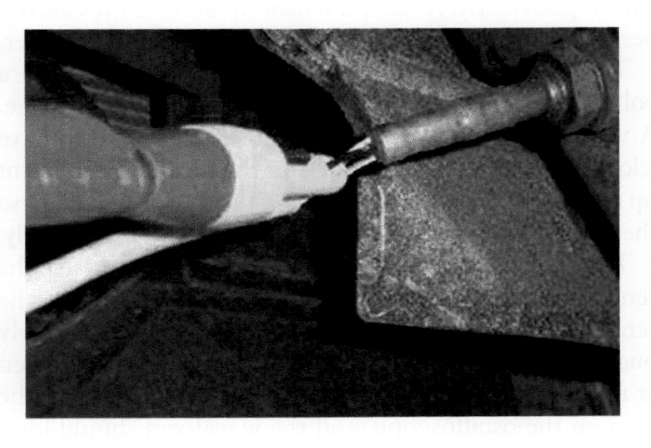

Figure 4.23 Titania lambda sensor in position.

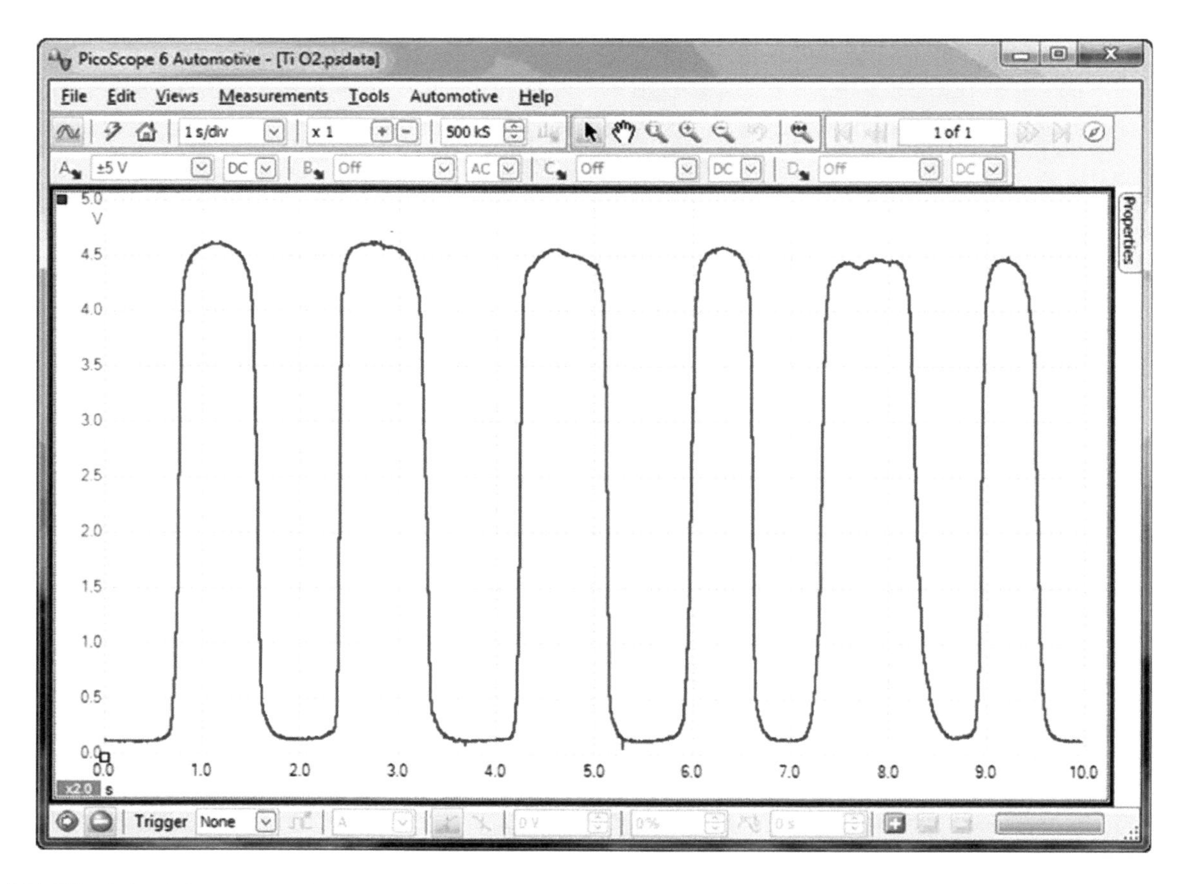

Figure 4.24 Titania lambda sensor output.

KEY FACT

Most lambda sensors operate best at temperatures over 300°C.

The main lambda sensor is fitted into the exhaust pipe before the catalytic converter. The sensor will have four electrical connections. It reacts to the oxygen content in the exhaust system and will produce an oscillating voltage between 0.5 (lean) and 4.0 V, or above (rich) when running correctly. A second sensor to monitor the catalyst performance may be fitted downstream of the converter.

Titania sensors, unlike Zirconia sensors, require a voltage supply as they do not generate their own voltage. A vehicle equipped with a lambda sensor is said to have 'closed loop', this means that after the fuel has been burnt up during the combustion process, the sensor will analyse the emissions and adjust the engine's fuelling accordingly.

Titania sensors have a heating element to assist the sensor to reach its optimum operating temperature. The sensor when working correctly will switch approximately once per second (1 Hz) but will only start to switch when at normal operating temperature. This switching can be seen on the oscilloscope, and the waveform should look similar to the one in the example (Figure 4.24).

4.2.8.2 Oxygen sensor (Zirconia)

The sensor will have varying electrical connections and may have up to four wires. It reacts to the oxygen content in the exhaust system and will produce a small voltage depending on the air/fuel mixture seen at the time. The voltage range seen will, in most cases, vary between 0.2 and 0.8 V. The 0.2 V indicates a lean mixture and a voltage of 0.8 V shows a richer mixture (Figure 4.25).

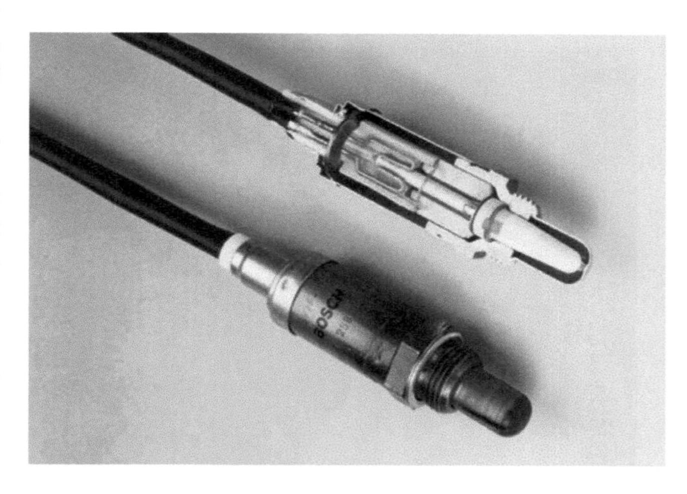

Figure 4.25 Zirconia-type oxygen sensor.

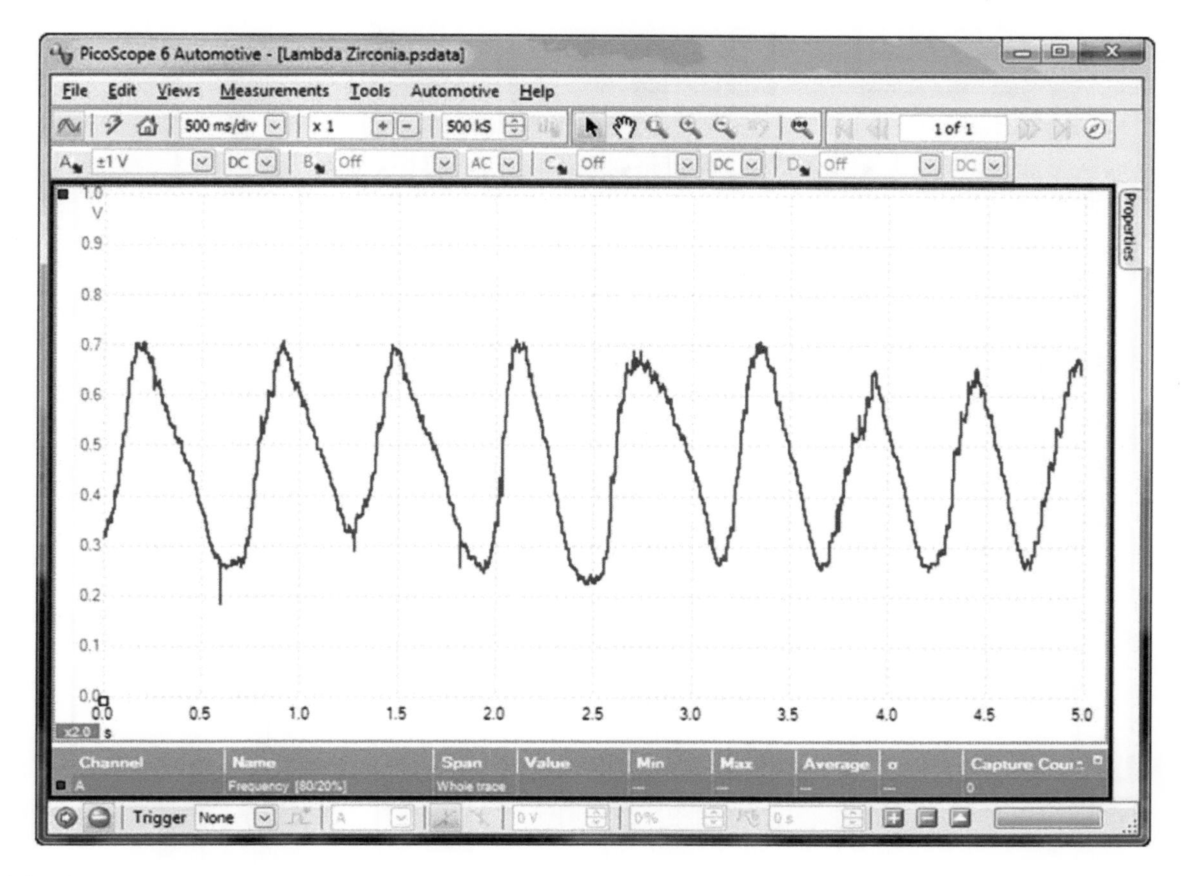

Figure 4.26 Zirconia oxygen sensor output.

Lambda sensors can have a heating element to assist the sensor reaching its optimum operating temperature. Zirconia sensors when working correctly will switch approximately once per second (1 Hz) and will only start to switch when at normal operating temperature. This switching can be seen on the oscilloscope, and the waveform should look similar to the one in the example waveform (Figure 4.26).

Many vehicles now have a pre- and post-cat lambda sensor. Comparing the outputs of these two sensors is a good indicator of catalyst operation and condition (Figure 4.27).

4.2.9 Pressure sensors

4.2.9.1 Strain gauges

When a strain gauge is stretched its resistance will increase, and when it is compressed its resistance decreases. Most strain gauges consist of a thin layer of film that is fixed to a flexible backing sheet. This in turn is bonded to the part where strain is to be measured. Most resistance strain gauges have a resistance of approximately 100 Ω.

Strain gauges are often used indirectly to measure engine manifold pressure. Figure 4.28 shows an arrangement of four strain gauges on a diaphragm forming part of an aneroid chamber used to measure pressure. When changes in manifold pressure act on the diaphragm, the gauges detect the strain. The output of the circuit is via a differential amplifier, which must have a very high input resistance so as not to affect the bridge balance. The actual size of this sensor may be only a few millimetres in diameter. Changes in temperature are compensated for by using four gauges which when affected in a similar way cancel out any changes.

4.2.9.2 Fuel pressure

Common rail diesel system pressure signals can be tested (remember these systems operate at very high pressure). The pulse width modulation (PWM) signal should be at the same amplitude but the on/off ratio can vary (Figure 4.29).

DEFINITION

PWM: Pulse width modulation is an adjustment of the duty cycle of a signal or power source, to either convey information over a communications channel or control the amount of power sent to a load (e.g. an actuator).

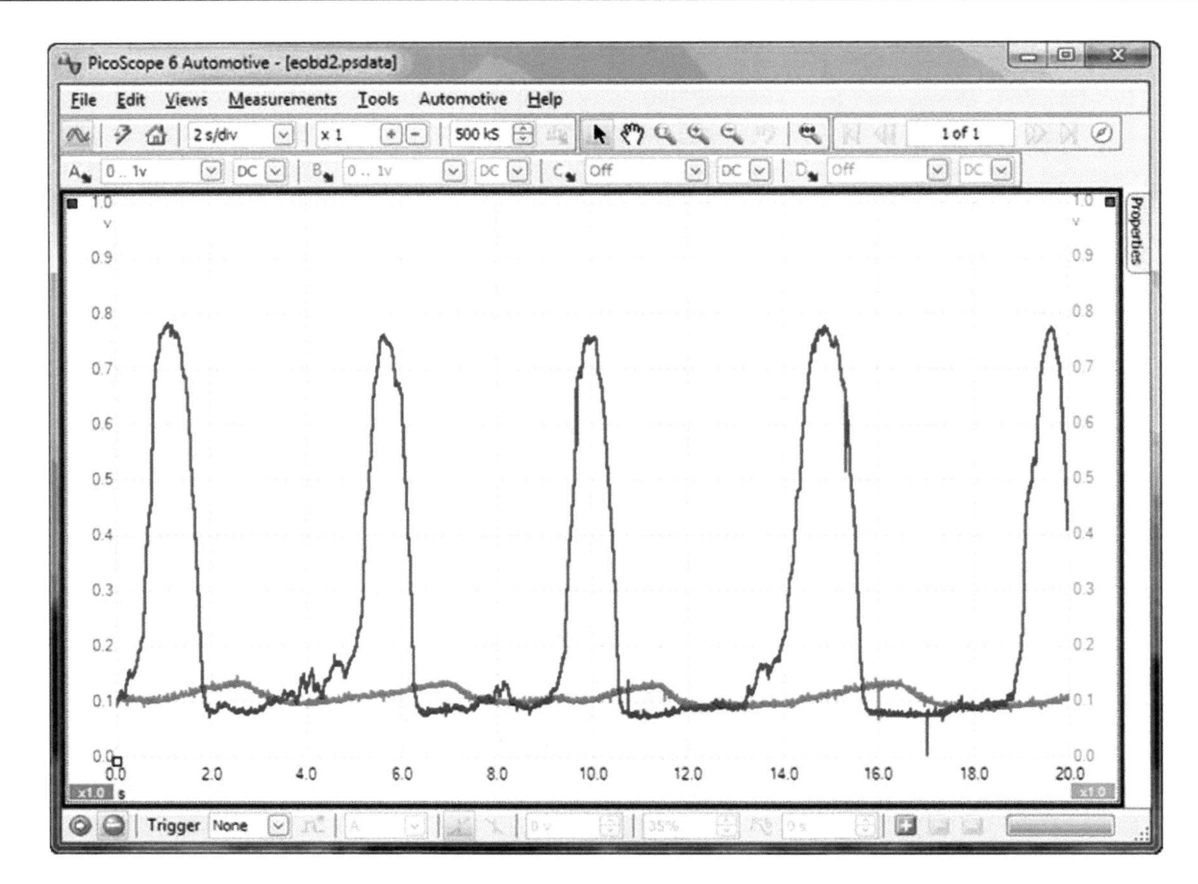

Figure 4.27 Pre-cat signal shown in blue and post-cat in red.

4.2.9.3 Manifold absolute pressure

Manifold absolute pressure (MAP) is a signal used to determine engine load. There are two main types: analogue and digital. The analogue signal voltage output varies with pressure whereas the digital signal varies with frequency. These sensors use a piezo crystal, strain gauges or variable capacitance sensors or similar. The signals are also processed internally (Figures 4.30 and 4.31).

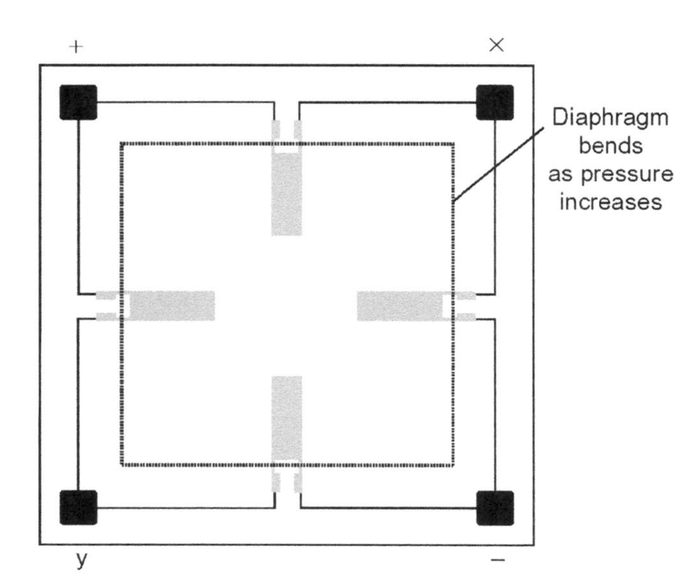

Figure 4.28 Strain gauge pressure sensor (+ and – are the supply, x and y the output signal).

4.2.10 Variable capacitance

The value of a capacitor is determined by the:

- surface area of its plates;
- distance between the plates;
- dielectric (insulation between the plates).

Sensors can be constructed to take advantage of these properties. Three sensors, each using the variable capacitance technique, are shown in Figure 4.32. These are (a) liquid level sensor in which the change in liquid level changes the dielectric value; (b) pressure sensor similar to the strain gauge pressure sensor in which the distance between capacitor plates changes; and (c) position sensor which detects changes in the area of the plates.

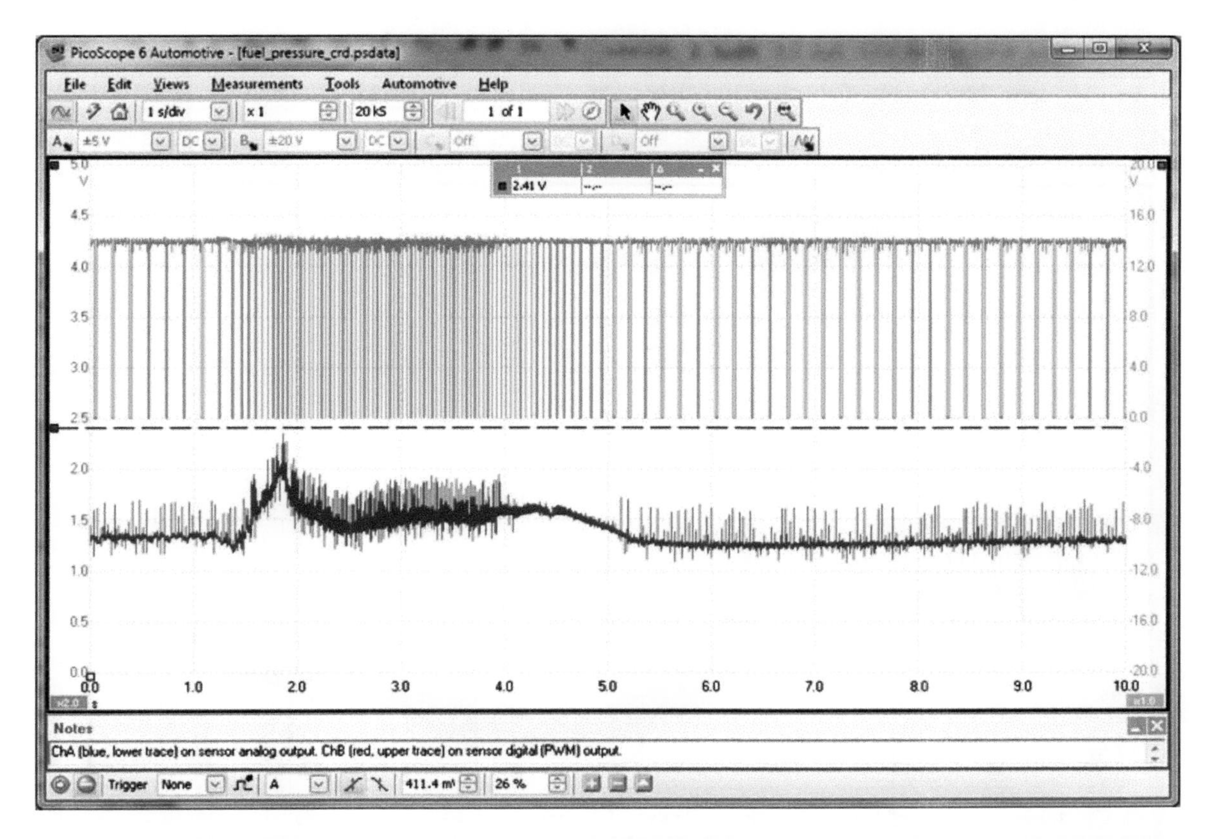

Figure 4.29 Common rail fuel pressure, the blue trace is from the sensor analogue output and the red trace is from sensor digital (PWM) output.

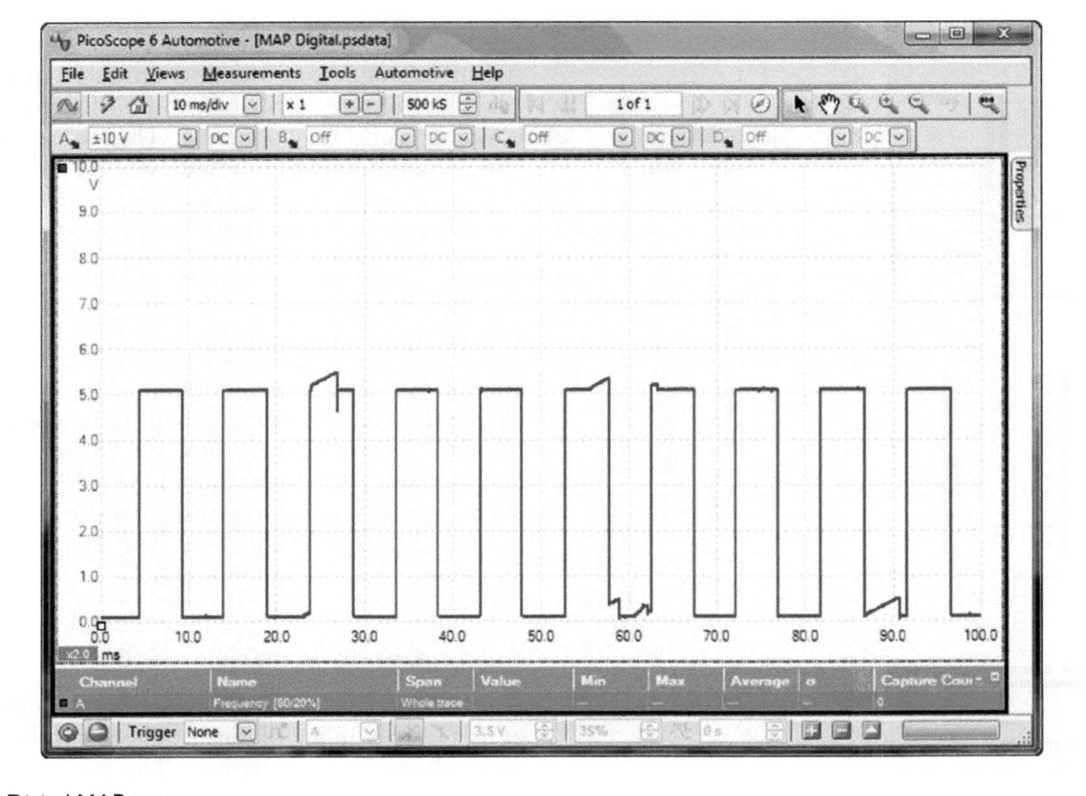

Figure 4.30 Digital MAP sensor.

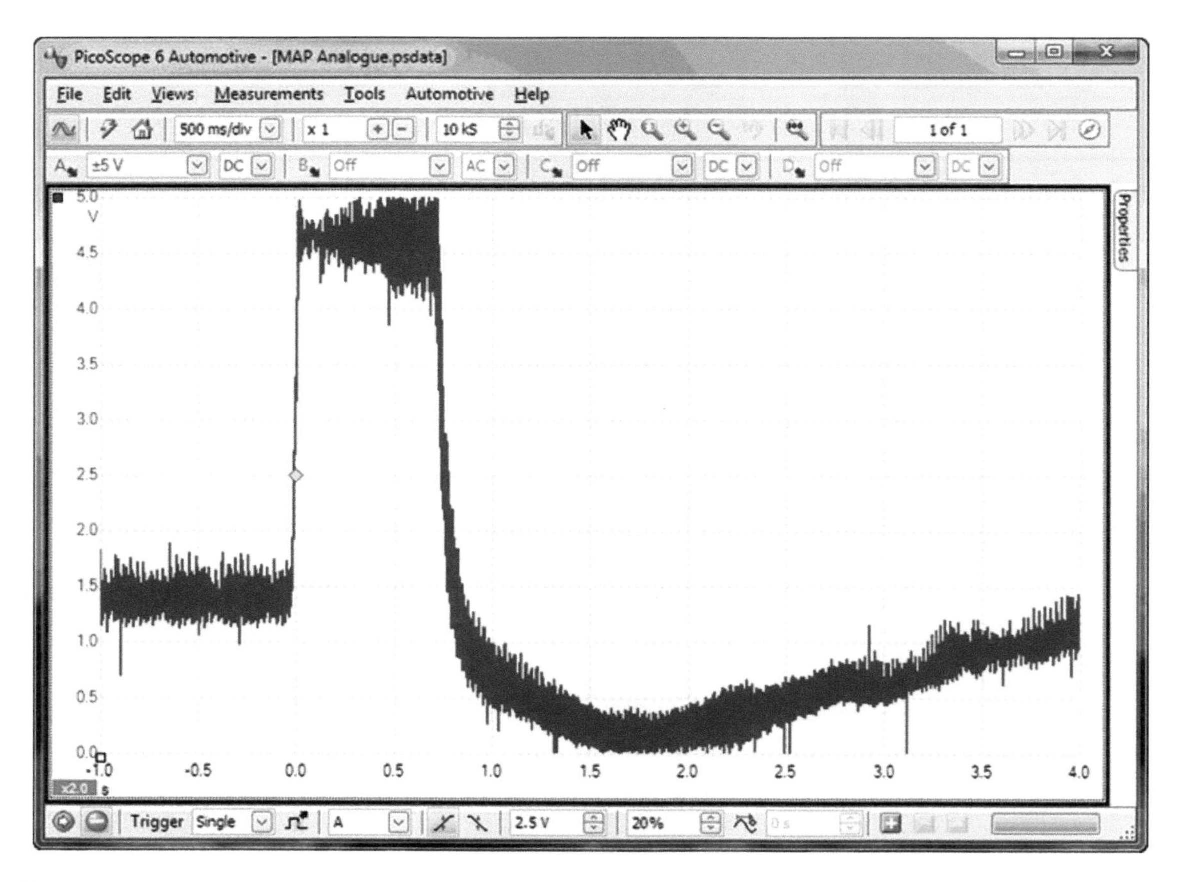

Figure 4.31 Analogue MAP sensor.

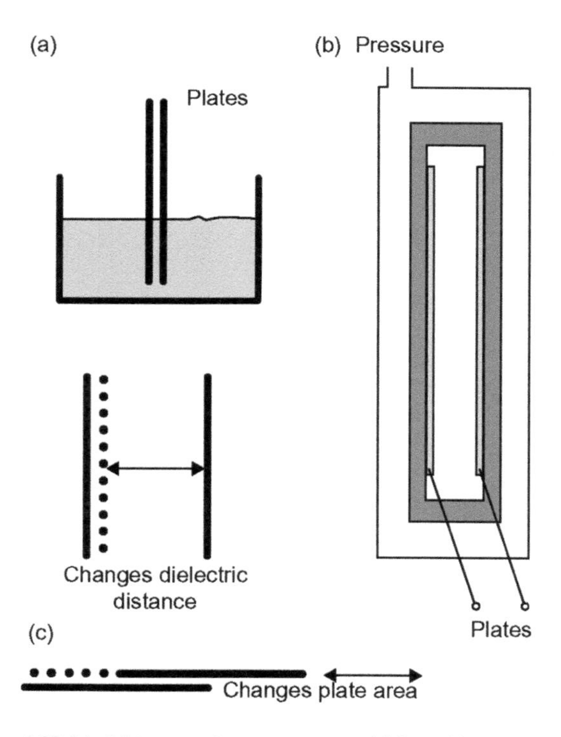

Figure 4.32 Variable capacitance sensors: (a) liquid level, (b) pressure and (c) position.

4.2.10.1 Oil quality sensor

An interesting sensor used to monitor oil quality is now available, which works by monitoring changes in the dielectric constant of the oil. This value increases as antioxidant additives in the oil deplete. The value rapidly increases if coolant contaminates the oil. The sensor output increases as the dielectric constant increases (Figure 4.33).

4.2.11 Optical sensors

An optical sensor for rotational position is a relatively simple device. The optical rotation sensor and circuit shown in Figure 4.34 consists of a phototransistor as a detector and a light-emitting diode (LED) light source. If the light is focused to a very narrow beam then the output of the circuit shown will be a square wave with frequency proportional to speed.

4.2.12 Dynamic position sensors

A dynamic position or movement of crash sensor can take a number of forms; these can be described as mechanical or electronic. The mechanical system

Figure 4.33 Oil quality sensor.

Source: Bosch Media.

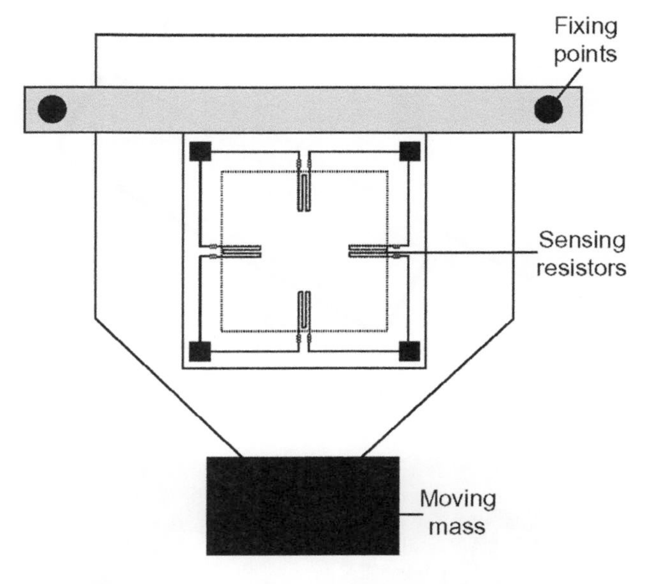

Figure 4.35 Strain gauge accelerometer.

works by a spring holding a roller in a set position until an impact (acceleration/deceleration) above a predetermined limit provides enough force to overcome the spring and the roller moves, triggering a micro switch. The switch is normally open with a resistor in parallel to allow the system to be monitored. Two switches similar to this may be used to ensure that an airbag is deployed only in the case of sufficient frontal impact.

Figure 4.35 is a further type of dynamic position sensor. Described as an accelerometer, it is based on strain gauges. There are two types of piezoelectric crystal accelerometer, one much like an engine knock sensor and the other using spring elements. A severe change in speed of the vehicle will cause an output from these sensors as the seismic mass moves or the springs bend. This sensor has been used by supplementary restraint systems (SRSs). Warning: For safety reasons, it is not recommended to test a sensor associated with an airbag circuit without specialist knowledge and equipment.

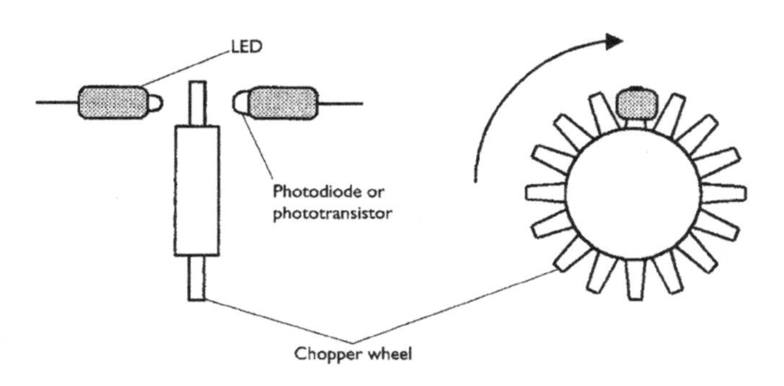

Figure 4.34 Optical sensor.

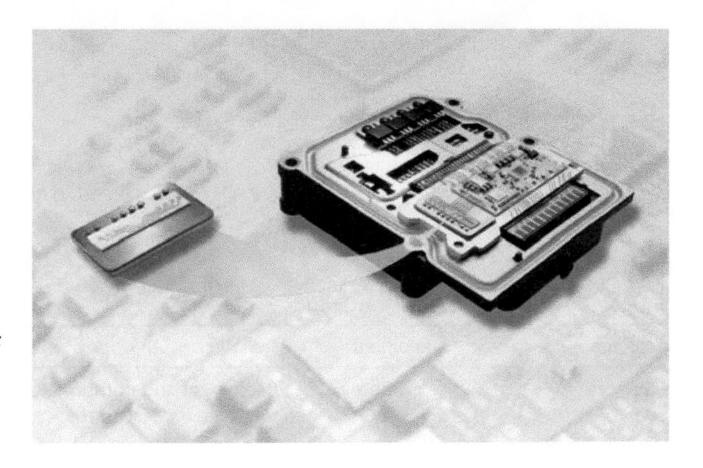

Figure 4.36 Yaw rate and acceleration sensor integrated into an electric stability program (ESP) control unit.

Source: Bosch Media.

Many sensors are now integrated into ECUs (Figure 4.36). This means that they are almost impossible to test – but fortunately have become very reliable!

4.2.13 Rain sensor

Rain sensors are used to switch on wipers automatically. Most work on the principle of reflected light. The device is fitted inside the windscreen and light from an LED is reflected back from the outer surface of the glass. The amount of light reflected changes if the screen is wet, even with a few drops of rain (Figures 4.37 and 4.38).

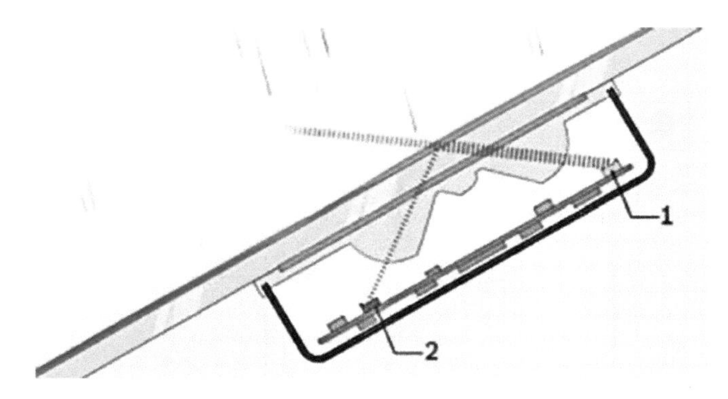

Figure 4.37 Rain sensor principle: 1 – LED; 2 – photo diode.

Figure 4.38 Rain sensor package.

Source: Bosch Media.

4.3 ACTUATORS

4.3.1 Introduction

There are many ways of providing control over variables in and around the vehicle. 'Actuators' is a general term used here to describe a control mechanism. When controlled electrically, they will work either by a thermal or by a magnetic effect. In this section, the term actuator will generally be used to mean a device which converts electrical signals into mechanical movement (Table 4.3).

4.3.2 Testing actuators

Testing actuators can be simple as many are operated by windings. The resistance can be measured with an ohmmeter. A good tip is that where an actuator has more than one winding (e.g. a stepper motor), the resistance of each should be about the same. Even if the expected value is not known, it is likely that if all the windings read the same then the device is in working order.

With some actuators, it is possible to power them up from the vehicle battery. A fuel injector should click, for example, and a rotary air bypass device should rotate about half a turn. Be careful with this method as some actuators could be damaged. At the very least, use a fused supply (jumper) wire.

Table 4.3 Actuator diagnostic methods

Actuator	*Equipment*	*Method(s)*	*Results*	*Scope waveform*
Solenoid Fuel injector Lock actuator	Ohmmeter	Disconnect the component and measure its resistance	The resistance of many injectors is approx. 16 Ω (but check data). Lock and other actuators may have two windings (e.g. lock and unlock). The resistance values are very likely to be the same	Figures 4.47 and 4.48
Motor See previous list	Battery supply (fused) Ammeter	Most 'motor' type actuators can be run from a battery supply after they are disconnected from the circuit. If necessary the current draw can be measured	Normal operation with current draw appropriate to the 'work' done by the device. For example, a fuel pump motor may draw up to 10 A, but an idle actuator will only draw 1 or 2 A	N/A
Solenoid actuator (idle speed control)	Duty cycle meter	Most types are supplied with a variable ratio square wave	The duty cycle will vary as a change is required	Figure 4.41
Stepper motor Idle speed air bypass Carburettor choke control Speedometer drivers	Ohmmeter	Test the resistance of each winding with the motor disconnected from the circuit	Winding resistances should be the same. Values in the region of 10–20 Ω are typical	Figure 4.44
Thermal Auxiliary air device Instrument display	Ohmmeter Fused battery supply	Check the winding for continuity; if OK, power up the device and note its operation (for instruments, power these but use a resistor in place of the sender unit)	Continuity and slow movement (several seconds to a few minutes) to close the valve or move as required	N/A
EGR valve	Ohmmeter Fused battery supply	Check the winding(s) for continuity; if OK, power up the device and note its operation	Continuity and rapid movement to close the valve	Figure 4.56

SAFETY FIRST

When powering a device, use a fused supply (jumper) wire.

4.3.3 Motorised and solenoid actuators

4.3.3.1 Motors

Permanent magnet electric motors are used in many applications and are very versatile. The output of a motor is of course rotation, and this can be used in many ways. If the motor drives a rotating 'nut' through which a plunger is fitted on which there is a screw thread, the rotary action can easily be converted to linear movement. In most vehicle applications, the output of the motor has to be geared down, this is to reduce speed and increase torque. Permanent magnet motors are almost universally used now in place of older and less practical motors with field windings. Some typical examples of the use of these motors are listed as follows:

- windscreen wipers;
- windscreen washers;
- headlight lift;
- electric windows;
- electric sunroof;
- electric aerial operation;
- seat adjustment;
- mirror adjustment;
- headlight washers;
- headlight wipers;
- fuel pumps;
- ventilation fans.

One disadvantage of simple motor actuators is that no direct feedback of position is possible. This is not required in many applications; however, in cases such as seat adjustment when a 'memory' of the position may be needed, a variable resistor–type sensor can be fitted to provide feedback. Three typical motor actuators are shown in Figure 4.39. The two motors on the right are used for window lift. Some of these use Hall effect sensors or an extra brush as a feedback device.

Figure 4.39 Window lift and wiper motors.

4.3.3.2 Rotary idle speed control valve

The rotary ISCV will have two or three electrical connections, with a voltage supply at battery voltage and either a single- or a double-switched earth path. The device is like a motor but only rotates about half a turn in each direction.

This device is used to control idle speed by controlling air bypass. There are two basic types in common use. These are single-winding types, which have two terminals, and double-winding types, which have three terminals. Under ECU, the motor is caused to open and close a shutter, controlling air bypass. These actuators only rotate approximately 90° to open and close the valve. As these are permanent magnet motors, the 'single or double windings' refer to the armature.

The single-winding type is fed with a square wave signal causing it to open against a spring and then close again, under spring tension. The on/off ratio or duty cycle of the square wave will determine the average valve open time and hence idle speed. With the double-winding type, the same square wave signal is sent to one winding but the inverse signal is sent to the other. As the windings are wound in opposition to each other, if the duty cycle is 50% then no movement will take place. Altering the ratio will now cause the shutter to move in one direction or the other (Figure 4.40).

The rate at which the earth path is switched is determined by the ECU to maintain a prerequisite idle speed according to its programming.

The valve will form an air bypass past the throttle butterfly to form a controlled air bleed within the induction tract. The rotary valve will have the choice of either single or twin earth paths, the single being pulled one way electrically and returned to its closed position via a spring; the double-switched earth system will switch the valve in both directions. This can be monitored on a dual trace oscilloscope. As the example waveform shows, the earth path is switched and the resultant picture is produced. The idle control device takes up a position determined by the on/off ratio (duty cycle) of the supplied signal.

Figure 4.40 Rotary idle control valve.

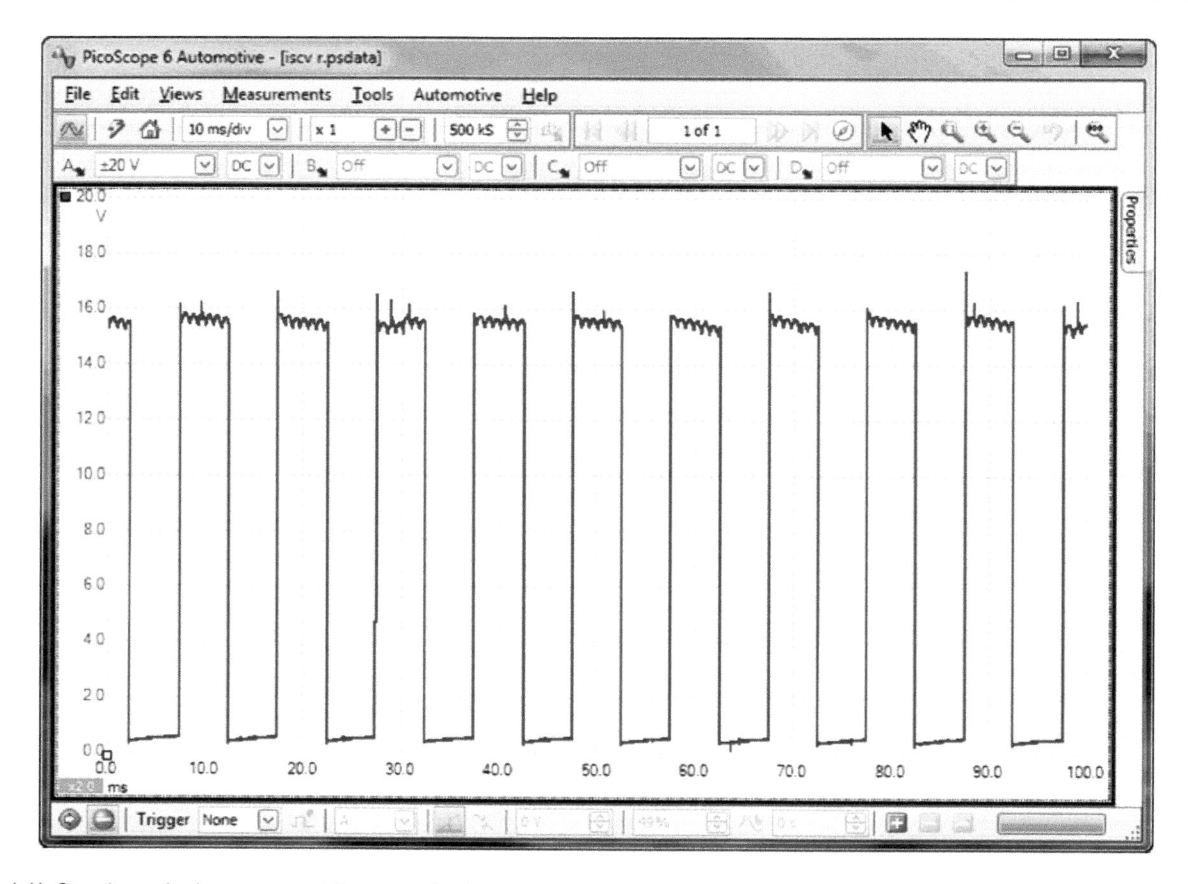

Figure 4.41 Signal supplied to a rotary idle control valve.

Probing onto the supply side will produce a straight line at system voltage, and when the earth circuit is monitored, a square wave will be seen (Figure 4.41). The frequency can also be measured as can the on/off ratio.

4.3.3.3 Stepper motors

Stepper motors are becoming increasingly popular as actuators in the motor vehicle.

This is mainly because of the ease with which they can be controlled by electronic systems. Stepper motors fall into the following three distinct groups, the basic principles of which are shown in Figure 4.42:

- variable reluctance motors;
- permanent magnet (PM) motors;
- hybrid motors.

The underlying principle is the same for each type. All of them have been and are being used in various vehicle applications. The basic design for a permanent magnet stepper motor comprises two double stators. The rotor is often made of barium-ferrite in the form of a sintered annular magnet. As the windings are energised in one direction then the other, the motor will rotate

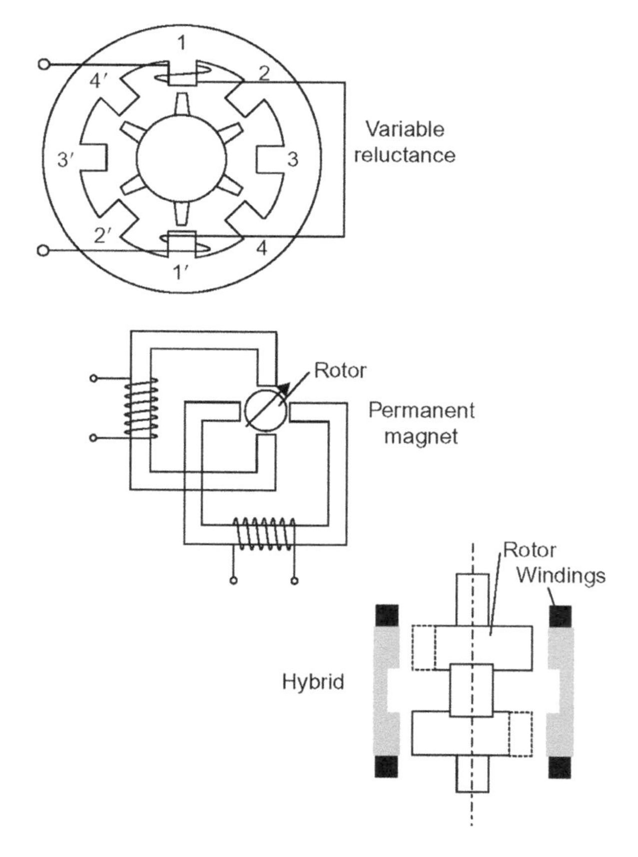

Figure 4.42 Stepper motor principle.

in 90° steps. Half step can be achieved by switching on two windings. This will cause the rotor to line up with the two stator poles and implement a half step of 45°. The direction of rotation is determined by the order in which the windings are switched on or off or reversed. The main advantages of a stepper motor are that feedback of position is not required. This is because the motor can be indexed to a known starting point and then a calculated number of steps will move the motor to any suitable position.

The stepper motor, when used to control idle speed, is a small electro-mechanical device that allows either an air bypass circuit or a throttle opening to alter in position depending on the amounts that the stepper is indexed (moved in known steps) (Figure 4.43).

Stepper motors are used to control the idle speed when an ISCV is not employed. The stepper may have four or five connections back to the ECU. These enable the control unit to move the motor in a series of 'steps' as the circuits are earthed to ground. These devices may also be used to control the position of control flaps, for example, as part of a heating and ventilation system.

The individual earth paths can be checked using the oscilloscope. The waveforms should be similar on each path. Variations to the example shown here may be seen between different systems (Figure 4.44 and Figure 4.45).

Figure 4.43 Stepper motor and throttle potentiometer on a throttle body.

4.3.4 Solenoid actuators

The basic operation of solenoid actuators is very simple. The term 'solenoid' actually means 'many coils of wire wound onto a hollow tube'. This is often misused but has become so entrenched that terms like 'starter

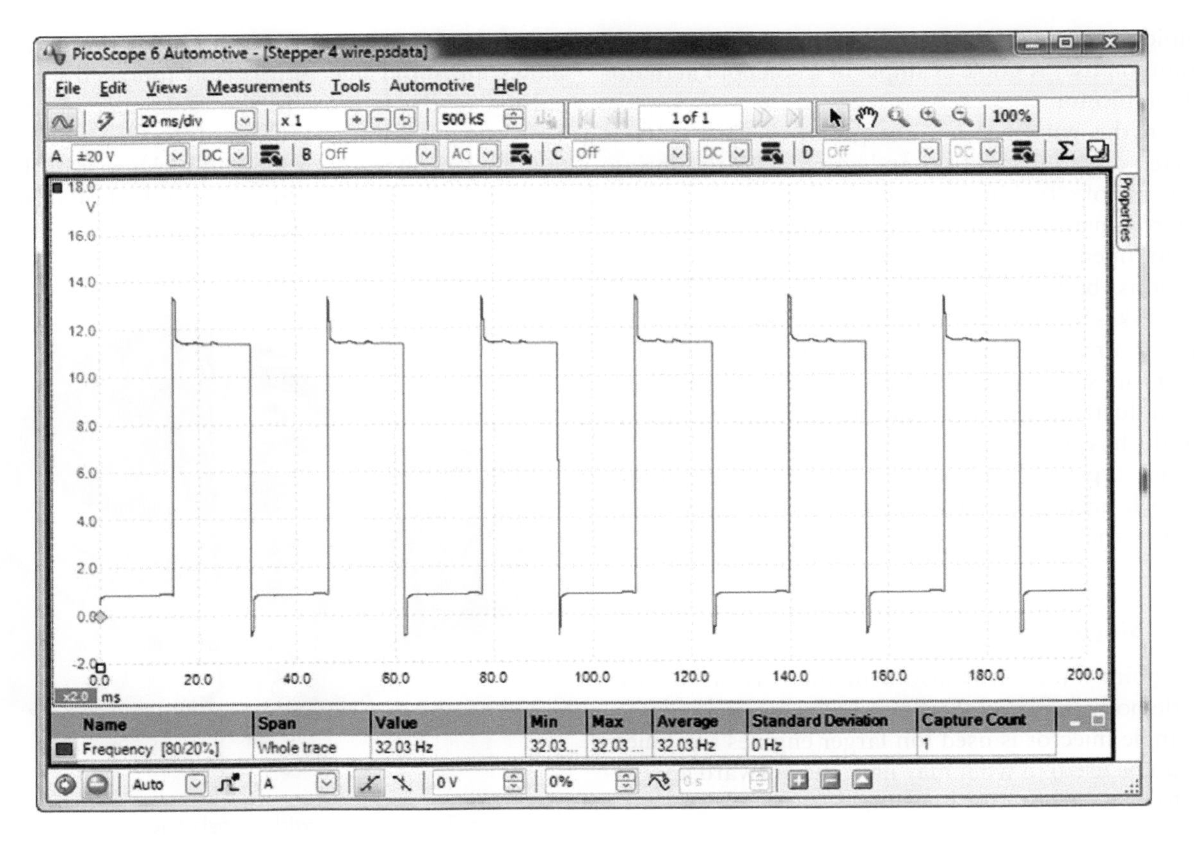

Figure 4.44 Stepper motor signals.

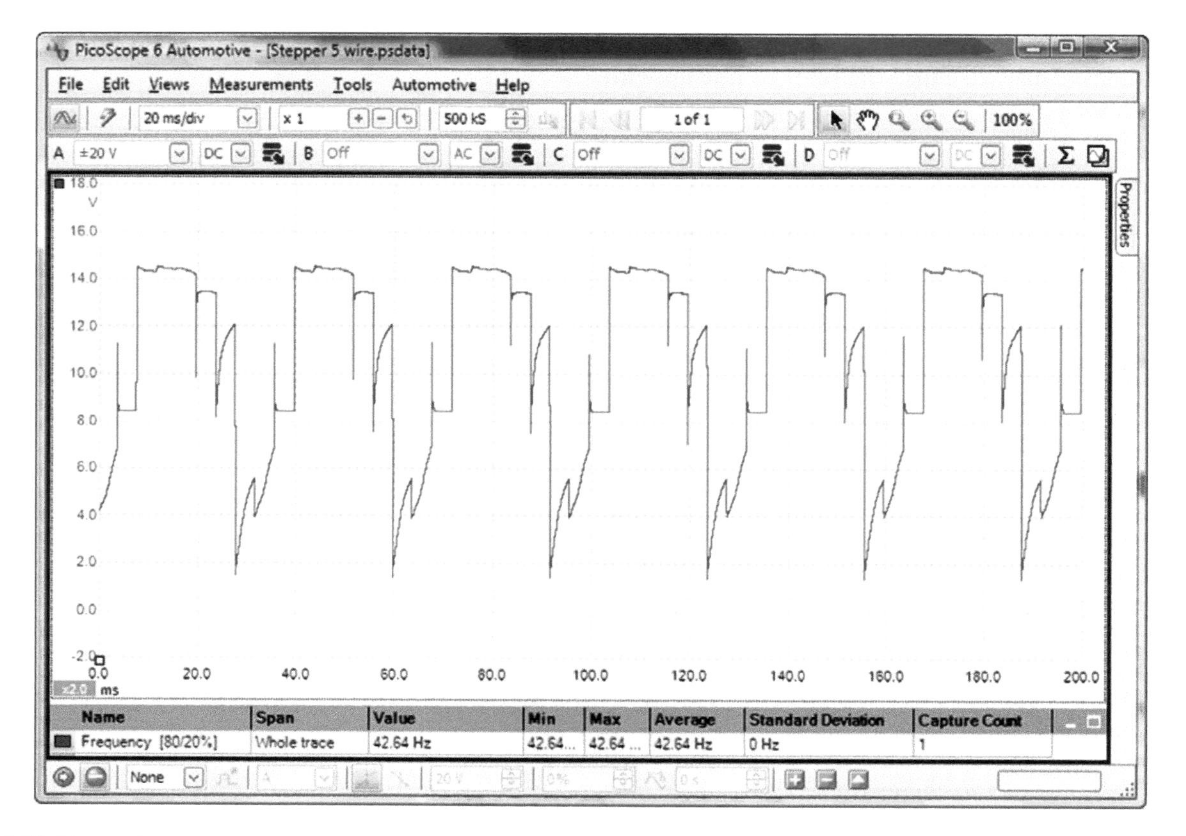

Figure 4.45 Alternative stepper motor signal.

solenoid', when really it is a starter actuator or relay, are in common use. A good example of a solenoid actuator is a fuel injector.

When the windings are energised, the armature is attracted due to magnetism and compresses the spring. In the case of a fuel injector, the movement is restricted to approximately 0.1 mm. The period that an injector remains open is very small; under various operating conditions, between 1.5 and 10 ms being typical. The time it takes an injector to open and close is also critical for accurate fuel metering. Some systems use ballast resistors in series with the fuel injectors. This allows lower inductance and resistance operating windings to be used, thus speeding up reaction time.

Other types of solenoid actuators, for example door lock actuators, have less critical reaction times. However, the basic principle remains the same.

4.3.4.1 Single-point injector

Single-point injection is also sometimes referred to as throttle body injection (Figure 4.46).

A single injector is used (on larger engines two injectors can be used) in what may have the outward appearance to be a carburettor housing.

The resultant waveform from the single-point system shows an initial injection period followed by voltage-pulsing of the injector in the remainder of the trace. This 'current limiting' section of the waveform is called the supplementary duration and is the part of the injection trace that expands to increase fuel quantity. This shows better in a current rather than voltage waveform (Figures 4.47 and 4.48).

Figure 4.46 Throttle body with a single injector.

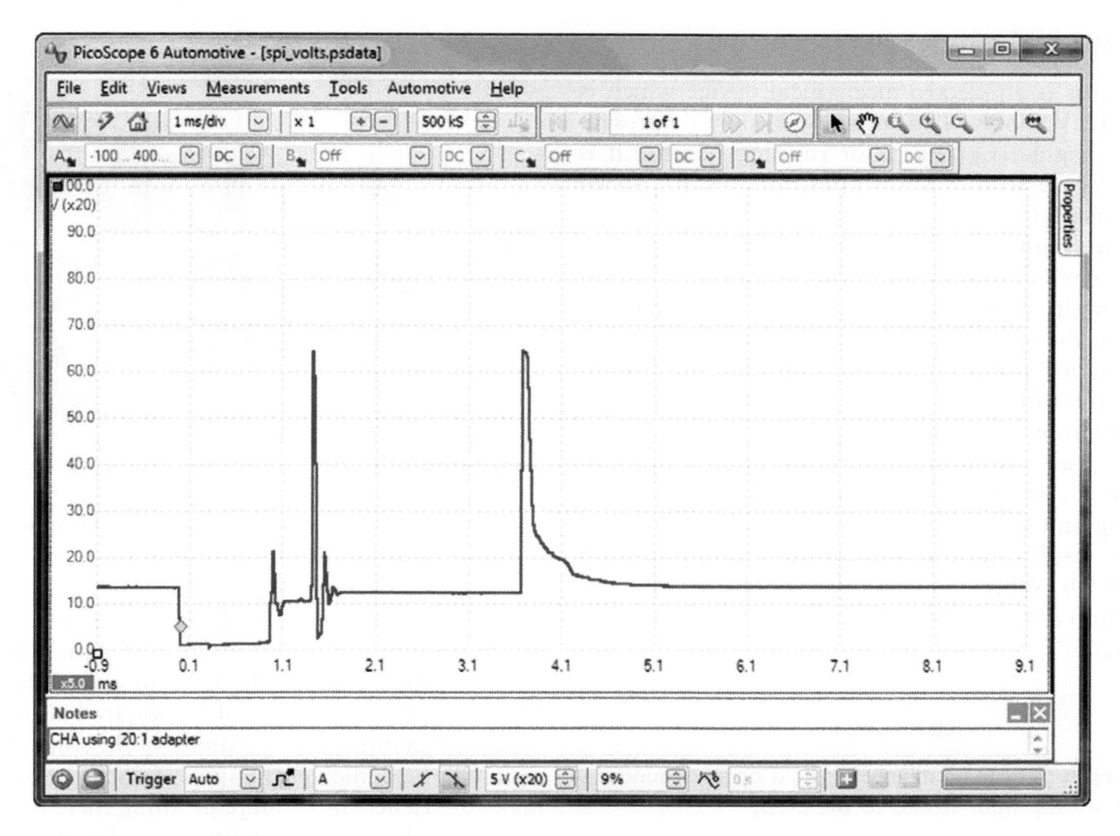

Figure 4.47 Single-point injector voltage waveform.

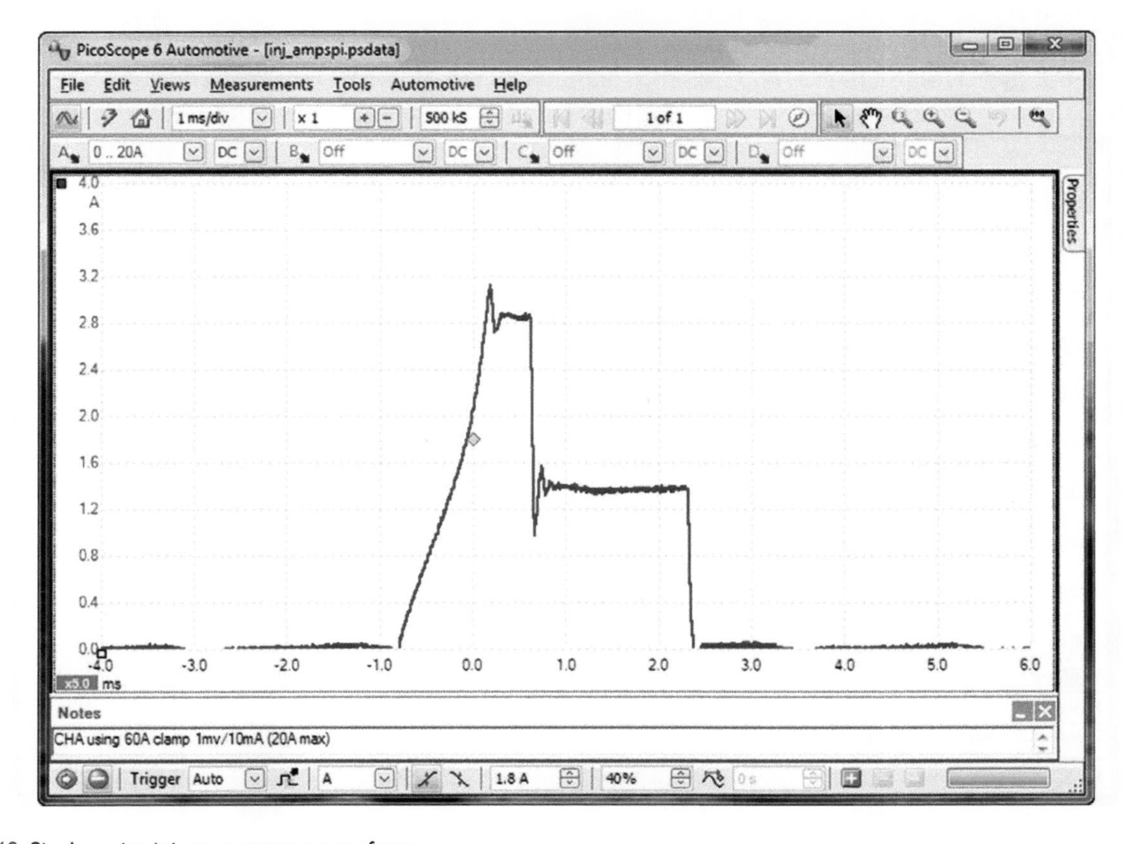

Figure 4.48 Single-point injector current waveform.

4.3.4.2 Multi-point injector

This injector is an electro-mechanical device which is fed by a 12 V supply. The voltage will only be present when the engine is cranking or running because it is controlled by a relay that operates only when a speed signal is available from the engine. Early systems had this feature built into the relay; most modern systems control the relay from the ECU (Figure 4.49).

The length of time the injector is held open will depend on the input signals seen by the ECU from its various engine sensors. The duration of open time or 'injector duration' will vary to compensate for cold engine starting and warm-up periods. The duration time will also expand under acceleration. The injector will have a constant voltage supply while the engine is running and the earth path will be switched via the ECU; the result can be seen in the example waveform (Figure 4.50). When the earth is removed, a voltage is induced into the injector and a spike approaching 60 V is recorded.

KEY FACT

The length of time an injector is held open depends on the sensor input signals to the ECU.

Figure 4.49 Multi-point injectors on the rail. Also shown are the pressure regulator and sensor.

The height of the spike will vary from vehicle to vehicle. If the value is approximately 35 V, it is because a Zener diode is used in the ECU to clamp the voltage. Make sure the top of the spike is squared off, indicating the Zener dumped the remainder of the spike. If it is not squared, this indicates the spike is not strong enough to make the Zener fully dump, meaning there is a problem with a weak injector winding. If a Zener diode is not

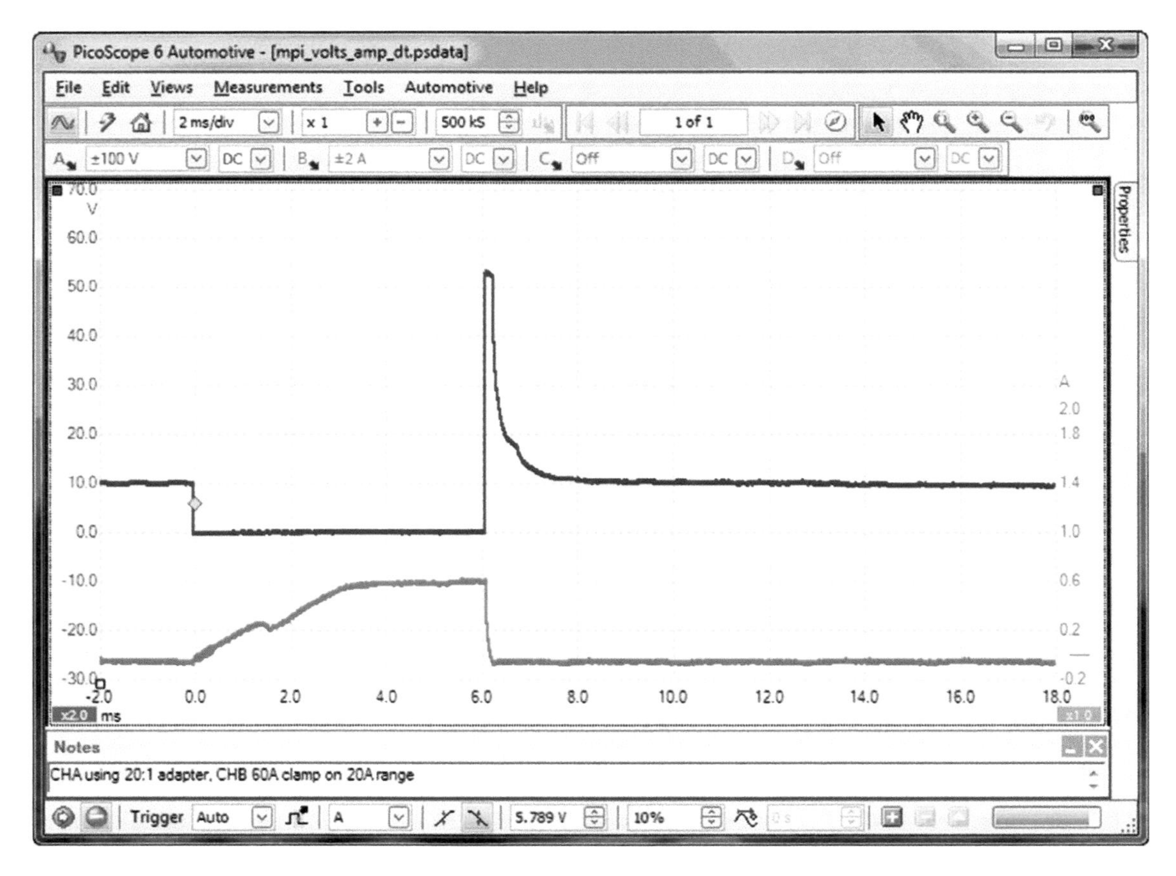

Figure 4.50 Multi-point injector waveform, red shows the current and blue the voltage signal.

used in the computer, the spike from a good injector will be 60 V or more.

Multi-point injection may be either sequential or simultaneous. A simultaneous system will fire all four injectors at the same time with each cylinder receiving two injection pulses per cycle (720° crankshaft rotation). A sequential system will receive just one injection pulse per cycle, which is timed to coincide with the opening of the inlet valve.

Monitoring the injector waveform using both voltage and amperage allows display of the 'correct' time that the injector is physically open. The current waveform (the one starting on the zero line) shows that the waveform is 'split' into two defined areas.

The first part of the current waveform is responsible for the electromagnetic force lifting the pintle; in this example, the time taken is approximately 1.5 ms. This is often referred to as the solenoid reaction time. The remaining 2 ms is the actual time the injector is fully open. This, when taken as a comparison against the injector voltage duration, is different to the 3.5 ms shown. The secret is to make sure you compare like with like!

4.3.4.3 Common rail diesel injector

Common rail diesel systems are becoming more, erm, *common*, particularly in Europe (Figure 4.51).

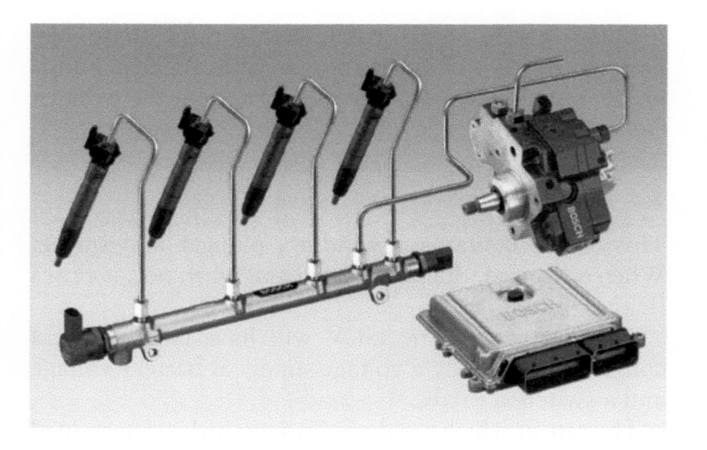

Figure 4.51 Common rail diesel pump, rail, injectors and ECU. **Source: Bosch Press.**

It can be clearly seen from the example waveform that there are two distinctive points of injection, the first being the 'pre-injection' phase, with the second pulse being the 'main' injection phase (Figure 4.52).

As the throttle is opened, and the engine is accelerated, the 'main' injection pulse expands in a similar way to a petrol injector. As the throttle is released, the 'main' injection pulse disappears until such time as the engine returns to just above idle.

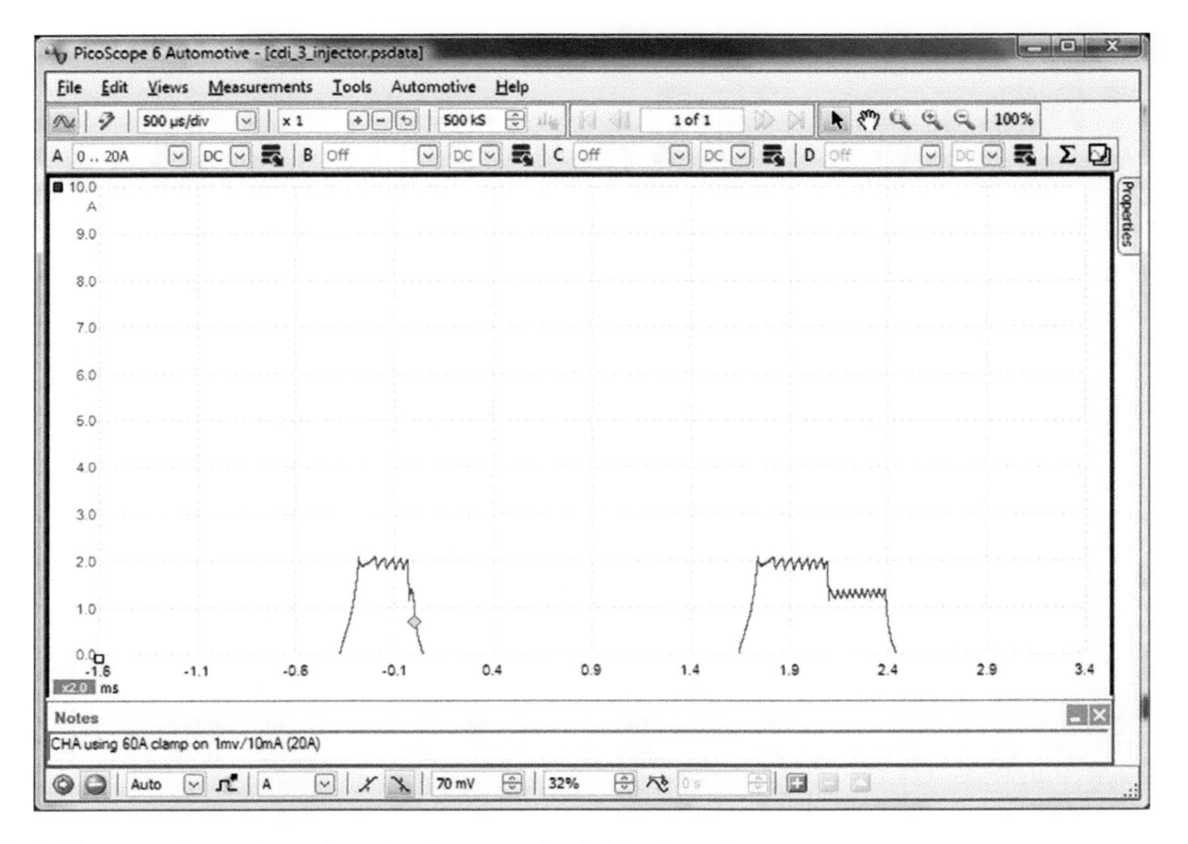

Figure 4.52 CR injector (current) waveform showing pre- and main injection pulses.

Under certain engine conditions, a third phase may be seen, this is called the 'post-injection' phase and is predominantly concerned with controlling the exhaust emissions.

4.3.4.4 Idle speed control valve

This device contains a winding, plunger and spring. When energised, the port opens, and when not, it closes (Figure 4.53).

The electromagnetic ISCV will have two electrical connections: usually a voltage supply at battery voltage and a switched earth.

The rate at which the device is switched is determined by the ECU to maintain a prerequisite speed according to its programming. The valve will form an air bypass around the throttle butterfly. If the engine has an adjustable air bypass and an ISCV, it may require a specific routine to balance the two air paths. The position of the valve tends to take up an average position determined by the supplied signal. Probing onto the supply side will produce a straight line at system voltage (Figure 4.54).

4.3.4.5 Exhaust gas recirculation valve

Various types of exhaust gas recirculation (EGR) valve are in use based on simple solenoid operation. One

Figure 4.53 Electromagnetic idle speed control valve.

development in actuator technology is the rotary electric exhaust gas recirculation (EEGR) valve for use in diesel engine applications. This device is shown in Figure 4.55. It has a self-cleaning action, accurate gas flow control and a fast reaction speed.

4.3.4.6 Carbon canister and other valves

There are a number of valves used that are effectively simple solenoid controlled devices. Measuring on one

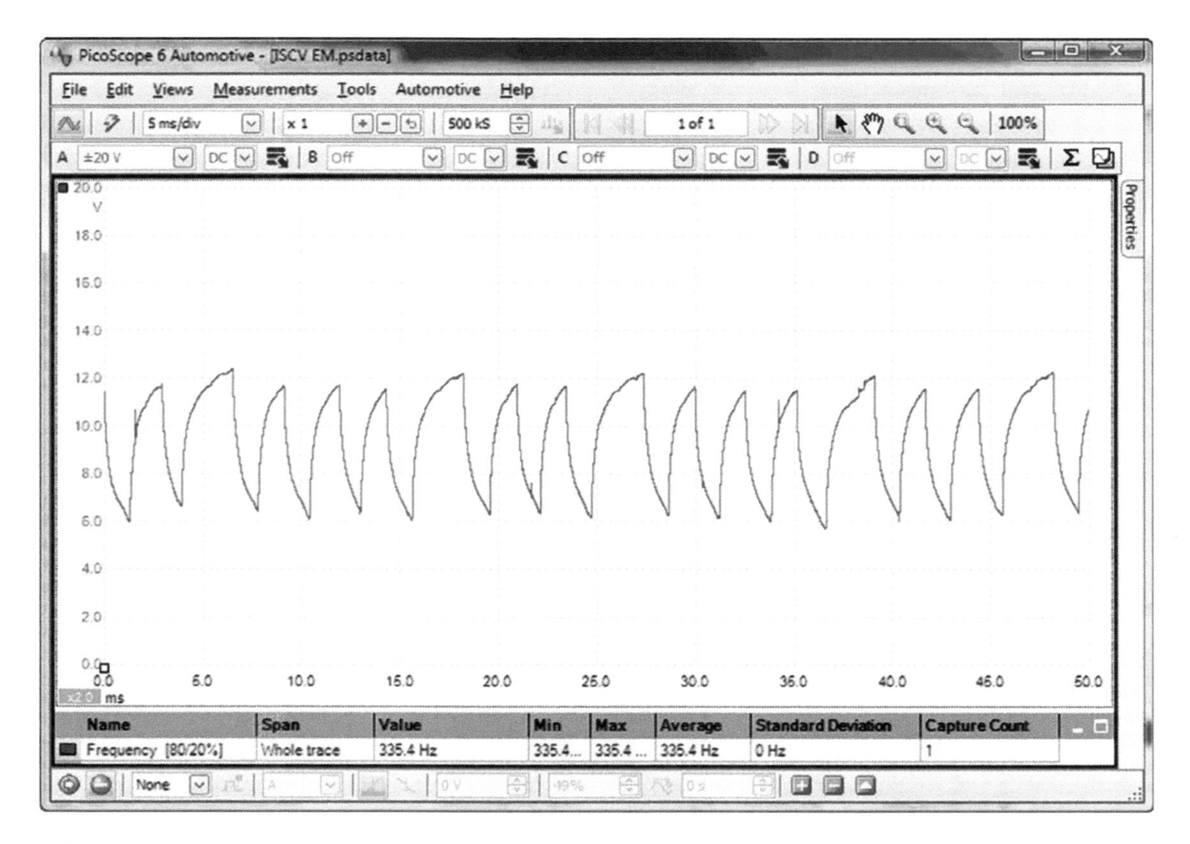

Figure 4.54 Signal produced by an electromagnetic idle speed control valve.

Figure 4.55 Rotary EGR valve.

Source: Delphi Media.

terminal will usually show battery supply voltage. The other terminal will show battery voltage when switched off and zero (ground or earth) voltage when the valve is switched on (Figure 4.56).

4.3.5 Thermal actuators

An example of a thermal actuator is the movement of a traditional type fuel or temperature gauge needle. A further example is an auxiliary air device used on many earlier fuel injection systems. The principle of the gauge is shown in Figure 4.57.

When current is supplied to the terminals, a heating element operates and causes a bimetallic strip to bend, which moves the pointer. The main advantage of this type of actuator, when used as an auxiliary device, apart from its simplicity, is that if it is placed in a suitable position, its reaction time will vary with the temperature of its surroundings. This is ideal for applications such as fast idle or cold starting control where, once the engine is hot, no action is required from the actuator.

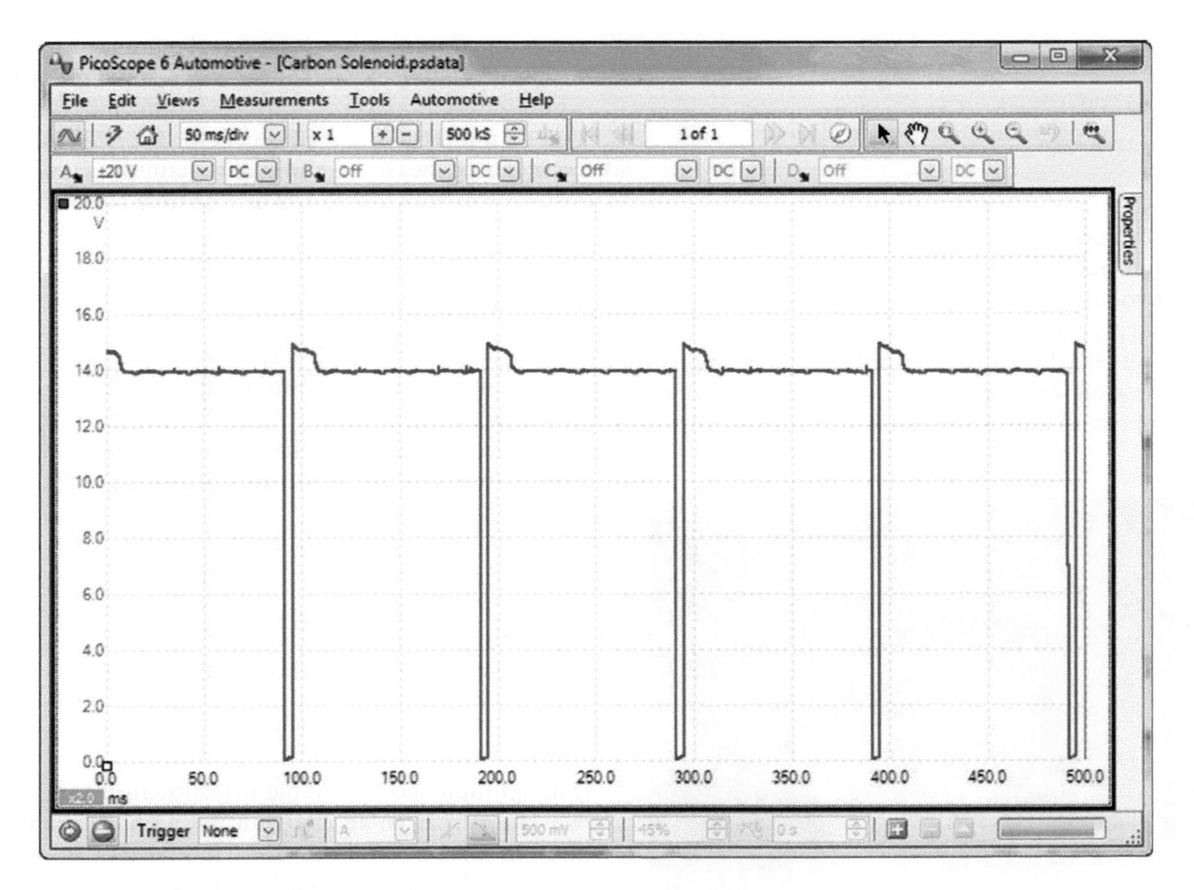

Figure 4.56 Carbon canister control valve signal.

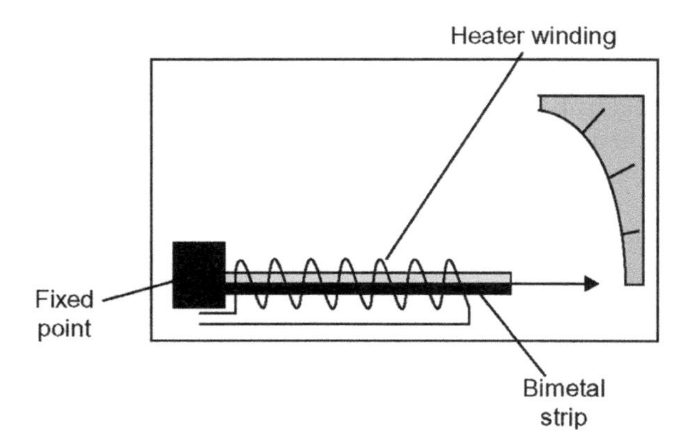

Figure 4.57 Thermal actuator used as a gauge.

4.4 ENGINE WAVEFORMS

4.4.1 Ignition primary

The ignition primary waveform is a measurement of the voltage on the negative side of the ignition coil. The earth path of the coil can produce over 350 V. Different types of ignition coils produce slightly different traces but the fundamental parts of the trace and principles are the same (Figure 4.58).

SAFETY FIRST

Even the earth path of the coil can produce over 350 V – take care.

In the waveform shown in Figure 4.59, the horizontal voltage line at the centre of the oscilloscope is at fairly constant voltage of approximately 30–40 V, which then drops sharply to what is referred to as the coil oscillation. The length of the horizontal voltage line is the 'spark duration' or 'burn time', which in this particular case is approximately 1 ms. The coil oscillation period should display a minimum of three to four peaks (both upper and lower). A loss of peaks would indicate a coil problem.

Figure 4.58 Coil on plug (COP) ignition.

There is no current in the coil's primary circuit until the dwell period. This starts when the coil is earthed and the voltage drops to zero. The dwell period is controlled by the ignition amplifier or ECU and the length of the dwell is determined by the time it takes to build up to approximately 6 A. When this predetermined current has been reached, the amplifier stops increasing the primary current and it is maintained until the earth is removed from the coil. This is the precise moment of ignition.

The vertical line at the centre of the trace is in excess of 300 V, this is called the 'induced voltage'. The induced voltage is produced by magnetic inductance. At the point of ignition, the coil's earth circuit is removed and the magnetic flux collapses across the coil's windings. This induces a voltage between 150 and 350 V. The coil's high-tension (HT) output will be proportional to this induced voltage. The height of the induced voltage is sometimes referred to as the primary peak volts.

From the example current waveform, the limiting circuit can be seen in operation. The current switches on as the dwell period starts and rises until the required value is achieved (usually 6–8 A). At this point, the current is maintained until it is released at the point of ignition.

The dwell will expand as the engine revs are increased to maintain a constant coil saturation time. This gives rise to the term 'constant energy'. The coil saturation time can be measured and will remain the same regardless of the engine speed. The example shows a charge time of approximately 3.5 ms.

4.4.2 Ignition secondary

The ignition secondary waveform is a measurement of the HT output voltage from the ignition coil. Some coils can produce over 50 000 V. Different types of ignition coils produce slightly different traces but the fundamental parts of the trace and principles are the same (Figure 4.60).

SAFETY FIRST

Some coils can produce over 50 000 V – take care.

The ignition secondary picture shown in the example waveform is from an engine fitted with electronic ignition. In this case, the waveform has been taken from the main coil lead (king lead). Suitable connection methods mean that similar traces can be seen for other types of ignition system (Figure 4.61).

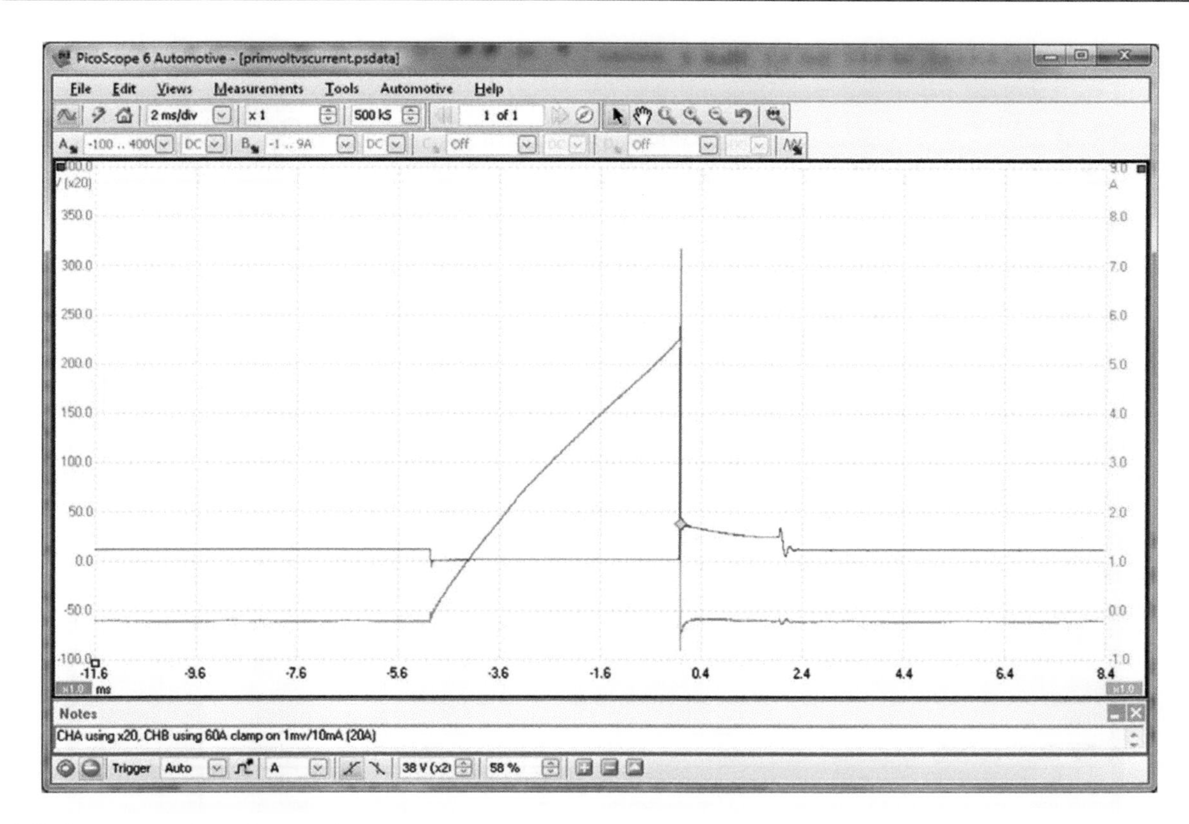

Figure 4.59 Primary ignition voltage and current traces.

The secondary waveform shows the length of time that the HT is flowing across the spark plug electrode after its initial voltage, which is required to initially jump the plug gap. This time is referred to as either the 'burn time' or the 'spark duration'. In the trace shown, it can be seen that the horizontal voltage line in the centre of the oscilloscope is at fairly constant voltage of approximately 4 or 5 kV, which then drops sharply into the 'coil oscillation' period.

The coil oscillation period should display a minimum of three or four peaks (same as for the primary trace).

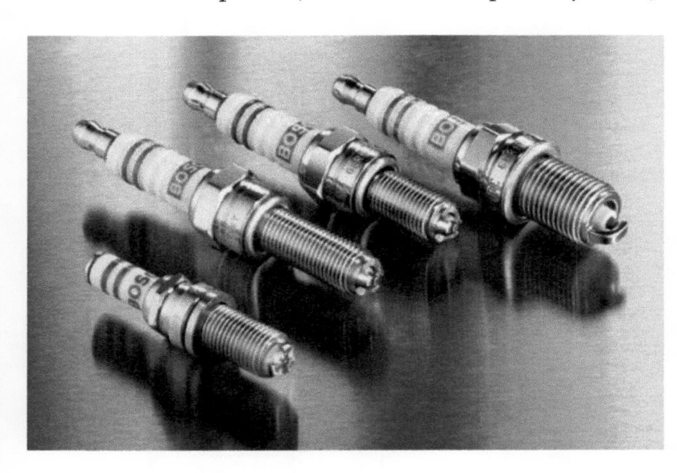

Figure 4.60 Spark plugs.

Source: Bosch Press.

A loss of peaks indicates that the coil may be faulty. The period between the coil oscillation and the next 'drop down' is when the coil is at rest and there is no voltage in the secondary circuit. The 'drop down' is referred to as the 'polarity peak' and produces a small oscillation in the opposite direction to the plug firing voltage. This is due to the initial switching on of the coil's primary current.

The plug firing voltage is the voltage required to jump and bridge the gap at the plug's electrode, commonly known as the 'plug kV'. In the example in Figure 4.62, the plug firing voltage is approximately 45 kV.

When the plug kVs are recorded on a distributorless ignition system (DIS) or coil per cylinder ignition system, the voltage seen on the waveform should be in the 'upright position'. If the trace is inverted, it would suggest that either the wrong polarity has been selected from the menu or in the case of DIS, the inappropriate lead has been chosen. The plug voltage, while the engine is running, is continuously fluctuating and the display will be seen to move up and down. The maximum voltage at the spark plug can be seen as the 'Ch A: Maximum (kV)' reading at the bottom of the screen.

It is a useful test to snap the throttle and observe the voltage requirements when the engine is under load. This is the only time that the plugs are placed under any strain and is a fair assessment of how they will perform on the road.

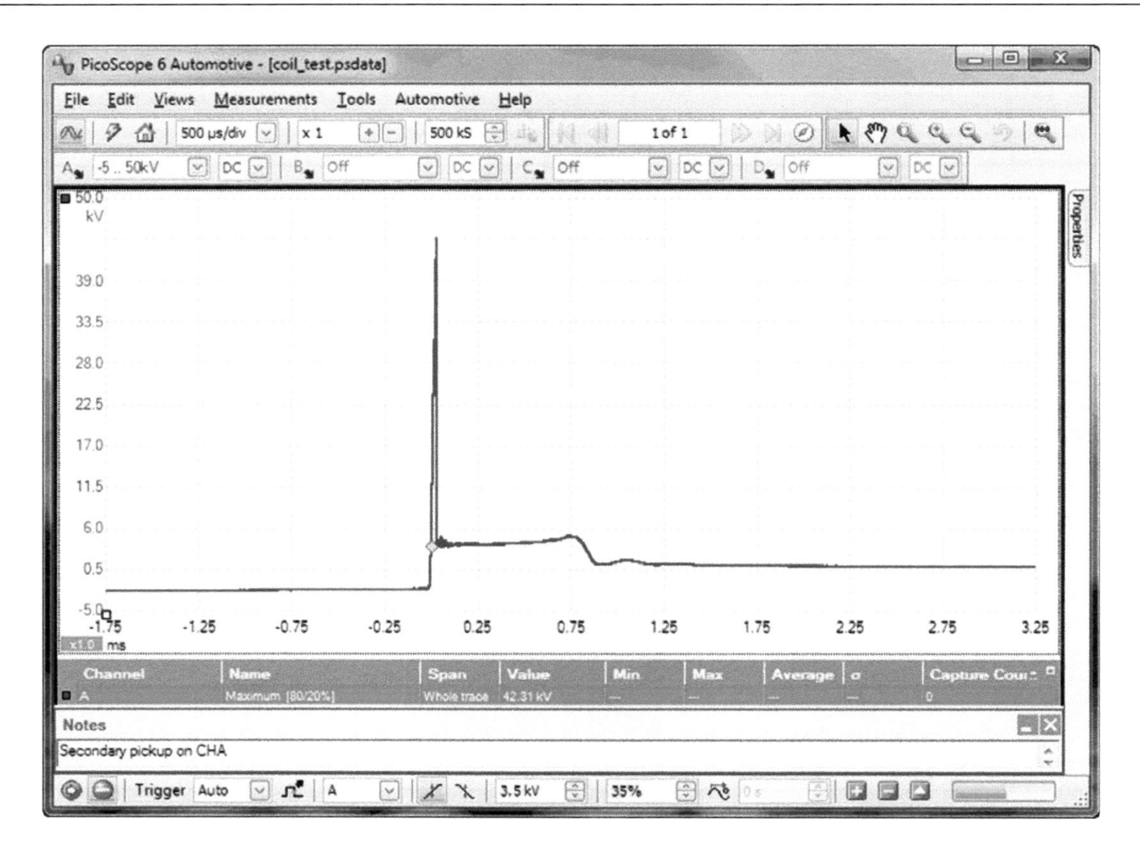

Figure 4.61 Ignition secondary trace.

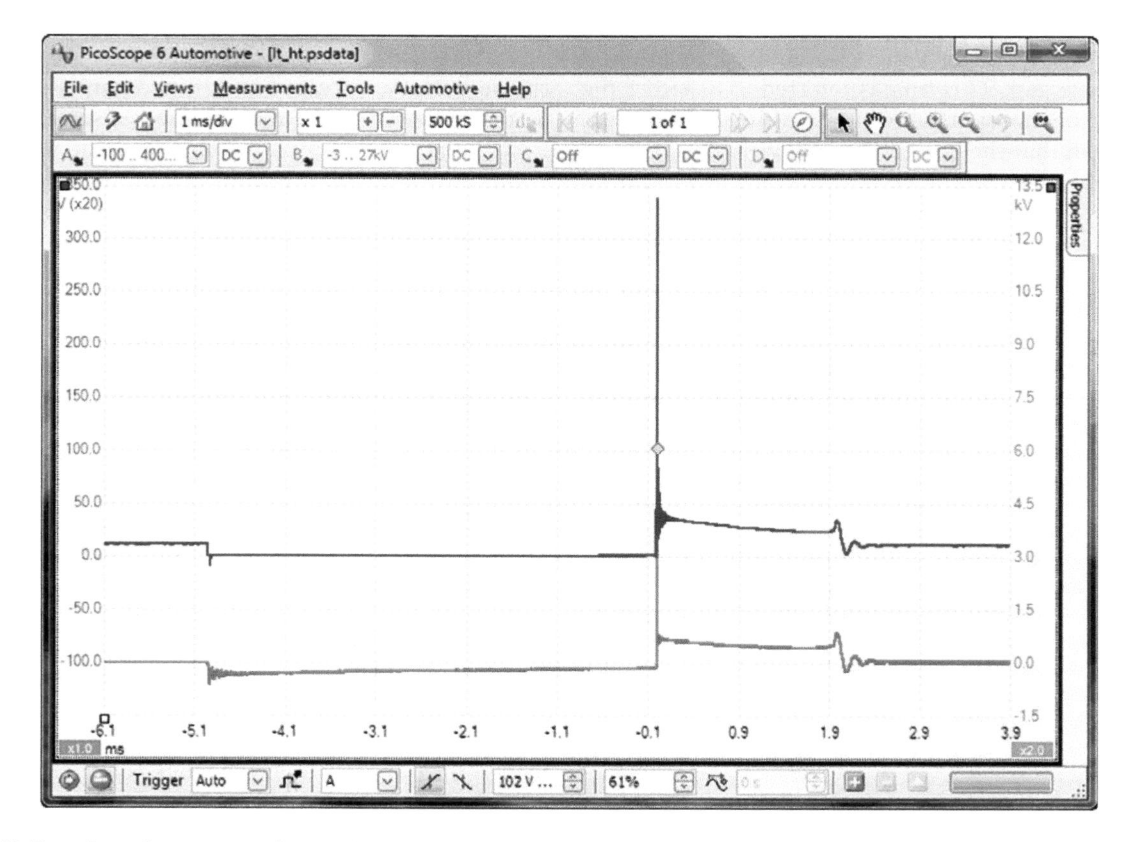

Figure 4.62 Distributorless ignition showing low and high tension (primary and secondary).

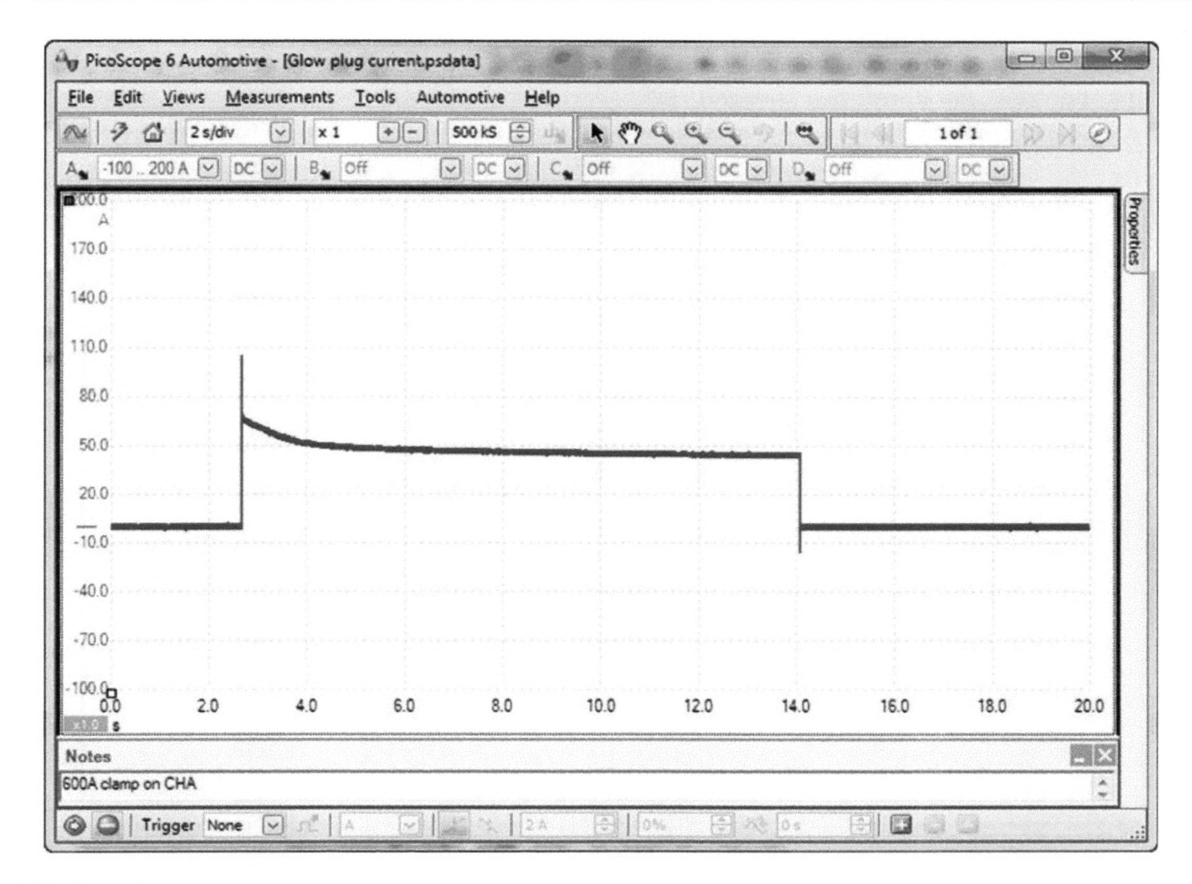

Figure 4.63 Glow plug current.

The second part of the waveform after the vertical line is known as the spark line voltage. This second voltage is the voltage required to keep the plug running after its initial spark to jump the gap. This voltage will be proportional to the resistance within the secondary circuit. The length of the line can be seen to run for approximately 2 ms.

4.4.3 Diesel glow plugs

A diesel glow plug is a simple heater. Measuring its current will indicate correct operation because as temperature increases in a glow plug so does resistance and therefore the current falls after an initial peak (Figure 4.63).

4.4.4 Alternator waveform

Checking the ripple voltage produced by an alternator (Figure 4.64) is a very good way of assessing its condition.

KEY FACT

Fluctuations in voltage on the spark line could indicate poor combustion.

The example waveform illustrates the rectified output from the alternator (Figure 4.65). The output shown is correct and there is no fault within the phase windings or the diodes (rectifier pack).

The three phases from the alternator have been rectified to DC from its original AC and the waveform shows that the three phases are all functioning.

Figure 4.64 Alternator.

Source: Bosch Media.

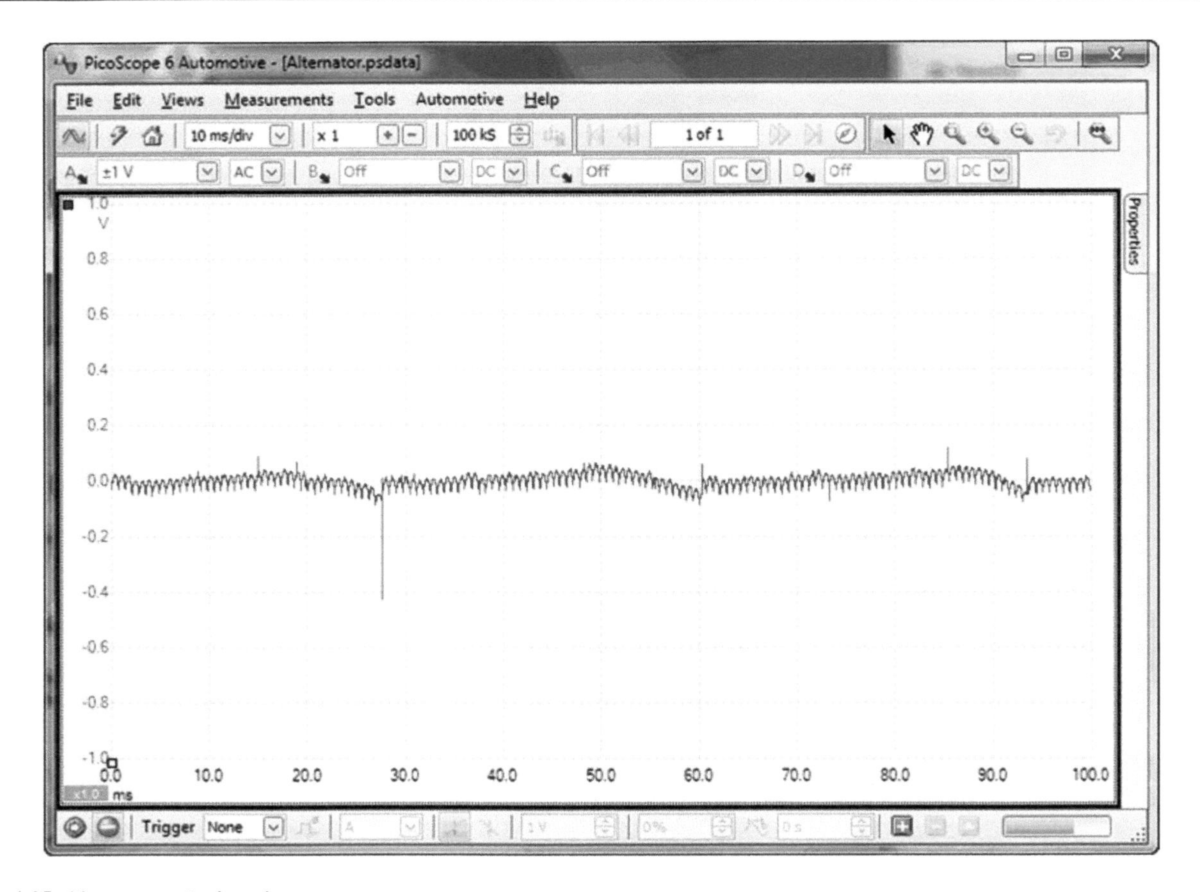

Figure 4.65 Alternator ripple voltage.

If the alternator is suffering from a diode fault, long downward 'tails' appear from the trace at regular intervals and 33% of the total current output will be lost. A fault within one of the three phases will show a similar picture to the one illustrated but is three or four times the height, with the base-to-peak voltage in excess of 1 V.

The voltage scale at the side of the oscilloscope is not representative of the charging voltage, but is used to show the upper and lower limits of the ripple. The 'amplitude' (voltage/height) of the waveform will vary under different conditions. A fully charged battery will show a 'flatter' picture, while a discharged battery will show an exaggerated amplitude until the battery is charged. Variations in the average voltage of the waveform are due to the action of the voltage regulator.

4.4.5 Relative compression petrol

Measuring the current drawn by the starter motor is useful to determine starter condition but it is also useful as an indicator of engine condition (Figure 4.66).

The purpose of this particular waveform is therefore to measure the current required to crank the engine and to evaluate the relative compressions.

The amperage required to crank the engine depends on many factors, such as the capacity of the engine, number of cylinders, viscosity of the oil, condition of the starter motor, condition of the starter's wiring circuit and compressions in the cylinders. Therefore, to evaluate the compressions, it is essential that the battery is charged and the starter and associated circuit are in good condition.

The current for a typical four cylinder petrol/gasoline engine is in the region of 100–200 A.

Figure 4.66 Starter and ring gear.

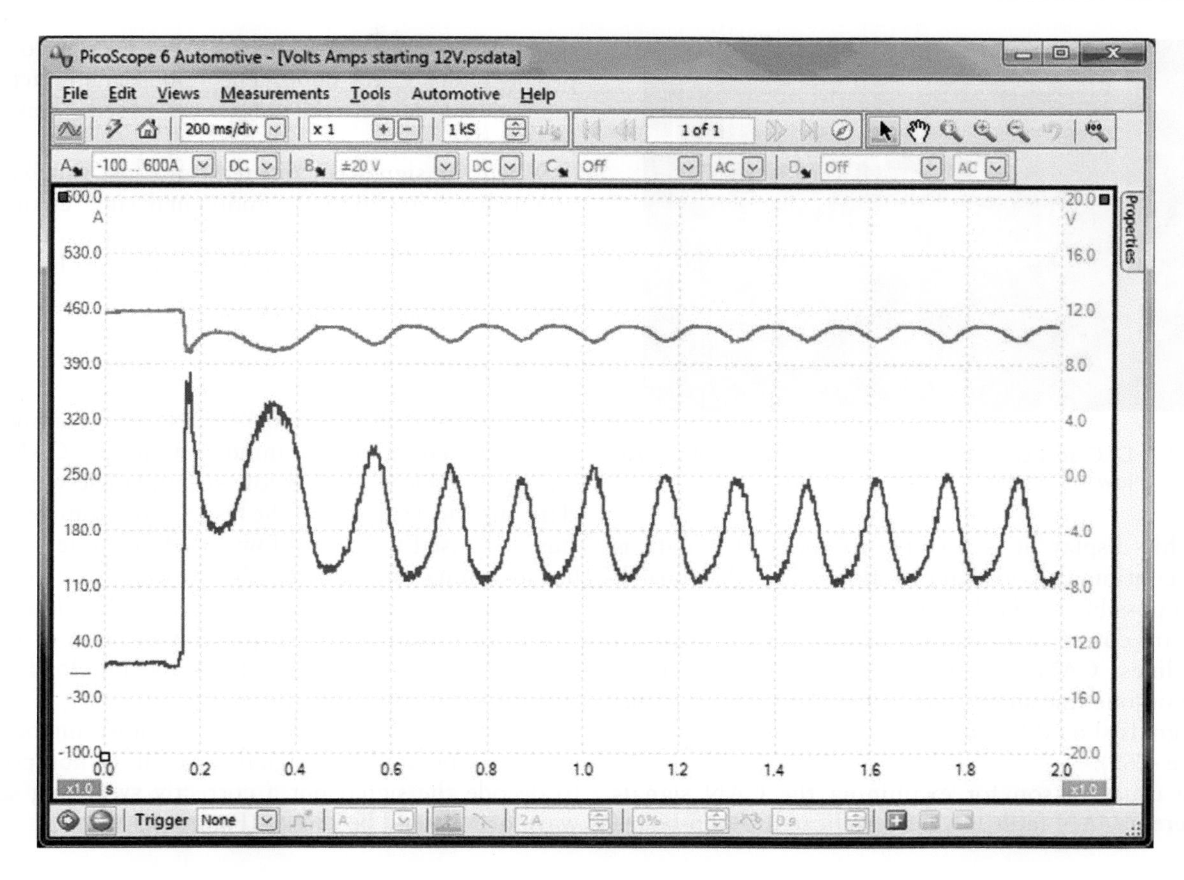

Figure 4.67 Spark ignition engine cranking amps.

In the waveform shown (Figure 4.67), the initial peak of current (approximately 300 A) is the current required to overcome the initial friction and inertia to rotate the engine. Once the engine is rotating, the current will drop. It is also worth mentioning the small step before the initial peak, which is being caused by the switching of the starter solenoid.

The compressions can be compared against each other by monitoring the current required to push each cylinder up on its compression stroke. The better the compression, the higher the current demand and vice versa. It is therefore important that the current draw on each cylinder is equal.

4.5 COMMUNICATION NETWORKS

4.5.1 Introduction

There are three common multiplexed communication systems currently in use. These systems reduce the number of wires needed and also allow information from sensors or different ECUs to be shared across a network. These systems are:

- CAN;
- LIN;
- FlexRay.

However, new protocols have been developed and are in use by many manufacturers. For example:

- Time-triggered CAN;
- Byteflight;
- Ethernet.

These systems are in use or being developed, such as MOST, time-triggered CAN (TT-CAN) and Byteflight (a protocol designed for use with safety-critical systems such as air bags and seat belt tensioners). However, one of the more interesting is ethernet.

4.5.2 CAN

Controller area network (CAN) is a protocol used to send information around a vehicle on data bus. It is made up of voltage pulses that represent ones and zeros, in other words, binary signals. The data is applied to two wires known as CAN high and CAN low (Figure 4.68).

DEFINITION

DLC: Diagnostic/Data link connector.

Figure 4.68 DLC socket – pin 6 is CAN high and pin 14 is CAN low.

In this display, it is possible to verify that data is being continuously exchanged along the CAN bus. It is also possible to check that the peak-to-peak voltage levels are correct and that a signal is present on both CAN lines. CAN uses a differential signal, and the signal on one line should be a coincident mirror image (the signals should line up) of the data on the other line (Figure 4.69).

The usual reason for examining the CAN signals is where a CAN fault has been indicated by on-board diagnostics, or to check the CAN connection to a suspected faulty CAN node. The vehicle manufacturer's manual should be referred to for precise waveform parameters.

When the signal is captured on a fast timebase (or zoomed in), it allows the individual state changes to be viewed. This enables the mirror image nature of the signals and the coincidence of the edges to be verified (Figure 4.70).

4.5.3 LIN

Local interconnect network (LIN) bus communication is becoming more common on modern CAN bus–equipped vehicles. It is a low-speed, single-wire serial data bus and a sub-bus of the faster, more complex CAN bus. It is used to control low-speed non-safety-critical housekeeping functions on the vehicle, especially windows, mirrors, locks, HVAC units and electric seats.

The LIN bus is proving popular because of its low cost and also because it reduces the bus load of the supervising CAN network.

LIN signals can be measured by connecting between earth/ground and the signal wire. It is not possible to decode the signal but a correctly switching square waveform should be shown (Figure 4.71).

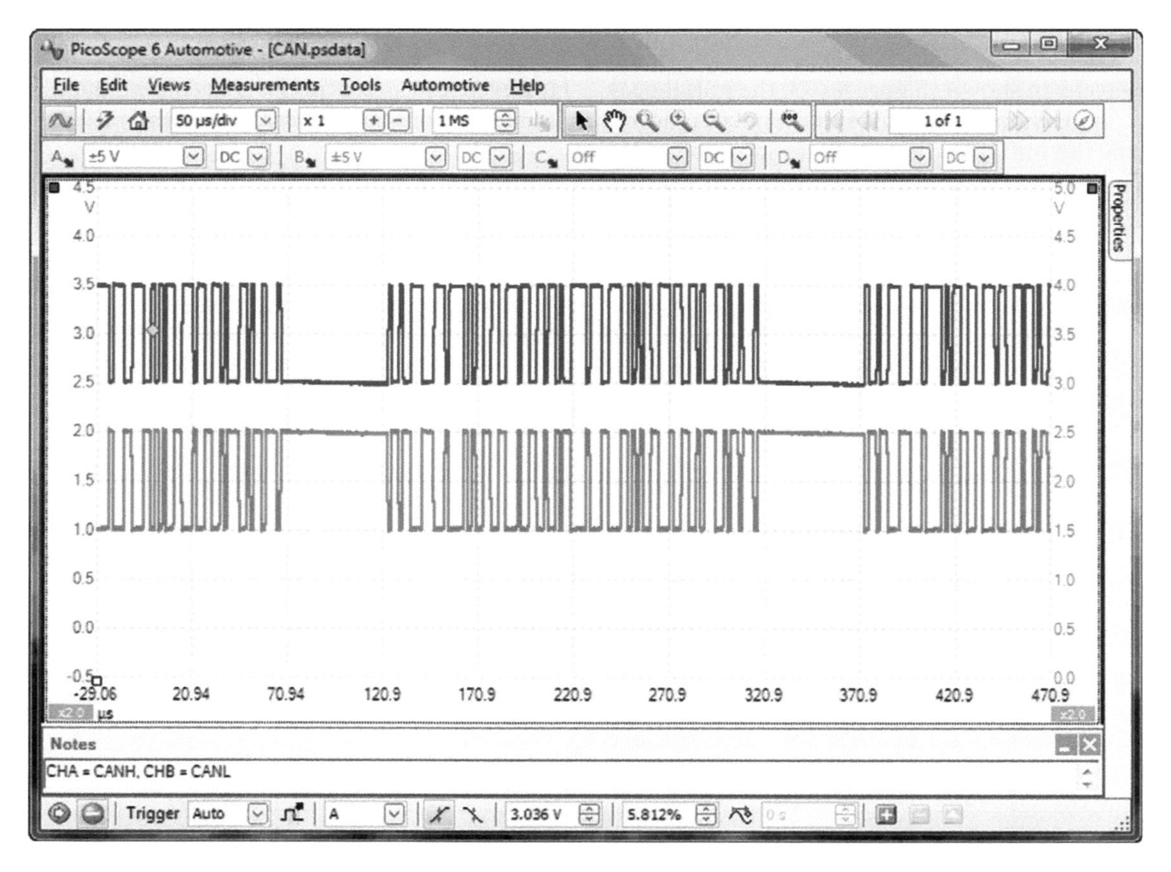

Figure 4.69 CAN high and low signals on a dual trace scope.

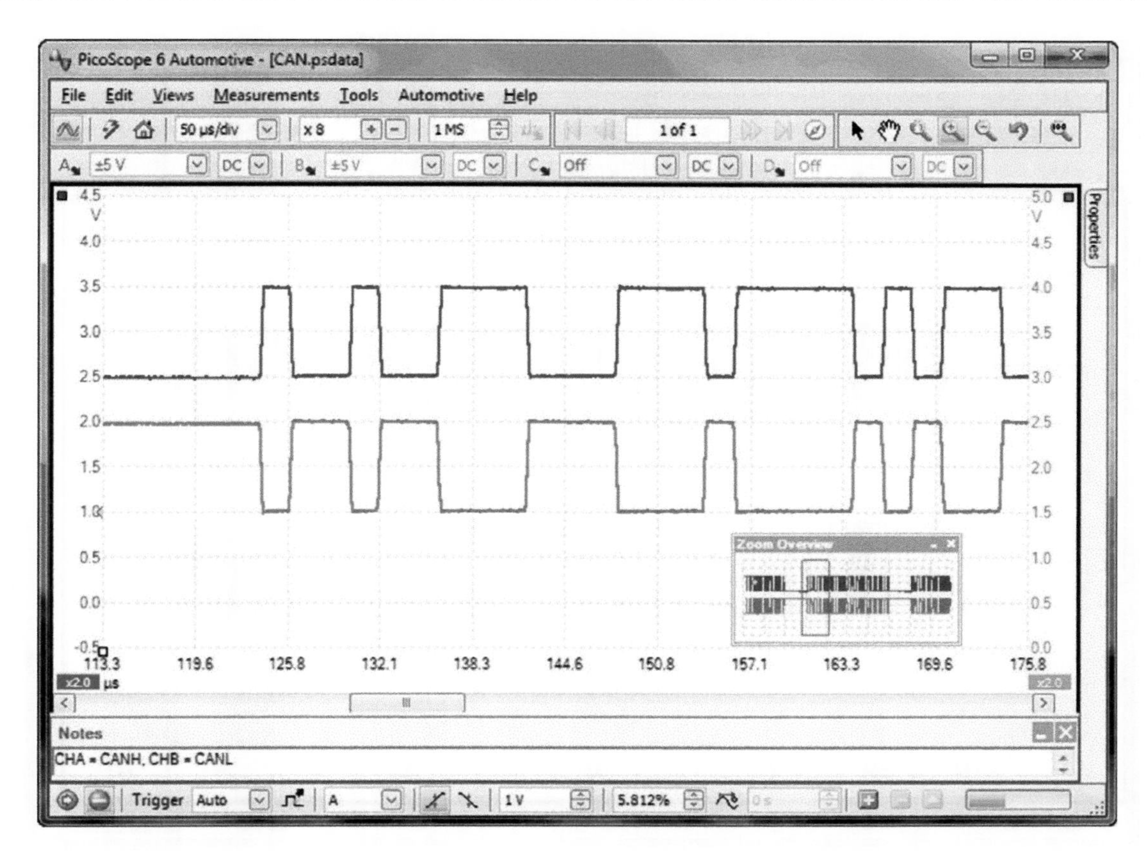

Figure 4.70 CAN signal zoomed in.

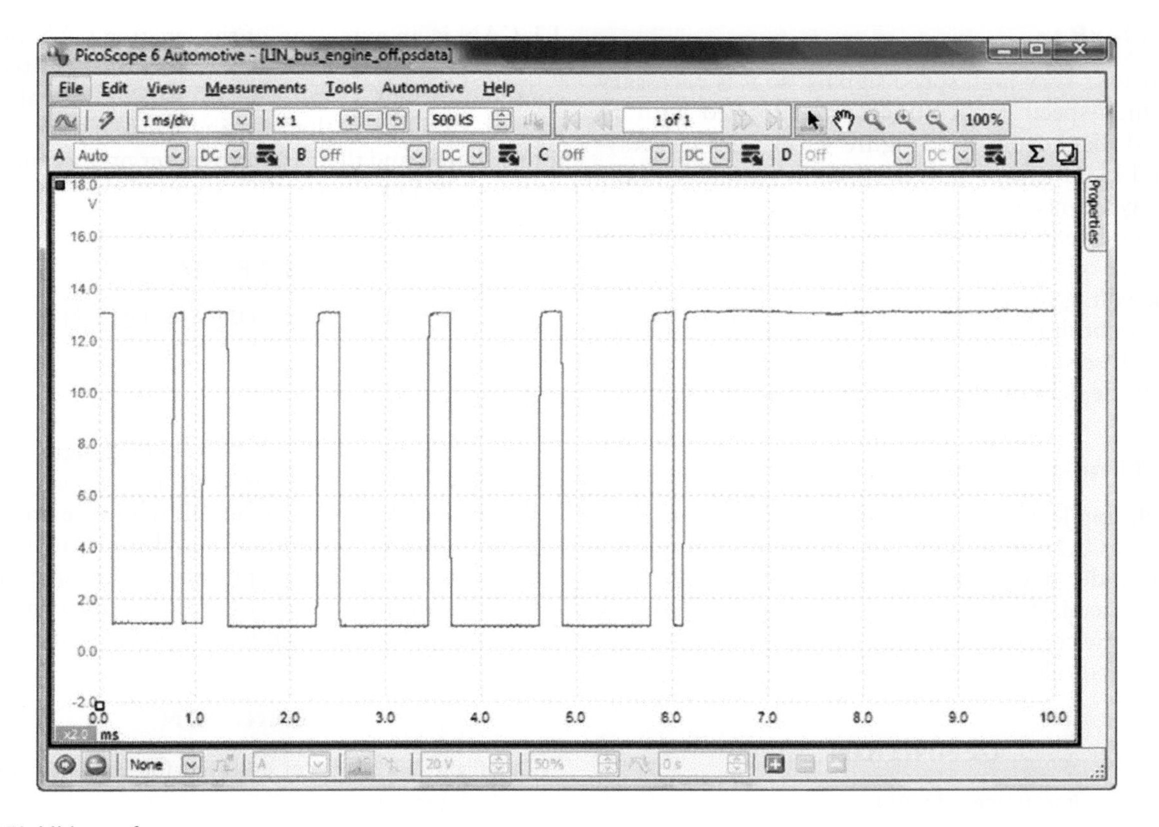

Figure 4.71 LIN waveform.

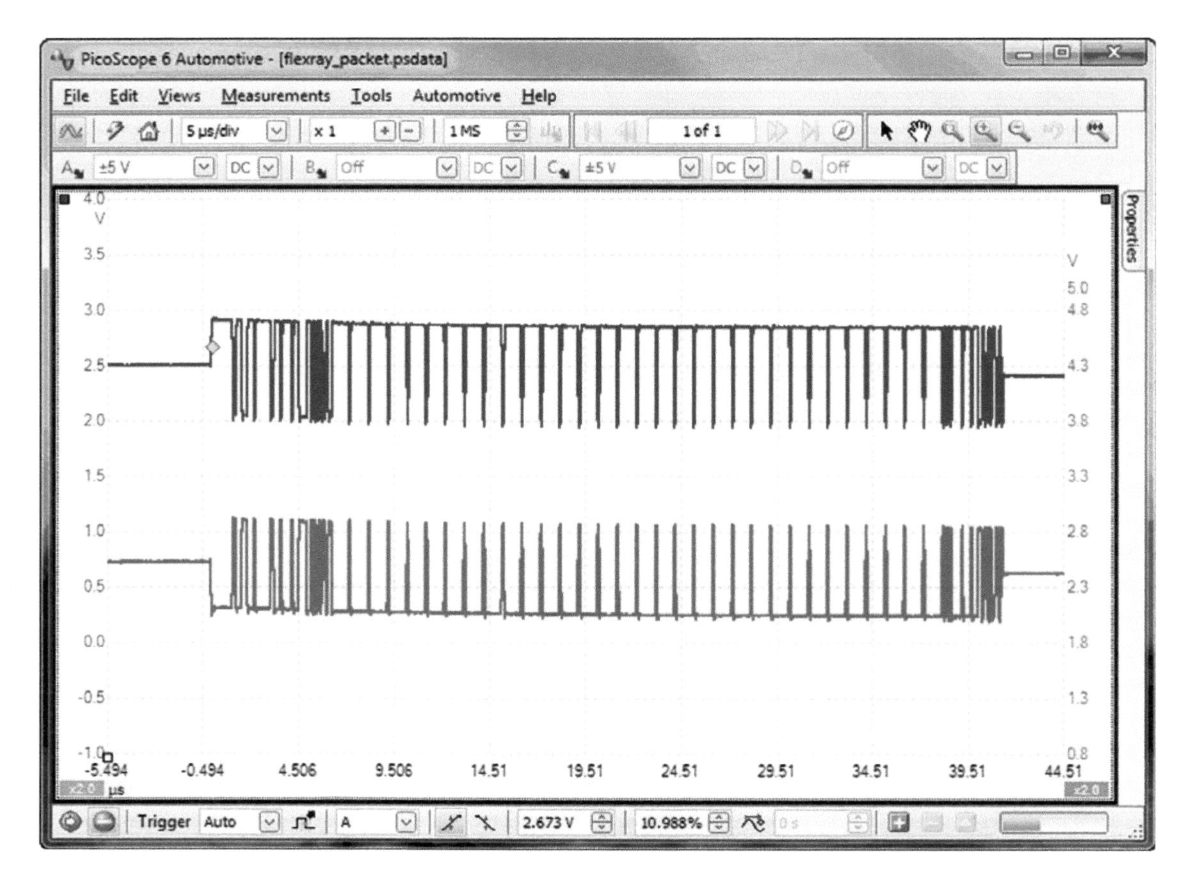

Figure 4.72 FlexRay signal.

4.5.4 FlexRay

FlexRay uses very high speed signals, so it is necessary to use high-speed probes (these are supplied with an advanced diagnostics kit) (Figure 4.72). The FlexRay-high and FlexRay-low pins are usually available at the multi-way connector at each ECU on the network.

It is possible to verify that data is being continuously exchanged on the FlexRay network, that the peak-to-peak voltage levels are correct and that a signal is present on both FlexRay lines. FlexRay uses a differential signal, so the signal on one line should be a mirror image of the data on the other line (Figure 4.73).

4.5.5 Time-triggered CAN

Network applications will soon need bit rates of 5 to 10 Mbps. CAN is limited to 1 Mbps so new protocols are under development by Bosch and others. One answer instead of using event-triggered transmission is to trigger the communication at a precise time. Because of this, speeds close to 'real time' can be achieved.

DEFINITION

TT-CAN: Time-triggered CAN.

TT-CAN is an extension of the existing CAN protocol so the system will still be able to use event-triggered communications (Figure 4.74). An additional advantage of TT-CAN is that missing messages are detected immediately, and there is much better protection against unauthorised bus access.

KEY FACT

TT-CAN is an extension of the existing CAN protocol.

4.5.6 Byteflight

Byteflight (Figure 4.75) is a protocol designed for use with safety-critical systems such as air bags and seat belt tensioners. It has a high fault tolerance and allows high speed data. It is known as a flexible time division multiple access (FTDMA) protocol. Because of its flexibility, it can also be used for body and convenience functions such as locking, windows and seat control.

DEFINITION

Byteflight is a protocol designed for use with safety-critical systems.

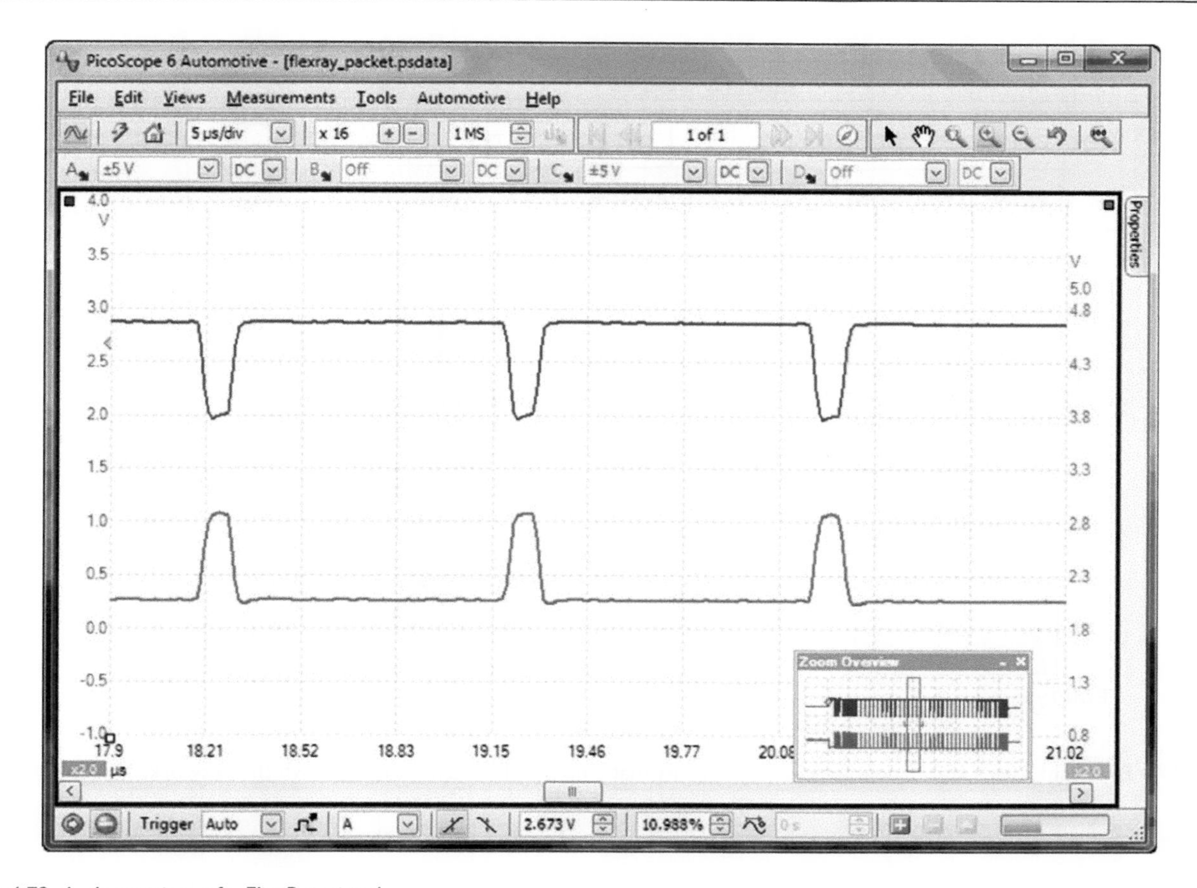

Figure 4.73 A closer view of a FlexRay signal.

KEY FACT

FTDMA: Flexible time division multiple access.

Star topography is used but the protocol also works on linear systems. Communication is by plastic optic fibre (POF) and the data rate, depending on loads, is 5 to 10 Mbps. One example of its use is the BMW 1 series where it links the air bag crash sensor satellite ECUs. The protocol was not designed for X-by-wire, but it is high performance and has many of the required features.

DEFINITION

POF: Plastic optic fibre.

4.5.7 Ethernet

Ethernet (Figures 4.76 and 4.77) is a family of computer networking technologies commonly used in local area networks, metropolitan area networks and wide area networks. It was commercially introduced in 1980 and first standardized in 1983 as IEEE 802.3.

Automotive Ethernet does more than support mobile connectivity and complex, high-bandwidth automotive applications; it is a key enabler for the fully connected autonomous vehicle (CAV). NXP, for example, has designed a broad portfolio of robust, flexible and cost-effective automotive Ethernet products to connect vehicle systems faster and more efficiently.

In an automotive Ethernet setup, multiple vehicle systems simultaneously access high bandwidth over a single cable (a twisted pair). The Ethernet is the backbone of the vehicle network and supports higher levels of

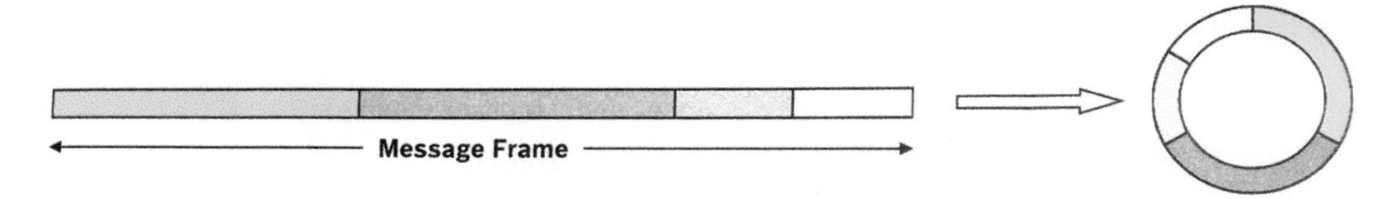

Figure 4.74 TT-CAN signal with the beginning and end joined to make a continuous cycle.

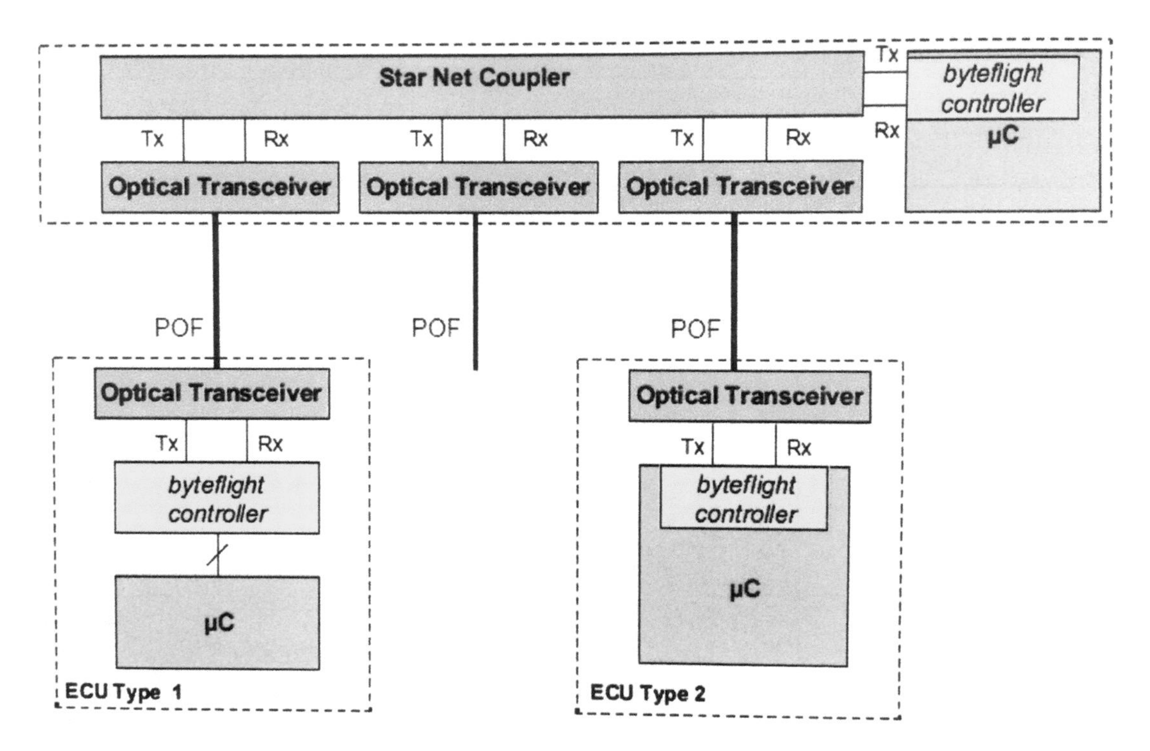

Figure 4.75 Byteflight uses a star topography as this ensures that a node failure cannot bring down the network.

data processing and more communication types. Each port receives dedicated bandwidth and the entire backbone is capable of IP connectivity. Instead of supporting individual high-bandwidth functions, the architecture supports all the high-bandwidth functions that reside on the same physical network but use logically separated virtual networks.

Real-time processing and in-vehicle networking technologies of the next decade will likely revolutionize how we move around, how we interact with our vehicles and how our vehicles react to the infrastructure and each other. NXP envisions automotive Ethernet as an essential part of this revolution.

Automotive Ethernet will connect the vehicle's internal digital devices, connect the vehicle with other vehicles and even make the car a productive part of the IoT. It supports the high bandwidth and real-time processing required for today's infotainment and ADAS systems and provides a platform for the development of truly autonomous driving.

Figure 4.76 Ethernet cable (CAT 5).

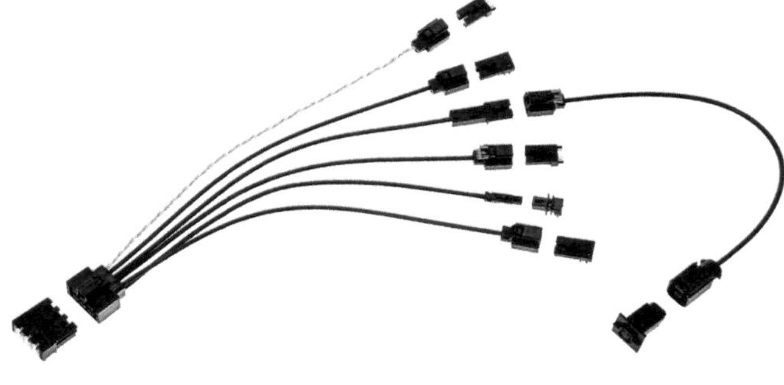

Figure 4.77 Ethernet cables.

Source: Delphi.

4.6 SUMMARY

'Scope' diagnostics, particularly for sensors and actuators, is now an essential skill for the technician to develop. As with all diagnostic techniques that use test equipment, it is necessary for the user to know how

1. the vehicle system operates;
2. to connect the equipment;
3. readings should be interpreted.

Remember that an oscilloscope is actually just a voltmeter or ammeter, but it draws a picture of the readings over a set period of time. Learn what good waveforms look like, and then you will be able to make good judgements about what is wrong when they are not so good.

ACKNOWLEDGEMENT

I am grateful to PicoTech for permission to use waveforms from their extensive library. Visit http://www.picotech.com for more information.

Chapter 5

On-board diagnostics

5.1 HISTORY

5.1.1 Introduction

Originating in the United States, and subsequently followed by Europe, Asia and many others, now governments around the globe have augmented vehicle emissions control legislation. This includes a requirement that all vehicles sold within their territories must support an on-board diagnostics (OBD) system that can be operated to determine the serviceability of the vehicle's emission control systems, sub-systems and components.

Enabled by the increasing advances in electronics and microprocessor software development, this system, now commonly termed as OBD, has been developed over recent years and is now implemented by all major motor vehicle manufacturers. Furthermore, this has been extended to allow diagnosis of non-emission-related vehicle systems.

5.1.2 Vehicle emissions and environmental health

From as early as 1930, the subject of vehicle engine emissions influencing environmental health was very topical in the state of California. Already with a population of 2 million vehicles, scores of people died and thousands became sick due to air pollution–related illnesses (Figure 5.1).

In 1943, following the outbreak of the Second World War, the population of California had risen to some 7 million people with 2.8 million vehicles travelling over a total of 24 billion miles. Already, smog was apparent and people suffered with stinging eyes, sore throat and breathing difficulties. The local government initiated a study into the cause of the problem. Scientists at Caltech and The University of California investigated the problem of smog.

DEFINITION

VMT: Vehicle miles travelled.

Figure 5.1 Early traffic jam.

In 1945, after the conclusion of the war, Los Angeles began its air pollution control program and established the Bureau of Smoke Control. On 10 June 1947, the then California governor Earl Warren signed the Air Pollution Control Act. By 1950, California's population had reached 11 million people. Total registered vehicles in California exceeded 4.5 million and vehicle miles travelled (VMT) was 44.5 billion. The search for the root cause of smog production went on (Figure 5.2). Reports of deaths in other countries became apparent, for example, thousands of people died in London of a 'mystery fog'.

In 1952, Dr Arie Haagen-Smit determined the root cause of smog production. He surmised that engine pollutants, carbon monoxide (CO), hydrocarbons (HC) and

Figure 5.2 Smog over Los Angeles.

various oxides of nitrogen (NO_x) combine to generate the smog, which consists of ozone and carbon dioxide.

Carbon dioxide is a pollutant, which is now said to contribute to global warming and climate change. Ozone, occupying a region of the lower atmosphere, is now known to cause respiratory ill health and lung disease and is also thought to make a much greater contribution to the greenhouse effect than even carbon dioxide.

The state became a centre for environmental activism. Naturally, amidst a public outcry to preserve the local environment, the state began to legislate for controls on motor vehicle emissions. So began an initiative that would span over 50 years, one that would drive change in a world industry and lead the world in the fight for clean air.

5.1.3 History of the emissions control legislation

In 1960, the Motor Vehicle Pollution Control Board was established with a mandate to certify devices proposed to be fitted on cars for sale in California. In addition, the Federal Motor Vehicle Act of 1960 was enacted, requiring federal research to combat motor vehicle engine pollution. Manufacturers made technology improvements, and during this period, California's population reached 16 million. Total registered vehicles approached 8 million and VMT was 71 billion.

In 1961, in an effort to control HC crankcase emissions, the first piece of vehicle emissions control legislation mandating the use of specific hardware was issued. Positive crankcase ventilation (PCV) controls HC crankcase emissions by extracting gases from the crankcase and recirculating them back into the fresh air/fuel charge in the cylinders.

A key turning point in history, in 1966, the California Motor Vehicle Pollution Control Board pioneered the adoption of vehicle tailpipe emissions standards for HC and CO and the California Highway Patrol began random roadside inspections of the smog control devices fitted to vehicles.

The following year, the governor of California, Ronald Reagan, signed the Mulford-Carrell Air Resources Act. This effectively allowed the state of California to set its own emissions standards. The same year saw the formation of the California Air Resources Board (CARB), which was created from the amalgamation of the Motor Vehicle Pollution Control Board and the Bureau of Air Sanitation.

In 1969, the first California State Ambient Air Quality Standards were extended by California for photochemical oxidants, suspended particulates, sulphur dioxide (SO_2), nitrogen dioxide (NO_2) and CO. California's population reached 20 million people. Total registered vehicles exceeded 12 million and VMT was 110 billion.

Total cumulative California vehicle emissions for HC and NO_x were estimated at 1.6 million tons/year (Table 5.1). In 1970, the US Environmental Protection Agency (EPA) was created. Its primary directive was, and still is, to protect all aspects of the environment. The next seven years witnessed further development of emissions control legislation and increasing employment of vehicle emissions control technology.

In 1971, CARB adopted the first vehicle NO_x standards. The EPA announced National Ambient Air Quality Standards for particulates, HC, CO, NO_2, photochemical oxidants (including ozone) and SO_2.

The first two-way catalytic converters came into use in 1975 as part of CARB's Motor Vehicle Emission Control Program followed by an announcement that CARB would limit lead in gasoline.

In 1977, Volvo introduced a vehicle marketed as 'Smog-Free'. This vehicle supported the first three-way catalytic (TWC) converter to control HC, NO_x and CO emissions.

In 1980, the California population reached 24 million people. Total registered vehicles were in the region of 17 million and VMT was 155 billion (Table 5.2).

Total cumulative California vehicle emissions for NO_x and HCs remained at 1970 levels of 1.6 million tons/year, despite a rise of 45 billion in VMT over those 10 years.

The legislative controls had clearly begun to have a positive effect. Spurred on by this victory, CARB began a program of compliance testing on 'in-use' vehicles in order to determine whether they continue to comply with

Table 5.1 California state-wide average emissions per vehicle, 1969

NO_x (g/mile)	HC (g/mile)
5.3	8.6

Table 5.2 California state-wide average emissions per vehicle, 1980

NO_x (g/mile)	HC (g/mile)
4.8	5.5

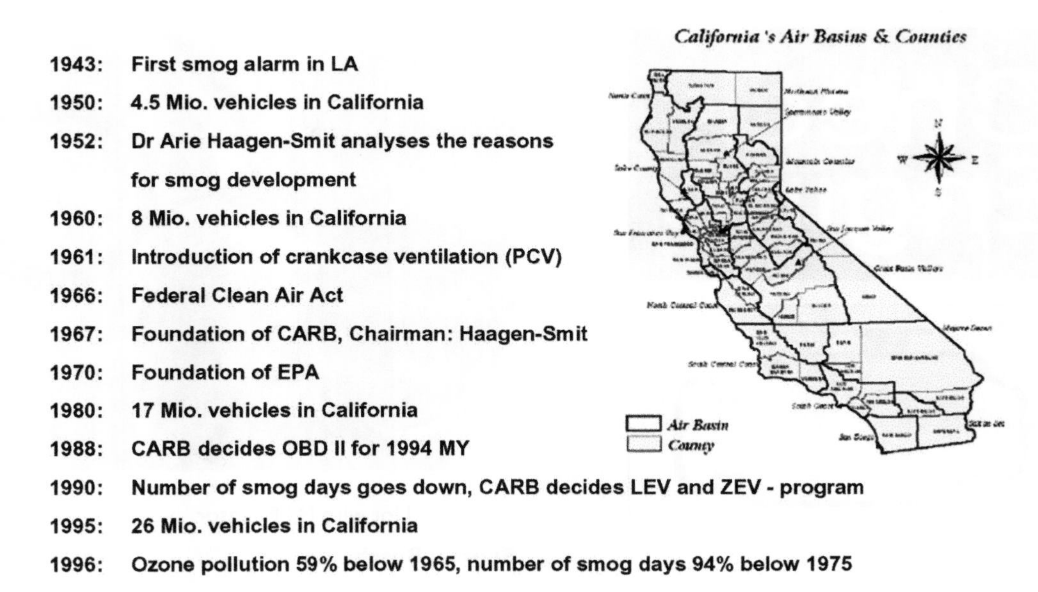

1943: First smog alarm in LA
1950: 4.5 Mio. vehicles in California
1952: Dr Arie Haagen-Smit analyses the reasons for smog development
1960: 8 Mio. vehicles in California
1961: Introduction of crankcase ventilation (PCV)
1966: Federal Clean Air Act
1967: Foundation of CARB, Chairman: Haagen-Smit
1970: Foundation of EPA
1980: 17 Mio. vehicles in California
1988: CARB decides OBD II for 1994 MY
1990: Number of smog days goes down, CARB decides LEV and ZEV - program
1995: 26 Mio. vehicles in California
1996: Ozone pollution 59% below 1965, number of smog days 94% below 1975
California's Air Basins & Counties
Air Basin
County

Figure 5.3 History of CARB emission legislation activity.

emission standards as vehicle mileage increases. Vehicle manufacturers commissioned the development of more durable emission control systems.

Introduction of the biennial California Smog Check Program was seen in 1984, the aim of which was to identify vehicles in need of maintenance and to confirm the effectiveness of their emissions control systems (Figure 5.3).

The mid-term period of emissions control legislation ended in 1988 with a key announcement, which saw the beginning of on-board diagnostics. The California Clean Air Act was signed and CARB adopted regulations that required that all 1994 and beyond model year cars were fitted with 'on-board diagnostic' systems. The task of these systems was, as it is now:

> *To monitor the vehicle emissions control systems performance and alert owners when there is a malfunction that results in the lack of function of an emissions control system/subsystem or component.*

DEFINITION

CARB: California Air Resources Board.

5.1.4 Introduction of vehicle emissions control strategies

To meet the ever increasing but justifiable and 'wanted' need of vehicle emissions control legislation, vehicle manufacturers were forced to invest heavily in the research and development of Vehicle Emission Control Strategies. Building upon the foundation laid by PCV, the two-way and three-way catalyst, manufacturers further developed emissions control hardware. Such systems included exhaust gas recirculation, secondary air injection, fuel tank canister purge, spark timing adjustment, air/fuel ratio (AFR) control biasing and fuel shut off under negative torque conditions (overrun or cruise down), to name but a few.

This development continued and expanded meaning that these systems demanded an ever-increasing array of sensors and actuators. The resolution of measurement, control of AFR, actuator displacement rates and accuracy of displacement, etc., was way beyond that which could be provided by traditional existing mechanical technologies.

At about this time, an enabler was provided in the form of recent advances in microprocessor technology. The path was clear, the drivers for OBD system monitoring were in force and the enablers were available. On-board diagnostics was born.

5.2 WHAT IS ON-BOARD DIAGNOSTICS?

Fundamentally, a contemporary microprocessor-based on-board diagnostics or OBD system is intended to self-diagnose and report when the performance of the vehicle's emissions control systems or components have degraded. This is to the extent that the tailpipe emissions have exceeded legislated levels or are likely to be exceeded in the long term.

When an issue occurs, the OBD system illuminates a warning lamp known as the malfunction indicator lamp (MIL) or malfunction indicator (MI) on the

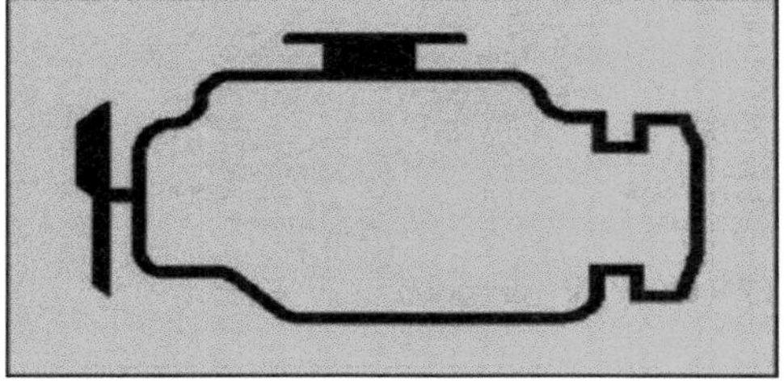

Figure 5.4 Malfunction indicator lamp symbols.

instrument cluster. In the United States, this symbol often appears with the phrase 'Check Engine', 'Check' or 'Service Engine Soon' contained within it. European vehicles tend to use the engine symbol on an orange background (Figure 5.4).

When the fault occurs, the system stores a diagnostic trouble code (DTC) that can be used to trace and identify the fault. The system will also store important information that pertains to the operating conditions of the vehicle when the fault was set. A service technician is able to connect a diagnostic scan tool or a code reader that will communicate with the microprocessor and retrieve this information. This allows the technician to diagnose and rectify the fault, make a repair/replacement, reset the OBD system and restore the vehicle emissions control system to a serviceable status.

As vehicles and their systems become more complex, the functionality of OBD is being extended to cover vehicle systems and components that do not have anything to do with vehicle emissions control. Vehicle body, chassis and accessories such as air conditioning or door modules can now also be interrogated to determine their serviceability as an aid to fault diagnosis.

5.2.1 OBD scenario example

While driving, a vehicle owner observes that the vehicle's engine 'lacks power' and 'jumps sometimes'. This is a problem often faced by technicians in that customers often have no engineering or automotive knowledge and use lay terms to describe what is happening with a very complex system. The driver does, however, report that the MIL has been illuminated.

The technician connects a scan tool that can communicate using an industry standard communications protocol. The OBD code memory is checked and data is presented in a way that also conforms to a standard. DTC P1101 with the description 'MAF sensor out of self-test range' is stored in memory, which means that the OBD system component monitor has identified the mass airflow (MAF) sensor circuit voltage as outside an acceptable range (Figure 5.5).

Figure 5.5 Hot wire MAF sensor.

Source: Bosch.

KEY FACT

When the fault occurs, the system stores a diagnostic trouble code (DTC).

Upon confirming the fault the system was smart, it defaulted to a 'safe' value of MAF, a concept known as failure mode effects management (FMEM), to allow the driver to take the vehicle to a place of repair. While this FMEM value was a good short-term solution, it is not a sufficient substitute for the full functionality of a serviceable MAF sensor.

DEFINITION

FMEM: Failure mode effects management allows the driver to take the vehicle to a place of repair. Also described as 'limp home' mode.

Since the MAF sensor determines the MAF going into the engine intake, it will be impossible for the system to run at the optimum AFR for efficient burning of the air/fuel charge within the cylinder. It may be that tailpipe emissions are likely to rise beyond legislated limits.

Also, the MAF sensor is used by other emissions control systems on the vehicle; now that its input is unreliable it follows that those systems are no longer working at their optimum levels and may not work at all. This is the reason for the MIL illumination, which says, in as many words, 'An emissions control system/sub system or component has become unserviceable!'

Visual inspection of the MAF sensor reveals that it has become damaged beyond repair and needs replacing. This is carried out, the technician clears the DTC from the OBD system memory, resets the system and takes a short test drive; later the diagnostic scan tool confirms that the DTC is no longer present. The road test also confirms that the previous drive issue is no longer apparent.

5.2.2 Origins of OBD in the USA

The previous example relates to the current situation, but when OBD was first introduced, standards and practices were less well defined. Manufacturers developed and applied their own systems and code descriptions. This state of affairs was obviously undesirable since non-franchised service and repair centres had to understand the various subtleties of each system; this meant having different scan tools, as well as a multitude of leads, manuals and connectors. This made diagnostics unwieldy and expensive. This stage became known as 'OBD1', the first stage of OBD introduction.

In the late 1980s, the Society of Automotive Engineers (SAE) defined a list of standard practices and recommended these to the EPA. The EPA acknowledged the benefits of these standards and recommendations, and adopted them. In combination, they changed the shape and application of OBD. The recommendations included having a standard diagnostic connector, a standard scan tool and a communications protocol that the standard scan tool could use to interface with the vehicle of any manufacturer.

The standard also included mandatory structures and descriptions for certain emission control system/component defects. These were called 'P0' codes. Manufacturers were still free to generate their own 'manufacturer-specific code descriptions' known as 'P1' codes. This phase of implementation became known as OBD2 and was adopted for implementation by 1 January 1996.

KEY FACT

OBD2 was adopted for implementation by 1 January 1996.

5.2.3 P code composition

The DTC is displayed as a five-character alphanumeric code. The first character is a letter that defines which vehicle system set the code, be it powertrain, body or chassis.

- P means powertrain system set the code.
- B means body system set the code.
- C means chassis system set the code.
- U is currently unused but has been 'stolen' to represent communication errors.

P codes are requested by the microprocessor controlling the powertrain or transmission and refer to the emissions control systems and their components.

B codes are requested by the microprocessor controlling the body control systems. Collectively these are grouped as lighting, air conditioning, instrumentation or even in-car entertainment or telematics.

C codes are requested by the microprocessor controlling the chassis systems that control vehicle dynamics such as ride height adjustment, traction control, etc.

The four numbers that follow the letter detail information pertaining to what sub-system declared the code. Using the example from before, see Figure 5.6.

An integral feature of the OBD system is its ability to store fault codes relating to problems that occur with the engine electronic control system, particularly faults that could affect the emission control system, and this is

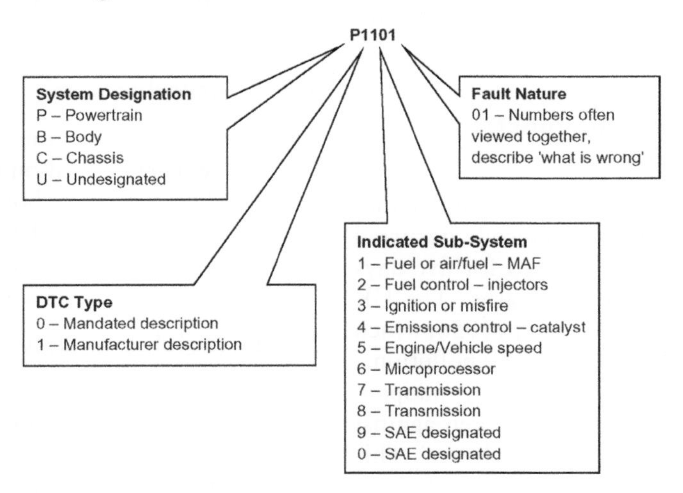

Figure 5.6 P-code composition.

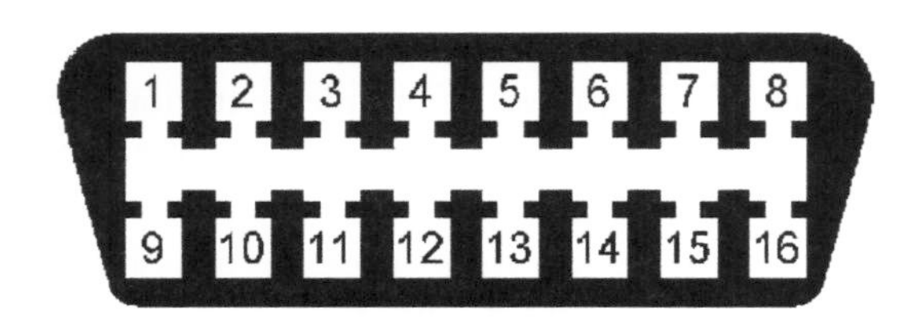

Figure 5.7 Sixteen pin DLC OBD2/EOBD connector.

one of its primary functions. For the diagnostic technician, this is a powerful feature which can clearly assist in locating and rectifying problems on the vehicle when they occur.

The diagnostic socket used by systems conforming to European OBD (EOBD)/OBD2 standards should have the following pin configuration (Figure 5.7):

1. ignition positive supply;
2. bus + line, SAE J1850 (PWM);
3. manufacturer's discretion;
4. chassis ground;
5. signal ground;
6. CAN bus H;
7. K-line;
8. manufacturer's discretion;
9. manufacturer's discretion;
10. bus – line (PWM);
11. manufacturer's discretion;
12. manufacturer's discretion;
13. manufacturer's discretion;
14. CAN bus L;
15. L-line or second K-line;
16. vehicle battery positive.

With the introduction of OBD2 and EOBD, this feature was made even more powerful by making it more accessible. Standardisation of the interface connector known as the diagnostic link connector (DLC) and communication protocol allowed the development of generic scan tools, which could be used on any OBD compliant vehicle.

5.2.4 European on-board diagnostics and global adoption

Europe was not immune to the environmental issues associated with smog. A major smog episode occurred in London in December 1952; this lasted for five days and resulted in approximately 4000 deaths. The UK government passed its first Clean Air Act in 1956, which aimed to control domestic sources of smoke pollution.

In 1970, the then European community adopted directive 70/220 EEC – 'Measures to be taken against Air Pollution by Emissions from Motor Vehicles'. This basically set the foundation for future legislation to curb motor vehicle pollution in Europe. This directive was amended over the next three decades when in October 1998 the amendment 98/69/EC 'On-Board Diagnostics (OBD) for Motor Vehicles' was adopted, which added Annex XI to the original 70/220 document. Annex XI details the functional aspects of OBD for motor vehicles in Europe and across the globe. This became known as EOBD.

5.2.5 Summary

A major contributing factor to environmental health issues in the United States was found to be motor vehicle emissions pollution. Scientific studies by government sponsored academic establishments and vehicle manufacturers then took place over several years. Legislative bodies were formed, which later developed and enacted vehicle emissions control legislation that forced vehicle manufacturers to develop control strategies and incorporate them within their production vehicles.

As microprocessor technologies became more advanced and commercially viable, the legislation was augmented to include a self-diagnosing OBD system, which would report when the emissions control system was unserviceable. First attempts by manufacturers to use such a system were applied unilaterally, which resulted in confusion, regenerative work and a poor reception of the OBD (now termed OBD1) concept. A revision of the legislation adopted SAE recommended standards, which resulted in the OBD (now termed OBD2) system becoming largely generic and applicable across the whole range of vehicle manufacturers.

As environmental activism spread across to Europe, vehicle manufacturers realised they had to support a philosophy of sustainable growth. Similar legislation was adopted and EOBD manifested itself in a form very similar to that observed in the United States.

5.3 PETROL/GASOLINE ON-BOARD DIAGNOSTIC MONITORS

5.3.1 Introduction

This section will cover the fundamentals of some of the OBD systems employed on mainstream petrol/gasoline vehicles. The concept of how the OBD system is divided into a series of software-based serviceability indicators, known as 'OBD monitors', is also covered.

5.3.2 Legislative drivers

In Europe, the European Directive 70/220 EEC was supplemented by European Directive 98/69/EC (Year: 1998: OJ Series: L – OJ Number: 350/1). This introduced legislation mandating the use of OBD systems in passenger vehicles manufactured and sold after 1 January 2001.

In the United States, legislation was first introduced in California by the California Air Resources Board (CARB) in 1988 and later federally by the Clean Air Act Amendments of 1990. This meant that the enforcing body, the EPA, requires that states have to develop state implementation plans (SIPs) that explain how each state will implement a plan to clean up pollution from sources including motor vehicles. One aspect of the requirement is the performance of OBD system checks as part of the required periodic inspection.

In order to be compliant with legislation and sell vehicles, manufacturers needed to engineer 'early warning' monitoring sub-systems that would determine when emission control systems had malfunctioned to the extent that tailpipe emissions had (or were likely to in the long term) exceeded a legislated level. OBD 'monitors' were derived for this purpose.

5.3.3 Component monitoring

The emission control systems integral with the vehicle employ many sensors and actuators. A software program housed within a microprocessor defines their actions.

The 'component monitor' is responsible for determining the serviceability of these sensors and actuators. Intelligent component drivers linked to the microprocessor have the ability to enable/disable sensors/actuators and to receive signals. The analogue inputs from the sensors are converted to digital values within the microprocessor.

In combination with these component drivers, the microprocessor possesses the functionality to detect circuit faults on the links between microprocessor and component. In addition, rationality tests can be performed to determine whether the sensor is operating out of range of its specification.

5.3.4 Rationality testing

Rationality tests can be performed on such sensors as the MAF sensor and throttle body. For example, the MAF is tested by observing its output value in comparison to a 'mapped' value normalised by throttle position and engine speed. The map or table contains expected MAF output values for the engine speed/throttle set point. Should the MAF output lie outside of an acceptable range (threshold) of values for that engine speed/throttle set point, then a fault is reported.

5.3.5 Circuit testing

The component monitor is capable of monitoring for circuit faults. Open circuits, short circuits to ground or voltage can be detected. Many manufacturers also include logic to detect intermittent errors.

5.3.6 Catalyst monitor

The purpose of the catalyst is to reduce tailpipe/exhaust emissions. The 'catalyst monitor' is responsible for determining the efficiency of the catalyst by inferring its ability to store oxygen. The method favoured most by the majority of manufacturers is to fit an oxygen sensor before and after the catalyst.

As the catalyst's ability to store oxygen (and hence perform three-way catalysis) deteriorates, the oxygen sensor downstream of the sensor will respond to the oxygen in the exhaust gas stream and its signal response will exhibit a characteristic similar to the upstream oxygen sensor (Figures 5.8 and 5.9).

An algorithm within the microprocessor analyses this signal and determines whether the efficiency of the catalyst has degraded beyond the point where the vehicle tailpipe emissions exceed legislated levels. If the microprocessor determines that this has occurred, then a malfunction and a DTC are reported. Repeat

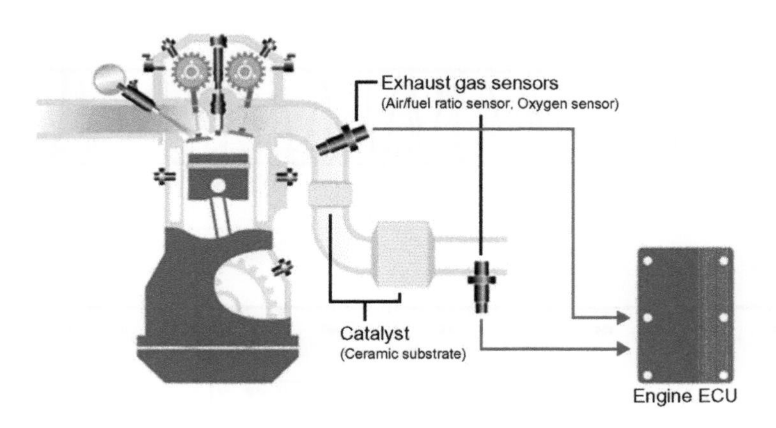

Figure 5.8 Exhaust gas oxygen sensors positioned pre- and post-catalyst.

Source: http://www.globaldensoproducts.com.

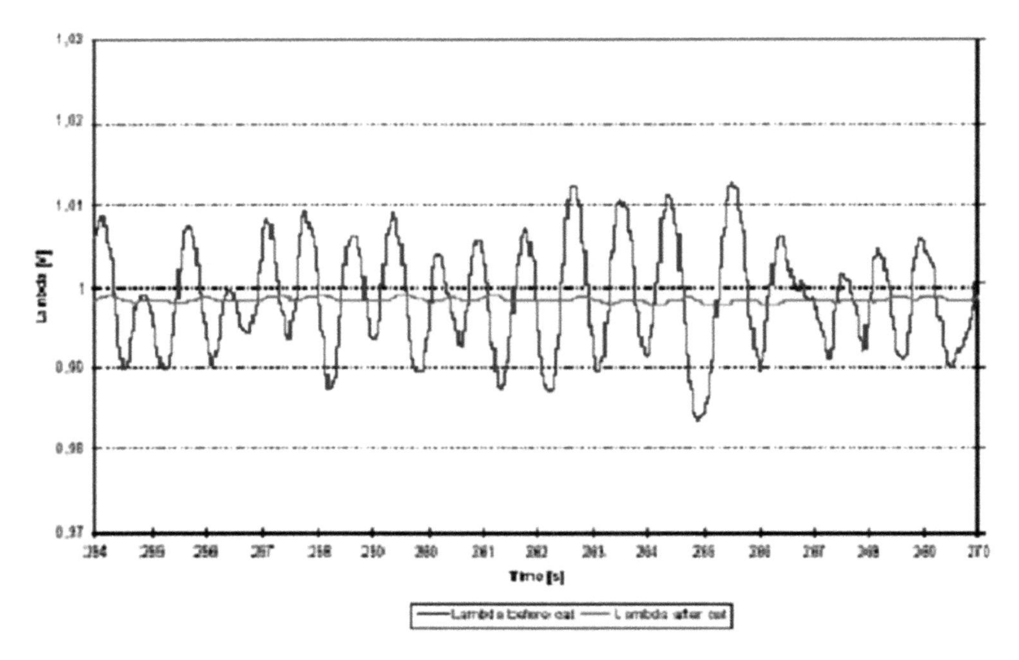

Figure 5.9 Upstream and downstream exhaust gas sensor activity – good catalyst.

Source: SAE 2001–01–0933 New Cat Preparation Procedure for OBD2 Monitoring Requirements.

detections of a failed catalyst will result in MIL illumination (Figure 5.10).

5.3.7 Evaporative system monitor

The purpose of the evaporative (EVAP) emissions control system is to store and subsequently dispose of unburned HC emissions, thus preventing them from entering the atmosphere. This is achieved by applying a vacuum across the fuel tank. The vacuum then causes fuel vapour to be drawn through a carbon canister in which the HC vapours are collected and stored.

During certain closed loop fuel control conditions, the microprocessor activates a solenoid-controlled 'vapour management valve'. This allows the manifold vacuum to draw vapour from the carbon canister along vapour

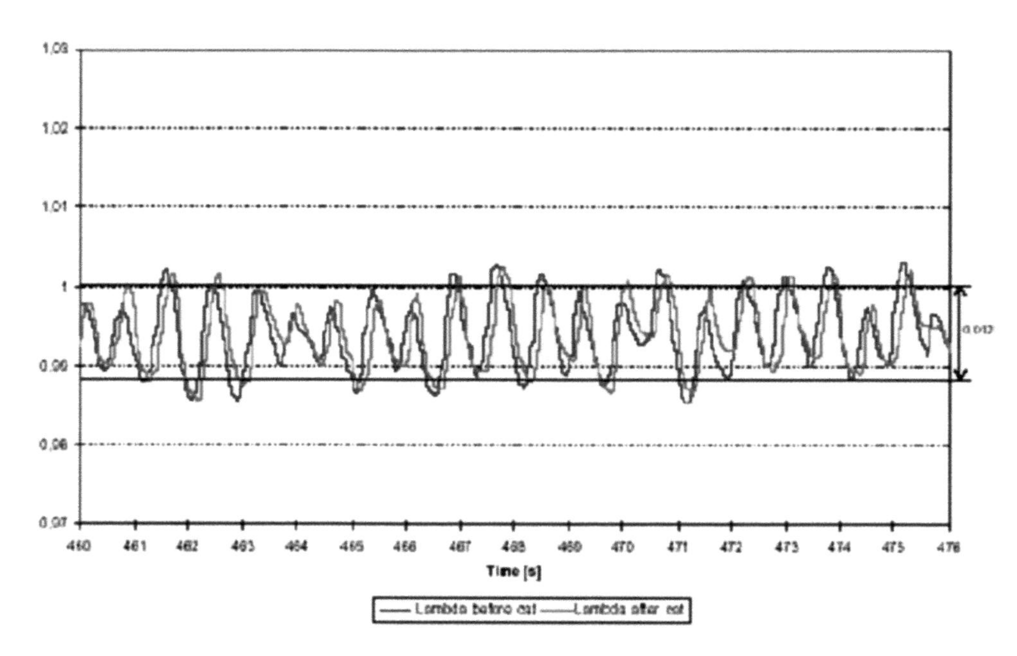

Figure 5.10 Upstream and downstream exhaust gas sensor activity – failed catalyst.

Source: SAE 2001–01–0933 New Cat Preparation Procedure for OBD2 Monitoring Requirements.

Evaporative emissions control system

1 Line from fuel tank to carbon canister. **2** Carbon canister. **3** Fresh air. **4** Canister-purge valve. **5** Line to intake manifold. **6** Throttle valve.

p_s Intake manifold pressure. p_u Atmospheric pressure. Δp Difference between intake manifold pressure and atmospheric pressure.

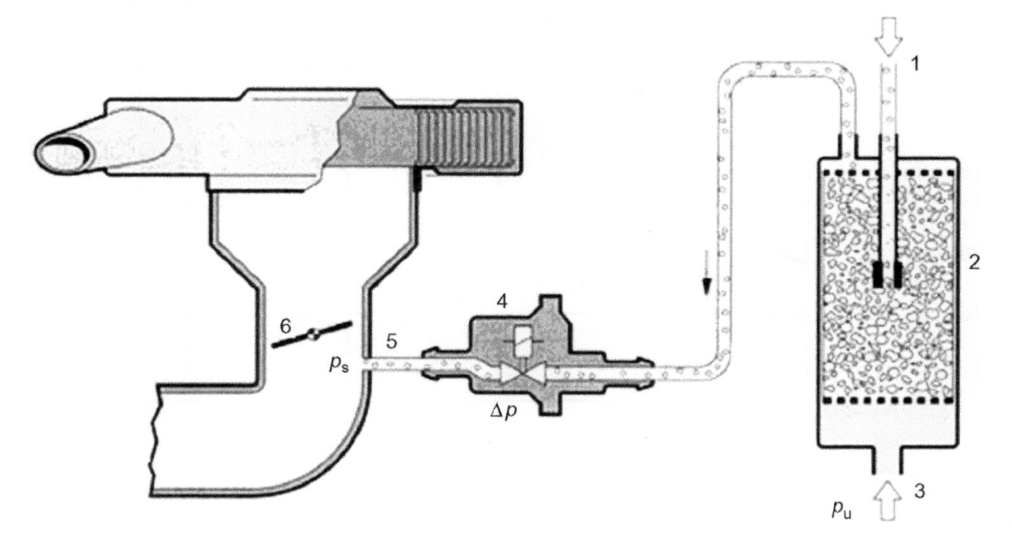

Figure 5.11 Evaporative emissions control system.

Source: Bosch.

lines, which terminate in the intake manifold. The fuel vapour is then combined and combusted with the standard air/fuel charge; the closed loop fuel control system caters for the additional AFR enrichment to ensure that stoichiometric fuelling continues (Figure 5.11).

The evaporative system monitor is responsible for determining the serviceability of the EVAP system components and detecting leaks in the vapour lines. Most manufacturers check for fuel vapour leaks by employing a diagnostic that utilises a pressure or vacuum test on the fuel system.

European legislation dictates that these checks are not required. However, vehicles manufactured in the United States after 1996 and before 1999 generally employ a system that uses a pressure or vacuum system. This must be able to detect a leak in a hose or filler cap that is equivalent to that generated by a hole, which is 0.040 inch (1 mm) in diameter. Vehicles manufactured after 2000 must support diagnostics that are capable of detecting a 0.020 inch (0.5 mm) hole.

5.3.8 Fuel system monitoring

As vehicles accumulate mileage so also do the components, sensors and actuators of the emissions control systems. MAF sensors become dirty and their response slows with age. Exhaust gas oxygen sensors also respond slower as they are subject to the in-field failure modes such as oil and fuel contamination, thermal stress and general ageing. Fuel pressure regulators perform outside of their optimum capacity; fuel injectors become slower in their response; and partial blockages mean that they deliver less and sometimes more fuel than requested.

If this component ageing were not compensated for, it would mean that the fuel system would not be able to maintain normal fuelling around stoichiometric AFR as shown in Figure 5.12. The end result would be the potential to exceed emission limits. A lambda value of 1 is required in order for the three-way catalyst to work. In addition to this, more severe fuelling errors would cause noticeable effects in the performance of the vehicle leading to customer complaints and potential to damage the manufacturer's brand image. This compensation strategy is known as adaptive learning. A dedicated piece of software contained within the electronic control unit (ECU) learns these deviations away from stoichiometry, while the fuel system is in closed loop control. They are stored in a memory that is only reset when commanded by a technician and which is also robust to battery changes.

These memory-stored corrections are often termed 'long-term' fuel corrections. They are often stored in memory as a function of air mass, engine speed or engine load.

An exhaust gas oxygen sensor detects the amount of oxygen in the catalyst feed gas and the sensor produces a voltage, which is fed back to the microprocessor. This is then processed to determine the instantaneous or 'short-term' fuel correction to be applied. This is

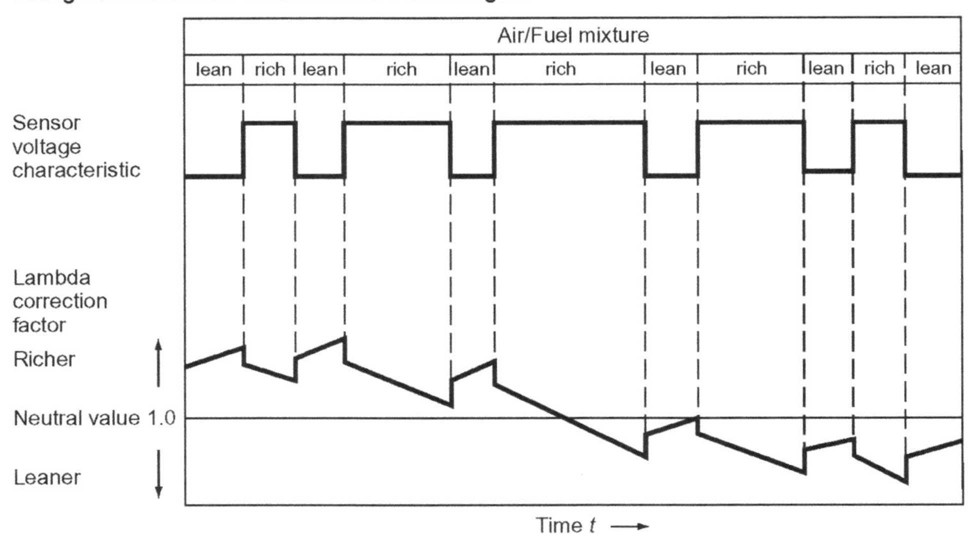

Figure 5.12 Rich AFR lambda sensor signal fuelling error.

Source: Bosch.

done in order to vary the fuel around stoichiometry and allow three-way catalysis to occur.

The microprocessor then calculates the amount of fuel required using an equation, which is shown here in its most basic form (Figure 5.13).

$$\text{Fuel mass} = \frac{\text{air mass} \times \text{long-term fuel trim}}{\text{short-term fuel trim} \times 14.64}$$

Referring to Figures 5.12 and 5.13, it can be seen that when there is a component malfunction, which causes the AFR in the exhaust stream to be rich, then there is a need to adapt to this to bring the AFR back into the region of stoichiometry. The value of the long-term fuel trim correction must decrease because less fuel is required.

Should the situation continue and the problem causing rich AFR become slowly worse, the error adaption will continue with an ever-decreasing value for the long-term fuel trim being applied, learned and stored in memory.

The purpose of the fuel monitor is to determine when the amount of long-term adaptive correction

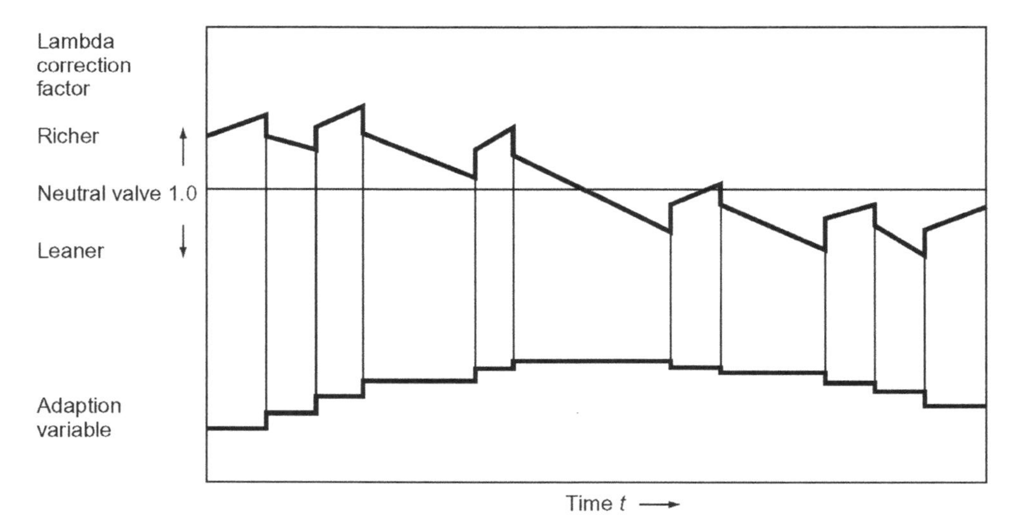

Figure 5.13 Adaptive fuel strategy in operation.

Source: Bosch.

has reached the point where the system can no longer cope. This is also where long-term fuel trim values reach a pre-defined or 'calibrated' limit at which no further adaption to error is allowed. This limit is calibrated to coincide with exhaust tailpipe emissions exceeding legislated levels. At this point and when a short-term fuelling error exceeds another 'calibrated' limit, a DTC is stored, and after consecutive drives, the MI is illuminated.

The opposite occurs, with extra fuel being added, via the long-term fuel trim parameter, should an error occur that causes the AFR at the exhaust gas oxygen sensor to be lean.

5.3.9 Exhaust gas recirculation monitor

As combustion takes place within the engine cylinders, nitrogen and oxygen combine to form various oxides of nitrogen, collectively termed as NO_x. NO_x emissions can be reduced up to a certain point by enriching the AFR, beyond the point at which HC and CO emissions begin to increase. NO_x emissions are generated as a function of combustion temperature, so another way to reduce these is to decrease the compression ratio which leads to other inefficiencies like poor fuel economy.

Most manufacturers employ an emissions control sub-system known as exhaust gas recirculation (EGR). This by definition recirculates some of the exhaust gases back into the normal intake charge. These 'combusted' gases cannot be burnt again so they act to dilute the intake charge. As a result, in-cylinder temperatures are reduced along with NO_x emissions (Figure 5.14).

The EGR system monitor is responsible for determining the serviceability of the sensors, hoses, valves and actuators that belong to the EGR system. Manufacturers employ systems that can verify that the requested amount of exhaust gas is flowing back into the engine intake. Methods can be both intrusive and non-intrusive, such as a change in manifold pressure as EGR flows and is then shut off.

One method monitors AFR excursions after the EGR valve is opened, and then closed as the AFR becomes lean. Another system employs a differential pressure scheme that determines the pressure both upstream and downstream of the exhaust to determine whether the requested flow rate is in effect. Yet another system employs a temperature sensor, which reports the change in temperature as EGR gases flow past the sensor. The temperature change will be mapped against the amount of EGR flowing, so when an amount of EGR

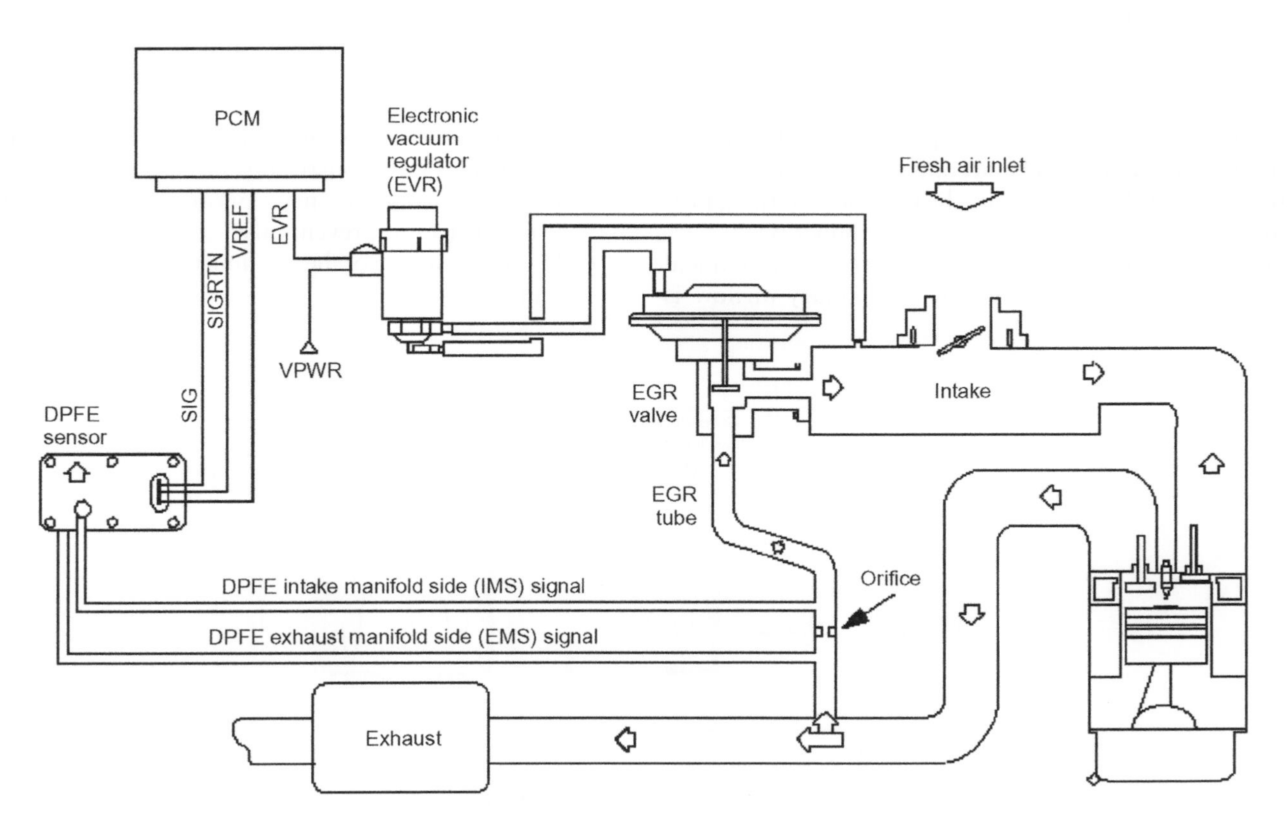

Figure 5.14 EGR system using differential pressure monitoring.

Source: Ford Motor Company.

is requested, the flow rate is inferred by measuring the change in temperature.

5.3.10 Secondary air monitor

The exhaust system catalyst is not immediately operative following a start where the engine and exhaust system is cool. Temperature thresholds above which the catalyst is working, and three-way catalysis is occurring, vary as a function of the exhaust gas system package. Typically, this 'light off' point occurs at temperatures of approximately 260°C/500°F. Some manufacturers employ electrically heated catalysts to reach this temperature rapidly, but these are expensive to manufacture and replace.

Most manufacturers rely on the exhaust gases as a source of heat in order to bring the catalyst up to light off temperature. When the vehicle is started from cold, the AFR is rich; this is required to ensure a stable engine start for cold pull-away. From an emissions perspective, the impact is observed in the production of HC and CO in the exhaust stream because the exhaust system catalyst has not reached light off.

DEFINITION

Catalyst light off temperature is the point at which it starts to operate fully.

The secondary air system uses a pump, which adds more air into the exhaust stream at a point before the catalyst follows a cold start. The secondary air combusts the HC in the catalyst, generating heat, which, in turn, promotes light off and further emissions reduction.

Older systems support a belt-driven mechanical pump with a bypass valve when secondary airflow is not required. Modern vehicles employ an electric air pump operated by the engine-management ECU [powertrain control module (PCM)] via relays.

The secondary air monitor is responsible for determining the serviceability of the secondary air system components. Most strategies monitor the electrical components and ensure the system pumps air when requested by the ECU. To check the airflow, the ECU observes the response of the exhaust gas oxygen sensor after it commands the fuel control system to enter open-loop control and force the AFR to become rich. The secondary air pump is then commanded on and the ECU determines the time taken for the exhaust gas oxygen sensor to indicate a lean AFR. If this time exceeds a calibrated threshold, a DTC is stored (Figure 5.15).

5.3.11 Monitors and readiness flags

An important part of any OBD system is the system monitors and associated readiness flags. These readiness flags indicate when a monitor is active. Certain monitors are continuous, for example, misfire and fuel system monitors.

Monitor status (ready/not ready) indicates if a monitor has completed its self-evaluation sequence. System monitors are set to 'not ready' if cleared by scan tool and/or the battery is disconnected. Some of the monitors must test their components under specific, appropriate preconditions:

- The evaporative system monitor has temperature and fuel fill level constraints.
- The misfire monitor may ignore input on rough road surfaces to prevent false triggers.
- The oxygen sensor heater must monitor from a cold start.

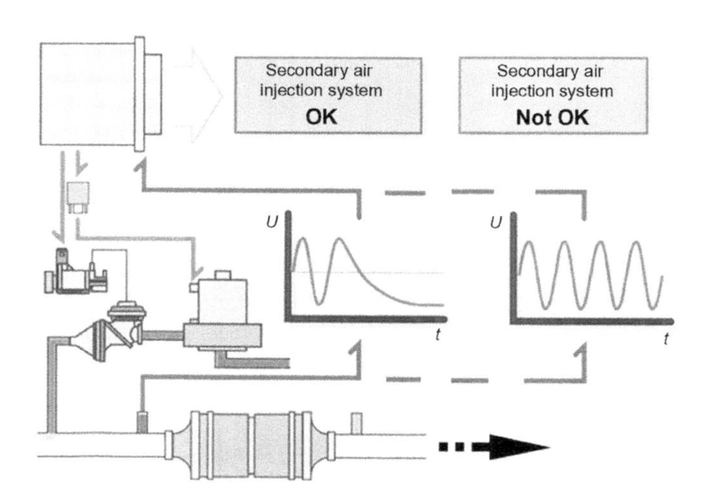

Figure 5.15 Secondary airflow diagnostic monitoring.

Insert | Remove | Advanced | ? | X

Parameter	Value	Unit
Calculated Engine Load	28.6	%
Catalyst Monitoring Status	Complete	
Comprehensive Component Monitoring Status	Complete	
EGR System Monitoring Status	Complete	
Engine Coolant Temperature	69	°C
Engine Speed	780	RPM
Fuel System Monitoring Status	Complete	
Fuel System Status Bank 1	CL-1	
Ignition Timing Advance	17.5	Deg
Intake Air Temperature	24	°C
Long Term Fuel Trim Bank 1	-3.1	%
Manifold Absolute Pressure (MAP)	27	kPA
Misfire Monitoring Status	Complete	
O2 Sensor - Bank 1 Sensor 1 (mV)	370	mV
O2 Sensor - Bank 1 Sensor 2 (mV)	150	mV
O2 Sensor Heater Monitoring Status	Complete	
O2 Sensor Monitoring Status	Complete	
OBD Requirements	EOBD	
Short Term Fuel Trim Bank 1	5.5	%
Short Term Fuel Trim from O2 Bank 1 Sensor 1	7	%
Short Term Fuel Trim from O2 Bank 1 Sensor 2	99.2	%
Throttle Position Angle	17.6	%
Vehicle Speed	0	MPH

Figure 5.16 System monitors (marked as 'Complete') and live data shown in scan tool.

Most other system monitors are not continuous and are only active under certain conditions. If these conditions are not fulfilled, then the readiness flag for that monitor is set to 'not ready'. Until the readiness flags are set appropriately, it is not possible to perform a test of the OBD system and its associated components (Figure 5.16).

There is no universal drive cycle that is guaranteed to set all the system monitors appropriately for a test of the OBD system. Most manufacturers and even cars have their own specific requirements, and irrespective of this, there are still some specific vehicles that have known issues when trying to set readiness flag status. To allow for this vehicles of model year 1996–2000 are allowed two readiness flags to be 'not ready'. After this, 2001 onwards, one readiness flag is allowed to be 'not ready' prior to a test.

5.4 MISFIRE DETECTION

5.4.1 Misfire monitor

When an engine endures a period of misfire, at best tailpipe emissions will increase and at worst catalyst damage and even destruction can occur. When misfire occurs, the unburned fuel and air is discharged direct to the exhaust system where it passes directly through the catalyst.

KEY FACT

When a misfire occurs, unburned fuel and air pass through the catalyst and can cause damage.

Subsequent normal combustion events can combust this air/fuel charge in something akin to a bellows effect, which causes catalyst temperatures to rise considerably. Catalyst damage failure thresholds are package specific but are in the region of 1000°C. The catalyst itself is a very expensive service item whether replaced by the customer or the manufacturer under warranty.

The misfire monitor is responsible for determining when misfire has occurred, calculating the rate of engine misfire and then initiating some kind of protective action in order to prevent catalyst damage.

The misfire monitor is in operation continuously within a 'calibrateable' engine speed/load window defined by the legislation. The United States requires misfire monitoring throughout the revs range but European legislation requires monitoring only up to 4500 rpm (Figure 5.17).

The crankshaft sensor generates a signal as the wheel rotates and the microprocessor processes this signal to determine the angular acceleration of the crankshaft produced by each engine cylinder when a firing event occurs. When a misfire occurs, the crankshaft decelerates and a cam position sensor identifies the cylinder that misfired.

Processing of the signal from the crank position sensor is not straightforward. A considerable amount of post-processing takes place to filter the signal and disable monitoring in unfavourable conditions. The misfire monitor must learn and cater for the differences in manufacturing tolerances of the crankshaft wheel and so has a specific sub-algorithm for learning these

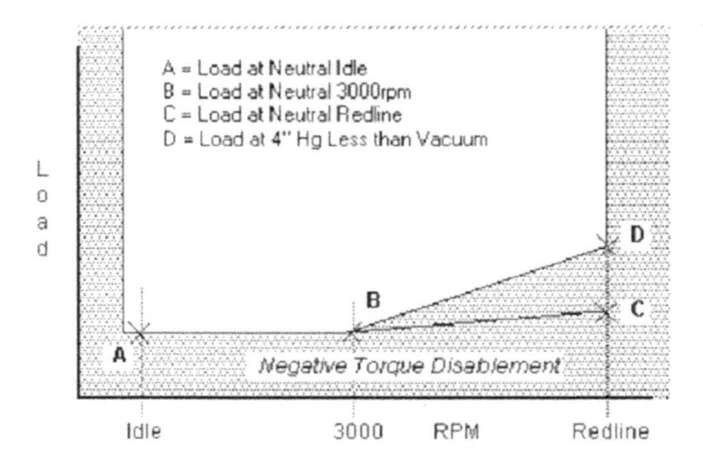

Figure 5.17 Misfire enablement window.

Source: Ford Motor Company.

differences and allowing for them when calculating the angular acceleration of the crankshaft (Figure 5.18). These correction factors are calculated during deceleration, with the injectors switched off. They should be re-learned following driveline component changes such as flywheel, torque converter, crankshaft sensor, etc.

The misfire monitor must be able to detect two types of misfire:

- Type A misfire;
- Type B misfire.

A type A misfire is defined as that rate of misfire which causes catalyst damage. When this occurs, the MI will flash at a rate of 1 Hz and is allowed to stop flashing should the misfire disappear. The MI will stay on steady state should the misfire reoccur on a subsequent drive and the engine operating conditions are 'similar', that is, engine speed is within 375 rpm, engine load is within 20% and the engine's warm-up status is the same as that under which the malfunction was first detected (and no new malfunctions have been detected).

The rate of misfire that will cause catalyst damage varies as a function of engine speed and load. Misfire rates in the region of 45% are required to damage a catalyst at neutral idle, while at 80% engine load and 4000 rpm, misfire rates in the region of only 5% are needed (Figure 5.19).

A type B misfire is defined as that rate of misfire which will cause the tailpipe emissions to exceed legislated levels. This varies from vehicle to vehicle and is dependent upon catalyst package. MI operation is the same as for standard DTCs.

The above is the most common method but misfires can be detected in a number of different ways as outlined in the following sections.

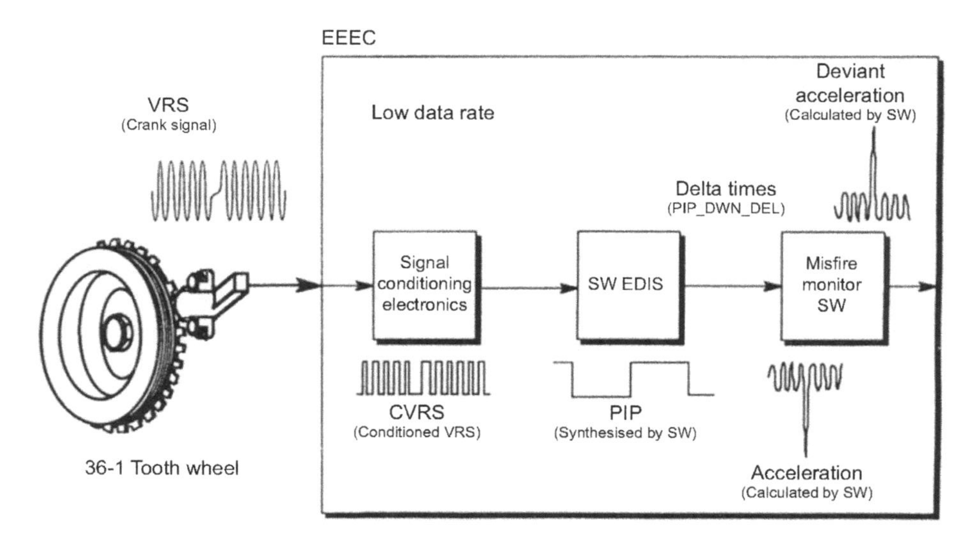

Figure 5.18 Crankshaft mounted wheel and sensor source of angular acceleration.

Source: Ford Motor Company.

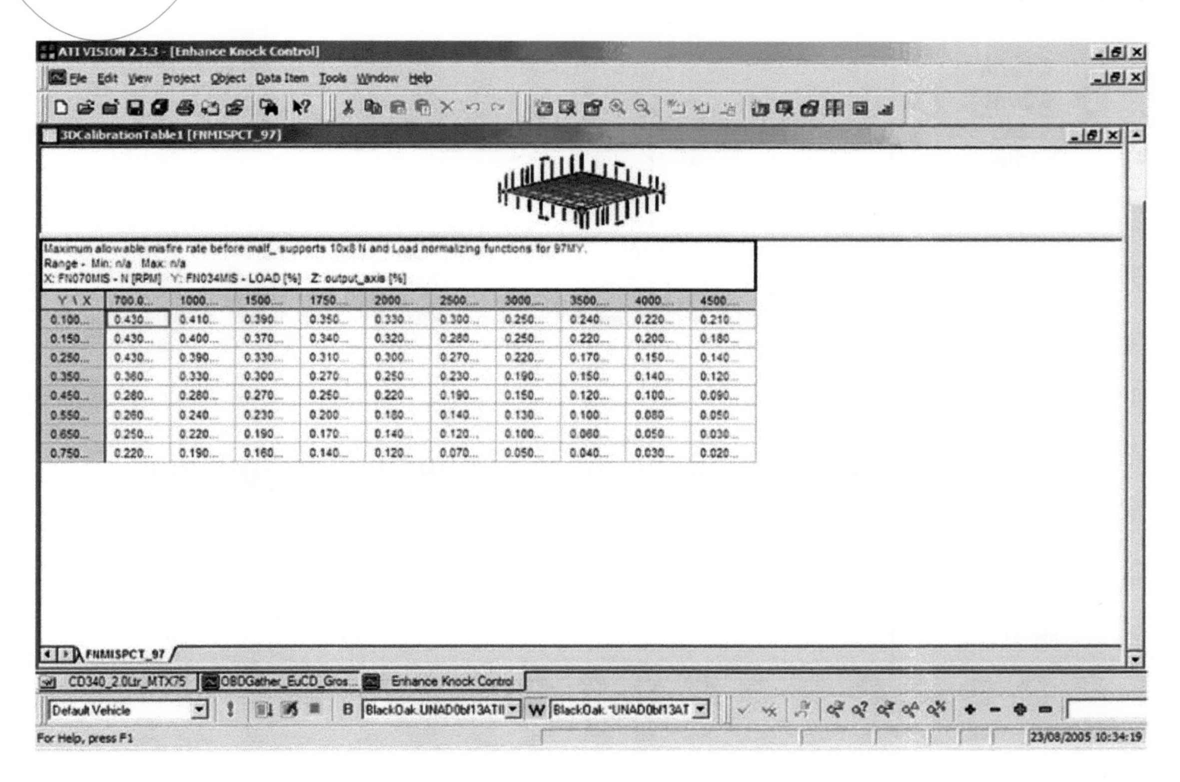

Y \ X	700.0...	1000....	1500....	1750....	2000....	2500....	3000....	3500....	4000....	4500....
0.100...	0.430...	0.410...	0.390...	0.350...	0.330...	0.300...	0.250...	0.240...	0.220...	0.210...
0.150...	0.430...	0.400...	0.370...	0.340...	0.320...	0.280...	0.250...	0.220...	0.200...	0.180...
0.250...	0.430...	0.390...	0.330...	0.310...	0.300...	0.270...	0.220...	0.170...	0.150...	0.140...
0.350...	0.360...	0.330...	0.300...	0.270...	0.250...	0.230...	0.190...	0.150...	0.140...	0.120...
0.450...	0.280...	0.280...	0.270...	0.250...	0.220...	0.190...	0.150...	0.120...	0.100...	0.090...
0.550...	0.260...	0.240...	0.230...	0.200...	0.180...	0.140...	0.130...	0.100...	0.080...	0.050...
0.650...	0.250...	0.220...	0.190...	0.170...	0.140...	0.120...	0.100...	0.060...	0.050...	0.030...
0.750...	0.220...	0.190...	0.160...	0.140...	0.120...	0.070...	0.050...	0.040...	0.030...	0.020...

Figure 5.19 System development screen showing type A misfire rates normalised by engine speed and load.

Source: Ford Motor Company.

5.4.2 Crank speed fluctuation

A misfire event in a cylinder results in a lost power stroke. The gap in the torque output of the engine and a consequential momentary deceleration of the crankshaft can be detected using the crankshaft position sensor. By closely monitoring the speed and acceleration of the crankshaft, misfiring cylinders can be detected; this technology is very commonly used in OBD systems to detect non-firing cylinders that can cause harmful emissions and catalyst damage (Figure 5.20).

There are a number of technical challenges that have to be overcome with this technique; the accuracy achieved and reliability of the system is very dependent on the algorithms used for signal processing and analysis. Under certain conditions, misfire detection can be difficult, particularly at light load with high engine speed. Under these conditions, the damping of firing pulses is low due to the light engine load, and this creates high momentary accelerations and decelerations of the crankshaft.

This causes speed variation which can be mistakenly taken by the OBD system as a misfire. With this method of misfire detection, careful calibration of the OBD system is necessary to avoid false detection. Another vehicle operation mode which can cause problems is operation of the vehicle on rough or poorly made roads. This also causes rapid crankshaft oscillation that could activate false triggers, and under these conditions the misfire detection must be disabled.

5.4.3 Ionising current monitoring

An ionisation current sensing ignition system consists of one ignition coil per cylinder, normally mounted directly above the spark plug. Eliminating the distributor and high-voltage leads helps promote maximum energy transfer to the spark plug to ignite the mixture. In this system, the spark plug is not only used as a device to ignite the air/fuel mixture but is also used as an in-cylinder sensor to monitor the combustion process. The operating principle used in this technology is that an electrical current flow in an ionised gas is proportional to the flame electrical conductivity. By placing a direct current bias across the spark plug electrodes, the conductivity can be measured. The spark current is used to create this bias voltage and this eliminates the requirement for any additional voltage source.

The ion current is monitored, and if no ion-generating flame is produced by the spark, no current flows through the measurement circuit during the working part of the cycle. The ion current versus time trace is very different from that of a cycle when normal combustion occurs, and this information can be used as a differentiator to detect misfire from normal combustion. This method

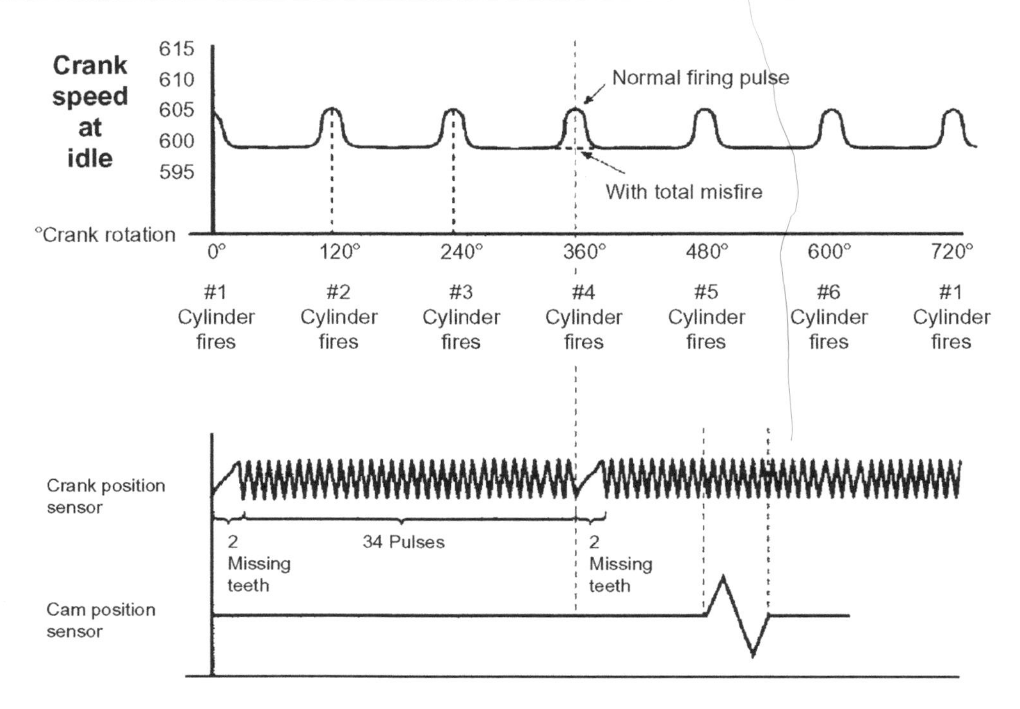

Figure 5.20 Misfire detection via crank sensor.

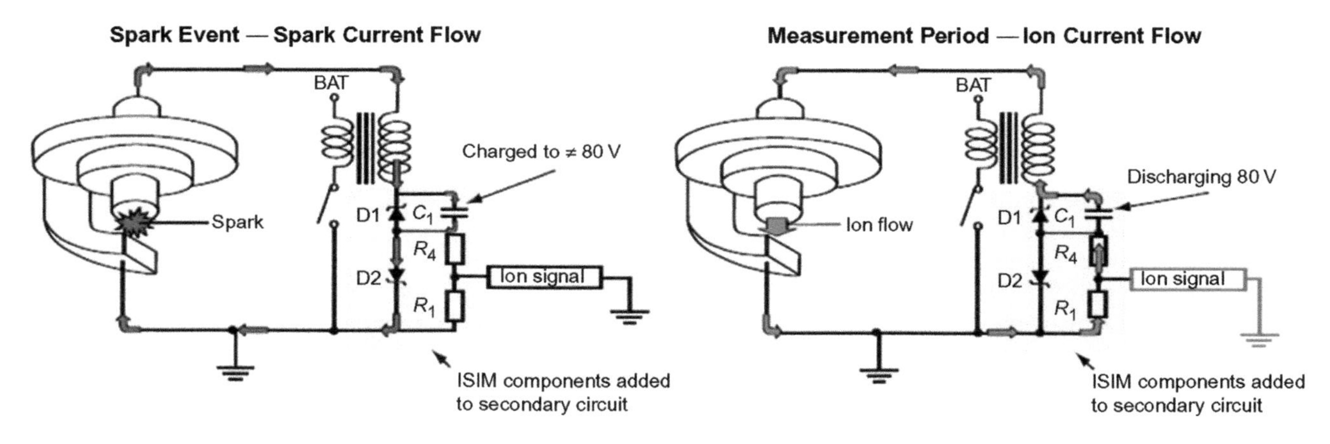

Figure 5.21 Ion-sensing circuit in direct ignition system.

has proven to be very effective at monitoring for misfires under test conditions and also in practice.

The signal the system produces contains misfire information and, in addition, can provide objective knock or detonation information. This can be used for engine control systems where knowledge of the actual combustion process is required (as mentioned above) (Figures 5.21 and 5.22).

5.4.4 Cylinder pressure sensing

This technology has great potential, not just for OBD applications but also for additional feedback to the engine-management system about the combustion process due to the direct measurement technique (Figure 5.23). This additional control dimension can be utilised to improve engine performance and reduce emissions further. With respect to misfire detection, this method provides reliable detection of a positive combustion event and can easily detect misfire with utmost reliability.

The major drawback is the availability of suitable sensors that could be installed into the engine at production and would be durable enough to last the life of the engine and provide the required performance expected of sensors in an OBD system. For certain

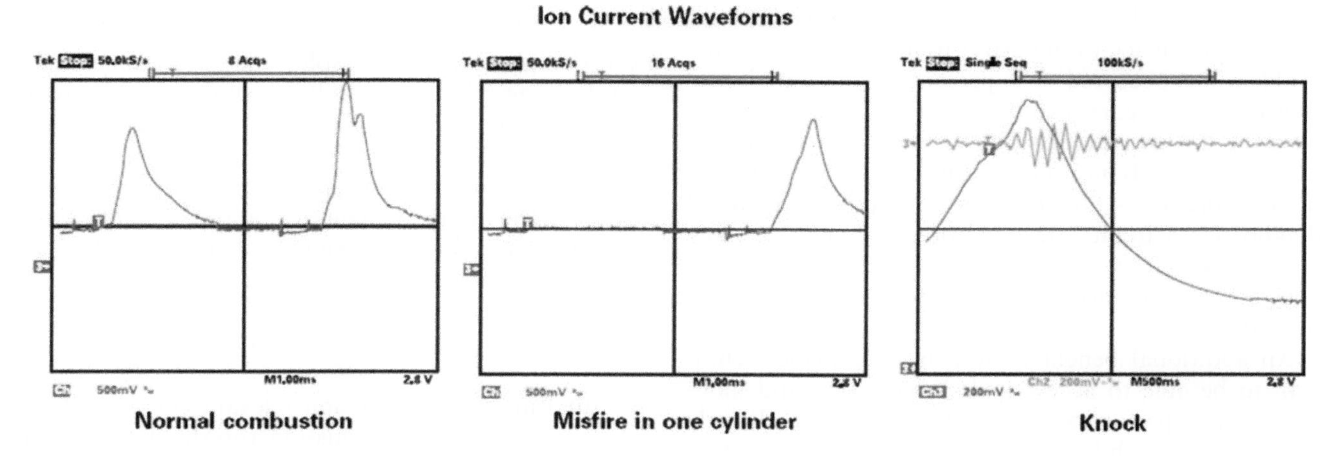

Figure 5.22 Resulting waveforms from the ion-sensing system.

engine applications, sensors are available, and currently combustion sensor technology is under rapid development such that this technical hurdle will soon be overcome.

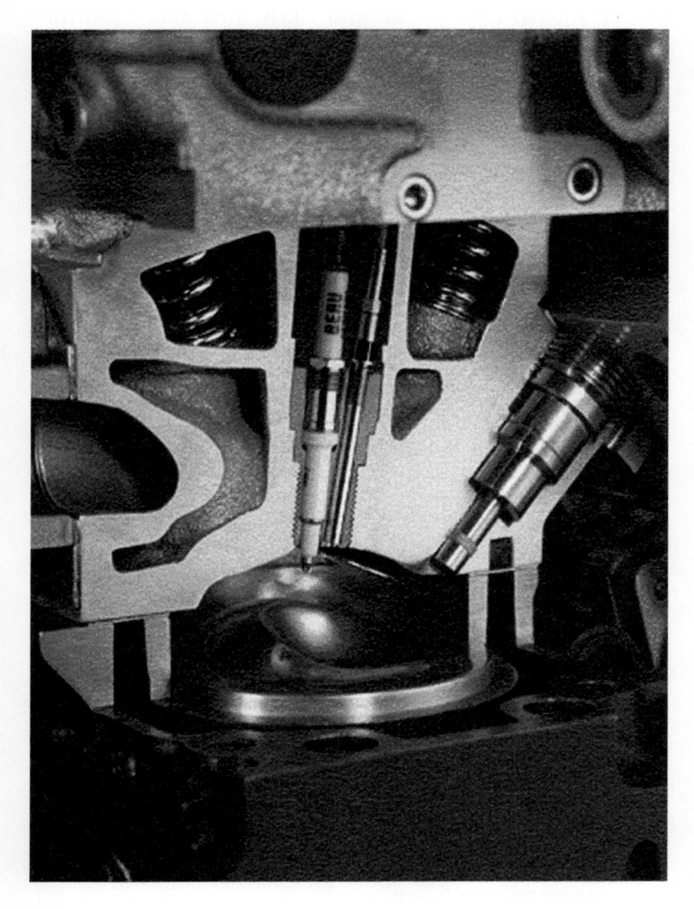

Figure 5.23 Cylinder pressure sensor mounted in the engine.

5.4.5 Exhaust pressure analysis

This solution involves using a pressure sensor in the exhaust manifold combined with a Fourier analysis as the first stage of the signal processing. Using a sensor to analyse the gas pulses in the exhaust manifold, it is possible to detect single misfires, and additionally, it is possible to identify which cylinder is misfiring. This method is less intrusive than the above and could potentially be retrofitted at the production stage. A sensor in the exhaust can detect misfiring cylinders but cannot give useful, qualitative information about the combustion process. This technique has been demonstrated as capable of detecting all misfires at engine speeds up to 6000 rpm, for all engine configurations, loads and fuels. Generally, a ceramic capacitive–type sensor has been employed, which has a short response time and good durability.

5.5 OBD OVERVIEW

5.5.1 Introduction

OBD monitoring applies to systems which are most likely to cause an increase in harmful exhaust emission, namely:

- all main engine sensors;
- fuel system;
- ignition system;
- EGR system.

The system uses information from sensors to judge the performance of the emission controls, but these sensors do not directly measure the vehicle emissions.

An important part of the system, and the main driver information interface, is the 'check engine' warning light, also known as the MIL. This is the main source of feedback to the driver to indicate if an engine problem has occurred or is present. When a malfunction or fault occurs, the warning light illuminates to alert the driver. Additionally, the fault is stored in the ECU memory. If normal condition is reinstated, the light extinguishes but the fault remains logged to aid diagnostics. Circuits are monitored for open or short circuits as well as plausibility. When a malfunction is detected, information about the malfunctioning component is stored.

An additional benefit allows the diagnostic technician to be able to access fault information and monitor engine performance via data streamed directly from the ECU while the engine is running (on certain vehicles). This information can be accessed via various scan tools available on the market and is communicated in a standardised format, so one tool (more or less!) works with all vehicles. The data is transmitted in a digital form via this serial interface. Thus, data values are transmitted as data words and the protocol used for this data stream has to be known in order to evaluate the information properly.

The benefits of having an OBD system are that it

- encourages vehicle and engine manufacturers to have a responsible attitude to reducing harmful emissions from their engines via the development of reliable and durable emission control systems;
- aids diagnosis and repair of complex electronic engine and vehicle control systems;
- reduces global emissions by identifying and highlighting immediately to the driver or user emission control systems in need of repair;
- provides 'whole life' emission control of the engine.

On-board diagnostics, or OBD, was the name given to the early emission control and engine-management systems introduced in cars. There was no single standard – each manufacturer often uses quite different systems (even between individual car models). OBD systems have been developed and enhanced, in line with United States government requirements, into the current OBD2 standard. The OBD2 requirement applies to all cars sold in the United States from 1996. EOBD is the European equivalent of the American OBD2 standard, which applies to petrol cars sold in Europe from 2001 (and diesel cars three years later).

KEY FACT

OBD2 (also OBDII) was developed to address the shortcomings of OBD1 and make the system more user friendly for service and repair technicians.

5.5.2 OBD2

Even though new vehicles sold today are cleaner than they have ever been, the millions of cars on the road and the ever-increasing miles they travel each day make them our single greatest source of harmful emissions. While a new vehicle may start out with very low emissions, infrequent maintenance or failure of components can cause the vehicle emission levels to increase at an undesirable rate. OBD2 works to ensure that the vehicles remain as clean as possible over their entire life. The main features of OBD2 are, therefore, as follows:

- malfunction of emission relevant components to be detected when emission threshold values are exceeded;
- storage of failures and boundary conditions in the vehicle's fault memory;
- diagnostic light (MIL) to be activated in case of failures;
- readout of failures with generic scan tool.

The increased power of micro controllers (CPUs) in ECUs has meant that a number of important developments could be added with the introduction of OBD2. These include catalyst efficiency monitoring, misfire detection, canister purge and EGR flow rate monitoring. An additional benefit was the standardisation of diagnostic equipment interfaces.

For OBD1, each manufacturer applied specific protocols. With the introduction of OBD2, a standardised interface was developed with a standard connector for all vehicles and a standardised theory for fault codes relating to the engine and powertrain (more about this later). This meant that generic scan tools could be developed and used in the repair industry by diagnostic technicians to aid troubleshooting of vehicle problems.

Another feature of OBD2 is that the prescribed thresholds at which a fault is deemed to have occurred are in relation to regulated emission limits. The basic monitor function is as follows:

- monitoring of catalyst efficiency, engine misfire and oxygen sensors function such that crossing a threshold of 1.5 times the emission limit will record a fault;
- monitoring of the evaporation control system such that a leak greater than the equivalent leak from a 0.04 inch hole will record a fault.

The main features of an OBD2 compliant system (as compared to OBD1) are as follows (Figure 5.24):

- pre- and post-catalyst oxygen sensors to monitor conversion efficiency;
- much more powerful ECU with 32 bit processor;

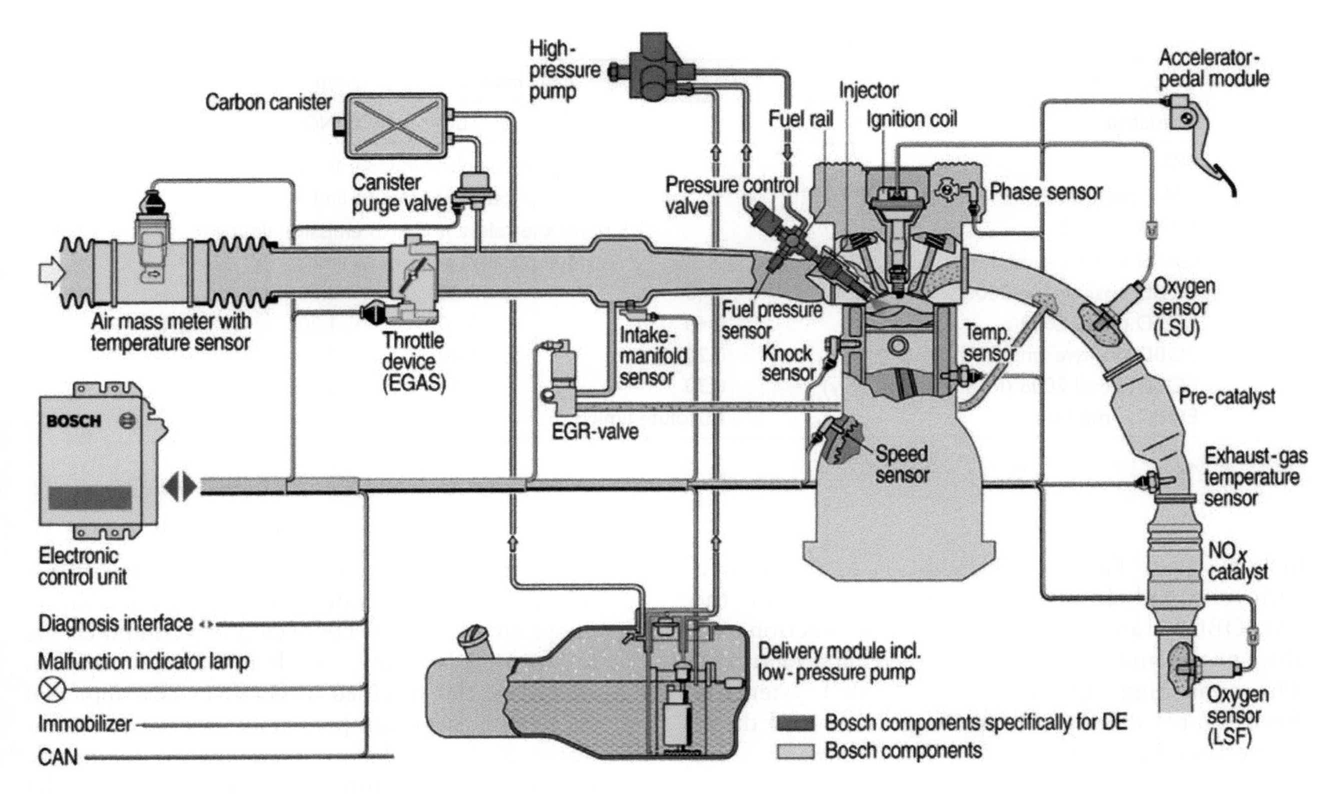

Figure 5.24 OBD2 system showing the main components of a gasoline direct injection system.

Source: Bosch Media.

- ECU map data held on EEPROMs such that they can be accessed and manipulated via an external link; no need to remove ECU from vehicle for software updates or tuning;
- more sophisticated EVAP system, can detect minute losses of fuel vapour;
- EGR systems with feedback of position/flow rate;
- sequential fuel injection with MAP and MAF sensing for engine load.

5.5.3 EOBD

EOBD is an abbreviation of European on-board diagnostics. All petrol/gasoline cars sold in Europe since 1 January 2001 and diesel cars manufactured from 2003 must have OBD systems to monitor engine emissions. These systems were introduced in line with European directives to monitor and reduce emissions from cars. All such cars must also have a standard EOBD diagnostic socket that provides access to this system. The EOBD standard is similar to the US OBD2 standard. In Japan, the JOBD system is used. The implementation plan for EOBD was as follows:

- January 2000 OBD for all new petrol/gasoline vehicle models
- January 2001 OBD for all new petrol/gasoline vehicles
- January 2003 OBD for all new diesel vehicle models PC/LDV
- January 2004 OBD for all new diesel vehicles PC/LDV
- January 2005 OBD for all new diesel vehicles HDV.

The EOBD system is designed, constructed and installed in a vehicle such as to enable it to identify types of deterioration or malfunction over the entire life of the vehicle. The system must be designed, constructed and installed in a vehicle to enable it to comply with the requirements during conditions of normal use.

DEFINITION

EOBD: European on-board diagnostics.

In addition, EOBD and OBD2 allow access to manufacturer-specific features available on some OBD2/EOBD compliant scan tools. This allows additional parameters or information to be extracted from the

Table 5.3 Emission limits table for comparison

Legislation	OBD malfunction limit (g/km)			
	HC	CO	NO_x	PM
EPA	≥1.5 times the applicable federal standard			
EPA – method	Multiplicative relative to limits			
CARB 1 and 2	≥1.5 times the relevant CARB emission limits			
CARB 1 and 2 – method	Multiplicative relative to limits			
EOBD positive ign. 2000	0.40	3.20	0.60	–
EOBD diesel 2003	0.40	3.20	1.20	0.18
EOBD positive ign. 2005	0.20	1.40	0.30	–
EOBD diesel 2008 (for indication only)	0.30	2.40	0.90	0.14
EOBD – method	Absolute limits			

vehicle systems. These are in addition to the normal parameters and information available within the EOBD/OBD2 standard. These enhanced functions are highly specific and vary widely between manufacturers.

The monitoring capabilities of the EOBD system are defined for petrol/gasoline (spark ignition) and diesel (compression ignition) engines. The following is an outline:

Spark ignition engines

- Detection of the reduction in the efficiency of the catalytic converter with respect to emissions of HC only.
- The presence of engine misfires in the engine operation region within the following boundary conditions.
- Oxygen sensor deterioration.
- Other emission control system components or systems, or emission-related powertrain components or systems which are connected to a computer, the failure of which may result in tailpipe emission exceeding the specified limits.
- Any other emission-related powertrain component connected to a computer must be monitored for circuit continuity.
- The electronic evaporative emission purge control must, at a minimum, be monitored for circuit continuity.

Compression ignition engines

- Where fitted, reduction in the efficiency of the catalytic converter.
- Where fitted, the functionality and integrity of the particulate trap.
- The fuel injection system electronic fuel quantity and timing actuator(s) is/are monitored for circuit continuity and total function failure.
- Other emission control system components or systems, or emission-related powertrain components or systems which are connected to a computer, the failure of which may result in tailpipe emission exceeding the specified limits given. Examples of such systems or components are those for monitoring and control of air mass flow, air volumetric flow (and temperature), boost pressure and inlet manifold pressure (and relevant sensors to enable these functions to be carried out).
- Any other emission-related powertrain component connected to a computer must be monitored for circuit continuity (Table 5.3).

5.5.4 Features and technology of current systems

To avoid false detection, the legislation allows verification and healing strategies. These are outlined as follows:

5.5.4.1 MIL activation logic for detected malfunctions

To avoid wrong detections, the legislation allows verification of the detected failure. The failure is stored in the fault memory as a pending code immediately after the first recognition but the MIL is not activated. The MIL will be illuminated in the third driving cycle in which the failure has been detected; the failure is then recognised as a confirmed fault.

5.5.4.2 MIL healing

The MIL may be deactivated after three subsequent sequential driving cycles during which the monitoring system responsible for activating the MIL ceases to detect the malfunction, and if no other malfunction has been identified that would independently activate the MIL.

5.5.4.3 Healing of the fault memory

The OBD system may erase a fault code, distance travelled and freeze frame information if the same fault is not re-registered in at least 40 engine warm-up cycles.

5.5.4.4 Freeze frame

This is a feature that can assist in the diagnosis of intermittent faults. Upon determination of the first malfunction of any component or system, 'freeze frame' engine conditions present at the time must be stored in the computer memory. Stored engine conditions must include, but are not limited to:

- calculated/derived load value;
- engine speed;
- fuel trim values (if available);
- fuel pressure (if available);
- vehicle speed (if available);
- coolant temperature;
- intake manifold pressure (if available);
- closed or open-loop operation (if available);
- the fault code which caused the data to be stored.

5.6 DRIVING CYCLES

5.6.1 Introduction

Even before a vehicle is subjected to OBD systems, it must pass stringent emissions tests. This is done by running the vehicle through test cycles and collecting the exhaust for analysis.

5.6.2 Europe

The New European Driving Cycle (NEDC) is a driving cycle consisting of four repeated ECE-15 driving cycles and an extra-urban driving cycle (EUDC). The NEDC is meant to represent the typical usage of a car in Europe, and is used, among other things, to measure emissions (Figure 5.25). It is sometimes referred to as MVEG (Motor Vehicle Emissions Group) cycle.

The *old* European ECE-15 driving cycle lies between 0 and 800 seconds and represented an urban drive cycle. The section from 800 seconds represents a suburban drive cycle, and is now called the New European Driving Cycle.

The cycle must be performed on a cold vehicle at 20°C (68°F). The cycles may be performed on a normal flat road, in the absence of wind. However, to improve repeatability, they are generally performed on a rolling road.

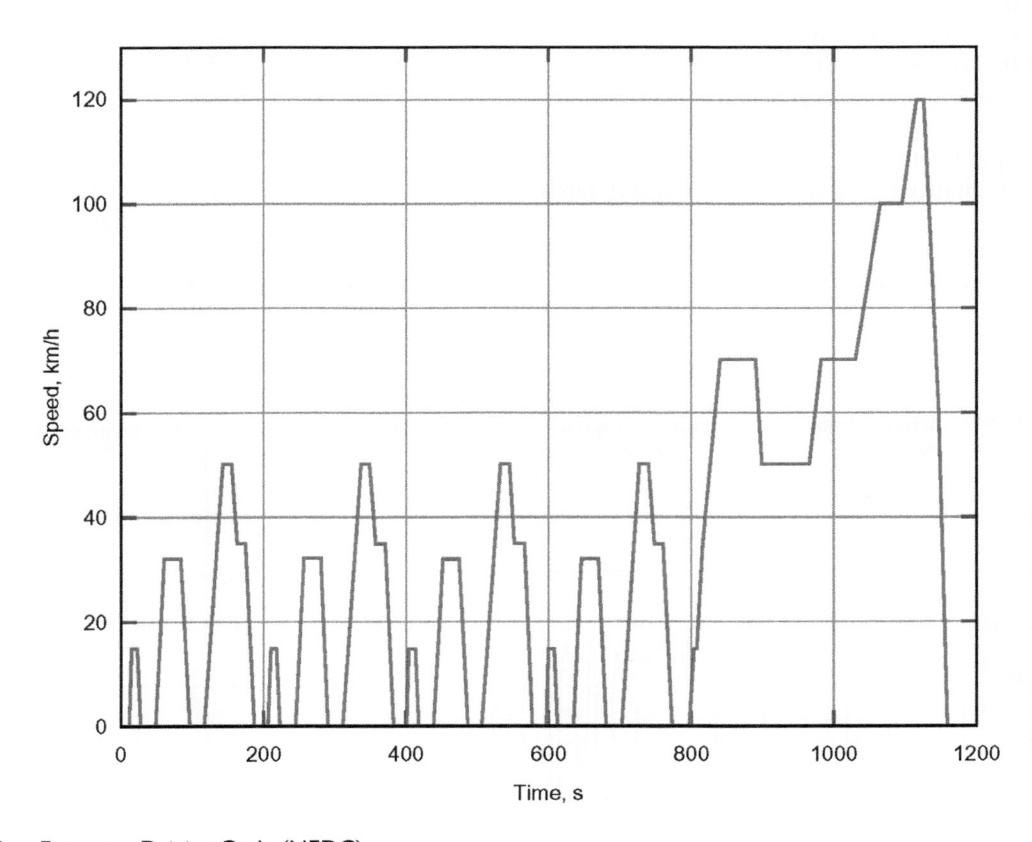

Figure 5.25 New European Driving Cycle (NEDC).

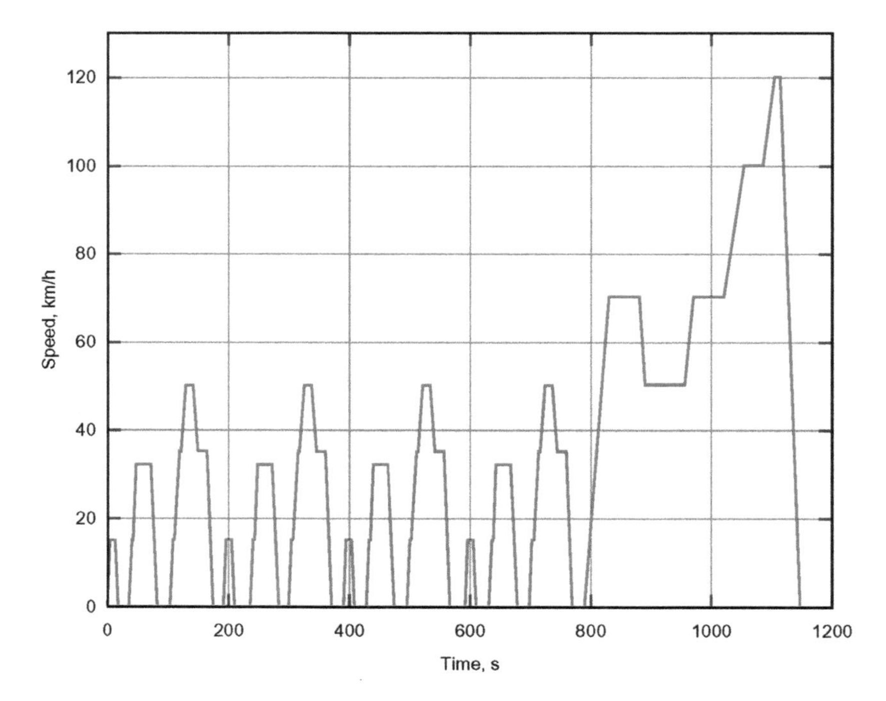

Figure 5.26 Modified New European Driving Cycle (MNEDC).

Several measurements are usually performed during the cycle. The figures made available to the general public are the following:

- urban fuel economy (first 800 seconds);
- extra-urban fuel economy (800–1200 seconds);
- overall fuel economy (complete cycle);
- CO_2 emission (complete cycle).

The following parameters are also generally measured to validate the compliance to European emission standards:

- carbon monoxide (CO);
- unburnt hydrocarbons (HC);
- nitrogen oxides (NO_x);
- particulate matter (PM).

A further tightening of the driving cycle is the Modified New European Driving Cycle (MNEDC), which is very similar to the NEDC except that there is no warm-up time at the start (Figure 5.26).

5.6.3 USA

In the United States, a cycle known as the Federal Test Procedure FTP-75 is used. This has been added to and became known as the Supplementary Federal Test Procedure (SFTP) (Figure 5.27).

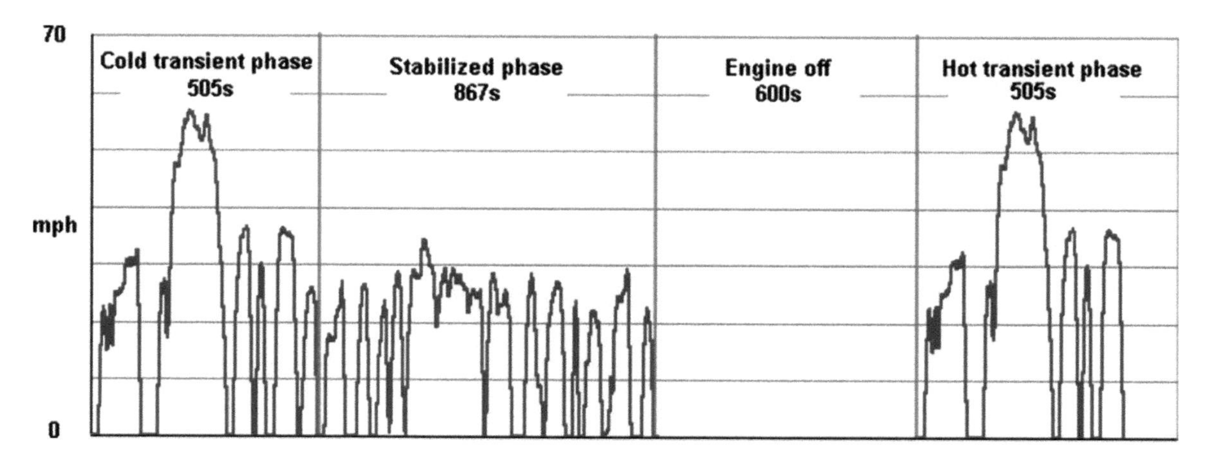

Figure 5.27 US Federal Test Procedure.

KEY FACT

OBD3 may take OBD2 further by adding remote data transfer.

5.6.4 Worldwide harmonized light vehicles test procedure

The worldwide harmonized light vehicles test procedure (WLTP) is a global standard for determining levels of pollutants and CO_2 emissions, fuel or energy consumption and electric range (Figure 5.28). It is used for passenger cars and light commercial vans. Experts from the EU, Japan and India, under guidelines of UNECE World Forum for Harmonization of Vehicle Regulations, developed the standard, which was released in 2015.

DEFINITION

WLTP: Worldwide harmonized light vehicles test procedure.

Like all previous test cycles, it has its drawbacks, but it is a good attempt to make the test more realistic. The key thing is that a standardised test allows vehicles to be compared accurately, even if the figures differ from real-world driving. It includes strict guidance regarding conditions of dynamometer tests and road load (motion resistance), gear changing, total car weight (by including optional equipment, cargo and passengers), fuel quality, ambient temperature, tyres and their pressure. Three different cycles are used depending on vehicle class defined by power/weight ratio PWr in kW/Tonne (rated engine power/kerb weight):

- Class 1 – low power vehicles with PWr <= 22;
- Class 2 – vehicles with 22 < PWr <= 34;
- Class 3 – high-power vehicles with PWr > 34.

Most modern cars, light vans and buses have a power-weight ratio of 40–100 kW/t, so belong to class 3, but some can be in class 2. In each class, there are several driving tests designed to represent real-world vehicle operation on urban roads, extra-urban roads, motorways and freeways.

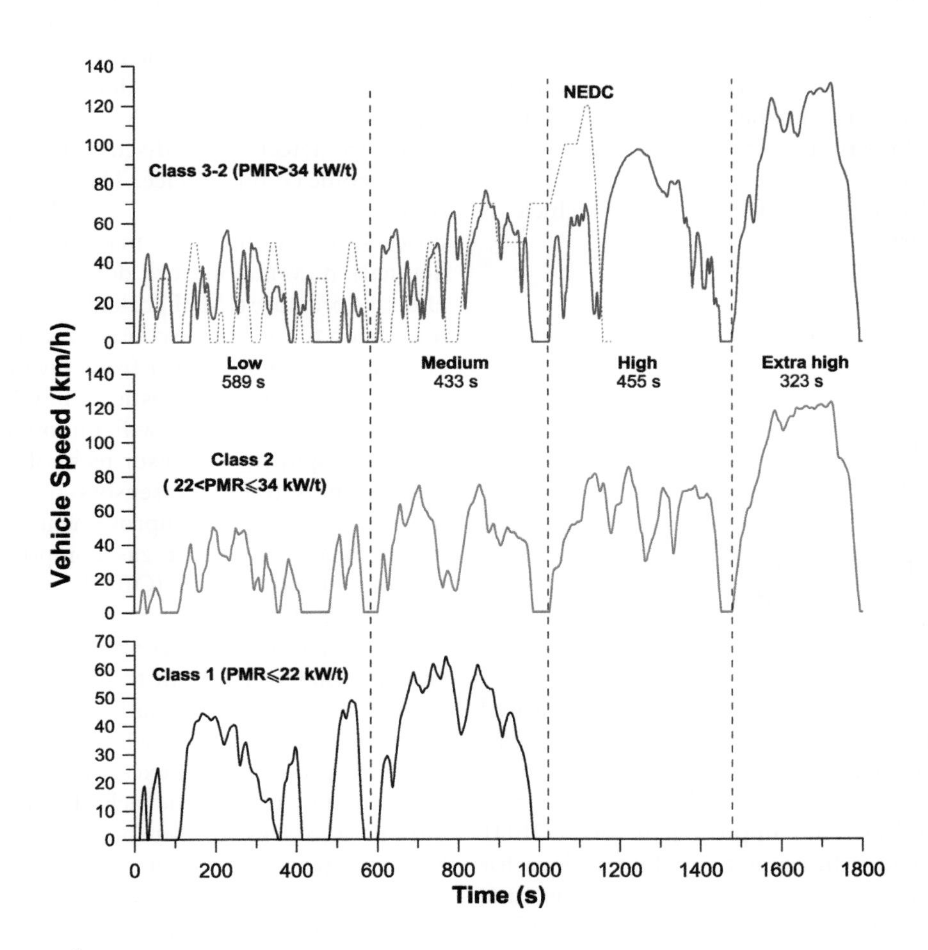

Figure 5.28 WLTP test cycles.

KEY FACT

Most modern cars, light vans and buses have a power-weight ratio of 40–100 kW/t, so belong to WLTP class 3.

The duration of each part of the tests is fixed between classes; the difference is that the acceleration and speed curves are shaped differently. The sequence of tests is also restricted by maximum vehicle speed (V_{max}). Because there is a wide range of manual gearboxes with 4, 5, 6, 7 or even 8 gears, it is impossible to specify fixed gear shift points.

To overcome this, the WLTP uses an algorithm for calculating optimal shift points, which considers total vehicle weight and full load power curves within normal engine speeds. This covers the wide range of rpm and engine power available on current vehicles. To reflect normal use and a fuel efficient driving style, gear changes are filtered out if they occur in less than 5 seconds.

WLTC is an improvement over the New European Driving Cycle (NEDC), but transitions are still very slow. For example, the most rapid 0 to 50 km/h (0 to 30 mph) time is 15 seconds. Most drivers in Western Europe accelerate from rest to 50 km/h in 5 to 10 seconds. There is no hill climbing in the cycle. Perhaps there should be as even modest gradients increase pollutant emissions because the engine load increases two to three times. Nonetheless, it is a step in the right direction.

The WLTC driving cycle for a class 3 vehicle is divided in four parts:

1. low;
2. medium;
3. high;
4. extra high speed.

If $V_{max} < 135$ km/h, the extra high speed part is replaced with low speed part.

5.7 FUTURE DEVELOPMENTS IN DIAGNOSTIC SYSTEMS

5.7.1 OBD3

The current generation of OBD is a very sophisticated and capable system for detecting emissions problems. However, it is necessary to get the driver of the vehicle to do something about the problem. With respect to this aspect, OBD2/EOBD is no improvement over OBD1 unless there is some enforcement capability. Plans for OBD3 have been under consideration for some time now. The idea is to take OBD2 a step further by adding remote data transfer.

An OBD3-equipped vehicle would be able to report emissions problems directly back to a regulatory authority. The transmitter, which would be similar to those currently used for automatic toll payments, would communicate the vehicle identification number (VIN) and any diagnostic codes that have been logged. The system could be set up to automatically report an emissions problem the instant the MIL light is on, or alternatively, the system could respond to answer a query about its current emissions performance status. It could also respond via a cellular or satellite link, reporting its position at the same time.

While somewhat 'Big Brother', this approach is very efficient. The need for periodic inspections could be eliminated because only those vehicles that reported problems would have to be tested. The regulatory authorities could focus their efforts on vehicles and owners who are actually causing a violation rather than just random testing. It is clear that with a system like this, much more efficient use of available regulatory enforcement resources could be implemented, with a consequential improvement in air quality.

An inevitable change that could come with OBD3 would be even closer scrutiny of vehicle emissions. The misfire detection algorithms currently required by OBD2 only look for misfires during driving conditions that occur during the prescribed driving cycles. It does not monitor misfires during other engine operating modes, like full load. More sophisticated methods of misfire detection (as discussed in Chapters 3 and 4) will become commonplace. These systems can feedback other information to the ECU about the combustion process, for example, the maximum cylinder pressure, detonation events or work done via an indicated mean effective pressure (IMEP) calculation. This adds another dimension to the engine control system allowing greater efficiency and more power from any given engine design by just using a more sophisticated ECU control strategy.

Future OBD systems will undoubtedly incorporate new developments in sensor technology. Currently, the evaluation is done via sensors monitoring emissions indirectly. Clearly an improvement would be the ability to measure exhaust gas composition directly via on-board measurement (OBM) systems. This is more in keeping with emission regulation philosophy and would overcome the inherent weakness of current OBD systems, that is, they fail to detect a number of minor faults that do not individually activate the MIL or cause excessive emissions but whose combined effect is to cause the production of excess emissions.

The main barrier is the lack of availability of suitably durable and sensitive sensors for CO, NO_x and HC. Some progress has been made with respect to this, and some vehicles are now being fitted with NO_x sensors. Currently, there does appear to be a gap between the laboratory-based sensors used in research and reliable mass produced units that could form the basis of an OBM system. The integration of combustion measurement in production vehicles produces a similar problem.

5.7.2 Diesel engines

Another development for future consideration is the further implementation of OBD for diesel engines. As diesel engine technology becomes more sophisticated, so does the requirement for OBD. In addition, emission legislation is driving more sophisticated requirements for after-treatment of exhaust gas. All of these subsystems are to be subjected to checking via the OBD system and present their own specific challenges; for example, the monitoring of exhaust after-treatment systems (particulate filters and catalysts) in addition to more complex EGR and air management systems.

5.7.3 Rate-based monitoring

Rate-based monitoring will be more significant for future systems which allow in-use performance ratio information to be logged. It is a standardised method of measuring monitoring frequency and filters out the effect of short trips, infrequent journeys, etc., as factors which could affect the OBD logging and reactions. It is an essential part of the evaluation where driving habits or patterns are not known and it ensures that monitors run efficiently in use and detect faults in a timely and appropriate manner. It is defined as

$$\text{Minimum frequency} = \frac{N}{D}$$

where N = number of times a monitor has run and D = number of times the vehicle has been operated.

5.7.4 Model-based development

A significant factor in the development of future systems will be the implementation of the latest technologies with respect to hardware and software development. Model-based development and calibration of the system will dramatically reduce the testing time by reducing the number of test iterations required. This technique is quite common for developing engine-specific calibrations for ECUs during the engine development phase (Figure 5.29).

Hardware-in-loop (HIL) simulation plays a part in rapid development of any hardware. New hardware can be tested and validated under a number of simulated conditions and its performance verified before it even goes near any prototype vehicle. The following tasks can be performed with this technology:

- full automation of testing for OBD functionality;
- testing of parameter extremes;
- testing of experimental designs;
- regression testing of new designs of software and hardware;
- automatic documentation of results.

5.7.5 OBD security

Researchers at the universities of Washington and California examined the security around OBD and found that they were able to gain control over many vehicle components via the interface. They were also able to upload different firmware into the engine control units. Vehicle-embedded systems are clearly not designed with security in mind!

Thieves have used specialist OBD reprogramming devices to steal cars without the use of a key. The main

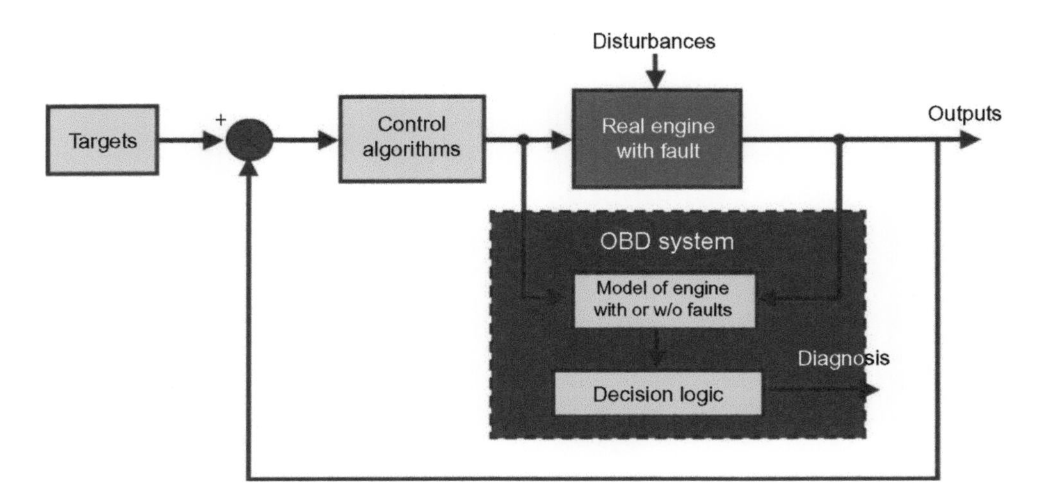

Figure 5.29 Model-based calibration of OBD system.

causes of this vulnerability are that vehicle manufacturers extend the data bus for purposes other than those for which it was designed. Lack of authentication and authorisation in the OBD specifications, which instead rely largely on security through obscurity, don't help. Of course, the physical locks on a vehicle are a deterrent.

5.7.6 Summary

Clearly, OBD is here to stay and will continue to be developed. It is a useful tool for the technician as well as a key driver towards cleaner vehicles. The creation of generic standards has helped those of us at the 'sharp end' of diagnostics significantly.

OBD has a number of key emission-related systems to 'monitor'. It saves faults in these systems in a standard form so that they can be accessed using a scan tool.

However, with the possibility of OBD3 using the navigation system to report where we are, speed and traffic light cameras everywhere and monitoring systems informing the authorities about the condition of our vehicles, whatever will be next?

Note: At the time of writing there was much controversy over 'defeat devices' and how some manufacturers have been cheating the emission tests. Expect to see legislation changes, in particular to the way in which the new vehicles are tested.

ACKNOWLEDGEMENT

I am most grateful to Dave Rogers (http://www.autoelex.co.uk) and Alan Malby (Ford Motor Company) for their excellent contributions to this chapter.

Chapter 6

Engine systems

6.1 INTRODUCTION

The main sections in this chapter that relate to an area of the vehicle start with an explanation of the particular system. The sections then conclude with appropriate diagnostic techniques and symptom charts. Extra tests and methods are explained where necessary.

6.2 ENGINE OPERATION

6.2.1 Four-stroke cycle

Figure 6.1 shows a modern vehicle engine. Engines like this can seem very complex at first but keep in mind when carrying out diagnostic work that, with very few exceptions, all engines operate on the four-stroke principle. The complexity lies in the systems around the engine to make it operate to its maximum efficiency or best performance. With this in mind then, back to basics: the engine components are combined to use the power of expanding gas to drive the engine.

When the term 'stroke' is used, it means the movement of a piston from top dead centre (TDC) to bottom dead centre (BDC) or the other way round. The following table explains the spark ignition (SI) and compression ignition (CI) four-stroke cycles – for revision purposes. Figure 6.2 shows the SI cycle.

Figure 6.1 Ford Focus engine.

Source: Ford Media.

KEY FACT

Stroke means the movement of a piston from top dead centre (TDC) to bottom dead centre (BDC) or the other way round.

6.2.2 Cylinder layouts

Great improvements can be made to the performance and balance of an engine by using more than one cylinder. Once this is agreed, the actual layout of the cylinders must be considered. The layout can be one of three possibilities as follows:

- **In-line or straight** – The cylinders are in a straight line. They can be vertical, inclined or horizontal.
- **Vee** – The cylinders are in two rows at a set angle. The actual angle varies but is often 60° or 90°.
- **Opposed** – The cylinders are in two rows opposing each other and are usually horizontal.

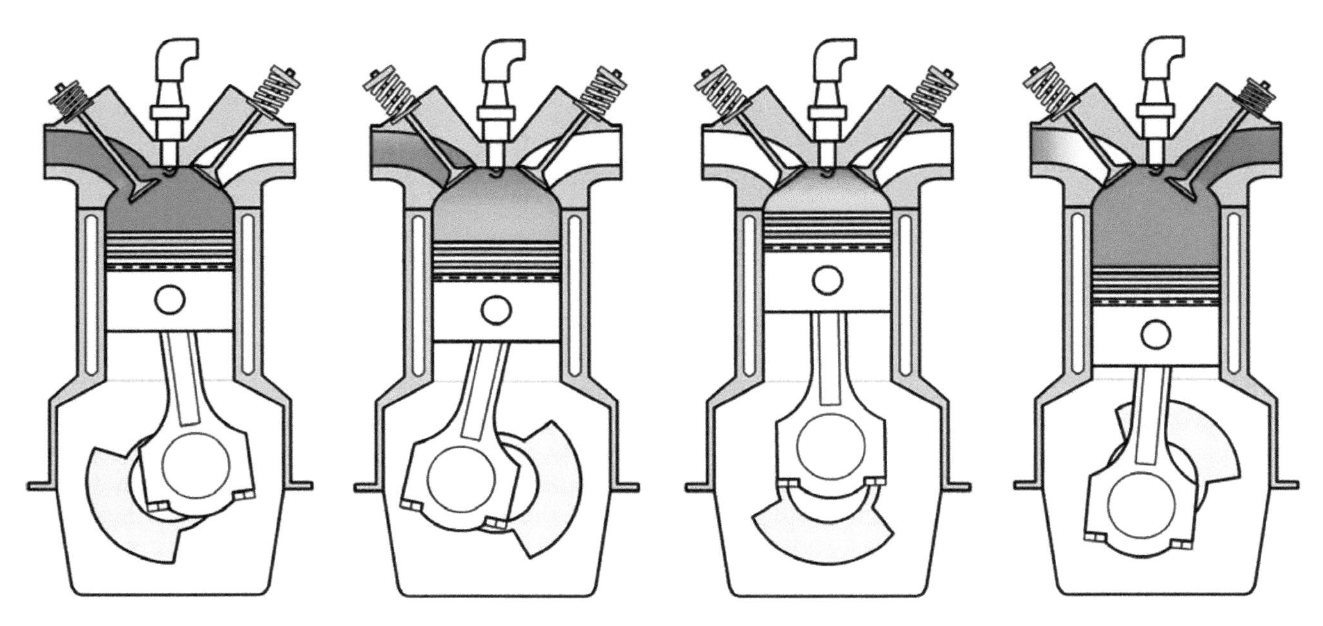

Figure 6.2 Four-stroke cycle: induction, compression, power and exhaust.

Stroke	*Spark ignition*	*Compression ignition*
Induction	The fuel air mixture is forced into the cylinder through the open inlet valve because as the piston moves down, it makes a lower pressure. It is acceptable to say the mixture is drawn into the cylinder	Air is forced into the cylinder through the open inlet valve because as the piston moves down, it makes a lower pressure. It is acceptable to say the air is drawn into the cylinder
Compression	As the piston moves back up the cylinder, the fuel air mixture is compressed to about an eighth of its original volume because the inlet and exhaust valves are closed. This is a compression ratio of 8:1, which is typical for many normal engines	As the piston moves back up the cylinder, the fuel air mixture is compressed in some engines to about a sixteenth of its original volume because the inlet and exhaust valves are closed. This is a compression ratio of 16:1, which causes a large build-up of heat
Power	At a suitable time before TDC, a spark at the plug ignites the compressed mixture. The mixture now burns very quickly and the powerful expansion pushes the piston back down the cylinder. Both valves are closed	At a suitable time before TDC, very high pressure atomised diesel fuel (at approximately 180 bar) is injected into the combustion chamber. The mixture burns very quickly and the powerful expansion pushes the piston back down the cylinder. The valves are closed
Exhaust	The final stroke occurs as the piston moves back up the cylinder and pushes the spent gases out of the now open exhaust valve	The final stroke occurs as the piston moves back up the cylinder and pushes the spent gases out of the now open exhaust valve

By far the most common arrangement is the straight four, and this is used by all manufacturers in their standard family cars. Larger cars do, however, make use of the 'Vee' configuration. The opposed layout although still used is less popular. Engine firing order is important. This means the order in which the power strokes occur. It is important to check in the workshop manual or data book when working on a particular engine.

6.2.3 Camshaft drives

The engine drives the camshaft in one of three ways: gear drive, chain drive or by a drive belt. The last of these is now the most popular, as it tends to be simpler and quieter. Note in all cases that the cam is driven at half the engine speed. This is done by the ratio of teeth between the crank and cam cogs which is 1:2, for example 20 crank teeth and 40 cam teeth.

KEY FACT

A camshaft is driven at half the speed of the crankshaft.

- **Camshaft drive gears** – Gears are not used very often on petrol engines but are used on larger diesel engines. They ensure a good positive drive from the crankshaft gear to the camshaft.

- **Camshaft chain drive** – Chain drive is still used but was even more popular a few years ago. The problems with it are that a way must be found to tension the chain and also provide lubrication.
- **Camshaft drive belt** – Camshaft drive belts have become very popular. The main reasons for this are that they are quieter, do not need lubrication and are less complicated. They do break now and then, but this is usually due to lack of servicing. Cam belts should be renewed at set intervals. Figure 6.3 shows an example of the data available relating to camshaft drive belt fitting. This is one of the many areas where data is essential for diagnostic checks.

BMW 5 Series 2.5 525i SE

Timing belt

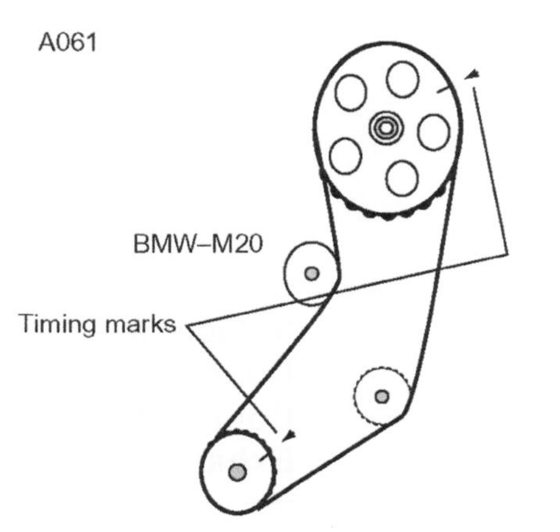

A061

BMW–M20

Timing marks

Replacement intervals:
at 36 000 miles

Engine setting position:
TDC at No. 1 cylinder

Special tools:

Viscous coupling tool	11.5.040
Pulley retaining tool	11.5.030

Torque settings:

Crank pulley	23 Nm
Tensioner	23 Nm
Fan coupling	43 Nm

Special notes:

The fan coupling has left hand thread. Set timing for camshaft and crankshaft and check TDC mark on the flywheel. Check for rotor arm alignment to distributor casing. After fitting belt, rotate engine two full turns and recheck the timing marks.

Figure 6.3 Timing belt data.

6.2.4 Valve mechanisms

A number of methods are used to operate the valves. Three common types are shown in Figure 6.4 and a basic explanation of each follows:

- Overhead valve with push rods and rockers – The method has been used for many years and although it is not used as much now, many vehicles still on the road are described as overhead valve (OHV). As the cam turns, it moves the follower, which in turn pushes the push rod. The push rod moves the rocker, which pivots on the rocker shaft and pushes the valve open. As the cam moves further, it allows the spring to close the valve.
- Overhead cam with followers – Using an overhead cam (OHC) reduces the number of moving parts. In the system shown here, the lobe of the cam acts directly on the follower which pivots on its adjuster and pushes the valve open.
- Overhead cam, direct acting and automatic adjusters – Most new engines now use an OHC with automatic adjustment. This saves on repair and service time and keeps the cost to the customer lower. Systems vary between manufacturers, some use followers and some have the cam acting directly on to the valve. In each case, though, the adjustment is by oil pressure. A type of plunger, which has a chamber where oil can be pumped under pressure, operates the valve. This expands the plunger and takes up any unwanted clearance.

Valve clearance adjustment is very important. If it is too large, the valves will not open fully and will be noisy. If the clearance is too small, the valves will not close and no compression will be possible.

When an engine is running, the valves become very hot and therefore expand. The exhaust valve clearance is usually larger than the inlet, because it gets hotter. Regular servicing is vital for all components but, in particular, the valve operating mechanism needs a good supply of clean oil at all times.

6.2.5 Valve and ignition timing

Valve timing is important. The diagram in Figure 6.5 shows accurately the degrees of rotation of the crankshaft where the inlet and exhaust valves open and close during the four-stroke cycle. The actual position in the cycle of operation when valves open and close depends on many factors and will vary slightly with different designs of engine. Some cars now control valve timing by electronics. The diagram is marked to show what is meant by valve lead, lag and overlap. Ignition timing

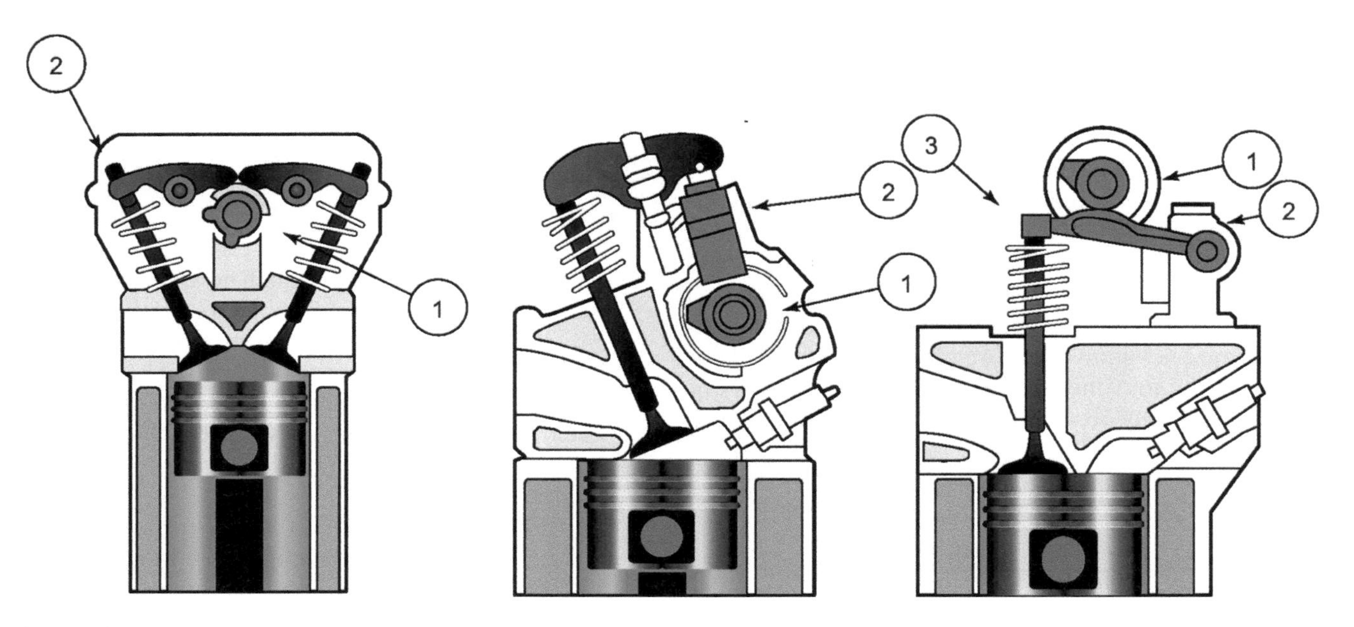

Figure 6.4 Valve operating mechanisms. Left: 1 – Cam; 2 – adjusting screw in direct acting rocker. Centre: 1 – Cam; 2 – hydraulic follower. Right: 1 – Cam; 2 – pivot and adjuster; 3 – finger follower.

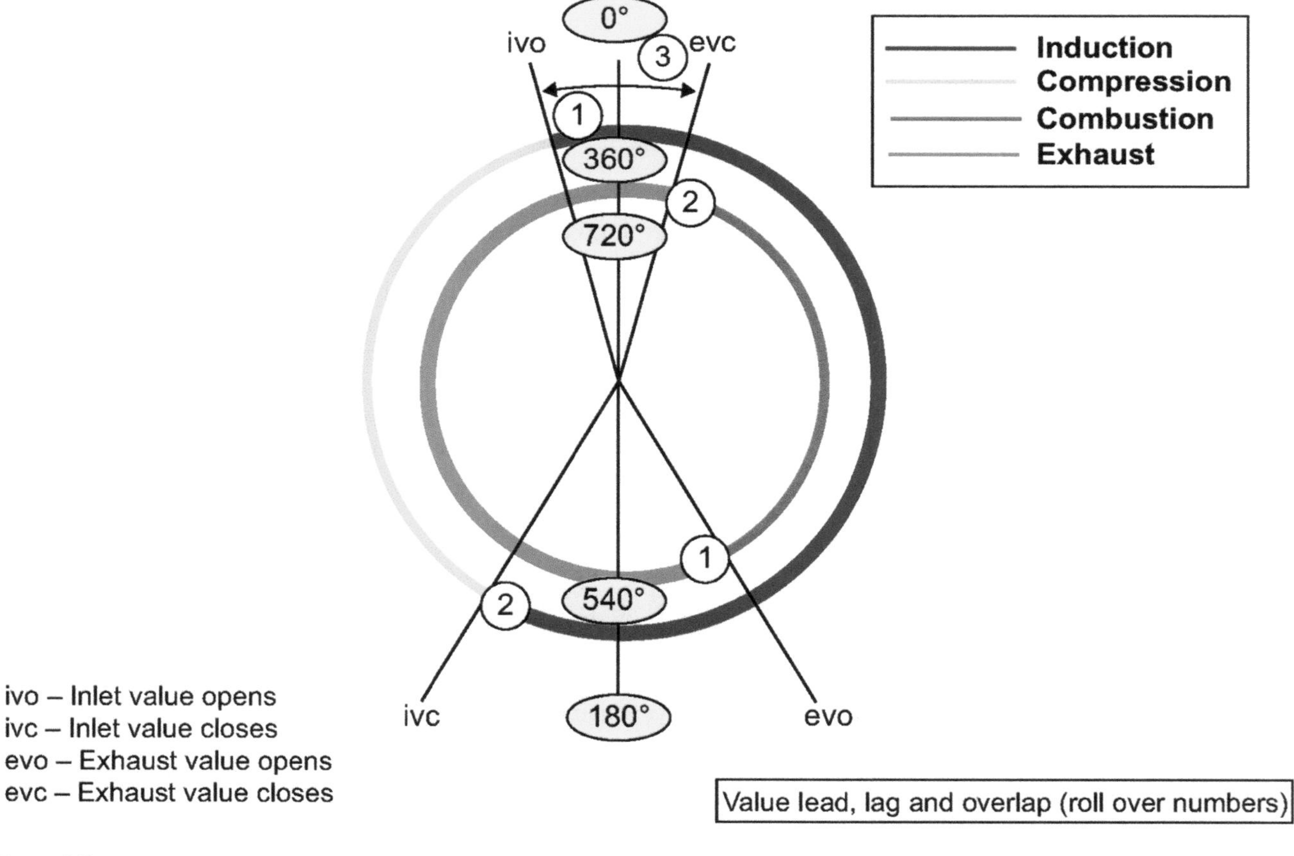

Figure 6.5 Valve timing diagrams (1 – lead; 2 – lag; 3 – overlap).

is marked on the diagram. Note how this changes as engine speed changes.

The valve timing diagram shows that the valves of a four-stroke engine open just before and close just after the particular stroke. Looking at the timing diagram, if you start at position IVO, the piston is nearly at the top of the exhaust stroke when the inlet valve opens (IVO). The piston reaches the top and then moves down on the intake stroke. Just after starting the compression stroke, the inlet valve closes (IVC). The piston continues upwards and, at a point several degrees before TDC, the spark occurs and starts the mixture burning.

The maximum expansion is 'timed' to occur after TDC; therefore, the piston is pushed down on its power stroke. Before the end of this stroke, the exhaust valve opens (EVO). Most of the exhaust gases now leave because of their very high pressure. The piston pushes the rest of the spent gases out as it moves back up the cylinder. The exhaust valve closes (EVC) just after the end of this stroke and the inlet has already opened, ready to start the cycle once again.

KEY FACT

Maximum expansion is 'timed' to occur after TDC on the power stroke.

The reason for the valves opening and closing like this is that it makes the engine more efficient by giving more time for the mixture to enter and the spent gases to leave. The outgoing exhaust gases in fact help to draw in the fuel air mixture from the inlet. Overall, this makes the engine have a better 'volumetric efficiency'.

6.3 DIAGNOSTICS – ENGINES

6.3.1 Systematic testing example

If the reported fault is excessive use of engine oil, proceed as follows:

1. Question the customer to find out how much oil is being used.
2. Examine the vehicle for oil leaks and blue smoke from the exhaust.
3. For example, oil may be leaking from a gasket or seal – if no leaks are found, the engine may be burning the oil.
4. A compression test, if the results were acceptable, would indicate a leak to be the most likely fault. Clean down the engine and run it for a while, the leak might show up.
5. For example, change the gasket or seals.
6. Run a thorough inspection of vehicle systems, particularly those associated with the engine. Double check that the fault has been rectified and that you have not caused any other problems.

SAFETY FIRST

Note: You should always refer to the manufacturer's instructions appropriate to the equipment you are using.

6.3.2 Test equipment

6.3.2.1 Compression tester

With this device the spark plugs are removed and the tester screwed or held in to each spark plug hole in turn. The engine is cranked over by the starter and the gauge will read the compression or pressure of each cylinder.

6.3.2.2 Cylinder leakage tester

A leakage tester uses compressed air to pressurise each cylinder in turn by a fitting to the spark plug hole. The cylinder under test is set to TDC compression. The percentage of air leaking out and where it is leaking from helps you determine the engine condition. For example, if air is leaking through the exhaust pipe, then the exhaust valves are not sealing. If air leaks into the cooling system, then a leak from the cylinder to the water jacket may be the problem (blown head gasket is possible). Figure 6.6 shows a selection of Snap-on diagnostic gauges – vacuum, compression and leakage.

Figure 6.6 Diagnostic gauges.

Table 6.1 Tests and information required

Test carried out	*Information required*
Compression test	Expected readings for the particular engine under test. For example, the pressure reach for each cylinder may be expected to read 800 kPa ±15%
Cylinder leakage test	The percentage leak that is allowed for the tester you are using – some allow approximately 15% leakage as the limit

6.3.3 Test results

Some of the information you may have to get from other sources such as data books or a workshop manual is listed in Table 6.1

6.3.4 Engine fault diagnosis table 1

Symptom	*Possible causes or faults*	*Suggested action*
Oil consumption	Worn piston rings and/or cylinders Worn valve stems, guides or stem oil seals	Engine overhaul Replace valves (guides if possible) and oil seals
Oil on engine or floor	Leaking gaskets or seals Build-up of pressure in the crankcase	Replace appropriate gasket or seal Check engine breather system
Mechanical knocking noises	Worn engine bearings (big ends or mains for example) Incorrect valve clearances or defective automatic adjuster Piston slap on side of cylinder	Replace bearings or overhaul engine, good idea to also check the oil pressure Adjust clearances to correct settings or replace defective adjuster Engine overhaul required now or quite soon
Vibration	Engine mountings loose or worn Misfiring	Secure or renew Check engine ancillary systems such as fuel and ignition

6.3.5 Engine fault diagnosis table 2

Please note that this section covers related engine systems as well as the engine itself.

Symptom	*Possible cause*
Engine does not rotate when trying to start	Battery connection loose or corroded Battery discharged or faulty Broken, loose or disconnected wiring in the starter circuit Defective starter switch or automatic gearbox inhibitor switch Starter pinion or flywheel ring gear loose Earth strap broken, loose or corroded
Engine rotates but does not start	No fuel in the tank! Discharged battery (slow rotation) Battery terminals loose or corroded Air filter dirty or blocked Low cylinder compressions Broken timing belt Damp ignition components Fuel system fault Spark plugs worn to excess Ignition system open circuit
Difficult to start when cold	Discharged battery (slow rotation) Battery terminals loose or corroded Air filter dirty or blocked Low cylinder compressions Fuel system fault Spark plugs worn to excess Enrichment device not working (choke or injection circuit)
Difficult to start when hot	Discharged battery (slow rotation) Battery terminals loose or corroded Air filter dirty or blocked Low cylinder compressions Fuel system fault
Starter noisy	Starter pinion or flywheel ring gear loose Starter mounting bolts loose Starter worn (bearings, etc.) Discharged battery (starter may jump in and out)
Starter turns engine slowly	Discharged battery (slow rotation) Battery terminals loose or corroded Earth strap or starter supply loose or disconnected Internal starter fault
Engine starts but then stops immediately	Ignition wiring connection intermittent Fuel system contamination Fuel pump or circuit fault (relay) Intake system air leak Ballast resistor open circuit (older cars)
Erratic idle	Air filter blocked Incorrect plug gaps Inlet system air leak Incorrect CO setting Uneven or low cylinder compressions (maybe valves) Fuel injector fault Incorrect ignition timing Incorrect valve timing
Misfire at idle speed	Ignition coil or distributor cap tracking Poor cylinder compressions Engine breather blocked Inlet system air leak Faulty plugs
Misfire through all speeds	Fuel filter blocked Fuel pump delivery low Fuel tank ventilation system blocked Poor cylinder compressions Incorrect plugs or plug gaps HT leads breaking down
Engine stalls	Idle speed incorrect CO setting incorrect Fuel filter blocked Air filter blocked Intake air leak Idle control system not working

(*Continued*)

Symptom	Possible cause
Lack of power	Fuel filter blocked Air filter blocked Ignition timing incorrect Low fuel pump delivery Uneven or low cylinder compressions (maybe valves) Fuel injectors blocked Brakes binding or clutch slipping
Backfires	Incorrect ignition timing Incorrect valve timing (cam belt not fitted correctly) Fuel system fault (airflow sensor on some cars)
Oil pressure gauge low or warning light on	Low engine oil level Faulty sensor or switch Worn engine oil pump and/or engine bearings Engine overheating Oil pick-up filter blocked Pressure relief valve not working
Runs on when switched off	Ignition timing incorrect Idle speed too high Anti-run on device not working Carbon build-up in engine Engine overheating
Pinking or knocking under load	Ignition timing incorrect Ignition system fault Carbon build-up in engine Knock sensor not working
Sucking or whistling noises	Leaking exhaust manifold gasket Leaking inlet manifold gasket Cylinder head gasket Inlet air leak Water pump or alternator bearing
Rattling or tapping	Incorrect valve clearances Worn valve gear or camshaft Loose component
Thumping or knocking noises	Worn main bearings (deep knocking/rumbling noise) Worn big-end bearings (heavy knocking noise under load) Piston slap (worse when cold) Loose component
Rumbling noises	Bearings on ancillary component

6.4 FUEL SYSTEM

Author's note: Even though carburettor fuel systems are now very rare, they are still used on some specialist vehicles. For this reason, and because it serves as a good introduction to fuel systems, I decided to include this section.

6.4.1 Introduction

All vehicle fuel systems consist of a fuel/air mixing stage (usually fuel injectors), fuel tank, fuel pump and fuel filter, together with connecting pipes. An engine works by the massive expansion of an ignited fuel air mixture acting on a piston. The job of the fuel system is to produce this mixture at just the right ratio to run the engine under all operating conditions. There are three main ways this is achieved:

- Petrol/gasoline is mixed with air in a carburettor;
- Petrol/gasoline is injected into the manifold, throttle body or cylinder or to mix with the air;
- Diesel is injected under very high pressure directly into the air already in the engine combustion chamber.

KEY FACT

A fuel system should produce the mixture at the right ratio to run the engine under all operating conditions.

6.4.2 Engine management

'Engine management' is a general term that describes the control of engine operation. This can range from a simple carburettor to control or manage the fuel, together with an ignition distributor with contact breakers for the ignition, to a very sophisticated electronic engine management control system. The fundamental tasks of an engine management system are to manage the ignition and fuelling, as well as other aspects. Very accurate control of fuelling and ignition means improved performance, better consumption and reduced emission.

KEY FACT

'Engine management' is a general term that describes the control of engine operation.

Many of the procedures and explanations in this chapter are generic. In other words, the ignition system explained in the next sections may be the same as the system used by a combined ignition and fuel control system.

6.5 DIAGNOSTICS – BASIC FUEL SYSTEM

6.5.1 Systematic testing example

If the reported fault is excessive fuel consumption, proceed as follows:

1. Check that the consumption is excessive for the particular vehicle. Test it yourself if necessary.
2. Are there any other problems with the vehicle, misfiring, for example, or difficult starting?

3. For example, if the vehicle is misfiring as well, this may indicate that an ignition fault is the cause of the problem.
4. Remove and examine spark plugs, check HT lead resistance and ignition timing. Check CO emissions.
5. Renew plugs and set fuel mixture.
6. Road-test the vehicle for correct engine operation.

6.5.2 Test equipment

6.5.2.1 Exhaust gas analyser

This is a sophisticated piece of test equipment used to measure the component gases of the vehicle's exhaust. The most common requirement is the measuring of carbon monoxide (CO). A sample probe is placed in the exhaust tailpipe or a special position before the catalytic converter (if fitted), and the machine reads out the percentage of certain gases produced. A digital readout is most common. The fuel mixture can then be adjusted until the required readings are obtained.

SAFETY FIRST

Note: You should always refer to the manufacturer's instructions appropriate to the equipment you are using.

6.5.2.2 Fuel pressure gauge

The output pressure of the fuel pump can be tested to ensure adequate delivery. The device is a simple pressure gauge but note the added precautions necessary when dealing with petrol (Figure 6.7).

6.5.3 Test results

Some of the information you may have to get from other sources such as data books or a workshop manual is listed in Table 6.2.

Figure 6.7 Exhaust gas analyser.

Table 6.2 Tests and information required

Test carried out	*Information required*
Exhaust gas analysis	CO setting. Most modern vehicles will have settings of approximately 1% or less. If a 'cat' is fitted, then the readings will be even lower when measured at the tailpipe.
Fuel pressure	The expected pressure readings will vary depending on the type of fuel system. Fuel injection pressure will be approximately 2.5 bar, whereas fuel pressure for a carburettor will be approximately 0.3 bar.
Fuel delivery	How much fuel the pump should move in a set time will again vary with the type of fuel system. One litre in 30 seconds is typical for some injection fuel pumps.

6.5.4 Fuel fault diagnosis table 1

Symptom	*Possible faults*	*Suggested action*
No fuel at carburettor or injection fuel rail	Empty tank!	Fill it!
	Blocked filter or line	Replace filter, renew/repair line
	Defective fuel pump	Renew/check it is being driven
	No electrical supply to pump	Check fuses/trace fault
Engine will not or is difficult to start	Choke or enrichment device not working	Check linkages or automatic actuator
Engine stalls or will not idle smoothly	Idle speed incorrectly set	Look up correct settings and adjust
	Mixture setting wrong	Look up correct settings and adjust
	Ignition problem	Check ignition system
Poor acceleration	Blockage in carburettor accelerator pump	Strip down and clean out or try a carburettor cleaner first
	Partially blocked filter	Renew
	Injection electrical fault	Refer to specialist information
Excessive fuel consumption	Incorrect mixture settings	Look up correct settings and adjust
	Driving technique!	Explain to the customer – but be diplomatic!
Black smoke from exhaust	Excessively rich mixture	Look up correct settings and adjust
	Flooding	Check and adjust carburettor float settings and operation

6.5.5 Fuel fault diagnosis table 2

Symptom	*Possible cause*
Excessive consumption	Blocked air filter
	Incorrect CO adjustment
	Fuel injectors leaking
	Ignition timing incorrect
	Temperature sensor fault
	Load sensor fault
	Low tyre pressures
	Driving style!
Fuel leakage	Damaged pipes or unions
	Fuel tank damaged
	Tank breathers blocked
Fuel smell	Fuel leak
	Breather incorrectly fitted
	Fuel cap loose
	Engine flooding
Incorrect emissions	Incorrect adjustments
	Fuel system fault
	Air leak into inlet
	Blocked fuel filter
	Blocked air filter
	Ignition system fault

6.6 IGNITION

6.6.1 Basics

The purpose of the ignition system is to supply a spark inside the cylinder, near the end of the compression stroke, to ignite the compressed charge of air fuel vapour. For a spark to jump across an airgap of 0.6 mm under normal atmospheric conditions (1 bar), a voltage of 2–3 kV is required. For a spark to jump across a similar gap in an engine cylinder having a compression ratio of 8:1, a voltage of approximately 8 kV is required. For higher compression ratios and weaker mixtures, a voltage up to 20 kV may be necessary. The ignition system has to transform the normal battery voltage of 12 V to approximately 8–20 kV and, in addition, has to deliver this high voltage to the right cylinder, at the right time. Some ignition systems will supply up to 40 kV to the spark plugs.

Conventional ignition is the forerunner of the more advanced systems controlled by electronics. However, the fundamental operation of most ignition systems is very similar; one winding of a coil is switched on and off causing a high voltage to be induced in a second winding.

A coil ignition system is composed of various components and subassemblies; the actual design and construction of these depend mainly on the engine with which the system is to be used.

KEY FACT

The fundamental operation of most ignition systems is very similar; one winding of a coil is switched on and off causing a high voltage to be induced in a second winding.

6.6.2 Advance angle (timing)

For optimum efficiency, the ignition advance angle should be such as to cause the maximum combustion pressure to occur approximately 10° after TDC. The ideal ignition timing is dependent on two main factors: engine speed and engine load. An increase in engine speed requires the ignition timing to be advanced. The cylinder charge, of air fuel mixture, requires a certain time to burn (normally approximately 2 ms). At higher engine speeds, the time taken for the piston to travel the same distance reduces. Advancing the time of the spark ensures that full burning is achieved.

A change in timing due to engine load is also required, as the weaker mixture used in low-load conditions burns at a slower rate. In this situation, further ignition advance is necessary. Greater load on the engine requires a richer mixture, which burns more rapidly. In this case, some retardation of timing is necessary. Overall, under any condition of engine speed and load, an ideal advance angle is required to ensure maximum pressure is achieved in the cylinder just after TDC. The ideal advance angle may also be determined by engine temperature and any risk of detonation.

Spark advance is achieved in a number of ways. The simplest of these is the mechanical system comprising a centrifugal advance mechanism and a vacuum (load sensitive) control unit. Manifold depression is almost inversely proportional to the engine load. I prefer to consider manifold pressure, albeit less than atmospheric pressure; the absolute manifold pressure (MAP) is proportional to engine load. Digital ignition systems may adjust the timing in relation to the temperature as well as speed and load. The values of all ignition timing functions are combined either mechanically or electronically in order to determine the ideal ignition point.

The energy storage takes place in the ignition coil. The energy is stored in the form of a magnetic field. To ensure that the coil is charged before the ignition point, a dwell period is required. Ignition timing is at the end of the dwell period.

6.6.3 Electronic ignition

Electronic ignition is now fitted to all spark ignition vehicles. This is because the conventional mechanical system has some major disadvantages:

- Mechanical problems with the contact breakers not least of which is the limited lifetime.
- Current flow in the primary circuit is limited to approximately 4 A, otherwise damage will occur to the contacts – or at least the lifetime will be seriously reduced.
- Legislation requires stringent emission limits which means the ignition timing must stay in tune for a long period of time.
- Weaker mixtures require more energy from the spark to ensure successful ignition, even at very high engine speed.

These problems can be overcome by using a power transistor to carry out the switching function and a pulse generator to provide the timing signal. Very early forms of electronic ignition used the existing contact breakers as the signal provider. This was a step in the right direction but did not overcome all the mechanical limitations such as contact bounce and timing slip. All systems nowadays are constant-energy systems ensuring high-performance ignition even at high engine speed. Figure 6.8 shows the circuit of a standard electronic ignition system.

The term 'dwell' when applied to ignition is a measure of the time during which the ignition coil is charging; in other words, when primary current is flowing. The dwell in conventional systems was simply the time during which the contact breakers were closed. This is now often expressed as a percentage of one charge-discharge cycle. Constant-dwell electronic ignition systems have now been replaced almost without exception by constant-energy systems discussed in the next section.

KEY FACT

The term 'dwell' when applied to ignition is a measure of the time during which the ignition coil is charging – in other words, when primary current is flowing.

Although this was a very good system in its time, constant dwell still meant that at very high engine speeds, the time available to charge the coil could only produce a lower-power spark. Note that as engine speed increases, dwell angle or dwell percentage remains the same but the actual time is reduced.

In order for a constant-energy electronic ignition system to operate, the dwell must increase with engine speed. This

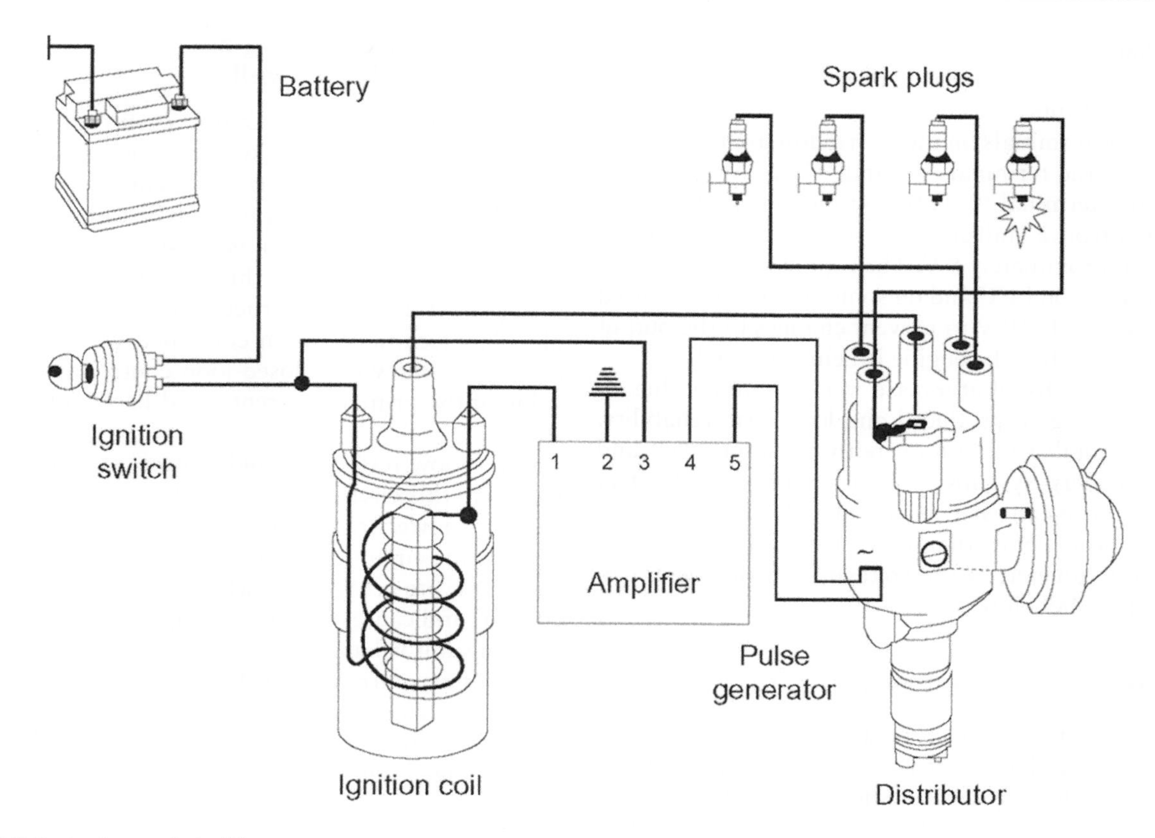

Figure 6.8 Early electronic ignition system.

will only be of benefit, however, if the ignition coil can be charged up to its full capacity in a very short time (the time available for maximum dwell at the highest expected engine speed). To this end, constant-energy coils are very low resistance and low inductance. Typical resistance values are less than 1 Ω (often 0.5 Ω). Constant energy means that, within limits, the energy available to the spark plug remains constant under all operating conditions.

KEY FACT

In order for a constant-energy electronic ignition system to operate, the dwell must increase with engine speed.

Owing to the high-energy nature of constant-energy ignition coils, the coil cannot be allowed to remain switched on for more than a certain time. This is not a problem when the engine is running, as the variable-dwell or current-limiting circuit prevents the coil from overheating. Some form of protection must be provided, however, for when the ignition is switched on but the engine is not running. This is known as stationary engine primary current cut-off.

6.6.4 Hall effect distributor

The Hall effect distributor has become very popular with many manufacturers. Figure 6.9 shows a typical example. As the central shaft of the distributor rotates, the chopper plate attached under the rotor arm alternately covers and uncovers the Hall chip. The number of vanes corresponds with the number of cylinders. In constant-dwell systems, the dwell is determined by the width of the vanes. The vanes cause the Hall chip to be alternately in and out of a magnetic field. The result of

Figure 6.9 Hall effect distributor.

this is that the device will produce almost a square wave output, which can then easily be used to switch further electronic circuits.

The three terminals on the distributor are marked '–', '0' and '+'; the terminals '–' and '+' are for a voltage supply and terminal '0' is the output signal. Typically, the output from a Hall effect sensor will switch between 0 V and approximately 8 V. The supply voltage is taken from the ignition ECU and on some systems is stabilised at approximately 10 V to prevent changes to the output of the sensor when the engine is being cranked.

Hall effect distributors are very common due to the accurate signal produced and long-term reliability. They are suitable for use on both constant-dwell and constant-energy systems. Operation of a Hall effect pulse generator can easily be tested with a DC voltmeter or a logic probe. Note that tests must not be carried out using an ohmmeter, as the voltage from the meter can damage the Hall chip.

6.6.5 Inductive distributor

Many forms of inductive-type distributors exist and all are based around a coil of wire and a permanent magnet. The example distributor shown in Figure 6.10 has the coil of wire wound on the pick-up, and as the reluctor rotates, the magnetic flux varies due to the peaks on the reluctor. The number of peaks or teeth on the reluctor corresponds to the number of engine cylinders. The gap between the reluctor and pick-up can be important and manufacturers have recommended settings.

KEY FACT

Many forms of inductive-type distributors exist and all are based around a coil of wire and a permanent magnet.

6.6.6 Current-limiting and closed-loop dwell

Primary current limiting not only ensures that no damage can be caused to the system by excessive primary current but also forms a part of a constant-energy system. The primary current is allowed to build up to its pre-set maximum as soon as possible and is then held at this value. The value of this current is calculated and then pre-set during construction of the amplifier module. This technique, when combined with dwell angle control, is known as closed-loop control as the actual value of the primary current is fed back to the control stages.

A very low resistance, high power, precision resistor is used in this circuit. The resistor is connected in series with the power transistor and the ignition coil. A voltage-sensing circuit connected across this resistor is activated at a pre-set voltage (which is proportional to the current) and causes the output stage to hold the current at a constant value.

Stationary current cut-off is for when the ignition is on but the engine is not running. This is achieved in many cases by a simple timer circuit, which will cut the output stage after about one second (Figure 6.11).

6.6.7 Programmed ignition/ electronic spark advance

Programmed ignition is the term used by some manufacturers; others call it electronic spark advance (ESA). Constant-energy electronic ignition was a major step forwards and is still used on countless applications. However, its limitations lay in still having to rely upon mechanical components for speed and load advance characteristics. In many cases, these did not match ideally the requirements of the engine.

Figure 6.10 Inductive distributor.

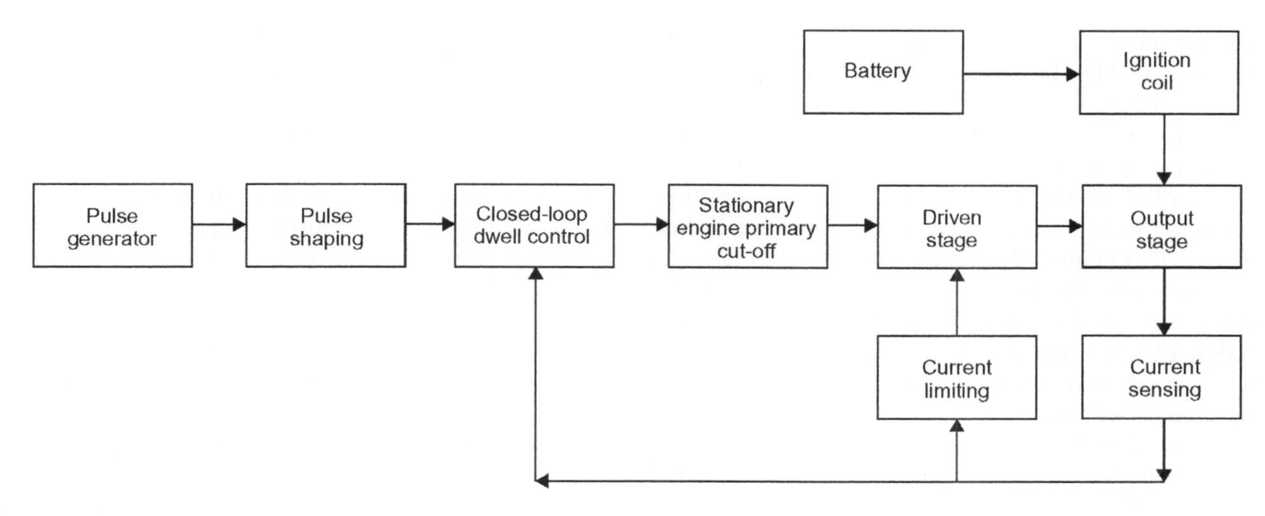

Figure 6.11 Closed-loop dwell control system.

DEFINITION

ESA: Electronic spark advance.

ESA systems have a major difference compared with earlier systems in that they operate digitally. Information about the operating requirements of a particular engine is programmed into memory inside the ECU. The data for storage in ROM is obtained from rigorous testing on an engine dynamometer and further development work in the vehicle under various operating conditions. Programmed ignition has several advantages:

- The ignition timing can be accurately matched to the individual application under a range of operating conditions.
- Other control inputs can be utilised such as coolant temperature and ambient air temperature.
- Starting is improved, fuel consumption and emissions are reduced, and idle control is better.
- Other inputs can be taken into account such as engine knock.
- The number of wearing components in the ignition system is considerably reduced.

Programmed ignition or ESA can be a separate system or included as part of the fuel control system. In order for the ECU to calculate suitable timing and dwell outputs, certain input information is required.

The crankshaft sensor consists of a permanent magnet, a winding and a soft iron core. It is mounted in proximity to a reluctor disc. The disc has 34 teeth spaced at 10° intervals around to periphery. It has two teeth missing 180° apart, at a known position BTDC. Many manufacturers use this technique with minor differences. As a tooth from the reluctor disc passes the core of the sensor, the reluctance of the magnetic circuit is changed. This induces a voltage in the winding, the frequency of the waveform being proportional to the engine speed. The missing tooth causes a 'missed' output wave and hence engine position can be determined.

Engine load is proportional to manifold pressure in that high-load conditions produce high pressure and lower-load conditions, such as cruise, produce lower pressure. Load sensors are therefore pressure transducers. They are either mounted in the ECU or as a separate unit and are connected to the inlet manifold with a pipe. The pipe often incorporates a restriction to damp out fluctuations and a vapour trap to prevent petrol fumes reaching the sensor.

KEY FACT

Engine load is proportional to manifold pressure in that high-load conditions produce high pressure and lower-load conditions, such as cruise, produce lower pressure.

Coolant temperature measurement is carried out by a simple thermistor. In many cases, the same sensor is used for the operation of the temperature gauge and to provide information to the fuel control system. A separate memory map is used to correct the basic timing settings. Timing may be retarded when the engine is cold to assist in more rapid warm-up.

Combustion knock can cause serious damage to an engine if sustained for long periods. This knock or detonation is caused by over advanced ignition timing. At variance with this is that an engine in general will run at its most efficient when the timing is advanced as far as possible. To achieve this, the data stored in the basic timing map will be as close to the knock limit of the engine as possible. The knock sensor provides a margin for error. The sensor itself is an accelerometer often of the piezoelectric type. It is fitted in the engine block between cylinders 2 and 3 on in-line four-cylinder engines. Vee engines require two sensors, one on each side. The ECU responds to signals from the knock sensor in the engine's knock window for each cylinder; this is often just a few degrees each side of TDC. This prevents clatter from the valve mechanism being interpreted as knock. The signal from the sensor is also filtered in the ECU to remove unwanted noise. If detonation is detected, the ignition timing is retarded on the fourth ignition pulse after detection (four-cylinder engine), in steps until knock is no longer detected. The steps vary between manufacturers, but approximately 2° is typical. The timing is then advanced slowly in steps of say 1° over a number of engine revolutions, until the advance required by memory is restored. This fine control allows the engine to be run very close to the knock limit without risk of engine damage.

KEY FACT

Combustion knock can cause serious damage to an engine if sustained for long periods.

Correction to dwell settings is required if the battery voltage falls, as a lower voltage supply to the coil will require a slightly larger dwell figure. This information is often stored in the form of a dwell correction map.

As the sophistication of systems has increased, the information held in the memory chips of the ECU has also increased. The earlier versions of programmed ignition system produced by Rover achieved accuracy in ignition timing of ±1.8°, whereas a conventional distributor is ±8°. The information, which is derived from dynamometer tests as well as running tests in the vehicle, is stored in ROM. The basic timing map consists of the correct ignition advance for 16 engine speeds and 16 engine load conditions.

A separate three-dimensional map is used which has eight speed and eight temperature sites. This is used to add corrections for engine coolant temperature to the basic timing settings. This improves driveability and can be used to decrease the warm-up time of the engine. The data is also subjected to an additional load correction below 70°C. Figure 6.12 shows a flow chart representing the logical selection of the optimum ignition setting. Note that the ECU will also make corrections to the dwell angle, both as a function of engine speed to provide constant energy output and due to changes in battery voltage. A lower battery voltage will require a slightly longer dwell and a higher battery voltage will require a slightly shorter dwell. A Windows® shareware program that simulates the ignition system (as well as many other systems) is available for download from my website.

The output of a system such as this programmed ignition is very simple. The output stage, in common with most electronic ignition, consists of a heavy-duty transistor which forms part of, or is driven by, a Darlington pair. This is simply to allow the high ignition primary current to be controlled. The switch-off point of the coil will control ignition timing and the switch-on point will control the dwell period.

The high-tension distribution is similar to a more conventional system. The rotor arm, however, is mounted on the end of the camshaft with the distributor cap positioned over the top. Figure 6.13 shows an early programmed ignition system.

6.6.8 Distributorless ignition

Distributorless ignition has all the features of ESA ignition systems but, by using a special type of ignition coil, outputs to the spark plugs without the need for an HT distributor. The system is generally only used on four-cylinder engines as the control system becomes too complex for higher numbers. The basic principle is that of the 'lost spark'. The distribution of the spark is achieved by using double-ended coils, which are fired alternately by the ECU. The timing is determined from a crankshaft speed and position sensor as well as load and other corrections. When one of the coils is fired, a spark is delivered to two engine cylinders, either 1 and 4 or 2 and 3. The spark delivered to the cylinder on the compression stroke will ignite the mixture as normal. The spark produced in the other cylinder will have no effect, as this cylinder will be just completing its exhaust stroke.

Because of the low compression and the exhaust gases in the 'lost spark' cylinder, the voltage used for the spark to jump the gap is only approximately 3 kV. This is similar to the more conventional rotor arm to cap voltage. The spark produced in the compression cylinder is therefore not affected.

KEY FACT

Because of the low compression and the exhaust gases in the 'lost spark' cylinder, the voltage used for the spark to jump the gap is only approximately 3 kV.

An interesting point here is that the spark on one of the cylinders will jump from the earth electrode to the spark plug centre. Many years ago, this would

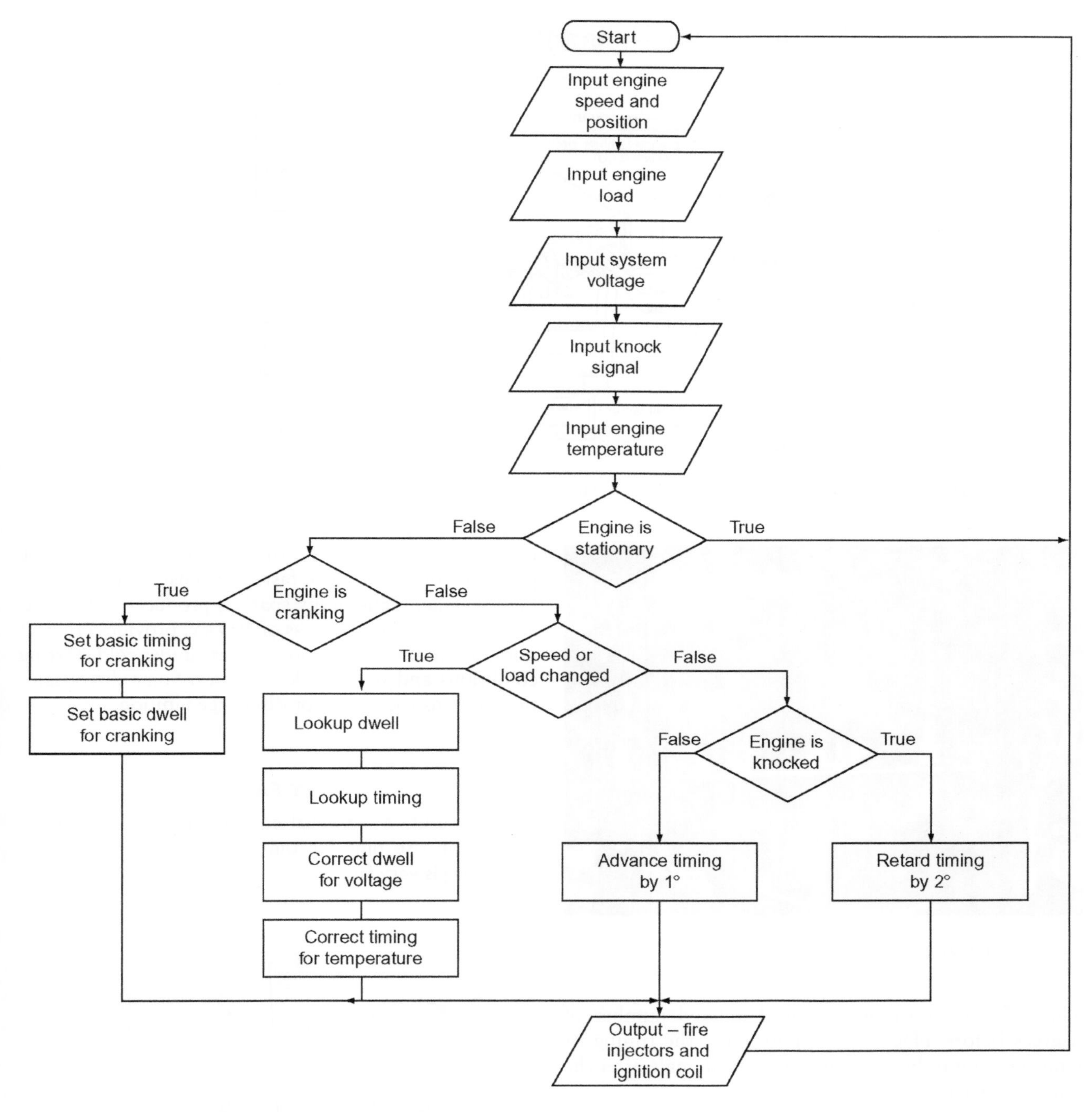

Figure 6.12 Electronic spark advance: representation of the process.

not have been acceptable, as the spark quality when jumping this way would not have been as good as when it jumps from the centre electrode. However, the energy available from modern constant-energy systems will produce a spark of suitable quality in either direction.

The direct ignition system (DIS) consists of three main components: the electronic module, a crankshaft position sensor and the DIS coil. In many systems, a MAP sensor is integrated in the module. The module functions in much the same way as has been described for the ESA system.

The crankshaft position sensor is similar in operation to the one described in the previous section. It is again a reluctance sensor and is positioned against the front of the flywheel or against a reluctor wheel just behind the front crankshaft pulley. The tooth pattern consists of 35 teeth. These are spaced at 10° intervals with a gap where the 36th tooth would be. The missing tooth is positioned at 90° BTDC for numbers 1 and 4 cylinders.

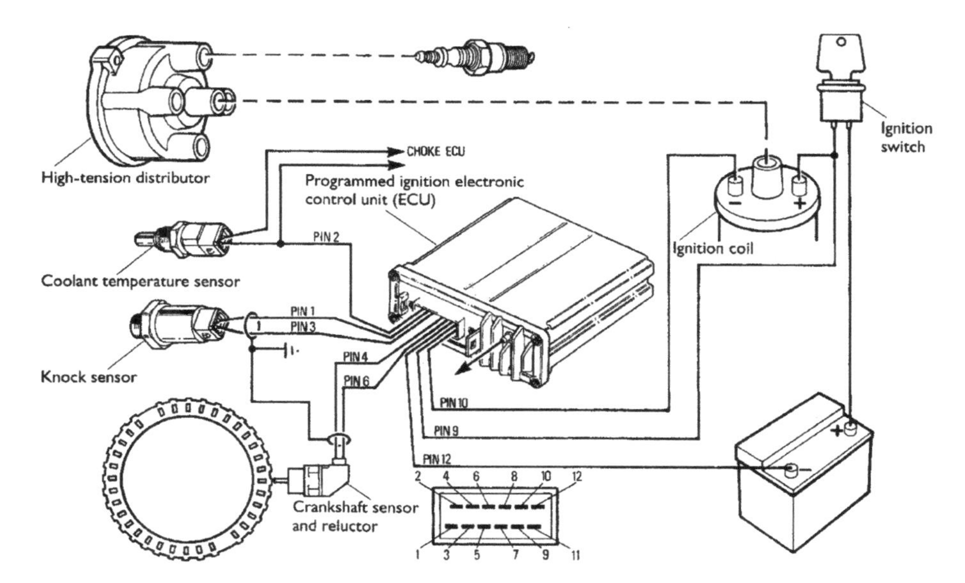

Figure 6.13 Programmed ignition system electronic spark advance (ESA).

Figure 6.14 DIS coil on a four-cylinder engine.

This reference position is placed a fixed number of degrees before TDC, in order to allow the timing or ignition point to be calculated as a fixed angle after the reference mark (Figure 6.14).

The low-tension winding is supplied with battery voltage to a centre terminal. The appropriate half of the winding is then switched to earth in the module. The high-tension windings are separate and are specific to cylinders 1 and 4, or 2 and 3.

6.6.9 Direct ignition

Direct ignition is in a way the follow-on from distributorless ignition. This system utilises an inductive coil for each cylinder. These coils are mounted directly on the spark plugs. Figure 6.15 shows a cross-section of the direct ignition coil. The use of an individual coil for each plug ensures that the rise time for the low-inductance primary winding is very fast. This ensures that a very high voltage, high-energy spark is produced. This voltage, which can be in excess of 400 kV, provides efficient initiation of the combustion process under cold starting conditions and with weak mixtures. Some direct ignition systems use capacitor discharge ignition.

KEY FACT

The use of an individual coil for each plug ensures that the rise time for the low-inductance primary winding is very fast.

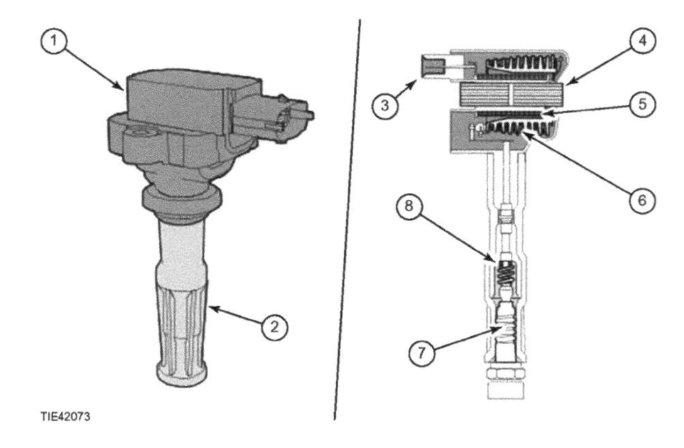

Figure 6.15 Direct ignition coil features: 1 – direct ignition coil; 2 – spark plug connector; 3 – low-voltage connection; outer: 4 – laminated iron core; 5 – primary winding; 6 – secondary winding; 7 – spark plug; 8 – highvoltage connection; inner: via spring contact.

Source: Ford Motor Company.

In order to switch the ignition coils, igniter units may be used. These can control up to three coils and are simply the power stages of the control unit but in a separate container. This allows less interference to be caused in the main ECU due to heavy current switching and shorter runs of wires carrying higher currents.

Ignition timing and dwell are controlled in a manner similar to the previously described programmed system. The one important addition to this on some systems is a camshaft sensor to provide information as to which cylinder is on the compression stroke. A system which does not require a sensor to determine which cylinder is on compression (engine position is known from a crank sensor) determines the information by initially firing all the coils. The voltage across the plugs allows measurement of the current for each spark and will indicate which cylinder is on its combustion stroke. This works because a burning mixture has a lower resistance. The cylinder with the highest current at this point will be the cylinder on the combustion stroke.

A further feature of some systems is the case when the engine is cranked over for an excessive time making flooding likely. The plugs are all fired with multisparks for a period of time after the ignition is left in the 'on' position for five seconds. This will burn away any excess fuel.

During difficult starting conditions, multisparking is also used by some systems during 70° of crank rotation before TDC. This assists with starting and then once the engine is running, the timing will return to its normal calculated position.

KEY FACT

Direct ignition is often described as coil on plug (COP).

6.6.10 Spark plugs

Figure 6.16 shows a standard spark plug. The centre electrode is connected to the top terminal by a stud. The electrode is constructed of a nickel-based alloy. Silver and platinum are also used for some applications. If a copper core is used in the electrode, this improves the thermal conduction properties.

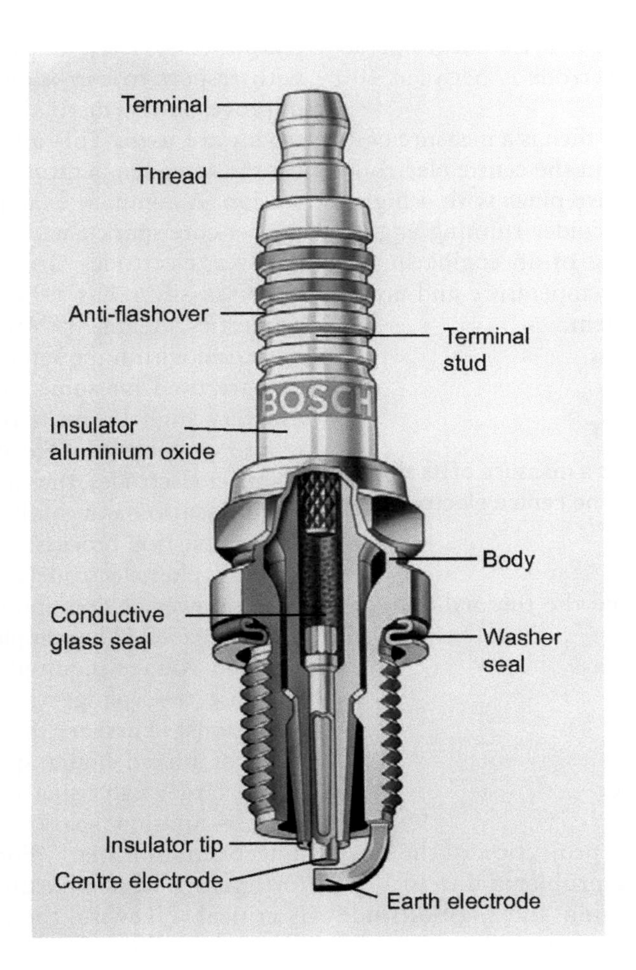

Figure 6.16 Construction of a copper-cored spark plug.

The insulating material is ceramic based and of a very high grade. The electrically conductive glass seal between the electrode and terminal stud is also used as a resistor. This resistor has two functions: first to prevent burn-off of the centre electrode and second to reduce radio interference. In both cases, the desired effect is achieved because the resistor damps the current at the instant of ignition.

Flashover or tracking down the outside of the plug insulation is prevented by ribs. These effectively increase the surface distance from the terminal to the metal fixing bolt, which is of course earthed to the engine.

Owing to the many and varied constructional features involved in the design of an engine, the range of temperatures a spark plug is exposed to can vary significantly. The operating temperature of the centre electrode of a spark plug is critical. If the temperature becomes too high then pre-ignition may occur, as the fuel air mixture may become ignited due to the incandescence of the plug electrode. On the other hand, if the electrode temperature is too low then carbon and oil fouling can occur, as deposits are not burnt off. Fouling of the plug nose can cause shunts (a circuit in parallel with the spark gap). It has been shown through experimentation and experience that the ideal operating temperature of the plug electrode is between 400 and 900°C.

The heat range of a spark plug then is a measure of its ability to transfer heat away from the centre electrode. A hot running engine will require plugs with a higher thermal loading ability than a colder running engine. Note that hot and cold running of an engine in this sense refers to the combustion temperature and not to the efficiency of the cooling system.

KEY FACT

The heat range of a spark plug is a measure of its ability to transfer heat away from the centre electrode.

The following factors determine the thermal capacity of a spark plug:

- insulator nose length;
- electrode material;
- thread contact length;
- projection of the electrode.

It has been found that a longer projection of the electrode helps to reduce fouling problems due to low-power operation, stop-go driving and high-altitude conditions. To use greater projection of the electrode, better-quality thermal conduction is required to allow suitable heat transfer at higher power outputs. Figure 6.17 shows the heat conducting paths of a spark plug together with changes in design for heat ranges.

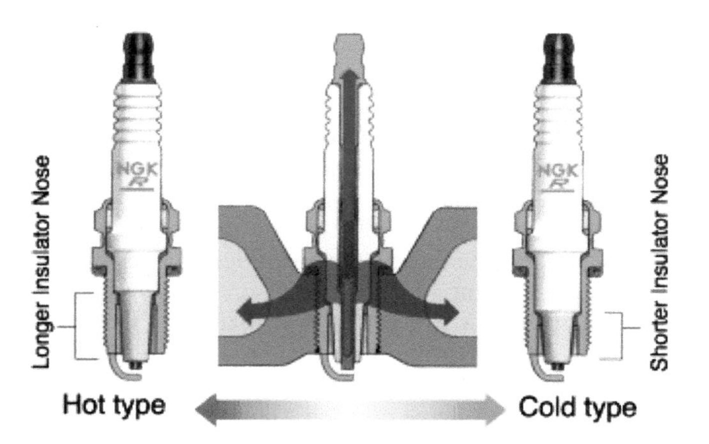

Figure 6.17 Hot and cold plugs. The cold plug is able to transfer heat more easily so is suitable for a hot engine.

For normal applications, alloys of nickel are used for the electrode material. Chromium, manganese, silicon and magnesium are examples of the alloying constituents. These alloys exhibit excellent properties with respect to corrosion and burn-off resistance. To improve on the thermal conductivity, compound electrodes are used. This allows a greater nose projection for the same temperature range as discussed in the last section. A common example of this type of plug is the copper core spark plug.

Silver electrodes are used for specialist applications, as silver has very good thermal and electrical properties. Again, with these plugs nose length can be increased within the same temperature range. Platinum tips are used for some spark plug applications due to the very high burn-off resistance of this material. It is also possible because of this to use much-smaller-diameter electrodes, thus increasing mixture accessibility. Platinum also has a catalytic effect, further accelerating the combustion process.

Spark plug electrode gaps in general have increased as the power of the ignition systems driving the spark has increased. The simple relationship between plug gap and voltage required is that as the gap increases, so must the voltage (leaving aside engine operating conditions). Further, the energy available to form a spark at a fixed engine speed is constant, which means that a larger gap using higher voltage will result in a shorter-duration spark. A smaller gap will allow a longer-duration spark. For cold starting an engine and for igniting weak mixtures, the duration of the spark is critical. Likewise, the plug gap must be as large as possible to allow easy access for the mixture to prevent quenching of the flame.

KEY FACT

Spark plug electrode gaps in general have increased as the power of the ignition systems driving the spark has increased.

The final choice is therefore a compromise reached through testing and development of a particular application. Plug gaps in the region of 0.6–1.2 mm seem to be the norm at present.

SAFETY FIRST

Warning: Caution/Achtung/Attention – High voltages can seriously damage your health.

6.7 DIAGNOSTICS – IGNITION SYSTEM

6.7.1 Testing procedure

The following procedure is generic and with a little adaptation can be applied to any ignition system. Refer to manufacturer's recommendations if in any doubt (Figure 6.18).

6.7.2 Ignition fault diagnosis table

Symptom	*Possible fault*
Engine rotates but does not start	Damp ignition components Spark plugs worn to excess Ignition system open circuit
Difficult to start when cold	Spark plugs worn to excess High resistance in ignition circuit
Engine starts but then stops immediately	Ignition wiring connection intermittent Ballast resistor open circuit (older cars)
Erratic idle	Incorrect plug gaps Incorrect ignition timing
Misfire at idle speed	Ignition coil or distributor cap tracking Spark plugs worn to excess Dwell incorrect (old systems)
Misfire through all speeds	Incorrect plugs or plug gaps HT leads breaking down
Lack of power	Ignition timing incorrect HT components tracking
Backfires	Incorrect ignition timing Tracking
Runs on when switched off	Ignition timing incorrect Carbon build-up in engine Idle speed too high Anti-run on device inoperative
Pinking or knocking under load	Ignition timing incorrect Ignition system electronic fault Knock sensor not working

Figure 6.19 shows a typical ignition timing light, essential to ensure correct settings where these are adjustable, or to check programmed advance systems for correct operation.

6.7.3 Ignition components and testing

Component	*Description*	*Test method*
Spark plug	Seals electrodes for the spark to jump across in the cylinder. Must withstand very high voltages, pressures and temperatures.	Compare nose condition to a manufacturer's chart. Inspect ignition secondary waveform, particularly when the engine is under load.
Ignition coil	Stores energy in the form of magnetism and delivers it to the distributor via the HT lead. Consists of primary and secondary windings.	Resistance checks of the primary and secondary windings: Primary: 1.5 Ω (ballasted) to 3 Ω Secondary: 5–10 kΩ
Ignition switch	Provides driver control of the ignition system and is usually also used to cause the starter to crank.	Voltage drop across the contacts.
Ballast resistor	Is shorted out during the starting phase to cause a more powerful spark. Also contributes towards improving the spark at higher speeds.	Resistance (often approximately 1.5 Ω) or check voltage at coil supply (approximately 6 or 7 V when the contact breakers are closed).
Contact breakers (breaker points)	Switches the primary ignition circuit on and off to charge and discharge the coil.	Voltage drop across them should not exceed approximately 0.2 V. General condition.
Capacitor (condenser)	Suppresses most of the arcing as the contact breakers open. This allows for a more rapid break of primary current and hence a more rapid collapse of the coil magnetism, which produces a higher voltage output.	Charge the capacitor up across a 12 V battery. Connect a digital meter and watch the voltage discharge from 12 V to almost 0 V over approximately five seconds.
HT distributor	Directs the spark from the coil to each cylinder in a pre-set sequence.	Visual inspection for signs of tracking (conducting lines) and contamination.
Centrifugal advance (engine speed)	Changes the ignition timing with engine speed. As speed increases, the timing is advanced.	Measure the timing at certain speeds using an 'advance' timing light. Refer to data.
Vacuum advance (engine load)	Changes timing depending on engine load. On conventional systems, the vacuum advance is most important during cruise conditions.	Apply a known vacuum and note timing changes (often just sucking on the pipe and noting movement is adequate).

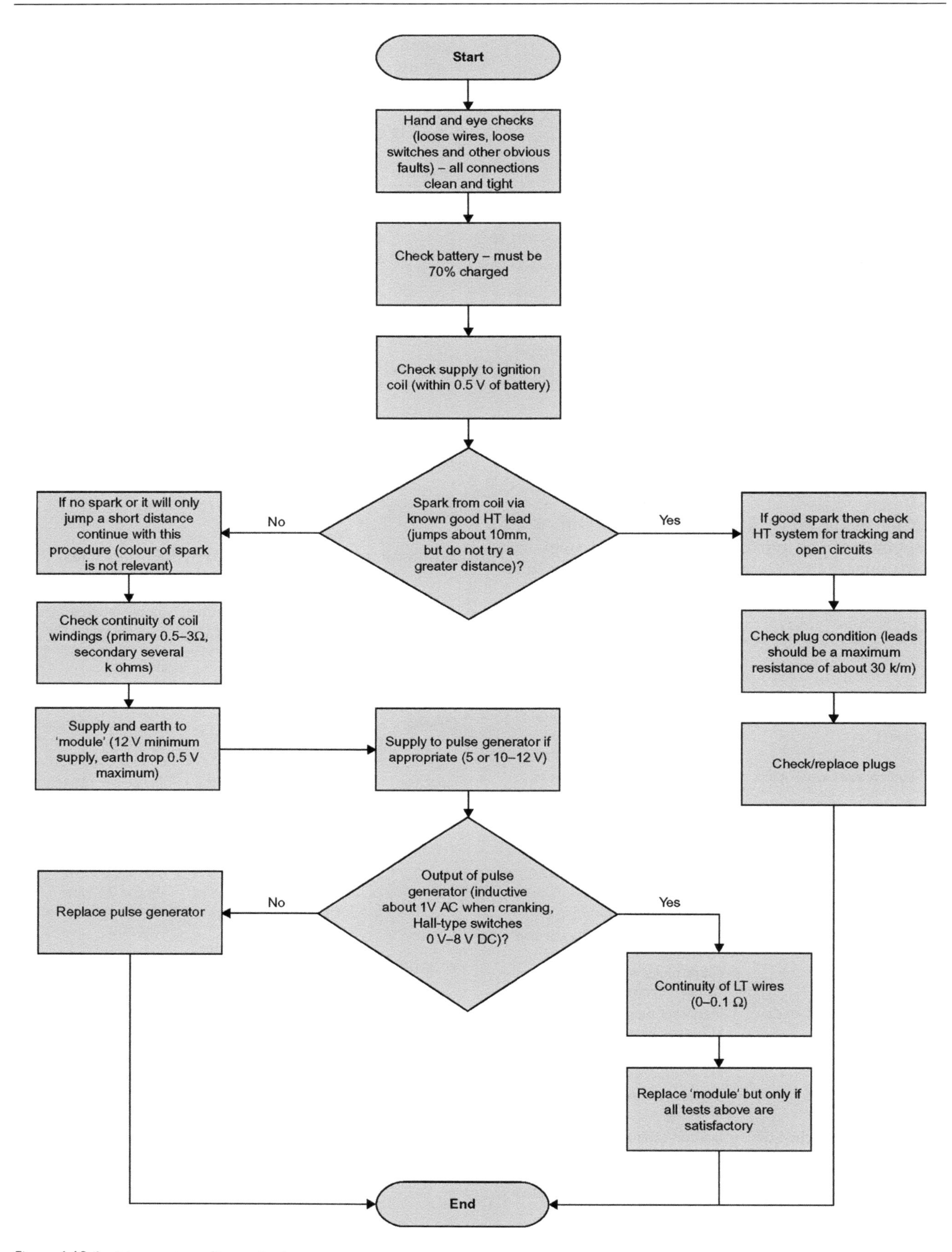

Figure 6.18 Ignition system diagnosis chart.

Figure 6.19 Timing light (used on earlier cars).

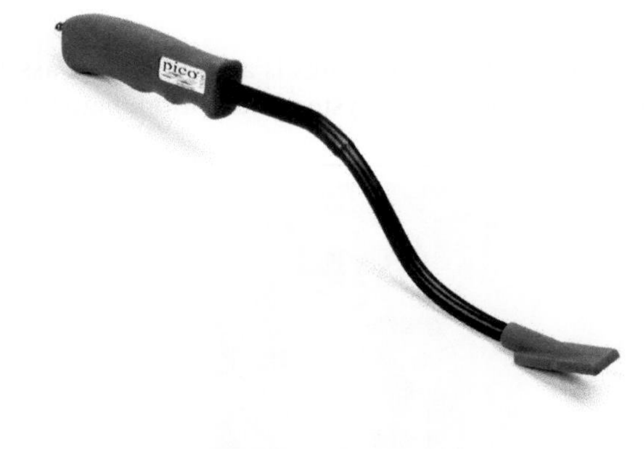

Figure 6.20 COP ignition probe.

Source: PicoScope.

6.7.4 DIS diagnostics

The DIS system is very reliable due to the lack of any moving parts. Some problems can be experienced when trying to examine HT oscilloscope patterns due to the lack of a king lead. This can often be overcome with a special adapter, but it is still necessary to move the sensing clip to each lead in turn.

The DIS coil can be tested with an ohmmeter. The resistance of each primary winding should be 0.5 Ω and the secondary windings between 11 and 16 kΩ. The coil will produce in excess of 37 kV in an open circuit condition. The plug leads have integral retaining clips to prevent water ingress and vibration problems. The maximum resistance for the HT leads is 30 kΩ per lead.

No service adjustments are possible with this system, with the exception of octane adjustment on some models. This involves connecting two pins together on the module for normal operation, or earthing one pin or the other to change to a different fuel. The actual procedure must be checked with the manufacturer for each particular model.

6.7.5 Coil on plug (COP) diagnostics

Connecting a normal probe to a COP ignition system is very difficult. It is sometimes possible to back probe the primary circuit but not always. A signal probe is the easiest and fastest non-intrusive way to check coil-on-plug ignition coils and spark plugs. The probe shown in Figure 6.20 will display a scope pattern of the secondary, faster than scoping the primary. It is simply held on top of the ignition coil.

A typical known good waveform is show in Figure 6.21.

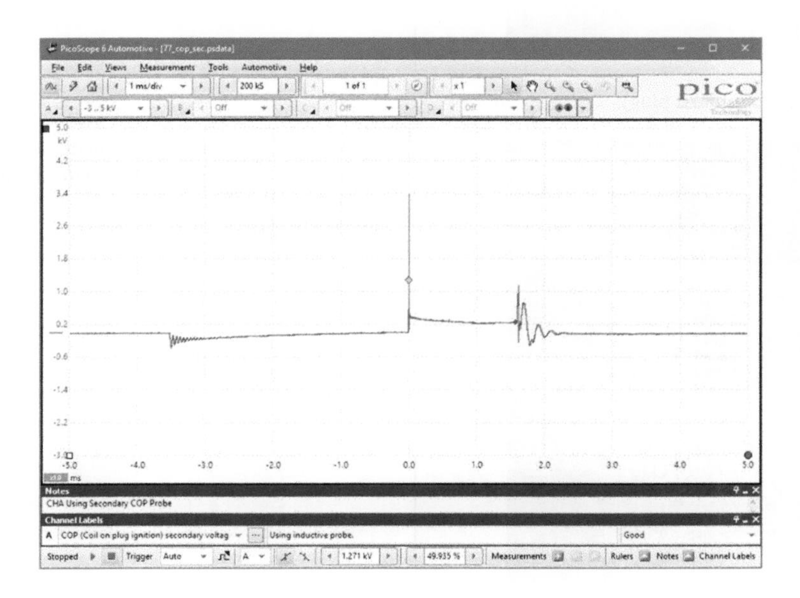

Figure 6.21 Coil-on-plug ignition secondary waveform.

Source: PicoScope.

6.7.6 Spark plugs

Examination of the spark plugs is a good way of assessing engine and associated systems condition. Figure 6.22 shows a new plug and Figures 6.23–6.27 show various conditions with diagnostic notes added.

Figure 6.22 New spark plug.

Use this image to compare with used spark plugs. Note in particular, on this standard design, how the end of the nose is flat and that the earth/ground electrode has a consistent size and shape.

Figure 6.23 Carbon fouled (standard plug).

This plug has black deposits over the centre electrode and insulator in particular. It is likely that this engine was running too rich – or on an older vehicle the choke was used excessively. However, carbon fouling may also be due to:

- poor-quality spark due to ignition fault;
- incorrect plug gap;
- overretarded timing;
- loss of cylinder compression;
- prolonged low-speed driving;
- incorrect (too cold) spark plug fitted.

Figure 6.24 Deposits.

The deposits on this plug are most likely to be caused by oil leaking into the cylinder. Alternatively, poor-quality fuel mixture supply or very short, cold engine operation could result in a similar condition.

Figure 6.25 Damaged insulation.

A plug that is damaged in this way is because of either overheating or impact damage. Impact is most likely in this case. The damage can of course be caused as the plug is being fitted! However, in this case a possible cause would be that the reach was too long for the engine and the piston hit the earth/ground electrode, closing up the gap and breaking the insulation.

Figure 6.26 Carbon fouled (platinum plug).

The carbon build-up on this plug would suggest an incorrect mixture. However, before diagnosing a fault based on spark plug condition, make sure the engine has been run up to temperature – ideally by a good road test. The engine from which this plug was removed is in good condition – it had just been started from cold and only run for a few minutes.

Figure 6.27 Overheating.

When a plug overheats, the insulator tips becomes glossy and/or they are blistered or melted away. The electrodes also wear quickly. Excessive overheating can result in the electrodes melting and serious piston damage is likely to occur. Causes of overheating are:

- overadvanced ignition;
- mixture too lean;
- cooling system fault;
- incorrect plug (too hot);
- incorrect fuel (octane low).

KEY FACT

Examination of the spark plugs is a good way of assessing engine and associated systems condition.

6.8 EMISSIONS

6.8.1 Introduction

Table 6.3 lists the four main exhaust emissions which are hazardous to health together with a short description of each.

Table 6.4 describes two further sources of emissions from a vehicle.

6.8.2 Exhaust gas recirculation (EGR)

Exhaust gas recirculation (EGR) is used primarily to reduce peak combustion temperatures and hence the production of nitrogen oxides (NO_x). EGR can be either internal due to valve overlap, or external via a simple arrangement of pipes and a valve (Figure 6.28 shows an

Figure 6.28 EGR valve.

Source: Delphi Media.

Table 6.3 Exhaust emissions

Substance	*Description*
Carbon monoxide (CO)	This gas is very dangerous even in low concentrations. It has no smell or taste and is colourless. When inhaled, it combines in the body with the red blood cells preventing them from carrying oxygen. If absorbed by the body, it can be fatal in a very short time.
Nitrogen oxides (NOx)	Oxides of nitrogen are colourless and odourless when they leave the engine, but as soon as they reach the atmosphere and mix with more oxygen, nitrogen oxides are formed. They are reddish brown and have an acrid and pungent smell. These gases damage the body's respiratory system when inhaled. When combined with water vapour nitric acid can be formed, which is very damaging to the windpipe and lungs. Nitrogen oxides are also a contributing factor to acid rain.
Hydrocarbons (HC)	A number of different hydrocarbons are emitted from an engine and are part of unburnt fuel. When they mix with the atmosphere, they can help to form smog. It is also believed that hydrocarbons may be carcinogenic.
Particulate matter (PM)	This heading in the main covers lead and carbon. Lead was traditionally added to petrol to slow its burning rate to reduce detonation. It is detrimental to health and is thought to cause brain damage, especially in children. Lead will eventually be phased out as all new engines now run on unleaded fuel. Particles of soot or carbon are more of a problem on diesel-fuelled vehicles and these now have limits set by legislation.

Table 6.4 Emission sources

Source	*Comments*
Fuel evaporation from the tank and system	Fuel evaporation causes hydrocarbons to be produced. The effect is greater as temperature increases. A charcoal canister is the preferred method for reducing this problem. The fuel tank is usually run at a pressure just under atmospheric by a connection to the intake manifold drawing the vapour through the charcoal canister. This must be controlled by the management system, however, as even a 1% concentration of fuel vapour would shift the lambda value by 20%. This is done by using a 'purge valve', which under some conditions is closed (full-load and idle for example) and can be progressively opened under other conditions. The system monitors the effect by use of the lambda sensor signal.
Crankcase fumes (blow by)	Hydrocarbons become concentrated in the crankcase mostly due to pressure blowing past the piston rings. These gases must be conducted back into the combustion process. This is usually via the air intake system. This is described as positive crankcase ventilation.

example) connecting the exhaust manifold back to the inlet manifold. A proportion of exhaust gas is simply returned to the inlet side of the engine.

KEY FACT

Exhaust gas recirculation (EGR) is used primarily to reduce peak combustion temperatures and hence the production of nitrogen oxides (NO_x).

This process is controlled electronically as determined by a ROM in the ECU. This ensures that driveability is not affected and also that the rate of EGR is controlled. If the rate is too high, then the production of hydrocarbons increases.

One drawback of EGR systems is that they can become restricted by exhaust residue over a period of time, thus changing the actual percentage of recirculation. However, valves that reduce this particular problem are now available.

6.8.3 Catalytic converters

Stringent regulations in most parts of the world have made the use of a catalytic converter almost indispensable. The three-way catalyst (TWC) is used to great effect by most manufacturers. It is a very simple device and looks similar to a standard exhaust box. Note that in order to operate correctly, however, the engine must be run at or very near to stoichiometry. This is to ensure that the right 'ingredients' are available for the catalyst to perform its function.

KEY FACT

For a three-way catalyst (TWC) to operate correctly, the engine must be run at or very near to stoichiometry.

Figure 6.29 shows some new metallic substrates for use inside a catalytic converter. There are many types of hydrocarbons, but the example illustrates the main reaction. Note that the reactions rely on some CO being produced by the engine in order to reduce the NO_x. This is one of the reasons that manufacturers have been forced to run the engine at stoichiometry. The legislation has tended to stifle the development of lean burn techniques. The fine details of the emission regulations can in fact have a very marked effect on the type of reduction techniques used. The main reactions in the 'cat' are as follows:

- Reduction of nitrogen oxides to nitrogen and oxygen:

$$2NOx \rightarrow xO_2 + N_2$$

Figure 6.29 Catalytic converter metal substrates.

- Oxidation of carbon monoxide to carbon dioxide:

$$2CO + O_2 \rightarrow 2CO_2$$

- Oxidation of unburnt hydrocarbons (HC) to carbon dioxide and water:

$$C_xH_{2x+2} + [(3x + 1)/2]O_2 \rightarrow xCO_2 + (x + 1)\ H_2O$$

Noble metals are used for the catalysts; platinum promotes the oxidation of HC and CO, and rhodium helps the reduction of NO_x. The whole three-way catalytic converter contains only about 3–4 g of the precious metals.

The ideal operating temperature range is from approximately 400 to 800°C. A serious problem to counter is the delay in the catalyst reaching this temperature. This is known as catalyst light-off time. Various methods have been used to reduce this time as significant emissions are produced before light-off occurs. Electrical heating is one solution, as is a form of burner which involves lighting fuel inside the converter. Another possibility is positioning the converter as part of the exhaust manifold and down pipe assembly. This greatly reduces light-off time, but gas flow problems, vibration and excessive temperature variations can be problems that reduce the potential life of the unit.

Catalytic converters can be damaged in two ways. The first is by the use of leaded fuel which causes lead compounds to be deposited on the active surfaces (Figure 6.30), thus reducing effective area. The second is engine misfire which can cause the catalytic converter to overheat due to burning inside the unit. BMW, for example, use a system on some vehicles where a sensor monitors output of the ignition HT system and will not allow fuel to be injected if the spark is not present.

For a catalytic converter to operate at its optimum conversion rate to oxidise CO and HC while reducing NO_x, a narrow band within 0.5% of lambda value 1 is essential. Lambda sensors in use at present tend to operate within approximately 3% of the lambda mean value. When a catalytic converter is in prime condition,

Figure 6.30 Catalytic converter ceramic substrates.

this is not a problem due to storage capacity within the converter for CO and O_2. Damaged converters, however, cannot store sufficient quantity of these gases and hence become less efficient. The damage as suggested earlier in this section can be due to overheating or by 'poisoning' due to lead or even silicon. If the control can be kept within 0.5% of lambda, the converter will continue to be effective even if damaged to some extent. Sensors which can work to this tolerance are becoming available. A second sensor fitted after the converter can be used to ensure ideal operation.

6.9 DIAGNOSTICS – EMISSIONS

6.9.1 Testing procedure

If the reported fault is incorrect exhaust emissions, the procedure shown in Figure 6.31 should be utilised.

6.9.2 Emissions fault diagnosis table

Symptom	*Possible cause*
EGR valve sticking	Build-up of carbon Electrical fault
High CO and high HC	Rich mixture Blocked air filter Damaged catalytic converter Engine management system fault
Low CO and high HC	Misfire Fouled plug(s) Weak mixture
Low CO and low or normal HC	Exhaust leak Fouled injector

6.10 FUEL INJECTION

6.10.1 Introduction

The ideal air/fuel ratio is approximately 14.7:1. This is the theoretical amount of air required to completely burn the fuel. It is given a 'lambda (λ)' value of 1.

DEFINITION

λ = actual air quantity ÷ theoretical air quantity.

Air/fuel ratio is altered during the following operating conditions of an engine to improve its performance, driveability, consumption and emissions:

- **cold starting** – richer mixture is needed to compensate for fuel condensation and improve driveability;
- **load or acceleration** – richer to improve performance;
- **cruise or light loads** – weaker for economy;
- **overrun** – very weak (if any) fuel, to improve emissions and economy.

The more accurately the air/fuel ratio is controlled to cater for external conditions, the better the overall operation of the engine.

The major advantage, therefore, of a fuel injection system is accurate control of the fuel quantity injected into the engine. The basic principle of fuel injection is that if petrol is supplied to an injector (electrically controlled valve), at a constant differential pressure, then the amount of fuel injected will be directly proportional to the injector open time.

KEY FACT

The major advantage of a fuel injection system is accurate control of the fuel mixture.

Most systems are now electronically controlled even if containing some mechanical metering components. This allows the operation of the injection system to be very closely matched to the requirements of the engine. This matching process is carried out during development on test beds and dynamometers, as well as development in the car. The ideal operating data for a large number of engine operating conditions is stored in a ROM in the ECU. Close control of fuel quantity injected allows the optimum setting for mixture strength when all operating factors are taken into account.

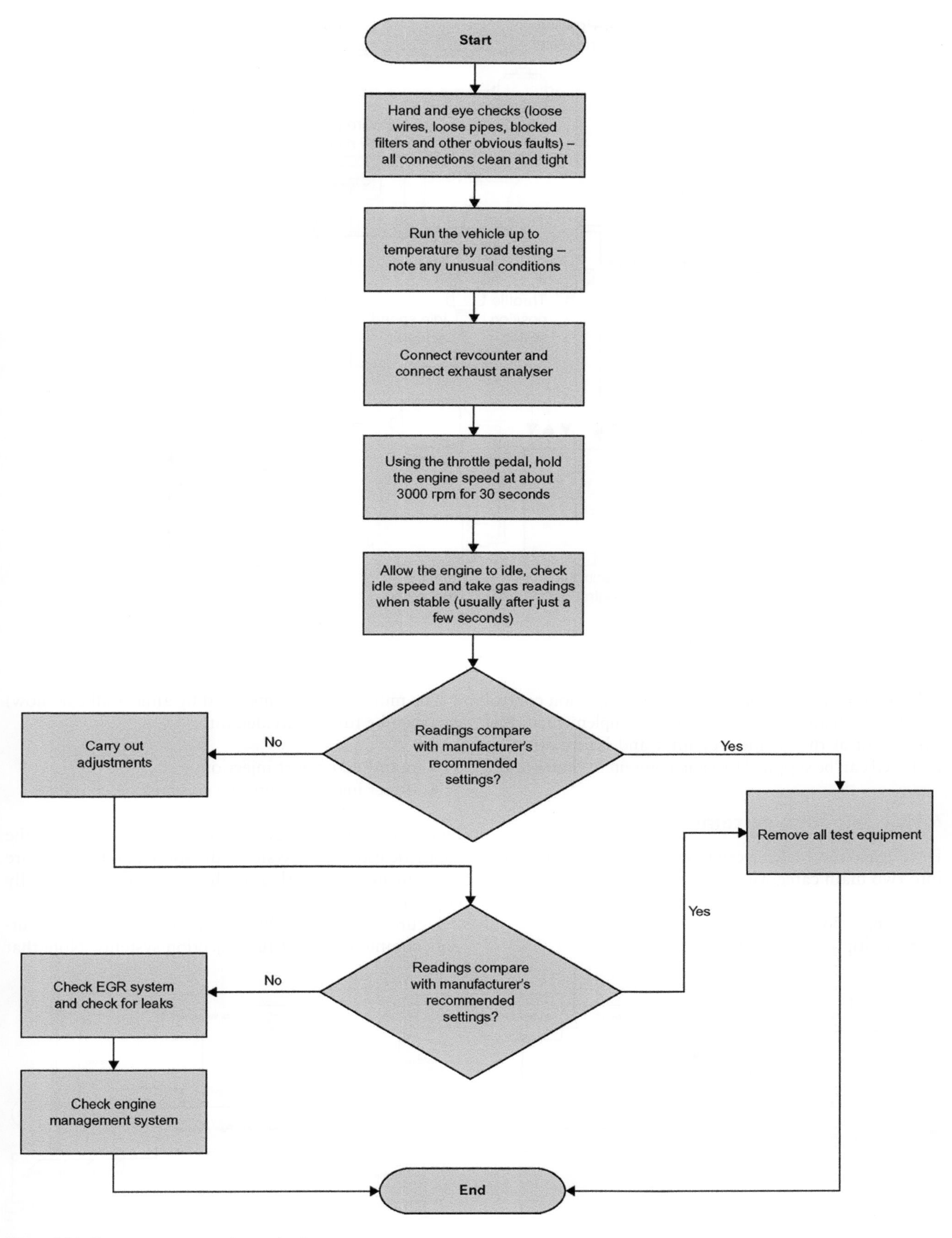

Figure 6.31 Emissions systems diagnosis chart.

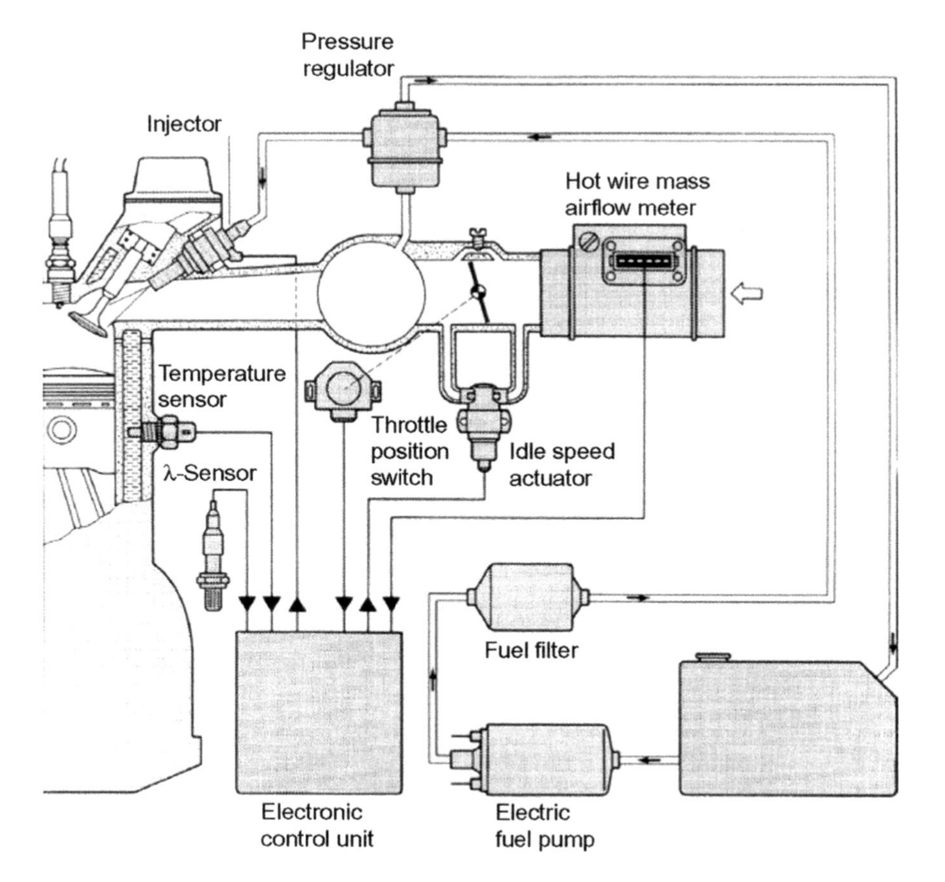

Figure 6.32 Fuel injection system layout.

Further advantages of electronic fuel injection control are that overrun cut-off can easily be implemented, fuel can be cut at the engines rev limit and information on fuel used can be supplied to a trip computer (Figure 6.32).

6.10.2 Injection systems

Petrol/gasoline fuel injection systems can be classified into two main categories:

- single point;
- multipoint injection.

The multipoint systems (used by almost all cars now) can then be further divided into:

- manifold or port injection;
- direct injection (into the combustion chamber).

Figure 6.33 shows these techniques. Depending on the sophistication of the system, idle speed and idle mixture adjustment can be either mechanically or electronically controlled.

Figure 6.34 shows a block diagram of inputs and outputs common to most fuel injection systems. Note that

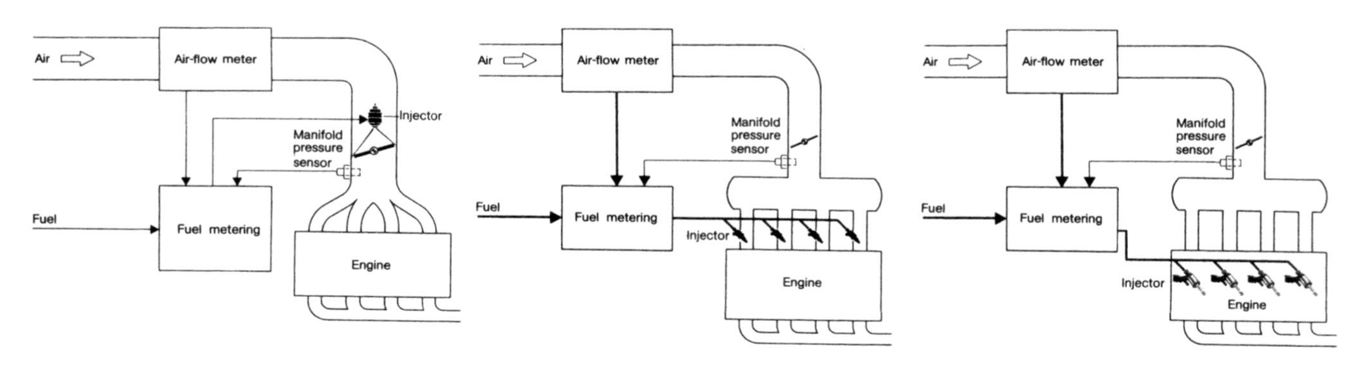

Figure 6.33 Figure Injection methods (single point, multipoint port, multipoint direct).

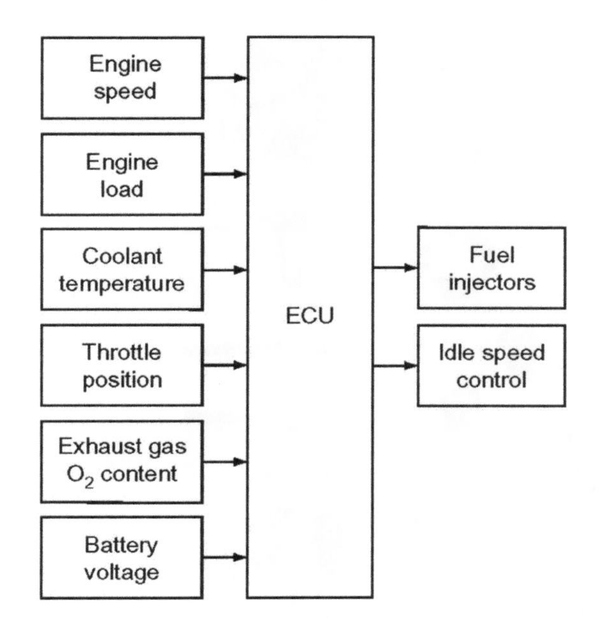

Figure 6.34 Fuel injection simplified block diagram.

the two most important input sensors to the system are speed and load. The basic fuelling requirement is determined from these inputs in a similar way to the determination of ignition timing.

An engine's fuelling requirements are stored as part of a ROM chip in the ECU. When the ECU has determined the 'lookup value' of the fuel required (injector open time), corrections to this figure can be added for battery voltage, temperature, throttle change or position and fuel cut-off. Figure 6.35 shows an injection system ECU.

Idle speed and fast idle are also generally controlled by the ECU and a suitable actuator. It is also possible to have a form of closed-loop control with electronic fuel injection. This involves a lambda sensor to monitor exhaust gas oxygen content. This allows very accurate control of the mixture strength, as the oxygen content

Figure 6.35 Engine management ECU.

of the exhaust is proportional to the air/fuel ratio. The signal from the lambda sensor is used to adjust the injector open time.

KEY FACT

A lambda sensor monitors exhaust gas oxygen content.

6.10.3 Fuel injection components

Many of the sensors and actuators associated with fuel injection are covered in Chapter 4. Figure 6.36 shows those associated with an earlier Motronic injection system. The main fuel components are outlined below.

6.10.3.1 Airflow meter

The type shown is a hot-wire meter. This allows direct measurement of air mass as temperature compensation is built in. The air quantity helps to determine the fuel required.

6.10.3.2 Electronic control unit (ECU)

This is also referred to as the electronic control module (ECM). The circuitry to react to the sensor signals by controlling the actuators is in the ECU. The data is stored in ROM.

6.10.3.3 Fuel pump

Pressurised fuel is supplied to the injectors. Most pumps work on the centrifugal roller principle. The pump ensures a constant supply of fuel to the fuel rail. The volume in the rail acts as a swamp to prevent pressure fluctuations as the injectors operate. The pump must be able to maintain a pressure of approximately 3 bar.

6.10.3.4 Fuel filter

The fuel supplied to the injectors must be free from any contamination or else the injector nozzle will be damaged or blocked.

6.10.3.5 Lambda sensor

The quantity of oxygen in the exhaust, when accurately measured, ensures that the fuel air mixture is kept within the lambda window (0.97–1.03).

6.10.3.6 Temperature sensor

A simple thermistor is used to determine the engine coolant temperature.

Figure 6.36 Earlier Motronic system components.

6.10.3.7 Fuel injectors

These are simple solenoid-operated valves designed to operate very quickly and produce a finely atomised spray pattern.

6.10.3.8 Idle or fast idle control actuator

The rotary actuator is used to provide extra air for cold fast idle conditions and to control idle speed. It is supplied with a variable duty cycle square wave.

6.10.3.9 Fuel pressure regulator

This device is to ensure a constant differential pressure across the injectors. It is a mechanical device and has a connection to the inlet manifold.

6.10.3.10 Throttle position switch

This is used to supply information as to whether the throttle is at idle, full load or somewhere in between.

6.10.4 Fuel mixture calculation

The quantity of fuel to be injected is determined primarily by the quantity of air drawn into the engine. This is dependent on two factors:

1. engine speed (rpm);
2. engine load (inlet manifold pressure).

This speed load characteristic is held in the ECU memory in ROM lookup tables.

KEY FACT

The quantity of fuel needed is determined by the mass of air drawn into the engine.

A sensor connected to the manifold by a pipe senses manifold absolute pressure. The sensor is fed with a stabilised 5 V supply and transmits an output voltage according to the pressure. The sensor is fitted away from the manifold and hence a pipe is required to connect it. The output signal varies between approximately 0.25 V at 0.17 bar to approximately 4.75 V at 1.05 bar. The density of air varies with temperature such that the information from the MAP sensor on air quantity will be incorrect over wide temperature variations. An air temperature sensor is used to inform the ECU of the inlet air temperature such that the quantity of fuel injected may be corrected. As the temperature of air decreases, its density increases and hence the quantity of fuel injected must also be increased. The other method of sensing engine load is direct measurement of air intake quantity using a hot-wire meter or a flap-type airflow meter.

To operate the injectors, the ECU needs to know, in addition to air pressure, the engine speed to determine the injection quantity. The same flywheel sensor used by the ignition system provides this information. All four injectors operate simultaneously once per engine revolution, injecting half of the required fuel. This helps to ensure balanced combustion. The start of injection varies according to ignition timing.

A basic open period for the injectors is determined by using the ROM information relating to manifold pressure and engine speed. Two corrections are then made, one relative to air temperature and another depending on whether the engine is idling, at full or partial load.

The ECU then carries out another group of corrections, if applicable:

- after-start enrichment;
- operational enrichment;
- acceleration enrichment;
- weakening on deceleration;
- cut-off on overrun;
- reinstatement of injection after cut-off;
- correction for battery voltage variation.

Under starting conditions, the injection period is calculated differently. This is determined mostly from a set figure varied as a function of temperature.

The coolant temperature sensor is a thermistor and is used to provide a signal to the ECU relating to engine coolant temperature. The ECU can then calculate any corrections to fuel injection and ignition timing. The operation of this sensor is the same as the air temperature sensor.

The throttle potentiometer is fixed on the throttle butterfly spindle and informs the ECU of throttle position and rate of change of throttle position. The sensor provides information on acceleration, deceleration and whether the throttle is in the full-load or idle position. It comprises a variable resistance and a fixed resistance. As is common with many sensors, a fixed supply of 5 V is provided and the return signal will vary approximately between 0 and 5 V. The voltage increases as the throttle is opened.

6.11 DIAGNOSTICS – FUEL INJECTION SYSTEMS

6.11.1 Testing procedure

The following procedure is generic and with a little adaptation can be applied to any fuel injection system. Refer to manufacturer's recommendations if in any doubt. It is assumed that the ignition system is operating correctly. Most tests are carried out while cranking the engine.

SAFETY FIRST

Warning: Caution/Achtung/Attention – Burning fuel can seriously damage your health.

6.11.2 Fuel injection fault diagnosis table

Symptom	*Possible fault*
Engine rotates but does not start	No fuel in the tank! Air filter dirty or blocked Fuel pump not running No fuel being injected
Difficult to start when cold	Air filter dirty or blocked Fuel system wiring fault Enrichment device not working Coolant temperature sensor short circuit
Difficult to start when hot	Air filter dirty or blocked Fuel system wiring fault Coolant temperature sensor open circuit
Engine starts but then stops immediately	Fuel system contamination Fuel pump or circuit fault (relay) Intake system air leak
Erratic idle	Air filter blocked Inlet system air leak Incorrect CO setting Idle air control valve not operating Fuel injectors not spraying correctly
Misfire through all speeds	Fuel filter blocked Fuel pump delivery low Fuel tank ventilation system blocked
Engine stalls	Idle speed incorrect CO setting incorrect Fuel filter blocked Air filter blocked Intake air leak Idle control system not working
Lack of power	Fuel filter blocked Air filter blocked Low fuel pump delivery Fuel injectors blocked
Backfires	Fuel system fault (airflow sensor on some cars) Ignition timing

6.12 DIESEL INJECTION

6.12.1 Introduction

The basic principle of the four-stroke diesel engine is very similar to the petrol system. The main difference is that the mixture formation takes place in the cylinder combustion chamber as the fuel is injected under very high pressure. The timing and quantity of the fuel injected is important from the usual issues of performance, economy and emissions (Figure 6.37).

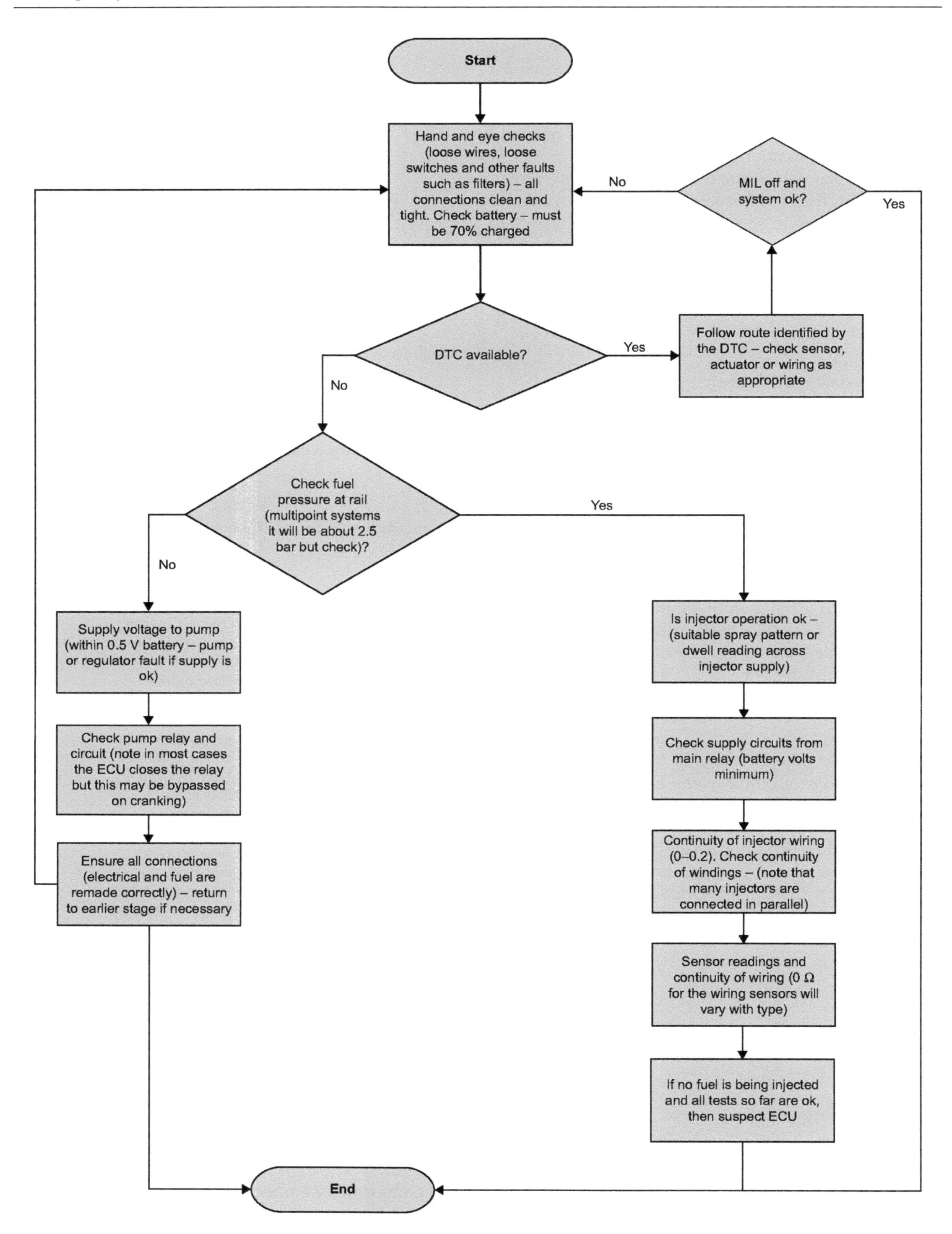

Figure 6.37 Fuel injection system diagnosis chart.

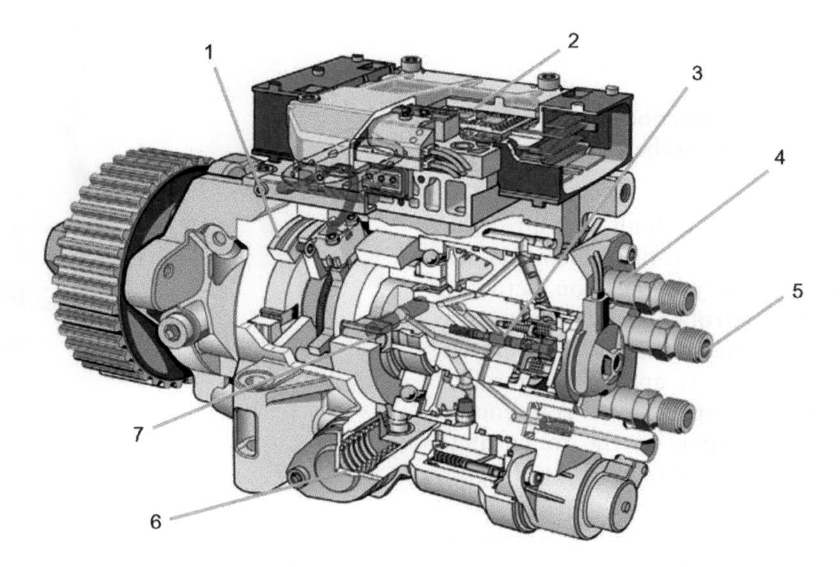

Figure 6.38 Solenoid-valve controlled radial-piston distributor pump: 1– sensor (position/timing); 2 – ECU; 3 – high-pressure solenoid valve needle; 4 – solenoid; 5 – outlets to injectors; 6 – timing device (ignition advance mechanism); 7 – radial-piston high-pressure pump.

Source: Bosch Media.

Fuel is metered into the combustion chamber by way of a high-pressure pump connected to injectors via heavy-duty pipes. When the fuel is injected, it mixes with the air in the cylinder and will self-ignite at approximately 800°C. The mixture formation in the cylinder is influenced by the following factors.

The timing of a diesel fuel injection pump to an engine is usually done using start of delivery as the reference mark. The actual start of injection, in other words when fuel starts to leave the injector, is slightly later than the start of delivery, as this is influenced by the compression ratio of the engine, the compressibility of the fuel and the length of the delivery pipes. This timing has a great effect on the production of carbon particles (soot) if too early and increases the hydrocarbon emissions if too late.

The duration of the injection is expressed in degrees of crankshaft rotation in milliseconds. This clearly influences fuel quantity, but the rate of discharge is also important. This rate is not constant due to the mechanical characteristics of the injection pump.

Pressure of injection will affect the quantity of fuel, but the most important issue here is the effect on atomisation. At higher pressures, the fuel will atomise into smaller droplets with a corresponding improvement in the burn quality. Indirect injection systems use pressures up to approximately 350 bar and direct injection systems can be up to approximately 1000 bar. Emissions of soot are greatly reduced by higher-pressure injection.

The direction of injection must match very closely the swirl and combustion chamber design. Deviations of only 2° from the ideal can greatly increase particulate emissions.

Diesel engines do not in general use a throttle butterfly, as the throttle acts directly on the injection pump to control fuel quantity. At low speeds in particular the very high excess air factor ensures complete burning and very low emissions. Diesel engines operate where possible with an excess air factor even at high speeds.

Figure 6.38 shows a typical diesel fuel injection pump. Detailed operation of the components is beyond the scope of this book. The principles and problems are the issues under consideration, in particular the way electronics can be employed to solve some of the problems.

6.12.2 Electronic control of diesel injection

The advent of electronic control over the diesel injection pump has allowed many advances over the purely mechanical system. The production of high pressure for injection is, however, still mechanical with all current systems. The following advantages are apparent over the non-electronic control system:

- more precise control of fuel quantity injected;
- better control of start of injection;
- idle speed control;
- control of EGR;
- drive by wire system (potentiometer on throttle pedal);
- an anti-surge function;
- output to data acquisition systems, etc.;
- temperature compensation;
- cruise control.

KEY FACT

Electronic control of diesel injection has allowed many advances over the purely mechanical system.

Because fuel must be injected at high pressure, the hydraulic head, pressure pump and drive elements are still used. An electromagnetic moving iron actuator adjusts the position of the control collar, which in turn controls the delivery stroke and therefore the injected quantity of fuel. Fuel pressure is applied to a roller ring and this controls the start of injection. A solenoid-operated valve controls the supply to the roller ring. These actuators together allow control of start of injection and injection quantity.

Ideal values for fuel quantity and timing are stored in memory maps in the ECU. The injected fuel quantity is calculated from the accelerator position and the engine speed. Start of injection is determined from the following:

- fuel quantity;
- engine speed;
- engine temperature;
- air pressure.

The ECU is able to compare start of injection with actual delivery from a signal produced by the needle motion sensor in the injector.

Control of EGR is a simple solenoid valve. This is controlled as a function of engine speed, temperature and injected quantity. The ECU is also in control of the stop solenoid and glow plugs via a suitable relay.

6.12.3 Common rail (CR) diesel systems

The development of diesel fuel systems is continuing, with many new electronic changes to the control and injection processes. One of the latest developments is the 'common rail' system, operating at very high injection pressures. It also has piloted and phased injection to reduce noise and vibration (Figure 6.39).

KEY FACT

Common rail diesel injection operates at very high injection pressures and has piloted and phased injection to reduce noise and vibration.

The common rail system has made it possible, on small, high-speed diesel engines, to have direct injection, whereas previously they would have been of indirect injection design. These developments are showing improvements in fuel consumption and performance of up to 20% over the earlier indirect injection engines of a similar capacity. The common rail injection system can be used on the full range of diesel engine capacities.

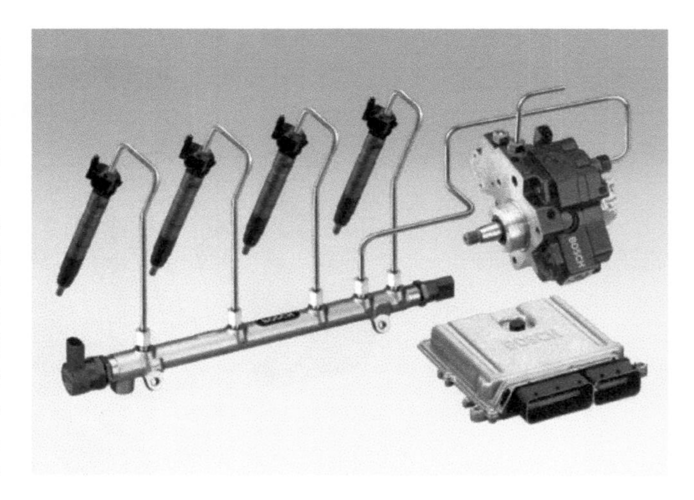

Figure 6.39 Common rail diesel system components.

Source: Bosch Media.

The combustion process, with common rail injection, is improved by a pilot injection of a very small quantity of fuel, at between 40° and 90° BTDC. This pilot fuel ignites in the compressing air charge so that the cylinder temperature and pressure are higher than in a conventional diesel injection engine at the start of injection. The higher temperature and pressure reduce ignition lag to a minimum, so that the controlled combustion phase during the main injection period is softer and more efficient (Figure 6.40).

Fuel injection pressures are varied – throughout the engine speed and load range – to suit the instantaneous conditions of driver demand and engine speed and load conditions. Data input from other vehicle system ECUs is used to further adapt the engine output to suit

Figure 6.40 Common rail diesel combustion.

changing conditions elsewhere on the vehicle. Examples are traction control, cruise control and automatic transmission gearshifts.

The electronic diesel control (EDC) module carries out calculations to determine the quantity of fuel delivered. It also determines the injection timing based on engine speed and load conditions.

The actuation of the injectors, at a specific crankshaft angle and for a specific duration, is made by signal currents from the EDC module. A further function of the EDC module is to control the accumulator (rail) pressure.

In summary, common rail diesel fuel injection systems consist of four main component areas:

- low-pressure delivery;
- high-pressure delivery with a high-pressure pump and accumulator (the rail);
- electronically controlled injectors (Figure 6.41);
- electronic control unit and associated sensors and switches.

The main sensors for calculation of the fuel quantity and injection advance requirements are the accelerator pedal sensor, crankshaft speed and position sensor, air mass meter and the engine coolant temperature sensor.

6.12.4 Diesel exhaust emissions

Exhaust emissions from diesel engines have been reduced considerably by changes in the design of combustion chambers and injection techniques.

More accurate control of start of injection and spill timing has allowed further improvements to be made. Electronic control has also made a significant contribution. A number of further techniques can be employed to control emissions.

Overall, the gas emissions from diesel combustion are far lower than those from petrol combustion. The CO, HC and NO_x emissions are lower mainly due to the higher compression ratio and excess air factor. The higher compression ratio improves the thermal efficiency and thus lowers the fuel consumption. The excess air factor ensures more complete burning of the fuel.

KEY FACT

Overall, the gas emissions from diesel combustion are lower than those from petrol combustion; the main problem area is that of particulates.

The main problem area is that of particulate emissions. These particle chains of carbon molecules can also contain hydrocarbons, aldehydes mostly. The dirt effect of this emission is a pollution problem, but the possible carcinogenic effect of this soot gives cause for concern. The diameter of these particles is only a few ten thousandths of a millimetre. This means they float in the air and can be inhaled.

In much the same way as with petrol engines, EGR is employed primarily to reduce NO_x emissions by reducing the reaction temperature in the combustion chamber. However, if the percentage of EGR is too high, increased hydrocarbons and soot are produced. This is appropriate to turbocharged engines such that if the air is passed through an intercooler and there are improvements in volumetric efficiency, lower temperature will again reduce the production of NO_x. The intercooler is fitted in the same way as the cooling system radiator.

6.12.5 Catalytic converter diesel

On a diesel engine, a catalyst can be used to reduce the emission of hydrocarbons but will have less effect on nitrogen oxides. This is because diesels are always run with excess air to ensure better and more efficient burning of the fuel. A normal catalyst therefore will not strip the oxygen off the NO to oxidise the hydrocarbons because the excess oxygen will be used. Special NO_x converters are becoming available.

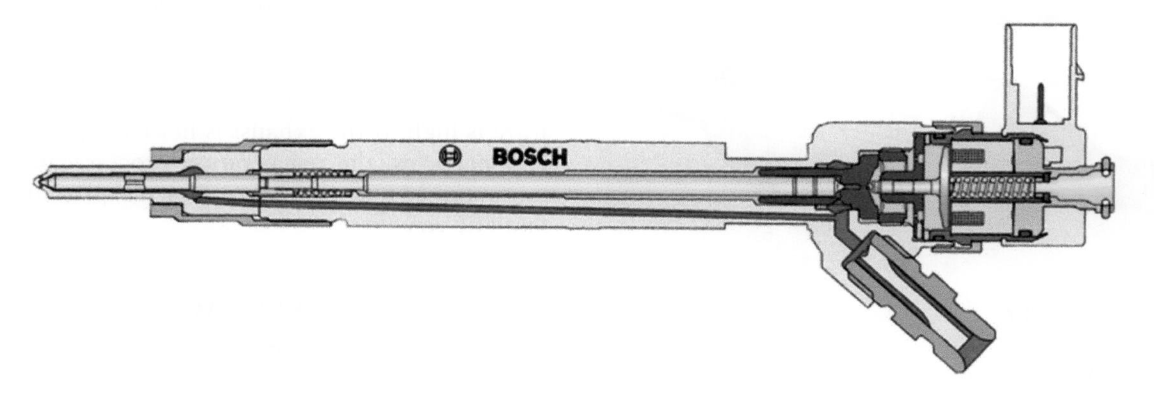

Figure 6.41 Electrically operated diesel fuel injector.

6.12.6 Diesel particulate filters

A diesel particulate filter (DPF) is a device designed to remove diesel particulate matter or soot from the exhaust gas of a diesel engine. Wall-flow diesel particulate filters usually remove 85% or more of the soot and under certain conditions can attain soot removal efficiencies of close to 100%.

DEFINITION

DPF: Diesel particulate filter.

The particulate filter shown in Figure 6.42 is made of sintered metal and lasts considerably longer than current ceramic models, since its special structure offers a high storage capacity for oil and additive combustion residues. The filter is designed in such a way that the filtered particulates are very evenly deposited, allowing the condition of the filter to be identified more reliably and its regeneration controlled far better than with other solutions. This diesel particulate filter is designed to last as long as the vehicle itself.

The two main DPF systems are those with additive and those without. To enable a vehicle to operate without an additive the particulate filter must be fitted close to the engine. Because the exhaust gases will not have travelled far from the engine, they will still be hot enough to burn off the carbon soot particles. In these systems, an oxidising catalytic converter will be integrated into the particulate filter. In other systems, the particulate filter is fitted some distance from the engine and as the exhaust gases travel along the exhaust they cool. The temperatures required for ignition of the exhaust gas can only be achieved using an additive.

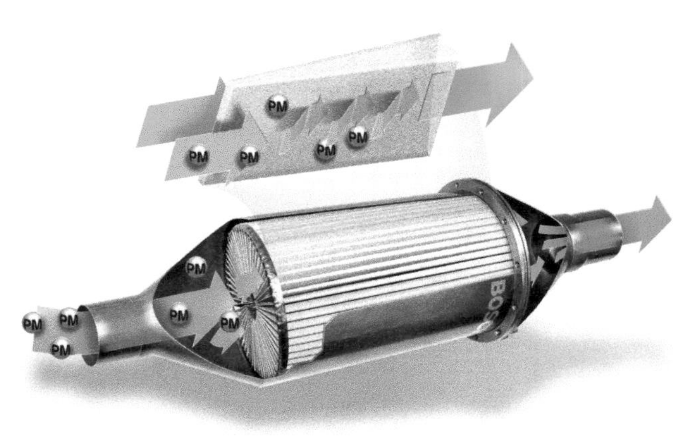

Figure 6.42 Diesel particulate filter.

Source: Bosch Media.

Use of an additive lowers the ignition temperature of the soot particles and, if the engine management ECU raises the temperature of the exhaust gas, the filter can be regenerated. Regeneration is usually necessary after 300–450 miles, depending on how the vehicle is driven. The process takes about 5–10 minutes, and the driver shouldn't notice it is occurring, although sometimes there may be a puff of white smoke from the exhaust during regeneration. The additive is stored in a separate tank and is used at a rate of about 1 litre of additive to 3000 litres of fuel. It works by allowing the carbon particles trapped in the particulate filter to burn at a significantly lower temperature than would usually be required (250–450°C rather than 600–650°C).

KEY FACT

Use of an additive lowers the ignition temperature of the soot particles.

On-board active filter management can use a variety of strategies:

- engine management to increase exhaust temperature through late fuel injection or injection during the exhaust stroke (the most common method);
- a fuel borne catalyst (the additive) to reduce soot burn-out temperature;
- a fuel burner after the turbo to increase the exhaust temperature;
- a catalytic oxidizer to increase the exhaust temperature, with after injection;
- resistive heating coils to increase the exhaust temperature;
- microwave energy to increase the particulate temperature.

Not running the regeneration cycle soon enough increases the risk of engine damage and/or uncontrolled regeneration (thermal runaway) and possible DPF failure.

There are two types of regeneration – passive and active. Passive regeneration takes place, automatically, on motorway-type runs in which the exhaust temperature is high. If the exhaust is hot enough to ignite the soot particles, the regeneration process can carry on continuously and steadily across the platinum coated catalytic converter.

KEY FACT

There are two types of DPF regeneration – passive and active.

Figure 6.43 Sectioned view of a new filter.

Once the storage capacity of the particulate filter (Figure 6.43) has been exhausted, the filter must be regenerated by passing hot exhaust gases through it which burn up the deposited particulates. To produce the necessary high exhaust gas temperatures, the EDC alters the amount of air fed to the engine as well as the amount of fuel injected and the timing of injection. In addition, some unburnt fuel can be fed to the oxidizing catalytic converter by arranging for extra fuel to be injected during the expansion stroke. The fuel combusts in the oxidizing catalytic converter and raises the exhaust temperature even further.

A significant number of people don't use motorways, so passive regeneration will be possible only occasionally. In the case of a filter without additive when the soot loading reaches about 45% the ECU switches off the EGR and increases the fuel injection period so there is a small injection after the main injection. These measures help to raise the engine exhaust temperature to over 600°C which is high enough to burn off the soot particles.

A warning light is triggered at a 55% soot loading. In such circumstances the car needs to be driven hard in a lower gear so the temperature in the particulate filter will be sufficient to burn off the soot. If the driver ignores the warning and continues to use the car as normal, the soot will continue to build until it reaches 75%. Additional warnings will then be given using the malfunction indication lamp (MIL). It will not now be possible to clear the DPF by driving and it may need to be replaced. If loading reaches 95% then the DPF will need to be replaced.

KEY FACT

The DPF regeneration process that involves heating and burning of the soot is controlled by the engine management system or as part of a service cleaning process.

6.12.7 Selective catalytic reduction

Selective catalytic reduction (SCR) is a method of converting nitrogen oxides (NO*x*) with the aid of a catalyst into nitrogen (N) and water (H_2O). A gaseous reductant, typically anhydrous ammonia, aqueous ammonia or urea, is added to a stream of exhaust gas and then into a catalyst (Figure 6.44). Carbon dioxide, CO_2, is a reaction product when urea is used as the reductant. A typical SCR fluid is known as AdBlue.

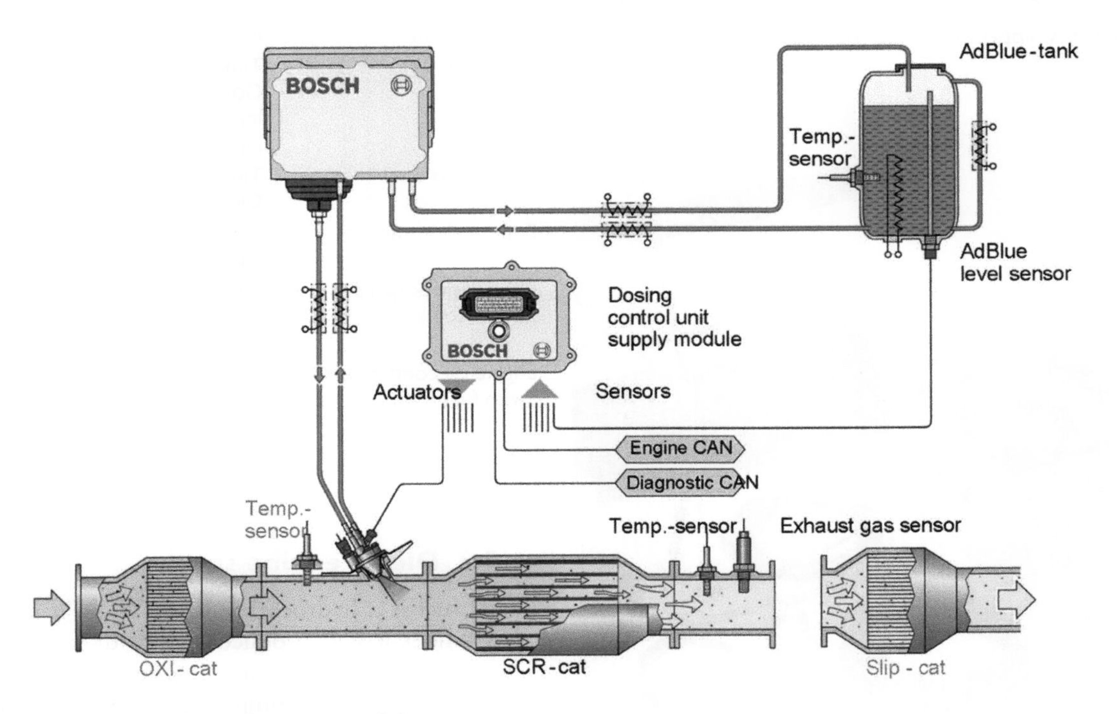

Figure 6.44 NO*x* reduction Bosch Denoxtronic 2.2.

Source: Bosch Media.

The use of SCR is now essential to meet the EURO6 diesel emissions standards for heavy trucks and also for cars and light commercial vehicles. In many cases, emissions of NO*x* and PM (particulate matter) have been reduced by upwards of 90% as compared with vehicles of the early 1990s.

DEFINITION

SCR: Selective catalytic reduction.

6.13 DIAGNOSTICS – DIESEL INJECTION SYSTEMS

6.13.1 Test equipment

6.13.1.1 Smoke meter

The smoke meter is an essential device in the United Kingdom and other countries where the level of smoke in the exhaust forms part of the annual test (Figure 6.45). Most devices use infrared light to 'count' the number of soot particles in the exhaust sample. This particulate matter is highly suspected of being carcinogenic.

SAFETY FIRST

Note: You should always refer to the manufacturer's instructions appropriate to the equipment you are using.

Figure 6.45 Gas analyser and smoke meter.

Figure 6.46 Injector pop tester.

6.13.1.2 Injector tester

The pressure required to 'crack' (lift the nozzle) on an injector can be tested (Figure 6.46).

6.13.2 Diesel injection fault diagnosis table

Symptom	*Possible fault*
Engine rotates but does not start	No fuel in the tank! Cam belt broken Fuel pump drive broken Open circuit supply to stop solenoid Fuel filter blocked
Excessive smoke	Refer to the next section
Lack of power	Timing incorrect Governor set too low Injector nozzles worn Injector operating pressure incorrect
Difficult to start	Timing incorrect Glow plugs not working
Fuel smell in the car	Fuel lines leaking Leak off pipes broken
Diesel knock (particularly when cold)	Timing incorrect Glow plug hold on for idle circuit not working
Engine oil contaminated with fuel	Piston broke (like me after a good holiday!) Work piston rings Excessive fuel injected

6.13.3 Diesel engine smoke

Diesel fuel is a hydrocarbon fuel. When it is burned in the cylinder it will produce carbon dioxide and water. There are, however, many circumstances under which the fuel may not be completely burned and one of the results is smoke. Despite the fact that diesel engines are designed to run under all conditions with an excess of air, problems still occur. Very often, these smoke problems are easily avoided by proper maintenance and servicing of the engine and its fuel system. The emission of smoke is usually due to a shortage of air (oxygen). If insufficient air is available for complete combustion, then unburnt fuel will be exhausted as tiny particles of fuel (smoke).

The identification of the colour of diesel smoke and under what conditions it occurs can be helpful in diagnosing what caused it in the first place. Poor-quality fuel reduces engine performance, increases smoke and reduces engine life. There are three colours of smoke: white, blue and black. All smoke diagnosis tests must be carried out with the engine at normal operating temperature.

6.13.3.1 *White or grey smoke*

White smoke is vaporised unburnt fuel and is caused by there being sufficient heat in the cylinder to vaporise it but not enough remaining heat to burn it. All diesel engines generate white smoke when starting from cold and it is not detrimental to the engine in any way – it is a diesel characteristic. Possible causes of white smoke are listed below:

- **Faulty cold starting equipment** – Cold engines suffer from a delay in the combustion process. A cold start unit is fitted to advance the injection timing to counteract this delay. This means that white smoke could be a cold start unit problem.
- **Restrictions in the air supply** – A partially blocked air cleaner will restrict the air supply – an easy cause to rectify but often overlooked. Incidentally, a blocked air cleaner element at light load in the workshop becomes a black smoke problem when the engine is under load. In both cases, there will not be sufficient air entering the cylinder for the piston to compress and generate full heat for combustion.
- **Cold running** – Check the cooling system thermostat to see if the correct rated thermostat is fitted.
- **Incorrect fuel injection pump timing** – If fuel is injected late (retarded timing), it may be vaporised but not burned.
- **Poor compressions** – Poor compressions may lead to leakage during the compression stroke and inevitably less heat would be generated.
- **Leaking cylinder head gasket** – If coolant were leaking into the combustion area, the result would be less temperature in the cylinder causing white smoke. Steam may also be generated if the leak is sufficient. All internal combustion engines have water as a by-product from burning fuel – you will have noticed your own car exhaust, especially on a cold morning.

6.13.3.2 *Blue smoke*

Blue smoke is almost certainly a lubricating oil burning problem. Possible causes of blue smoke are:

- incorrect grade of lubricating oil;
- worn or damaged valve stem oil seals, valve guides or stems where lubricating oil is getting into the combustion chamber;
- worn or sticking piston rings;
- worn cylinder bores.

6.13.3.3 *Black smoke*

Black smoke is partly burned fuel. Possible causes are listed below:

- **Restriction in air intake system** – A blocked air cleaner element will not let enough air in to burn all the fuel.
- **Incorrect valve clearances** – Excessive valve clearances will cause the valves to not fully open and to close sooner. This is another form of insufficient air supply.
- **Poor compressions** – Air required for combustion may leak from the cylinder.
- **Defective or incorrect injectors** – Check the injector to see if the spray is fully atomised and solid fuel is not being injected.
- **Incorrect fuel injection pump timing** – This is less likely because the timing would need to be advanced to the point where additional engine noise would be evident.
- **Low boost pressure** – If a turbocharger is fitted and is not supplying enough air for the fuel injected, this is another form of air starvation.

6.13.4 Glow plug circuit

Figure 6.47 shows a typical glow plug circuit controlled by an ECU. Most timer circuits put the glow plugs on for a few seconds before cranking. A warning light may be used to indicate the 'ready' condition to the driver.

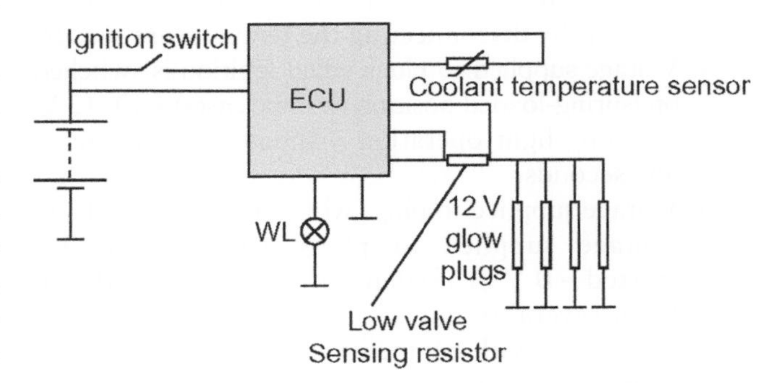

Figure 6.47 Glow plug circuit.

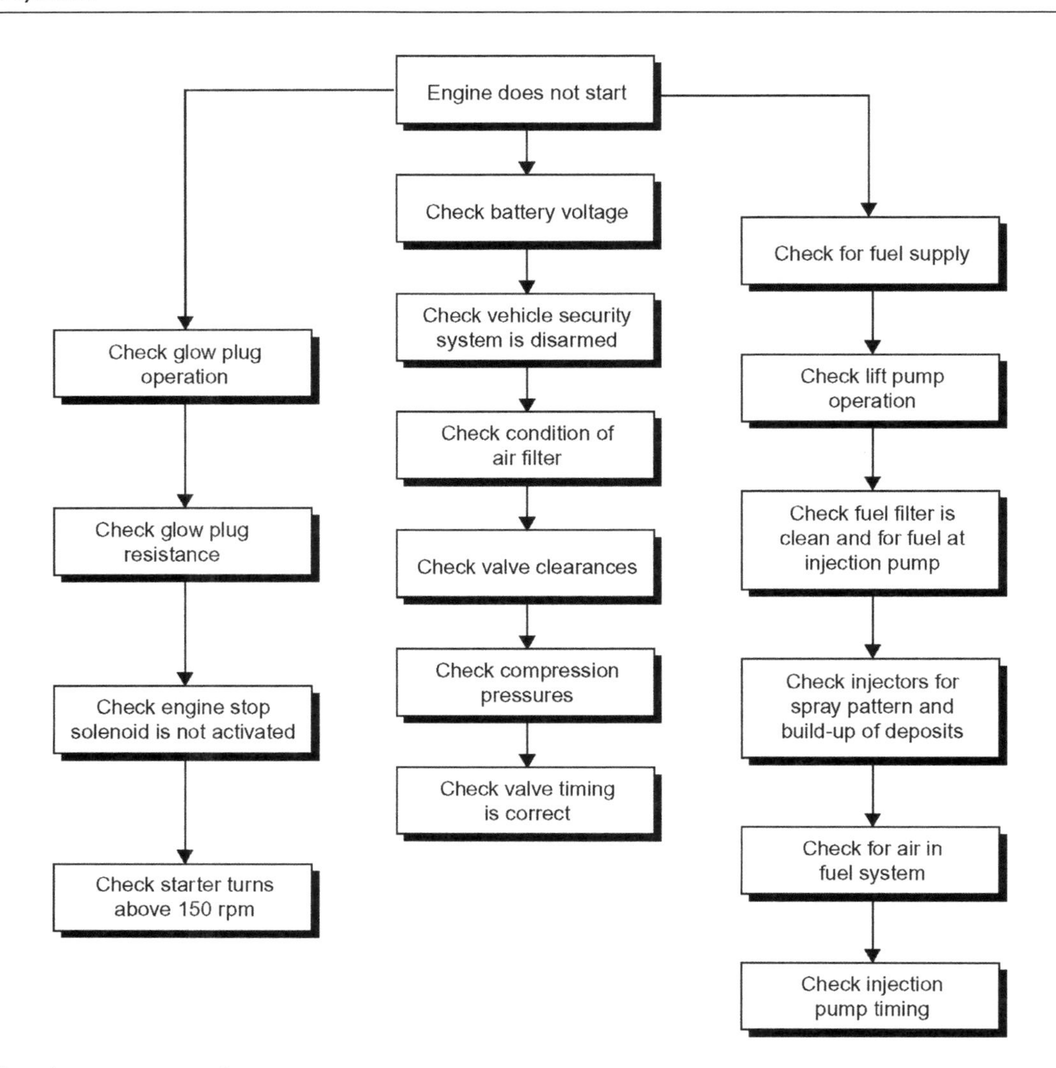

Figure 6.48 Diesel system generic diagnostic process.

Take care to note the type of glow plugs used; most are 12 V and connected in parallel, but some are connected in series (4 – 3 V plugs). To check the operation of most 12 V glow plug circuits, use the following steps:

1. Hand and eye checks.
2. Battery condition – at least 70%.
3. Engine must be cold – it may be possible to simulate this by disconnecting the temperature sensor.
4. Voltage supplied to plugs when ignition is switched on (spring-loaded position in some cases) – 10–12 V.
5. Warning light operation – should go out after a few seconds.
6. Voltage supplied to plugs while cranking – 9–11 V.
7. Voltage supplied to plugs after engine has started – 0 V or if silent idle system is used 5–6 V for several minutes.
8. Same tests with engine at running temperature – glow plugs may not be energised or only for the starting phase.

6.13.5 Diesel systems

It is recommended that when the injection pump or the injectors are diagnosed as being at fault, reconditioned units should be fitted. Other than basic settings of timing, idle speed and governor speed, major overhaul is often required. Figure 6.48 shows a general diagnosis pattern for diesel systems.

6.14 ENGINE MANAGEMENT

6.14.1 Introduction

As the requirement for lower and lower emissions continues together with the need for better performance, other areas of engine control are constantly being investigated. This is becoming even more important as the possibility of carbon dioxide emissions being included in the regulations increases. Some of the current and potential areas for further control of engine operation are included in this section. Most of the common aspects have been covered earlier in the 'Ignition' and 'Fuel injection' sections. The main areas of control are as follows:

- ignition timing;
- dwell angle;
- fuel quantity;
- idle speed.

Further areas of engine control may include the following:

- EGR;
- canister purge;
- valve timing;
- inlet manifold length;
- closed-loop lambda control.

It is not possible for an engine to operate at its best volumetric efficiency with fixed manifolds. This is because the length of the inlet tract determines the velocity of the intake air and in particular the propagation of the pressure waves set up by the pumping action of the cylinders. These standing waves can be used to improve the ram effect of the charge as it enters the cylinder but only if they coincide with the opening of the inlet valves. The length of the inlet tract has an effect on the frequency of these waves.

With the widespread use of twin-cam engines, one cam for the inlet valves and one for the exhaust valves, it is possible to vary the valve overlap while the engine is running. Honda has a system that improves the power and torque range by opening both of the inlet valves only at higher speed.

Many systems use oil pressure controlled by valves to turn the cam with respect to its drive gear. This alters the cam phasing or relative position. The position of the cams is determined from a suitable map held in ROM in the control unit.

6.14.2 Closed-loop lambda control

Current regulations have almost made closed-loop control of air fuel mixture in conjunction with a three-way catalytic converter mandatory. Lambda control is a closed-loop feedback system in that the signal from a lambda sensor in the exhaust can directly affect the fuel quantity injected. The lambda sensor is described in more detail in Chapter 4.

A graph to show the effect of lambda control in conjunction with a catalytic converter is shown in Figure 6.49.

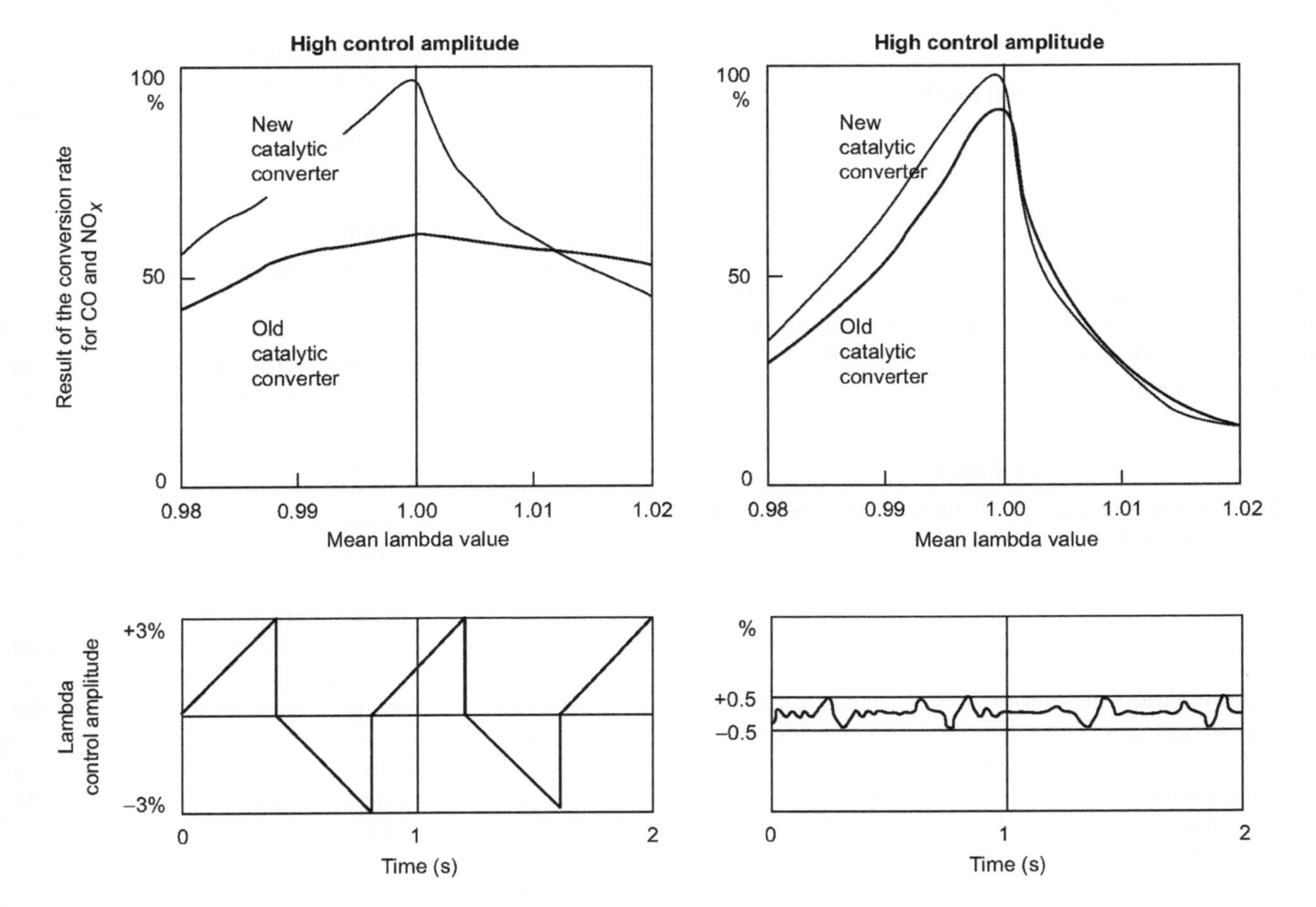

Figure 6.49 Effect of lambda control on catalytic converters.

The principle of operation is as follows: the lambda sensor produces a voltage which is proportional to the oxygen content of the exhaust which is in turn proportional to the air/fuel ratio. At the ideal setting, this voltage is approximately 450 mV. If the voltage received by the ECU is below this value (weak mixture), the quantity of fuel injected is increased slightly. If the signal voltage is above the threshold (rich mixture), the fuel quantity is reduced. This alteration in air/fuel ratio must not be too sudden, as it could cause the engine to buck. To prevent this, the ECU contains an integrator, which changes the mixture over a period of time.

KEY FACT

Lambda control is a closed-loop negative feedback system.

A delay also exists between the mixture formation in the manifold and the measurement of the exhaust gas oxygen. This is due to the engine's working cycle and the speed of the inlet mixture, the time for the exhaust to reach the sensor and the sensor's response time. This is sometimes known as dead time and can be as much as one second at idle speed but only a few hundred milliseconds at higher engine speeds.

DEFINITION

TWC: Three-way catalyst.

Owing to the dead time, the mixture cannot be controlled to an exact value of lambda equals 1. If the integrator is adjusted to allow for engine speed, then it is possible to keep the mixture in the lambda window (0.97–1.03), which is the region in which the TWC is at its most efficient.

KEY FACT

The lambda window (0.97–1.03) is the region in which the TWC is most efficient.

6.14.3 Engine management operation

The combination of ignition and injection control has several advantages. The information received from various sensors is used for computing both fuelling and ignition requirements. Perhaps more importantly, ignition and injection are closely linked. The influence they have on each other can be easily taken into account to ensure that the engine is working at its optimum under all operating conditions.

Overall, this type of system is less complicated than separate fuel and ignition systems and in many cases the ECU is able to work in an emergency mode by substituting missing information from sensors with pre-programmed values. This will allow limited but continued operation in the event of certain system failures.

The ignition system is integrated and is operated without a high-tension (HT) distributor. The ignition process is controlled digitally by the ECU. The data for the ideal characteristics are stored in ROM from information gathered during both prototyping and development of the engine. The main parameters for ignition advance are engine speed and load, but greater accuracy can be achieved by taking further parameters into account such as engine temperature. This provides both optimum output and close control of anti-pollution levels. Performance and pollution level control means that the actual ignition point must be in many cases a trade-off between the two.

The injection system shown in Figure 6.50 is multipoint and, as is the case for all fuel systems, the amount of fuel delivered is primarily determined by the amount of air 'drawn' into the engine. The method for measuring this data is indirect in the case of this system as a pressure sensor is used to determine the air quantity.

Electromagnetic injectors control fuel supply into the engine. The injector open period is determined by the ECU. This will obtain very accurate control of the air fuel mixture under all operating conditions of the engine. The data for this is stored in ROM in the same way as for the ignition.

KEY FACT

Injector open period is determined by the ECU.

The main source of reference for the ignition system is from the crankshaft position sensor. This is a magnetic inductive pick-up sensor positioned next to a flywheel ring containing 58 teeth. Each tooth takes up a 6° angle of the flywheel with one 12° gap positioned 114° before top dead centre (BTDC) for number 1 cylinder. The signal produced by the flywheel sensor is essentially a sine wave with a cycle missing corresponding to the gap in the teeth of the reluctor plate. The information provided to the ECU is engine speed from the frequency of the signal and engine position from the number of pulses before or after the missed pulses.

The basic ignition advance angle is obtained from a memorised cartographic map. This is held in a ROM chip within the ECU. The parameters for this are

- **engine rpm** – given by the flywheel sensor;
- **inlet air pressure** – given by the MAP sensor.

Figure 6.50 Engine management components.

Source: Bosch Media.

The above two parameters (speed and load) give the basic setting, but to ensure optimum advance angle the timing is corrected by

- coolant temperature;
- air temperature;
- throttle butterfly position.

The ignition is set to a predetermined advance during the starting phase. Figure 6.51 shows typical advance, fuelling and dwell maps used by an engine management system. This data is held in ROM.

For full ignition control, the electronic control unit has to first determine the basic timing for three different conditions:

- Under idling conditions, ignition timing is often moved very quickly by the ECU in order to control idle speed. When timing is advanced, engine speed will increase within certain limits.
- Full-load conditions require careful control of ignition timing to prevent combustion knock. When a full-load signal is sensed by the ECU (high manifold pressure), the ignition advance angle is reduced.
- Partial throttle is the main area of control and, as already stated, the basic timing is set initially by a programme as a function of engine speed and manifold pressure.

Corrections are added according to

- operational strategy;
- knock protection;
- phase correction.

The ECU will also control ignition timing variation during overrun fuel cut-off and reinstatement and also to ensure anti-jerk control. When starting, the ignition timing plan is replaced by a specific starting strategy.

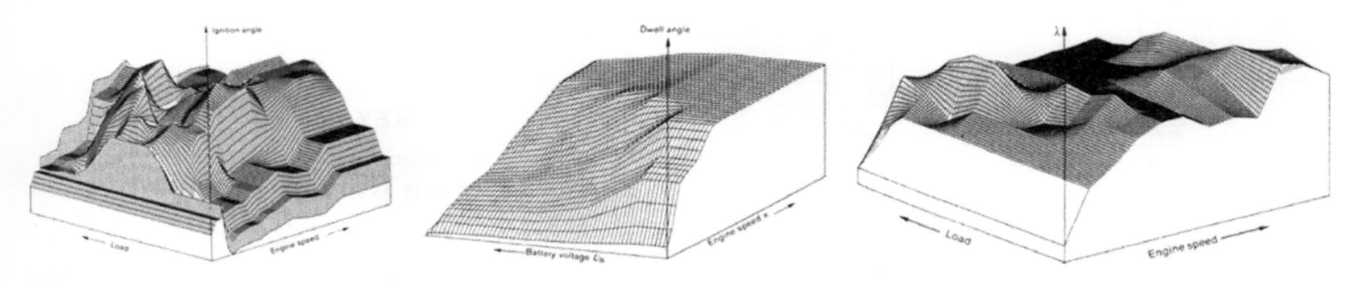

Figure 6.51 Ignition timing, dwell and lambda maps.

Phase correction is when the ECU adjusts the timing to take into account the time taken for the HT pulse to reach the spark plugs. To ensure good driveability, the ECU can limit the variations between the two ignition systems to a maximum value, which varies according to engine speed and the basic injection period.

An anti-jerk function operates when the basic injection period is less than 2.5 ms and the engine speed is between 720 and 3200 rpm. This function operates to correct the programmed ignition timing in relation to the instantaneous engine speed and a set filtered speed; this is done to stabilise the engine rotational characteristics as much as possible.

To maintain constant-energy HT, the dwell period must increase in line with engine speed. To ensure that the ignition primary current reaches its maximum at the point of ignition, the ECU controls the dwell by use of another memory map, which takes battery voltage into account.

Fuel is collected from the tank by a pump either immersed in it or outside, but near the tank. The immersed type is quieter in operation, has better cooling and has no internal leaks. The fuel is directed forwards to the fuel rail or manifold via a paper filter.

Fuel pressure is maintained at approximately 2.5 bar above manifold pressure by a regulator mounted on the fuel rail. Excess fuel is returned to the tank. The fuel is usually picked up via a swirl pot in the tank to prevent aeration of the fuel. Each of the four inlet manifold tracts has its own injector.

Most fuel pumps on manifold injection systems are similar to Figure 6.52 and the pump operates as shown in Figures 6.53 and 6.54.

The fuel enters the pump housing where it is pressurised by rotation of the pump and the reduction of the volume in the roller chambers. This pressure opens a residual valve and fuel passes to the filter. When the pump stops, pressure is maintained by this valve, which prevents the fuel returning. If, due to a faulty regulator or a blockage in the line, fuel pressure rises above 7 bar,

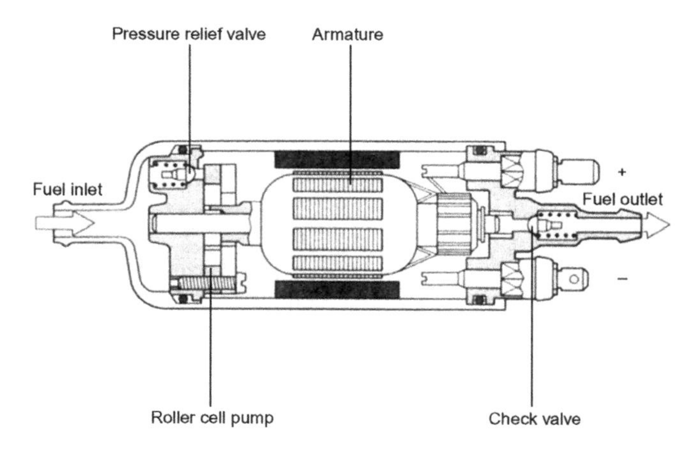

Figure 6.52 Roller cell pump.

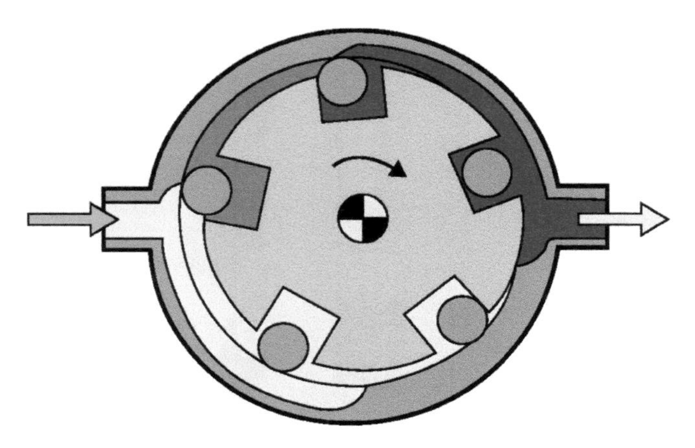

Figure 6.53 Roller cell pump (1).

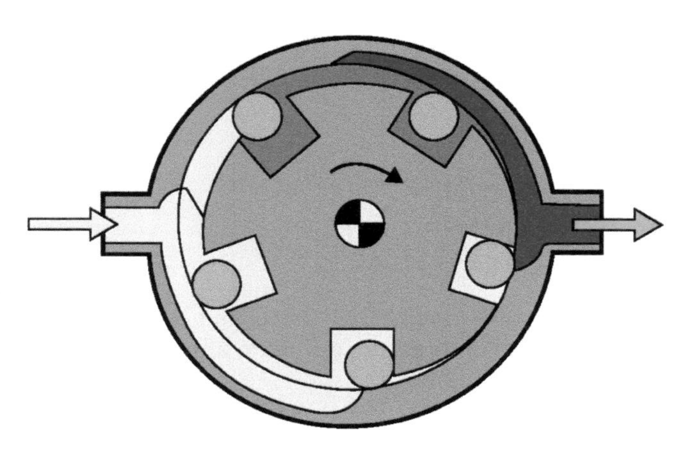

Figure 6.54 Roller cell pump (2).

an overpressure valve will open releasing fuel back to the tank.

The fuel filter is placed between the fuel pump and the fuel rail. It is fitted one way only to ensure the outlet screen traps any paper particles from the filter element. The filter will stop contamination down to between 8 and 10 μm. Replacement varies between manufacturers, but 80 000 km (50 000 miles) is often recommended.

The fuel rail, in addition to providing a uniform supply to the injectors, acts as an accumulator. Depending on the size of the fuel rail, some systems also use an extra accumulator. The volume of the fuel rail is large enough to act as a pressure fluctuation damper, ensuring that all injectors are supplied with fuel at a constant pressure.

KEY FACT

The fuel rail, in addition to providing a uniform supply to the injectors, acts as an accumulator.

Multipoint systems use one injector for each cylinder although very high-performance vehicles may use two.

The injectors are connected to the fuel rail by a rubber seal. The injector is an electrically operated valve manufactured to a very high precision. The injector is composed of a body and needle attached to a magnetic core. When the winding in the injector housing is energised, the core or armature is attracted and the valve opens, compressing a return spring. The fuel is delivered in a fine spray to wait behind the closed inlet valve until the induction stroke begins. Provided the pressure across the injector remains constant, the quantity of fuel admitted is related to the open period, which, in turn, is determined by the time the electromagnetic circuit is energised.

The purpose of the fuel pressure regulator is to maintain differential pressure across the injectors at a predetermined constant. This means the regulator must adjust the fuel pressure in response to changes in manifold pressure. It is made of two compressed cases containing a diaphragm, spring and a valve.

The calibration of the regulator valve is determined by the spring tension. Changes in manifold pressure vary the basic setting. When the fuel pressure is sufficient to move the diaphragm, the valve opens and allows the fuel to return to the tank. The decrease in pressure in the manifold, also acting on the diaphragm at say idle speed, will allow the valve to open more easily, hence maintaining a constant differential pressure between the fuel rail and the inlet manifold. This is a constant across the injectors, and hence the quantity of fuel injected is determined only by the open time of the injectors. The differential pressure is maintained at approximately 2.5 bar.

KEY FACT

Multipoint manifold injection differential pressure is usually maintained at approximately 2.5 bar – but always check data.

The air supply circuit will vary considerably between manufacturers, but an individual manifold from a collector housing, into which the air is fed via a simple butterfly, supplies essentially each cylinder. The air is supplied from a suitable filter. A supplementary air circuit is utilised during the warm-up period after a cold start and to control idle speed.

6.14.4 Gasoline direct injection (GDI)

High-pressure injection systems for petrol/gasoline engines are based on a pressure reservoir and a fuel rail, which a high-pressure pump charges to a regulated pressure of up to 120 bar. The fuel can therefore be injected directly into the combustion chamber via electromagnetic injectors.

KEY FACT

Gasoline direct injection (GDI) systems use a high-pressure pump to create a regulated pressure of up to 120 bar.

DEFINITION

Homogeneous: A substance that is uniform in composition.
Heterogeneous: A substance that is random and non-uniform in composition.

DEFINITION

Stratified charge: Fuel/air mixture is in layers.

The air mass drawn in can be adjusted through the electronically controlled throttle valve and is measured with the help of an air mass meter. For mixture control, a wide-band oxygen sensor is used in the exhaust, before the catalytic converters. This sensor can measure a range between a lambda value of 0.8 and infinity. The engine electronic control unit regulates the operating modes of the engine with gasoline direct injection (GDI) in three ways:

- stratified charge operation – with lambda values greater than 1;
- homogeneous operation – at lambda = 1;
- rich homogeneous operation – with lambda = 0.8.

Compared to the traditional manifold injection system, the Bosch DI-Motronic must inject the entire fuel amount in full-load operation in a quarter of the time. The available time is significantly shorter during stratified charge operation in part-load. Especially at idle, injection times of less than 0.5 ms are required due to the lower fuel consumption. This is only one-fifth of the available time for manifold injection (Figure 6.55).

The fuel must be atomised very finely in order to create an optimal mixture in the brief moment between injection and ignition (Figure 6.56). The fuel droplets for direct injection are on average smaller than 20 μm. This is only one-fifth of the droplet size reached with the traditional manifold injection and one-third of the diameter of a single human hair. This improves efficiency considerably.

Direct injection engines operate according to the stratified charge concept in the part-load range and function with high excess air. In return, very low fuel consumption is achieved.

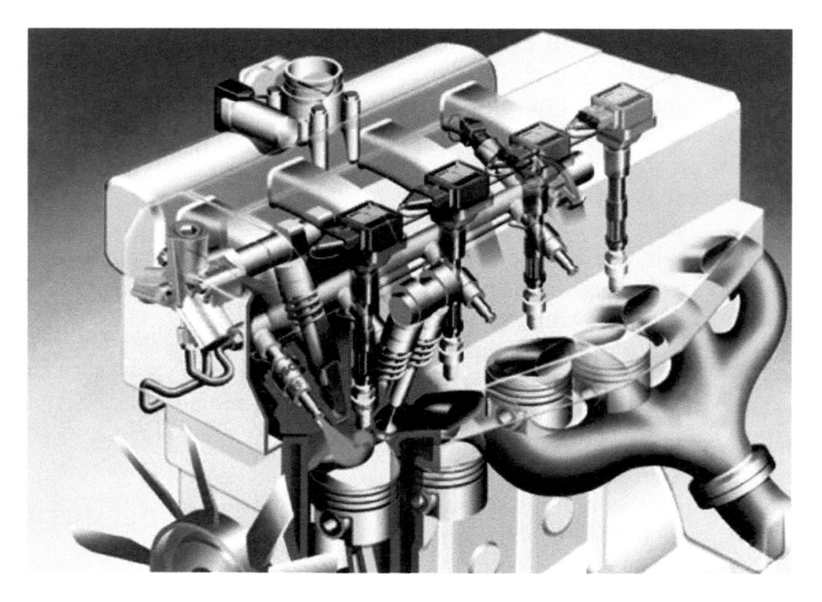

Figure 6.55 Gasoline direct injection (GDI).

KEY FACT

The fuel droplets for direct injection are on average smaller than 20 µm.

The engine operates with an almost completely opened throttle valve, which avoids additional alternating charge losses. With stratified charge operation, the lambda value in the combustion chamber is between approximately 1.5 and 3. In the part-load range, GDI achieves the greatest fuel savings with up to 40% at idle compared to conventional petrol injection processes. With increasing engine load, and therefore increasing injection quantities, the stratified charge cloud becomes even richer and emission characteristics become worse.

Because soot may form under these conditions, the DI-Motronic engine control converts to a homogeneous cylinder charge at a pre-defined engine load. The system injects very early during the intake process in order to achieve a good mixture of fuel and air at a ratio of lambda = 1. As is the case for conventional manifold injection systems, the amount of air drawn in for all operating modes is adjusted through the throttle valve according to the desired torque specified by the driver.

Diagnosing faults with a GDI system is little different from the manifold injection types. Extra care is needed because of the higher fuel pressures of course. Injector

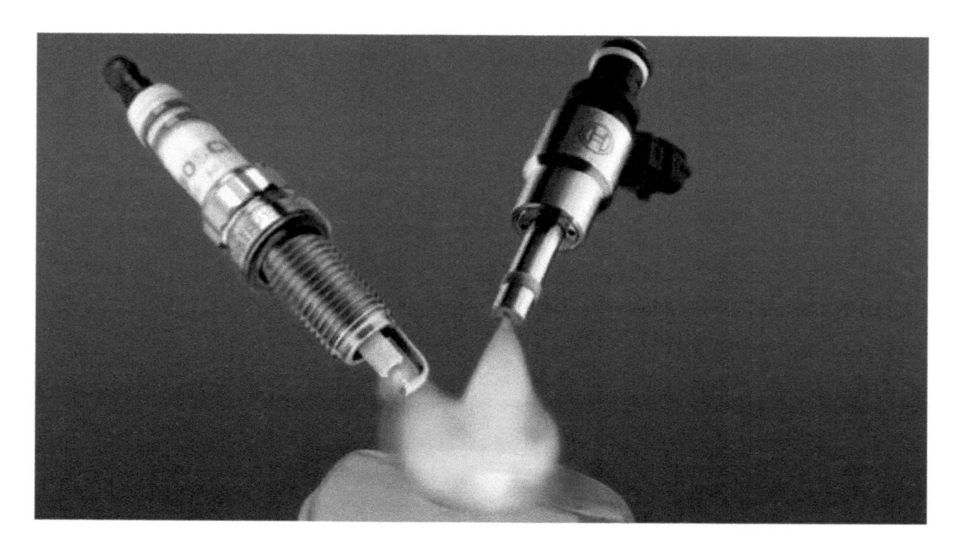

Figure 6.56 Atomisation.

waveforms can be checked as can those associated with the other sensors and actuators.

6.14.5 ECU calibration

With each new generation of vehicle, the sophistication of the powertrain, chassis and body control systems increases significantly. This is only possible due to the increasing number of electronic controllers around the vehicle (ECUs). With each vehicle generation, they become more powerful (in terms of memory and processing power), and they are increasingly more 'connected' – sharing data and information during vehicle operation. Having a complicated controller is now essential to achieve the required performance levels, as well as driver expectation. Developing these control units to function and behave in the correct way is a complex task. This is undertaken during vehicle and powertrain system development, and is known as 'calibration' (Figure 6.57).

An ECU is a microprocessor-based control unit that performs complex functions by running sophisticated control algorithms. The tuning of these to suit the application (e.g. engine, vehicle type, etc.) is known as calibration. The calibration engineer will adjust the parameters of the control function algorithms that are stored in permanent memory inside the ECU (maps, tables, etc.) during development. This will allow optimisation of the controller, based on targets (good fuel consumption, good driveability, low emissions, etc.) whilst keeping within boundary limits (noise, temperatures).

Within the ECU software, each physical data value is known as a 'label' – this could be a calibration value (i.e. a value that needs adjusting or 'calibrating') or a measurement value (i.e. an internal value used within the software, e.g. 'engine temperature' or 'desired ignition angle for spark advance'). The number of labels has grown massively with increasing complexity. In just 10 years the number has grown from 3000 to 5000 labels to something like 40 000 to 50 000 labels (Figure 6.58). This shows what a massive task the overall calibration of an ECU is.

In years gone by the 'calibrations' were held on memory chips. These chips could be replaced or rewritten offline and re-installed into the ECU, thus providing a new calibration. There were (and are) many providers in the aftermarket doing this, providing updated calibrations that added horsepower (at the expense of other things, such as fuel consumption, emissions and driveability). The technical term for this approach is 'offline' calibration.

The offline technology does not support an efficient calibration process in development – typically, there is a lot of 'round trip Engineering' when calibrating an ECU, for example, changing values, checking responses and comparing with measurement data online and offline. To support this task efficiently, an online calibration technology is needed. That is, seeing and changing what is going on inside the ECU, during operation. If you have used a scan tool to view and record data during a test drive, you will have seen online measurement labels. However, the engineer needs to see all the labels for the task, including being able to change some of them, and test and measure the response immediately. For this, a suitable software and hardware tool chain is needed (Figure 6.59). Typically, this consists of

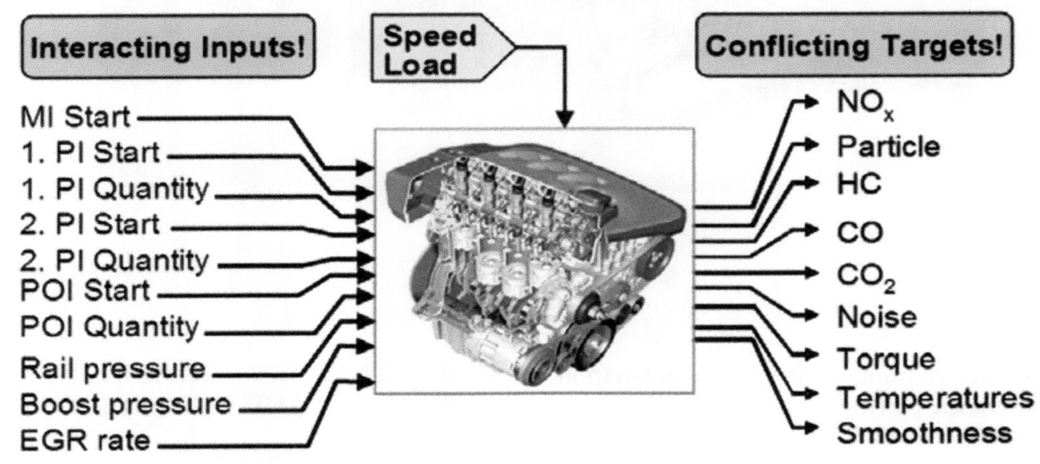

Figure 6.57 Typical ECU parameters needing to be adjusted in order to achieve the required targets for a modern diesel engine – this is ECU calibration.

Source: www.autoelex.co.uk.

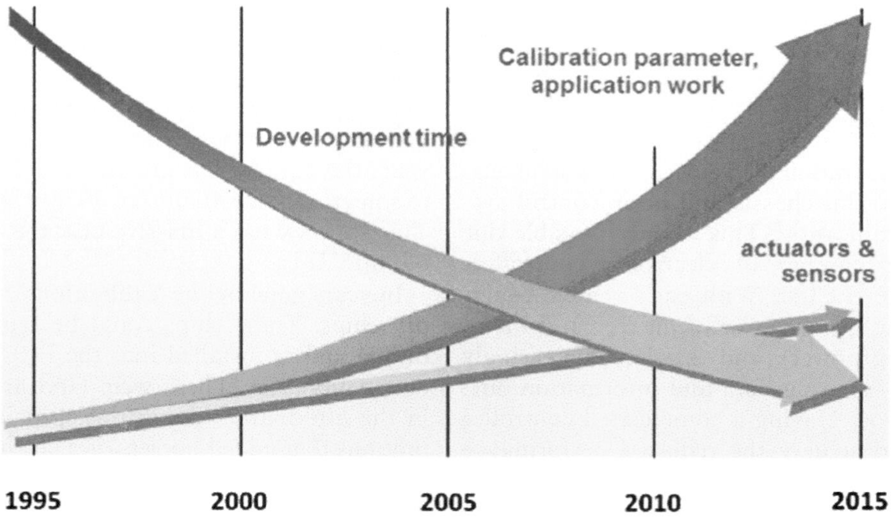

Figure 6.58 The increasing complexity of the ECU calibration task.

Source: www.autoelex.co.uk.

Figure 6.59 Motronic control unit.

Source: Bosch Media.

application software running on a PC in combination with an interface to the ECU.

Having a complicated ECU to achieve the required performance and emission levels, as well as driver expectation, is now essential. Developing these control units to function and behave in the correct way is a complex, but fascinating, task!

6.15 DIAGNOSTICS – COMBINED IGNITION AND FUEL SYSTEMS

6.15.1 Testing procedure

The following procedure is very generic but with a little adaptation can be applied to any system. Refer to manufacturer's recommendations if in any doubt (Figure 6.60).

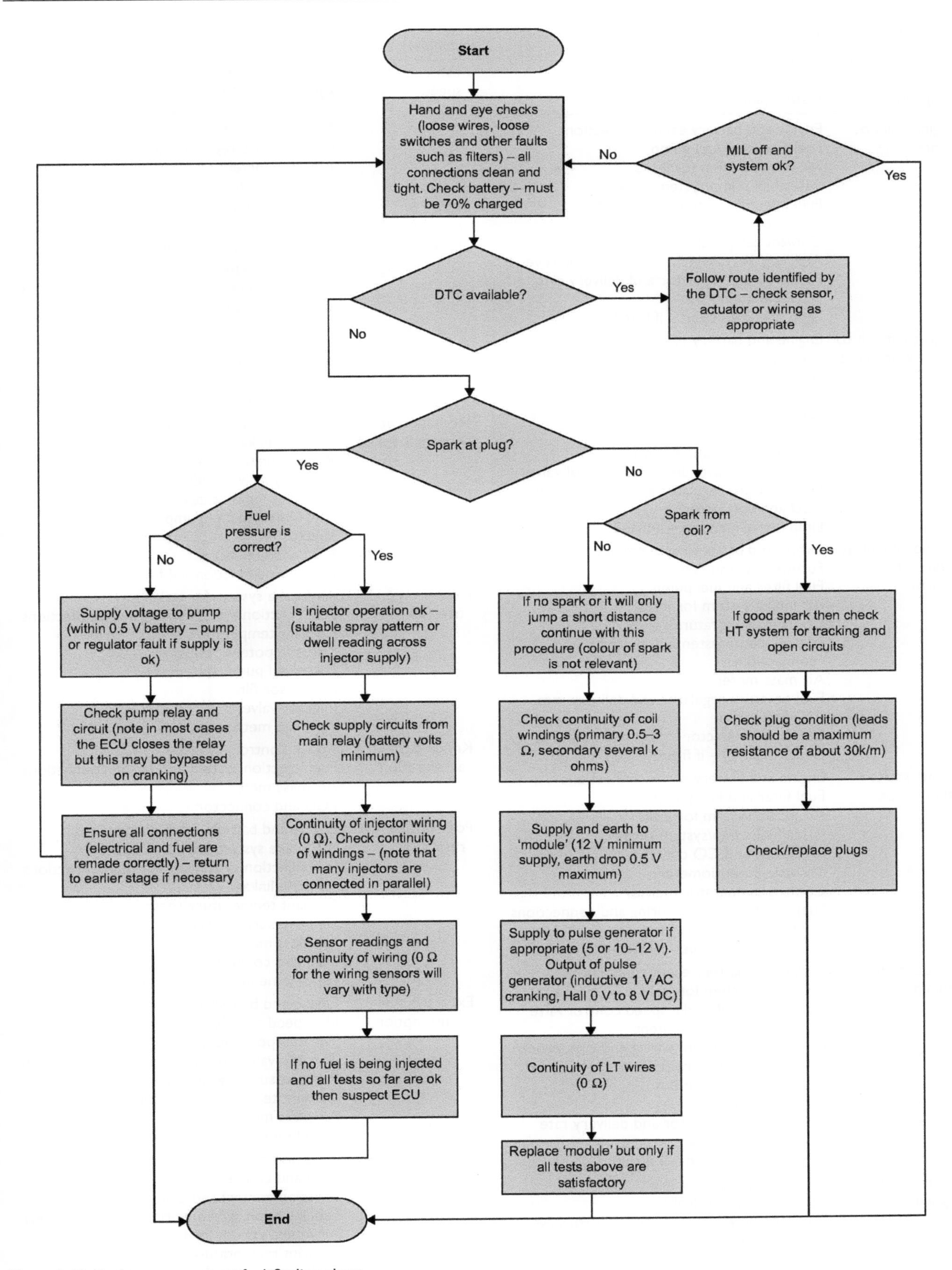

Figure 6.60 Engine management faultfinding chart.

6.15.2 Combined ignition and fuel control fault diagnosis table

Symptom	*Possible fault*
Engine will not start	Engine and battery earth connections Fuel filter and fuel pump Air intake system for leaks Fuses/fuel pump/system relays Fuel injection system wiring and connections Coolant temperature sensor Auxiliary air valve/idle speed control valve Fuel pressure regulator and delivery rate ECU and connector Limp home function – if fitted
Engine difficult to start when cold	Engine and battery earth connections Fuel injection system wiring and connections Fuses/fuel pump/system relays Fuel filter and fuel pump Air intake system for leaks Coolant temperature sensor Auxiliary air valve/idle speed control valve Fuel pressure regulator and delivery rate ECU and connector Limp home function – if fitted
Engine difficult to start when warm	Engine and battery earth connections Fuses/fuel pump/system relays Fuel filter and fuel pump Air intake system for leaks Coolant temperature sensor Fuel injection system wiring and connections Air mass meter Fuel pressure regulator and delivery rate Air sensor filter ECU and connector Knock control – if fitted
Engine starts then stops	Engine and battery earth connections Fuel filter and fuel pump Air intake system for leaks Fuses/fuel pump/system relays Idle speed and CO content Throttle potentiometer Coolant temperature sensor Fuel injection system wiring and connections ECU and connector Limp home function – if fitted
Erratic idling speed	Engine and battery earth connections Air intake system for leaks Auxiliary air valve/idle speed control valve Idle speed and CO content Fuel injection system wiring and connections Coolant temperature sensor Knock control – if fitted Air mass meter Fuel pressure regulator and delivery rate ECU and connector Limp home function – if fitted
Incorrect idle speed	Air intake system for leaks Vacuum hoses for leaks Auxiliary air valve/idle speed control valve Idle speed and CO content Coolant temperature sensor
Misfire at idle speed	Engine and battery earth connections Air intake system for leaks Fuel injection system wiring and connections Coolant temperature sensor Fuel pressure regulator and delivery rate Air mass meter Fuses/fuel pump/system relays
Misfire at constant speed	Airflow sensor
Hesitation when accelerating	Engine and battery earth connections Air intake system for leaks Fuel injection system wiring and connections Vacuum hoses for leaks Coolant temperature sensor Fuel pressure regulator and delivery rate Air mass meter ECU and connector Limp home function – if fitted
Hesitation at constant speed	Engine and battery earth connections Throttle linkage Vacuum hoses for leaks Auxiliary air valve/idle speed control valve Fuel lines for blockage Fuel filter and fuel pump Injector valves ECU and connector Limp home function – if fitted
Hesitation on overrun	Air intake system for leaks Fuel injection system wiring and connections Coolant temperature sensor Throttle potentiometer Fuses/fuel pump/system relays Air sensor filter Injector valves Air mass meter
Knock during acceleration	Knock control – if fitted Fuel injection system wiring and connections Air mass meter ECU and connector
Poor engine response	Engine and battery earth connections Air intake system for leaks Fuel injection system wiring and connections Throttle linkage Coolant temperature sensor Fuel pressure regulator and delivery rate Air mass meter ECU and connector Limp home function – if fitted
Excessive fuel consumption	Engine and battery earth connections Idle speed and CO content Throttle potentiometer Throttle valve/housing/sticking/initial position Fuel pressure regulator and delivery rate Coolant temperature sensor Air mass meter Limp home function – if fitted
CO level too high	Limp home function – if fitted ECU and connector Emission control and EGR valve – if fitted Fuel injection system wiring and connections Air intake system for leaks Coolant temperature sensor Fuel pressure regulator and delivery rate

(*Continued*)

Symptom	*Possible fault*
CO level too low	Engine and battery earth connections Air intake system for leaks Idle speed and CO content Coolant temperature sensor Fuel injection system wiring and connections Injector valves ECU and connector Limp home function – if fitted Air mass meter Fuel pressure regulator and delivery rate
Poor performance	Engine and battery earth connections Air intake system for leaks Throttle valve/housing/sticking/initial position Fuel injection system wiring and connections Coolant temperature sensor Fuel pressure regulator/fuel pressure and delivery rate Air mass meter ECU and connector Limp home function – if fitted

6.15.3 Fuel pump testing

Typical high-pressure fuel pump characteristics are as follows (Figure 6.61):

- **delivery** – 120 L/h (1 L in 30 s) at 3 bar;
- **resistance** – 0.8 Ω (static);
- **voltage** – 12 V;
- **current** – 10.5 A.

An ideal test for a fuel pump is its delivery. Using a suitable measuring receptacle, bypass the pump relay and check the quantity of fuel delivered in a set time (refer to manufacturer's specifications). A reduced amount would indicate either a fuel blockage, a reduced electrical supply to the pump or an inefficient pump.

6.15.4 Injector testing

Injectors typically have the characteristics as listed:

- supply voltage – 12 V;
- resistance – 16 Ω;
- static output – 150 cc/min at 3 bar.

As always, check with actual data before carrying out any tests (Figure 6.62).

Resistance checks (with the supply disconnected) are an ideal start to testing injectors. Further tests with the fuel pressurised by the pump and each injector in turn held in a suitable receptacle include the following:

- **spray pattern** – usually a nice cone shape with good atomisation;
- **delivery** – set quantity over a set time;
- **leakage** – any leakage of more than two drops a minute for standard non-direct injectors is considered excessive (zero is desirable).

6.15.5 ECU fuel trim diagnostics

Fuel trim is technology used to keep the fuelling control operating correctly over a vehicle's lifetime in order to comply with emissions regulations. Trim values are accessible via scan tools and this knowledge can help diagnostic procedures (Figure 6.63). Short-term fuel trim (STFT) and long-term fuel trim (LTFT) are expressed as a percentage. Positive fuel trim percentages indicate that the ECU is attempting to make the fuel mixture richer, to compensate for a lean condition. Negative fuel trim percentages indicate the opposite. Fuel trim values are generated and stored with respect to engine operating condition – generally speed or load.

Figure 6.61 Airflow meter under test.

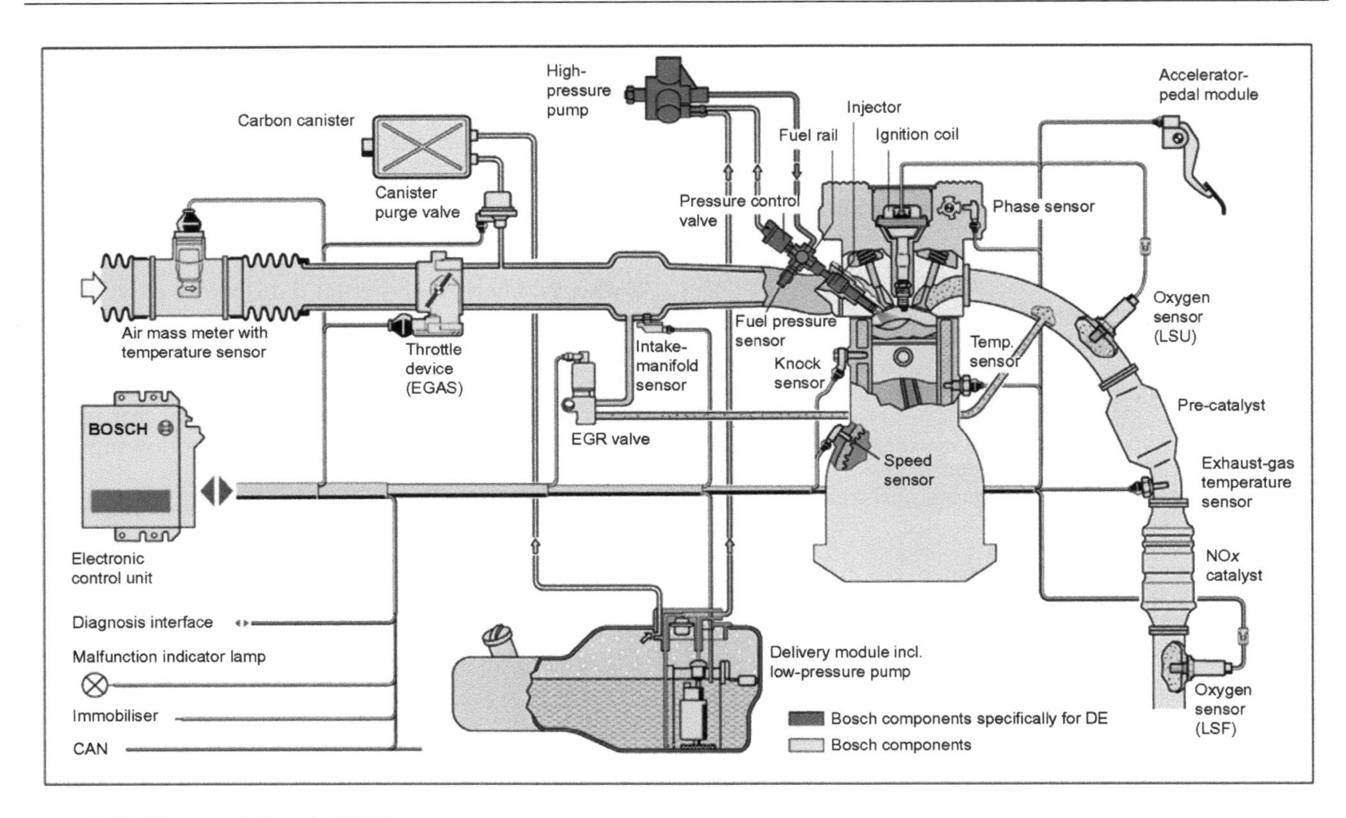

Figure 6.62 Motronic M5 with OBD2.

Source: Bosch Media.

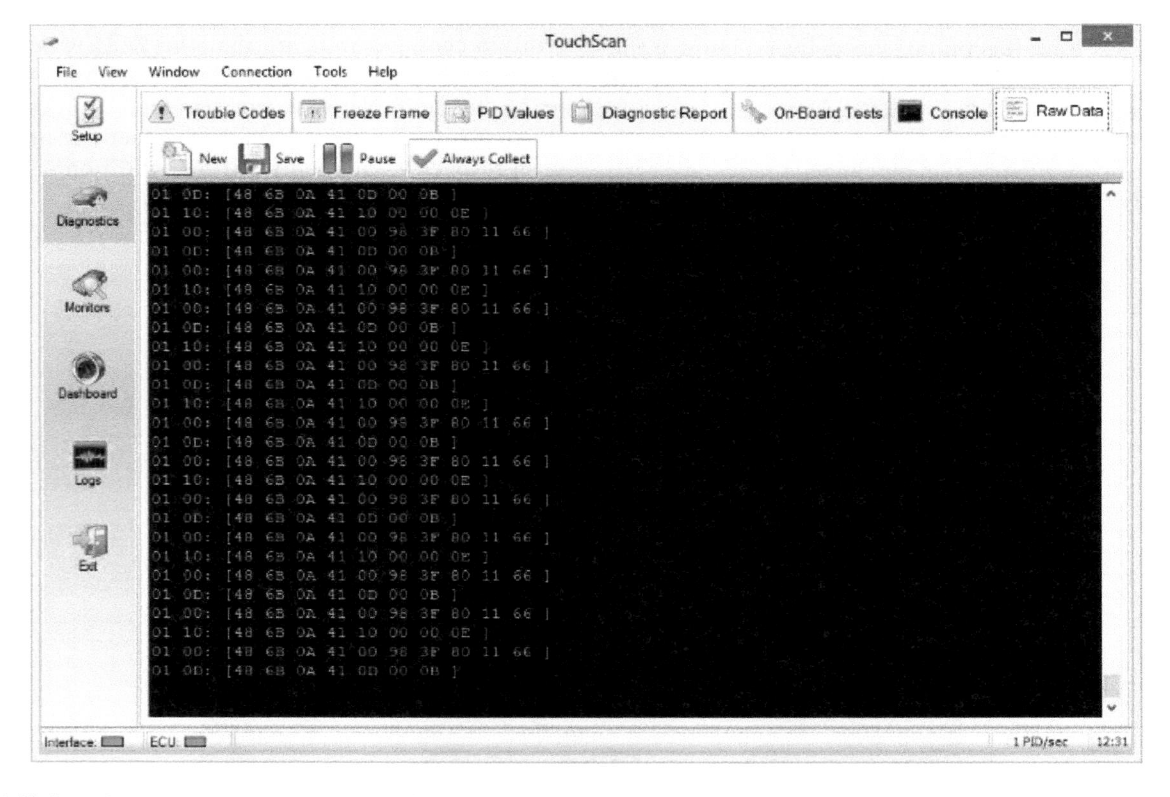

Figure 6.63 Raw data transmission to the TouchScan scan tool.

Figure 6.64 Manifold/port fuel injector.

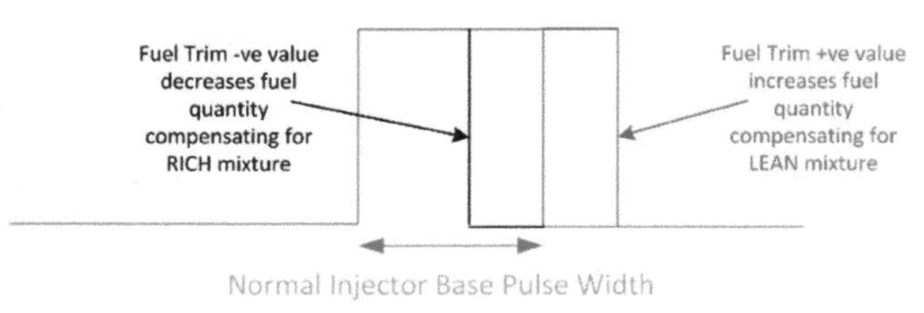

Figure 6.65 Fuel trim is effected by modifying the basic injector demand for the engine operating condition.

As vehicles operate over their lifetime, the engine system components experience wear and tear. With respect to the fuelling control system, airflow meters can become dirty and contaminated, and this has an effect on their initial calibration and response characteristic. Oxygen sensors can also become contaminated and this impacts on their accuracy and reliability. The fuel injection system also suffers from long-term fatigue effects – injectors can clog, thus affecting their spray pattern, and the ability to provide a homogeneous charge for combustion (Figure 6.64). Fuel pressure regulators lose their calibration and accuracy due to the continuous operation mode. Add to these factors general engine wear, due to loading and thermal effects, and it's clear that over a long period of time the accuracy and capability of the engine control system, with respect to correct fuelling, becomes compromised. Due to ever-tightening emissions regulations, this is no longer acceptable. The engine control system must maintain emissions within prescribed limits over vehicle life, and if it can't (perhaps due to component failure), then the system must inform the driver.

Long-term trim values are the result of an adaptive learning strategy within the ECU; this monitors the effectiveness of the control system during engine operation. For example, a gasoline engine ECU system will determine deviations from stoichiometry in the exhaust gas content over time. The information is stored in non-volatile memory and used to adjust or offset the fuelling value from the original stored fuel map value, to the required value, taking into account wear and tear factors. This is known as a long-term trim value – it is learned by the ECU and it is not lost when the battery is disconnected; it generally can only be reset with a scan tool.

Short-term trim values are offset values based on short-term effects in response to temporary changes. Typically the short-term trim values are a result of the closed-loop lambda control system; the main part of this loop being the pre-catalyst oxygen sensor. Short- and long-term fuel trim correction values work together to maintain the correct, required lambda value (Figure 6.65). The short-term trim will provide immediate correction, and if this correction is needed to maintain the required value on an ongoing basis, then the long-term fuel trim will learn this and provide a permanent offset to the basic fuelling map. Note though, this process is monitored and maintained within calibrated limits that correlate with tailpipe emissions. If the long-term trim value becomes greater than a certain, pre-set limit, a trigger in the diagnostic system will inform the driver of an emissions-related problem.

A useful aspect of fuel trim values is the potential to use them as part of a diagnostic procedure. Many fault code readers, including low-cost, generic devices, can provide fuel trim values. If you know and understand what these values are, and how they are generated, then you can use them to help you understand the nature and root cause of many fuel systems-related faults that you may come across.

During normal engine operation, the ECU records long (LTFT) and short (STFT) term fuel trim values as ECU labels. STFT normally cycles at the same frequency as the oxygen sensor switches; in effect it reflects the operation of the closed-loop lambda control. LTFT is normally quite stable, and any adjustment of the value is made over a long period of time, so is less noticeable. Remember that LTFT is held in non-volatile memory (also known as KAM – keep alive memory), but STFT is dynamic, thus constantly changes during run-time, but starting at zero, both values re-set when DTCs are cleared.

The cycle-to-cycle and cylinder-to-cylinder variations of an engine cause loss of efficiency and roughness. The ECU is therefore continually adjusting spark timing and fuelling to compensate. As injectors wear, the delivered quantity and spray pattern are compromised, and this can cause cylinder-specific fuelling faults that can be identified very efficiently using fuel trim values. The long- and short-term fuel trim value will reflect any variations in injector delivery or combustion efficiency for a specific cylinder, and can help diagnose faults such as leaking, blocked or dirty injectors (Figure 6.66).

Trim values should generally not exceed 10% either way. Positive means the ECU is trying to compensate for a weak mixture, it does this by extending the basic

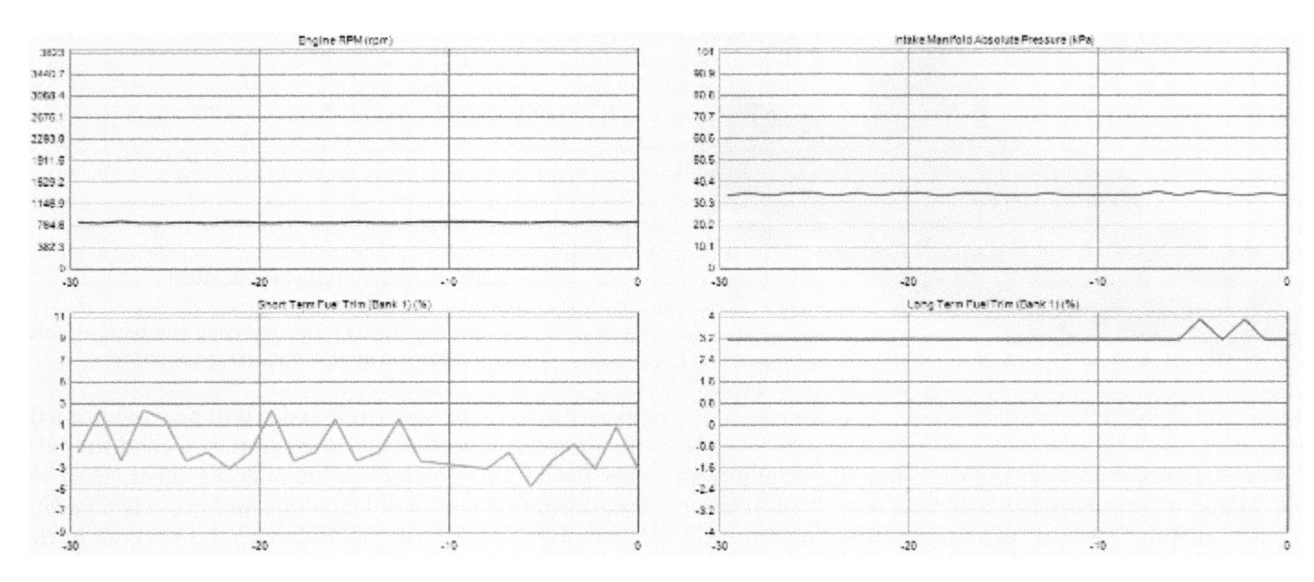

Figure 6.66 Typical fuel trim values at idle, as shown on a standard scan tool, connected to the vehicle OBD connector. Note that STFT is negative and follows closed-loop lambda operation. LTFT is much more stable over time, and is a positive value. Both are within acceptable limits.

injector pulse width (Figure 6.67). This could indicate faults caused by air leaks in the flow system, manifold or vacuum pipework. In addition, there may be MAF (mass airflow meter) or oxygen sensor issues. Lean mixture problems are the most common failure mode. Negative trims mean that the engine is running rich, and the base pulse width is reduced to compensate. This could indicate problems with EGR, MAF, fuel pressure regulator, leaking injectors or oxygen sensor problems. Individual cylinder contributions can also be evaluated. Disconnecting the injectors one by one and monitoring fuel trim values allows the technician to compare each cylinder with the others. Thus individual misfire or injector problems can be identified clearly.

Different ECUs respond differently to a given fault or diagnostic stimulus, in particular with respect to how LTFT and STFT work together. This is part of the ECU calibration strategy, so there are no hard and fast rules, but if you understand what LT and ST fuel trims are and what they do, then you can observe what you see and make an informed judgement. There is no doubt that being familiar with the concept will help your diagnostics of fuelling-related problems. If you have time, it's always worth making some observational

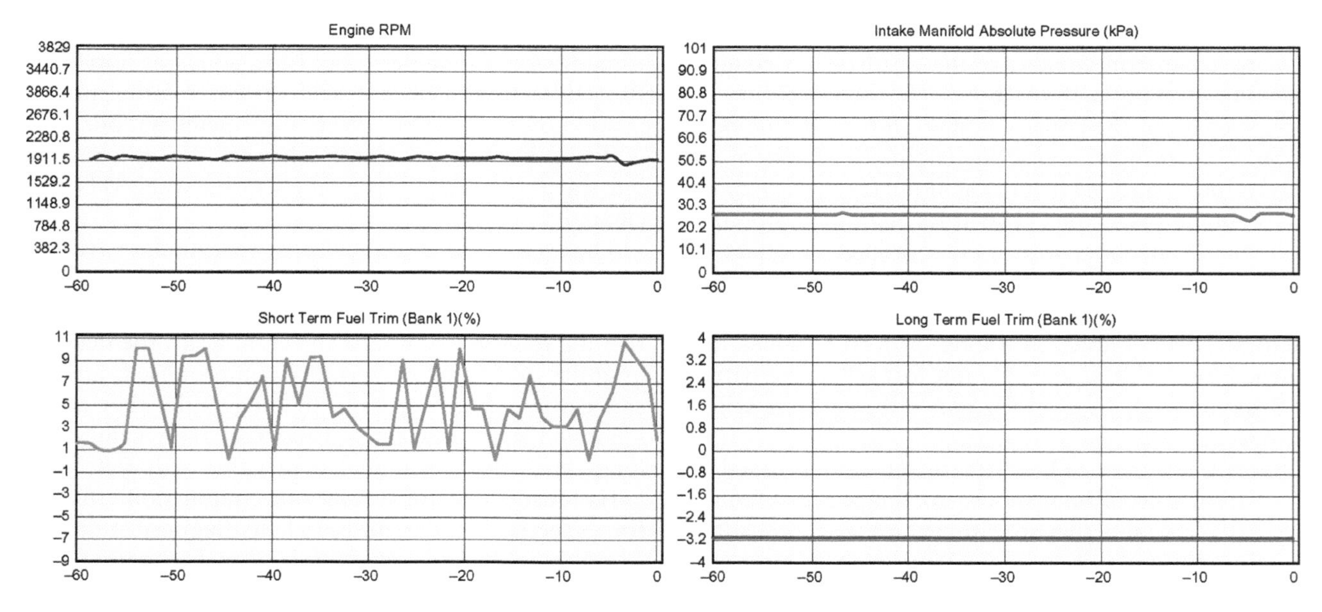

Figure 6.67 Fuel trim – high idle speed. STFT is now positive, LTFT has switched negative. This shows how both work together to achieve the required target of lambda 1.

measurements of a known 'good' vehicle, and record/note the data ready for the next faulty vehicle of that type or make to appear!

6.16 AIR SUPPLY AND EXHAUST SYSTEMS

6.16.1 Exhaust system

A vehicle exhaust system directs combustion products away from the passenger compartment, reduces combustion noise and, on most modern vehicles, reduces harmful pollutants in the exhaust stream. The main parts of the system are the exhaust manifold, the silencer or muffler, the pipes connecting them and a catalytic converter.

Most exhaust systems are made from mild steel, but some are made from stainless steel which lasts much longer. The system is suspended under the vehicle on rubber mountings. These allow movement because the engine is also rubber mounted, and they also reduce vibration noise.

An exhaust manifold links the engine exhaust ports to the down pipe and main system. It also reduces combustion noise and transfers heat downstream to allow the continued burning of hydrocarbons and carbon monoxide. The manifold is connected to the down pipe, which in turn can be connected to the catalytic converter. Most exhaust manifolds are made from cast iron, as this has the necessary strength and heat transfer properties.

The silencer's main function is to reduce engine noise to an acceptable level. Engine noise is a mixed-up collection of its firing frequencies (the number of times per second each cylinder fires). These range from approximately 100 to 400 Hz (cycles/s). A silencer reduces noise in two main ways:

- interior chambers using baffles, which are tuned to set up cancelling vibrations;
- absorptive surfaces function like sound-deadening wall and ceiling panels to absorb noise.

When the exhaust gases finally leave the exhaust system, their temperature, pressure and noise have been reduced considerably. The overall length of an exhaust system including the silencers can affect the smooth flow of gases. For this reason, do not alter the length or change the layout of an exhaust system (Figure 6.68).

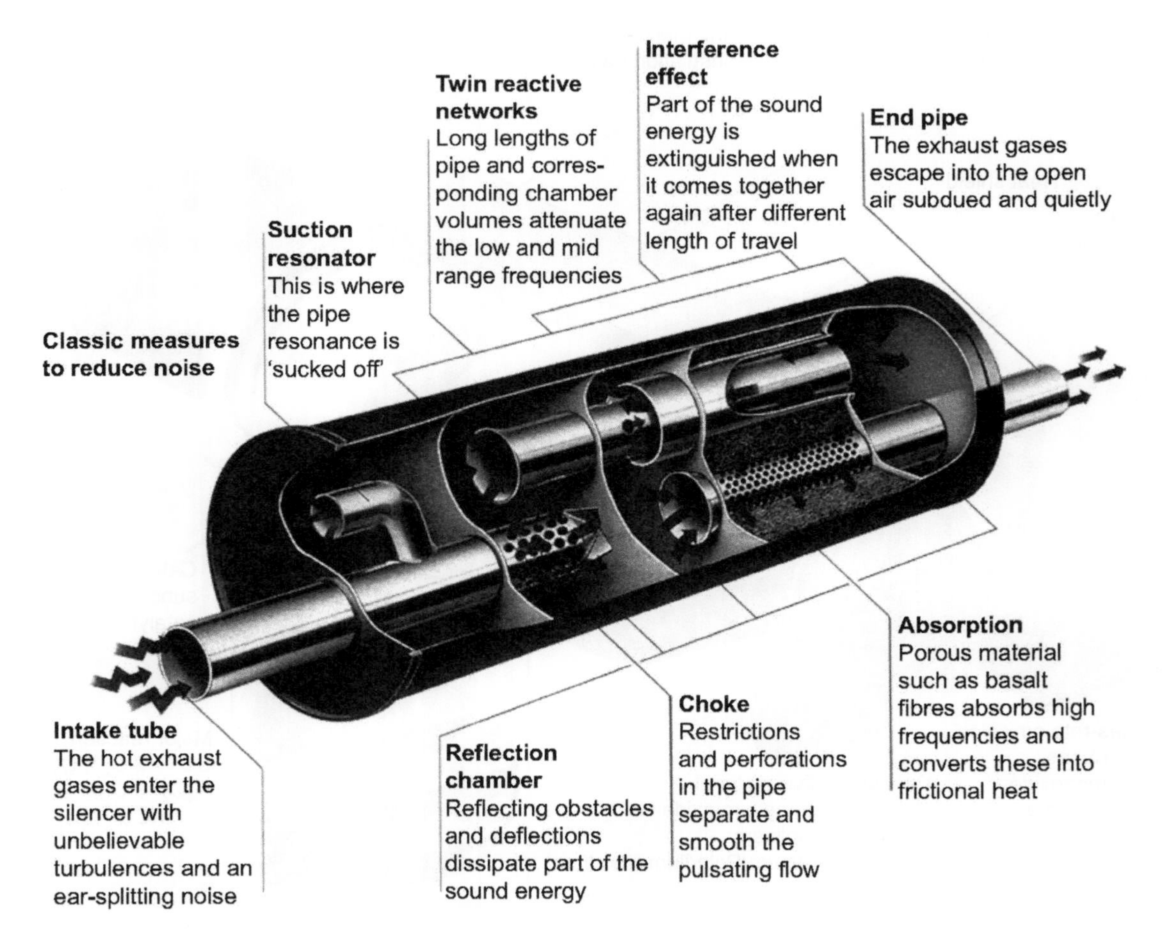
Classic measures to reduce noise
Suction resonator This is where the pipe resonance is 'sucked off'
Twin reactive networks Long lengths of pipe and corresponding chamber volumes attenuate the low and mid range frequencies
Interference effect Part of the sound energy is extinguished when it comes together again after different length of travel
End pipe The exhaust gases escape into the open air subdued and quietly
Intake tube The hot exhaust gases enter the silencer with unbelievable turbulences and an ear-splitting noise
Reflection chamber Reflecting obstacles and deflections dissipate part of the sound energy
Choke Restrictions and perforations in the pipe separate and smooth the pulsating flow
Absorption Porous material such as basalt fibres absorbs high frequencies and converts these into frictional heat

Figure 6.68 Exhaust noise reduction methods.

6.16.2 Catalytic converters

Stringent regulations in most parts of the world have made the use of a catalytic converter essential. The TWC is used to great effect by most manufacturers. It is in effect a very simple device; it looks similar to a standard exhaust silencer box. Note that in order for the 'cat' to operate correctly, the engine must be always well tuned. This is to ensure that the right 'ingredients' are available for the catalyst to perform its function. A catalytic converter works by converting the dangerous exhaust gases into gases which are non-toxic (Figure 6.69).

KEY FACT

Stringent regulations in most parts of the world have made the use of a catalytic converter essential.

The core has traditionally been made from ceramic of magnesium aluminium silicate. Owing to the several thousand very small channels, this provides a large surface area. It is coated with a wash coat of aluminium oxide, which again increases its effective surface area by about several thousand times. 'Noble' metals are used for the catalysts. Platinum helps to burn off the hydrocarbons (HC) and carbon monoxide (CO), and rhodium helps in the reduction of nitrogen oxides (NO_x). The whole three-way catalytic converter contains only about 3–4 g of these precious metals. Some converters now use metal cores (substrates).

The engine can damage a catalytic converter in one of two ways:

- first by the use of leaded fuel which can cause lead compounds to be deposited on the active surfaces;
- second by engine misfire which can cause the catalytic converter to overheat due to burning fuel inside the unit.

Some manufacturers use a system on some vehicles where a sensor checks the output of the ignition HT system and, if the spark is not present, will not allow fuel to be injected. Misfire detection is also part of current on-board diagnostic (OBD) legislation in some countries and future legislation in others.

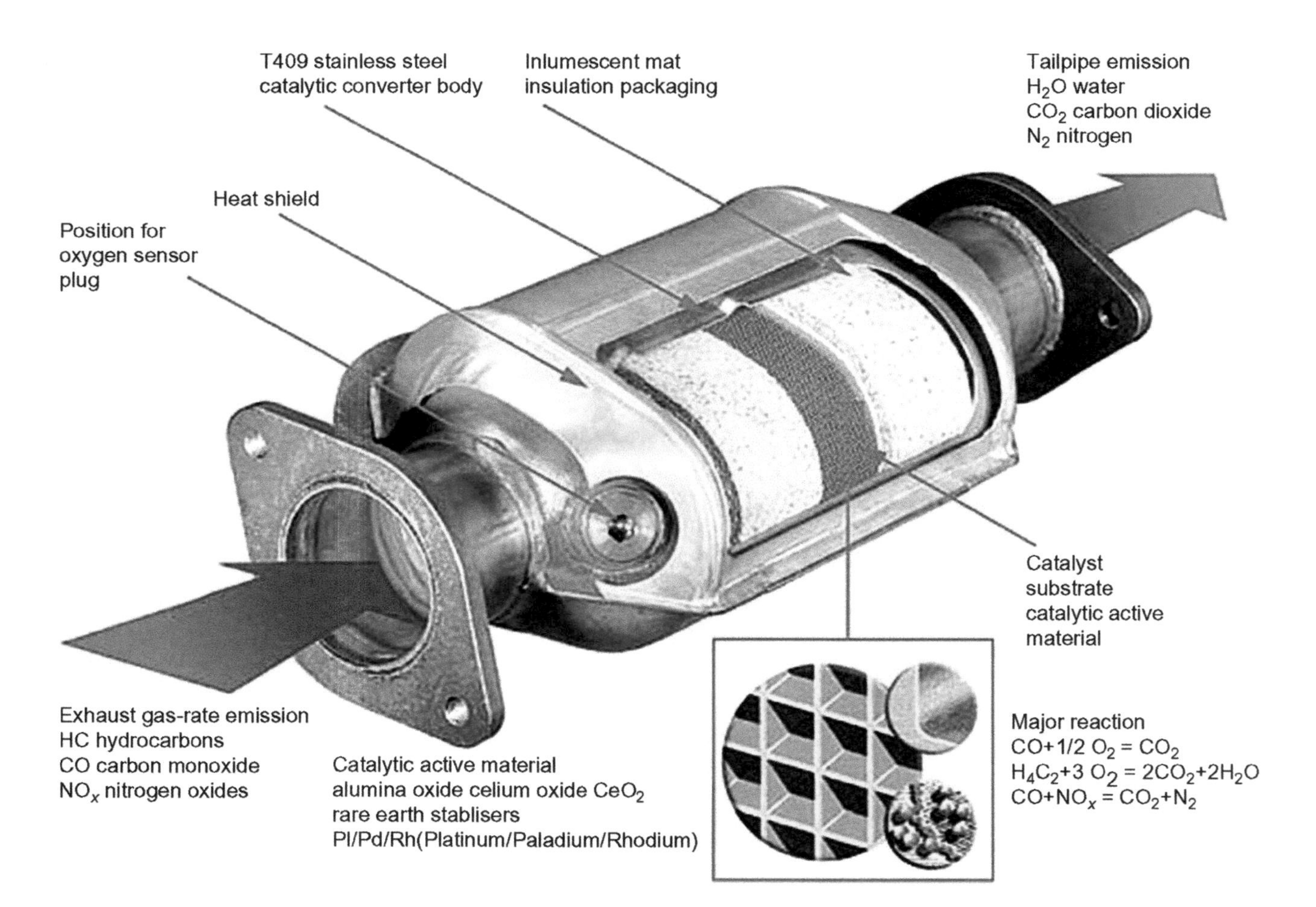

Figure 6.69 Catalytic converter components and operation.

6.16.3 Air supply system

There are three purposes of the complete air supply system:

1. clean the air;
2. control air temperature;
3. reduce noise.

A filter does the air cleaning and drawing air from around the exhaust manifold helps to control air temperature. When large quantities of air are drawn into the engine, it causes the air to vibrate and this makes it noisy. In the same way as with the exhaust system, baffles are used to stop resonance. Resonance means that when vibrations reach a natural level they tend to increase and keep going. A good example of how much noise is reduced by the air intake system is to compare the noise when an engine is run with the air filter removed.

Two types of air filter are in use, the first of these being by far the most popular:

- paper element;
- oil bath and mesh.

The paper element is made of resin-impregnated paper. Air filters using this type of replaceable element are used for both car and commercial vehicles. They provide a very high filtering efficiency and reasonable service life. They can be mounted in any position available under the bonnet. Service intervals vary, so check recommendations.

KEY FACT

A paper air filter element is made of resin-impregnated paper.

The oil bath and mesh type of air cleaner was widely used on non-turbocharged commercial vehicles. However, it is not very practical for modern low styled bonnets. Because it can be cleaned and fresh oil added, an oil bath air cleaner might still be used for vehicles operating in dusty conditions.

Air temperature control is used to help the vehicle conform to emission control regulations and for good driveability when the engine is cold. Good vaporisation of the fuel is the key. An automatic control is often fitted to make sure that the air intake temperature is always correct. The air cleaner has two intake pipes, one for cold air and the other for hot air from the exhaust manifold or hot box. The proportion of hot and cold air is controlled by a flap, which is moved by a diaphragm acted on by low pressure from the inlet manifold. The flap rests in the hot air pick-up position.

A thermo-valve in the air stream senses the temperature of the air going into the engine. When a temperature of approximately 25°C is reached, the valve opens. This removes the connection to the manifold, which in turn increases the pressure acting on the diaphragm. The flap is now caused to move and the pick-up is now from the cool air position. The flap is constantly moving, ensuring that the temperature of air entering the engine remains constant. Picking up hot air when the engine is very cold can also help to prevent icing.

6.17 DIAGNOSTICS – EXHAUST AND AIR SUPPLY

6.17.1 Systematic testing

If the reported fault is a noisy exhaust, proceed as follows:

1. Check if the noise is due to the exhaust knocking or blowing.
2. Examine the vehicle on the lift.
3. Check whether further tests are required or the fault is obvious.
4. Cover the end of the exhaust pipe with a rag for a second or two to highlight where the exhaust may be blowing.
5. Renew the exhaust section or complete system as appropriate.
6. Run and test for leaks and knocking.

6.17.2 Test results

Some of the information you may have to get from other sources such as data books or a workshop manual is listed in Table 6.5.

Table 6.5 Tests and information required

Test carried out	*Information required*
Air filter condition	Clearly a physical examination but note the required service intervals
Exhaust noise	An idea of the normal noise level – note that 'big-bore' exhausts will make more noise than the 'correct' type

6.17.3 Exhaust and air supply fault diagnosis table 1

Symptom	*Possible faults*	*Suggested action*
Exhaust noise	Hole in pipe, box or at joints	Renew as appropriate
Knocking noise	Exhaust incorrectly positioned	Reposition
	Broken mountings	Renew
Rich mixture/smoke	Blocked air filter	Replace
Noisy air intake	Intake trunking or filter box leaking or loose	Repair or secure as required
Poor cold driveability	Hot air pick-up not operating	Check pipe connections to inlet manifold for leaks
		Renew temperature valve or actuator

6.17.4 Exhaust fault diagnosis table 2

Symptom	*Possible cause*
Excessive noise	Leaking exhaust system or manifold joints
	Hole in exhaust system
Excessive fumes in car	Leaking exhaust system or manifold joints
Rattling noise	Incorrect fitting of exhaust system
	Broken exhaust mountings
	Engine mountings worn

6.18 COOLING

6.18.1 Air-cooled system

Air-cooled engines with multicylinders, especially under a bonnet, must have some form of fan cooling and ducting. This is to make sure that all cylinders are cooled evenly. The cylinders and cylinder heads are finned. Hotter areas, such as near the exhaust ports on the cylinders, have bigger fins.

Fan-blown air is directed by a metal cowling, so it stays close to the finned areas. A thermostatically controlled flap will control airflow. When the engine is warming up, the flap will be closed to restrict the movement of air. When the engine reaches its operating temperature, the flap opens and allows the air to flow over the engine. The cooling fan is a large device and is driven from the engine by a belt. This belt must not be allowed to slip or break, because serious damage will occur.

Car heating is not easy to arrange with an air-cooled engine. Some vehicles use a heat exchanger around the exhaust pipe. Air is passed through this device where it is warmed. It can then be used for demisting and heating with the aid of an electric motor and fan.

6.18.2 Water-cooled system

The main parts of a water-cooled system are as follows:

- water jacket;
- water pump;
- thermostat;
- radiator;
- cooling fan.

Water-cooled engines work on the principle of surrounding the hot areas inside the engine with a water jacket. The water takes on heat from the engine and, as it circulates through the radiator, gives it off to atmosphere. The heat concentration around the top of the engine means a water pump is needed to ensure proper circulation.

The water pump circulates water through the radiator and around the engine when the thermostat is open. Water circulates only round the engine when the thermostat is closed and not through the radiator. Forcing water around the engine prevents vapour pockets forming in very hot areas. This circulation is assisted by the thermo-siphon action. The thermo-siphon action causes the water to circulate because as the water is heated it rises and moves to the top of the radiator. This pushes down on the colder water underneath which moves into the engine. This water is heated, rises and so on.

Coolant from the engine water jacket passes through a hose to the radiator at the top. It then passes through thin pipes called the radiator matrix to the lower tank and then back to the lower part of the engine.

Many water passages between the top and bottom tanks of the radiator are used to increase the surface area. Fins further increase the surface area to make the radiator even more efficient. A cooling fan assists airflow. The heat from the coolant passes to the pipes and fins and then to the air as it is blown by a fan over the fins.

KEY FACT

Most modern radiators are made from aluminium pipes and fins with plastic tanks.

Many modern radiators are made from aluminium pipes and fins with plastic tanks top and bottom (down-flow), or at each end (cross-flow). The cross-flow radiators with tanks at each end are becoming the most popular. The more traditional method was to use copper and brass.

A thermostat is a temperature-controlled valve. Its purpose is to allow coolant to heat up more quickly and then be kept at a constant temperature. The total coolant volume in an engine takes time to heat up. Modern engines run more efficiently when at the correct operating temperature. The action of the thermostat is such as to prevent water circulation from the engine to the radiator until a set temperature is reached. When the valve opens, there is a full circuit for the coolant and a good cooling action occurs because of full flow through the radiator. The constant action of the thermostat ensures that the engine temperature remains at a constant level. The thermostat used by almost all modern engines is a wax capsule type. If the thermostat is faulty, ensure that the correct type for the engine is fitted, as some work at different temperatures.

The water pump is driven by a V-belt or multi-V-belt from the crankshaft pulley or by the cam belt. The pump is a simple impeller type and is usually fitted at the front of the engine (where the pulleys are). It assists with the thermo-siphon action of the cooling system, forcing water around the engine block and radiator.

The engine fan, which maintains the flow of air through the radiator, is mounted on the water-pump pulley on older systems. Most cooling fans now are electric. These are more efficient because they only work when needed. The forward motion of the car also helps the air movement through the radiator.

6.18.3 Sealed and semi-sealed systems

Cooling systems on most vehicles today are sealed or semi-sealed. This allows them to operate at pressures as much as 100 N/m^2 (100 Pa) over atmospheric pressure, raising the boiling point of the coolant to as much as 126.6°C (remember water boils at 100°C at atmospheric pressure). The system can therefore operate at a higher temperature and with greater efficiency.

The pressure build-up is made possible by the radiator pressure cap. The cap contains a pressure valve which opens at a set pressure and a vacuum valve which opens at a set vacuum. On a semi-sealed system, air is pushed out to atmosphere through the pressure valve as the coolant expands. Air is then drawn back into the radiator through the vacuum valve as the coolant cools and contracts. A sealed system has an expansion tank into which coolant is forced as it expands, and when the engine cools, coolant can flow from the tank back into the cooling system. Figure 6.70 shows a semi-sealed type cooling system.

SAFETY FIRST

Warning: If a pressure cap is removed from a hot system, hot water under pressure will boil the instant pressure is released. This can be very dangerous.

Correct levels in the expansion tank or in an unsealed radiator are very important. If too much coolant is used, it will be expelled on to the floor when the engine gets hot. If not enough is used, then the level could become low and overheating could take place.

Heat from the engine can be used to increase the temperature of the car interior. This is achieved by use of a heat exchanger, often called the heater matrix. Owing to the action of the thermostat in the engine cooling system, the water temperature remains nearly constant. The air being passed over the heater matrix is therefore heated to a set level.

A source of hot air is now available for heating the vehicle interior. Some form of control is required over how much heat is required. The method used on most modern vehicles is blending. This is a control flap that determines how much of the air being passed into the vehicle is directed over the heater matrix. Some systems use a valve to control the hot coolant flowing to the heater matrix.

By a suitable arrangement of flaps, it is possible to direct air of the chosen temperature to selected areas of

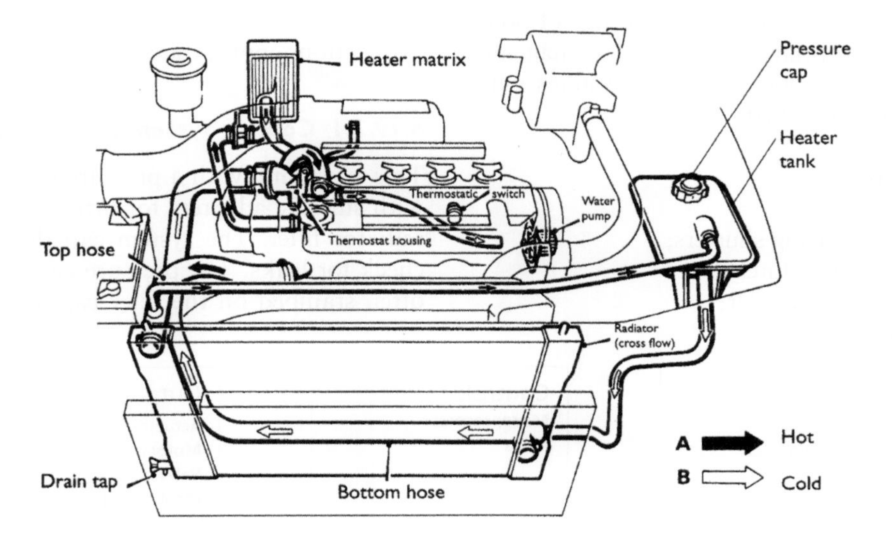

Figure 6.70 Semi-sealed cooling system.

the vehicle interior. In general, basic systems allow the warm air to be adjusted between the inside of the windscreen and the driver and passenger footwells. Fresh cool air outlets with directional nozzles are also fitted.

One final facility, which is available on many vehicles, is the choice between fresh or recirculated air. The primary reason for this is to decrease the time taken to demist or defrost the vehicle windows and simply to heat the car interior more quickly, and to a higher temperature. The other reason is that for example, in heavy congested traffic, the outside air may not be very clean.

6.18.4 Thermal management systems

Thermal management is about controlling air and coolant flows intelligently. However, it goes beyond cooling system technology because it considers all the heat flow systems on the vehicle. With this approach it is possible to improve efficiency, reduce emissions and improve passenger comfort.

Giving intelligence to the cooling system of a vehicle is the first step. This can be locally or in microprocessor-controlled systems (Figure 6.71). This intelligence will, for example, operate shutters (air control), coolant thermostats, bypass or mixing valves and electrically driven coolant pumps (water/glycol control). Warm-up times are faster and aerodynamic drag can be reduced by using radiator shutters. There is a reduction in energy consumed, passenger compartment warm-up time is reduced and temperature control of the engine is far more precise.

In addition, it is possible to warm up and control the temperature of transmission fluids using engine coolant, improving service life and efficiency of the transmission.

Thermal management systems need to be integrated with powertrain control systems, and a holistic view of the whole system should be taken. The range of potential applications appropriate for thermal management is considerable; here are some example systems where temperature can be controlled:

- coolant pump(s);
- thermostat;
- radiator air flow using shutters;
- cooling fan viscous clutch;
- coolant-cooled intercooler;
- exhaust gas cooling (EGR system);
- transmission oil temperature;
- HVAC;
- turbocharger;
- throttle body.

In passenger vehicles the potential fuel saving could be as much as 7.5%, along with the consequential reduction in CO_2 emissions.

Just as the use of electronic components has increased in the engine, the same is true of the TMS. Modern engines demand more extensive cooling, but in a controlled manner. This is delivered in part by adding electronic functionality to the TMS, and in part by redesigning TMS components.

6.19 DIAGNOSTICS – COOLING

6.19.1 Systematic testing

If the reported fault is loss of coolant, proceed as follows:

1. Check coolant level and discuss with customer how much is being lost.
2. Run the engine to see if it is overheating.
3. If the engine is not overheating, a leak would seem to be most likely.
4. Pressure-test the cooling system and check for leaks from hoses, gaskets and the radiator.
5. Renew a gasket or the radiator, clips or hoses as required. Top up the coolant and check antifreeze content.
6. Road-test the vehicle to confirm the fault is cured and that no other problems have occurred.

6.19.2 Test equipment

Note: You should always refer to the manufacturer's instructions appropriate to the equipment you are using.

6.19.2.1 Cooling system pressure tester

This is a pump with a pressure gauge built in, together with suitable adapters for fitting to the header tank or radiator filler. The system can then be pressurised to check for leaks. The pressure can be looked up or it is often stamped on the filler cap. A good way of doing

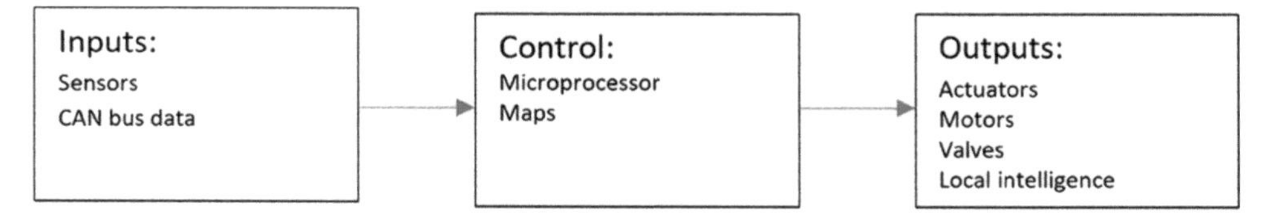

Figure 6.71 Simple representation of a thermal management control system.

this test is to pressurise the system when cold and then start the engine and allow it to warm up. You can be looking for leaks, but beware of rotating components.

6.19.2.2 *Antifreeze tester*

This piece of equipment is a hydrometer used to measure the relative density of the coolant. The relative density of coolant varies with the amount of antifreeze. A table can be used to determine how much more antifreeze should be added to give the required protection.

6.19.2.3 *Temperature meter/thermometer*

Sometimes the dashboard temperature gauge reading too high can create the symptoms of an overheating problem. A suitable meter or thermometer can be used to check the temperature. Note, though, that normal operating temperature is often well above 90°C (hot enough to burn badly) (Figure 6.72).

6.19.3 Test results

Some of the information you may have to get from other sources such as data books or a workshop manual is listed in Table 6.6.

Figure 6.72 Cooling system testing kit.

Source: Sykes Pickavant.

Table 6.6 Tests and information required

Test carried out	*Information required*
Leakage test	System pressure. Printed on the cap or from data books. Approximately 1 bar is normal.
Antifreeze content	Cooling system capacity and required percentage of antifreeze. If the system holds 6L, then for a 50% antifreeze content you will need to add 3 L of antifreeze. Do not forget you will need to drain out 3 L of water to make room for the antifreeze.
Operating temperature	This is about the same as the thermostat opening temperature. 88–92°C is a typical range.

6.19.4 Cooling fault diagnosis table 1

Symptom	*Possible faults*	*Suggested action*
Overheating	Lack of coolant	Top up but then check for leaks
	Thermostat stuck closed	Renew
	Electric cooling fan not operating	Check operation of thermal switch
	Blocked radiator	Renew
	Water pump/fan belt slipping	Check, adjust/renew
Loss of coolant	Leaks	Pressure test when cold and hot, look for leaks and repair as required
Engine does not reach normal temperature or it takes a long time	Thermostat stuck in the open position	Renew

6.19.5 Cooling fault diagnosis table 2

Symptom	*Possible cause*
Overheating	Low coolant level (maybe due to a leak)
	Thermostat stuck closed
	Radiator core blocked
	Cooling fan not operating
	Temperature gauge inaccurate
	Airlock in system (some systems have a complex bleeding procedure)
	Pressure cap faulty
Overcooling	Thermostat stuck open
	Temperature gauge inaccurate
	Cooling fan operating when not needed
External coolant leak	Loose or damaged hose
	Radiator leak
	Pressure cap seal faulty
	Water pump leak from seal or bearing
	Boiling due to overheating or faulty pressure cap
	Core plug leaking

(*Continued*)

Symptom	*Possible cause*
Internal coolant leak	Cylinder head gasket leaking
	Cylinder head cracked
Corrosion	Incorrect coolant (antifreeze, etc.)
	Infrequent flushing
Freezing	Lack of antifreeze
	Incorrect antifreeze

6.20 LUBRICATION

6.20.1 Lubrication system

From the sump reservoir under the crankshaft oil is drawn through a strainer into the pump. Oil pumps have an output of tens of litres per minute and operating pressures of more than 5 bar at high speeds. A pressure relief valve limits the pressure of the lubrication system to between 2.5 and 4 bar. This control is needed because the pump would produce excessive pressure at high speeds. After leaving the pump, oil passes into a filter and then into a main oil gallery in the engine block or crankcase.

Drillings connect the gallery to the crankshaft bearing housings and, when the engine is running, oil is forced under pressure between the rotating crank journals and the main bearings. The crankshaft is drilled so that the oil supply from the main bearings is also to the big-end bearing bases of the connecting rods.

The con rods are often drilled near the base so that a jet of oil sprays the cylinder walls and the underside of the pistons. In some cases, the con rod may be drilled along its entire length so that oil from the big-end bearing is taken directly to the gudgeon pin (small end). The surplus then splashes out to cool the underside of the piston and cylinder.

The camshaft operates at half crankshaft speed, but it still needs good lubrication because of the high-pressure loads on the cams. It is usual to supply pressurised oil to the camshaft bearings and splash or spray oil on the cam lobes. On overhead camshaft engines, two systems are used. In the simplest system, the rotating cam lobes dip into a trough of oil. Another method is to spray the cam lobes with oil. This is usually done by an oil pipe with small holes in it alongside the camshaft. The small holes in the side of the pipe aim a jet of oil at each rotating cam lobe. The surplus splashes over the valve assembly and then falls back into the sump.

On cars where a chain drives the cam, a small tapping from the main oil gallery sprays oil on the chain as it moves past or the chain may simply dip in the sump oil.

6.20.2 Oil filters

Even new engines can contain very small particles of metal left over from the manufacturing process or grains of sand which have not been removed from the crankcase after casting. Old engines continually deposit tiny bits of metal worn from highly loaded components such as the piston rings. To prevent any of these lodging in bearings or blocking oil ways, the oil is filtered.

The primary filter is a wire mesh strainer that stops particles of dirt or swarf from entering the oil pump. This is normally on the end of the oil pick-up pipe. An extra filter is also used that stops very fine particles. The most common type has a folded, resin-impregnated paper element. Pumping oil through it removes all but smallest solids from the oil.

Most engines use a full-flow system to filter all the oil after it leaves the pump. The most popular method is to pump the oil into a canister containing a cylindrical filter. From the inner walls of the canister the oil flows through the filter and out from the centre to the main oil gallery. Full-flow filtration works well, provided the filter is renewed at regular intervals. If it is left in service too long, it may become blocked. When this happens, the build-up of pressure inside the filter forces open a spring-loaded relief valve in the housing and the oil bypasses the filter. This valve prevents engine failure, but the engine will be lubricated with dirty oil until the filter is renewed. This is better than no oil.

A bypass filtration system was used on older vehicles. This system filters only a proportion of the oil pump output. The remainder is fed directly to the oil gallery. At first view, this seems a strange idea, but all the oil does eventually get filtered. The smaller amount through the filter allows a higher degree of filtration (Figure 6.73).

6.20.3 Oil pumps

In its simplest form, an oil pump consists of two gear wheels meshed together in a tight space so that oil cannot escape past the sides. The engine drives one wheel. As the gears rotate in opposite directions, the gap between each tooth in each wheel traps a small quantity of oil from an inlet port. The trapped oil is carried round by each wheel towards an outlet port on the opposite side where it is forced out by the meshing teeth.

The principle of squeezing oil from an ever-decreasing space is also used in the rotor-type pump. An inner and outer rotor are mounted on different axes in the same cylinder. The inner rotor, which commonly has four lobes, is driven by the engine. It meshes with an outer rotor, which has five lobes. As they rotate, the spaces between them change size. The inlet port is at a point where the space between the rotor lobes is increasing. This draws the oil into the pump. The oil is then carried round the pump. As rotation continues, the space between the lobes gets smaller. This compresses the oil out of the outlet port.

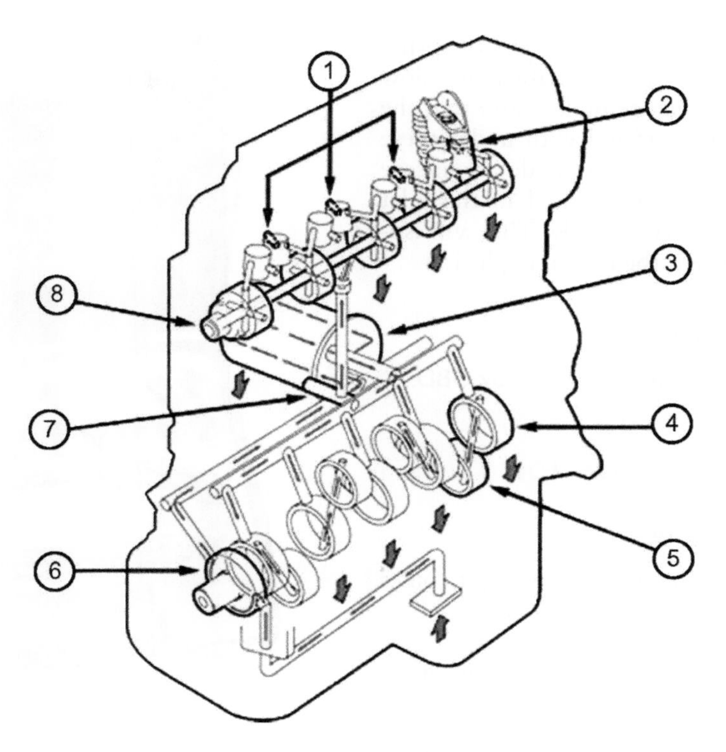

Figure 6.73 Oil flow: 1 – oil to rocker arms; 2 – hydraulic tappets; 3 – filter; 4 – crank main bearings; 5 – big-end bearings; 6 – crank driven oil pump; 7 – oil under pressure; 8 – camshaft.

Oil pumps can produce more pressure than is required. A valve is used to limit this pressure to a set value. The pressure relief valve is a simple device, which in most cases works on the ball and spring principle. This means that when the pressure on the ball is greater than the spring, the ball moves. The pressure relief valve is placed in the main gallery so that excess pressure is prevented. When the ball moves, oil is simply returned to the sump.

KEY FACT

Oil pumps can produce more pressure than is required, so a pressure relief valve is used.

6.20.4 Crankcase ventilation engine breather systems

Breathing is very important; without being able to breathe, we would die. It is almost as important for an engine breathing system to work correctly. There are two main reasons for engine breathers:

1. Prevent pressure build-up inside the engine crankcase due to combustion gases blowing past the pistons. The build-up of pressure will blow gaskets and seals but also there is a high risk of explosion.
2. Prevent toxic emissions from the engine. Emission limits are now very strict, for good reason – our health.

Crankcase breathing or ventilation of the engine was first achieved by what is known as an open system, but this has now been completely replaced by the closed system. The gases escaping from an engine with open crankcase ventilation as described above are very toxic. Legislation now demands a positive closed system of ventilation. This makes the pollution from cylinder blow-by gases negligible. Positive crankcase ventilation is the solution to this problem.

In early types of closed-system crankcase ventilation, the lower pressure at the carburettor air cleaner was used to cause an airflow through the inside of the engine. The breather outlet was simply connected by a pipe to the air cleaner. This caused the crankcase gases to be circulated and then burned in the engine cylinders. A flame trap was included in the system to prevent a crankcase explosion if the engine backfired.

In modern closed systems, the much lower pressure within the inlet manifold is used to extract crankcase gases. This has to be controlled in most cases by a variable regulator valve or pressure conscious valve (PCV). The valve is fitted between the breather outlet and the inlet manifold. It consists of a spring-loaded plunger, which opens as the inlet manifold pressure reduces. When the engine is stationary, the valve is closed. Under normal running conditions, the valve opens to allow crankcase gases to enter the inlet manifold with minimum restriction. At low manifold pressures during idling and overrun (pressure is less than atmospheric), further travel of the valve plunger against its spring closes it in the opposite direction. This reduces gas flow to the inlet manifold. This feature makes sure that the fuel control process is not interfered with under these conditions. The valve also acts as a safety device in case of a backfire. Any high pressure created in the inlet manifold will close the valve completely. This will isolate the crankcase and prevent the risk of explosion.

KEY FACT

Crankcase emission systems are monitored by OBD.

6.21 DIAGNOSTICS – LUBRICATION

6.21.1 Systematic testing

If the reported fault is that the oil pressure light comes on at low speed, proceed as follows:

1. Run the engine and see when the light goes off or comes on.
2. Is the problem worse when the engine is hot? Check the oil level. When was it last serviced?
3. If oil level is correct, then you must investigate further.
4. Carry out an oil pressure test to measure the actual pressure.
5. If pressure is correct, then renew the oil pressure switch. If not, engine strip down is likely.
6. Run and test for leaks.

6.21.2 Test equipment

6.21.2.1 Oil pressure test gauge

This is a simple pressure gauge that can be fitted with suitable adapters into the oil pressure switch hole. The engine is then run and the pressure readings compared to data.

6.21.2.2 Vacuum gauge

A simple 'U' tube full of water is often used. This is connected to the oil dip stick tube and the engine is run.

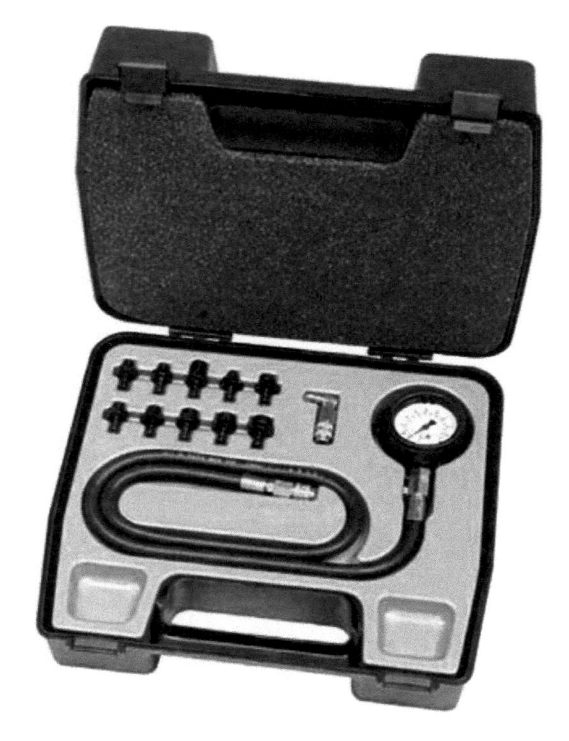

Figure 6.74 Oil pressure testing kit.

The gauge should show a pressure less than atmospheric (a partial vacuum). This checks the operation of the crankcase ventilation system (Figure 6.74).

6.21.3 Test results

Some of the information you may have to get from other sources such as data books or a workshop manual is listed in Table 6.7.

SAFETY FIRST

Note: You should always refer to the manufacturer's instructions appropriate to the equipment you are using.

Table 6.7 Tests and information required

Test carried out	*Information required*
Oil pressure	Oil pressure is measured in bars. A typical reading would be approximately 3 bar.
Crankcase pressure	By tradition pressures less than atmosphere are given in strange ways, such as, inches of mercury or inches of water! This is why I like to stick to absolute pressure and the bar! 0 bar is no pressure, 1 bar is atmospheric pressure and so on. 2–3 bar is more than atmospheric pressure like in a tyre. The trouble is standards vary, so make sure you compare like with like! Back to crankcase pressure – it should be less than atmospheric; check data.
Oil condition	Recommended type of lubricant.

6.21.4 Lubrication fault diagnosis table 1

Symptom	*Possible faults*	*Suggested action*
Low oil pressure	Lack of oil	Top up
	Blocked filter	Renew oil and filter
	Defective oil pump	Renew after further tests
	Defective oil pressure relief valve	Adjust if possible or renew
High crankcase pressure	Blocked crankcase breather	Clean or replace
	Blocked hose	Clean or renew hose
	Pressure blowing by pistons	Engine overhaul may be required
Loss of oil	Worn piston rings	Engine overhaul may be required
	Leaks	Renew seals or gaskets

6.21.5 Lubrication fault diagnosis table 2

Symptom	*Possible cause*
Oil leaks	Worn oil seal (check breather system)
	Gasket blown
	Cam or rocker cover loose
	Oil filter seal
Blue smoke	Piston rings
	Valve stem seals
	Head gasket

6.22 BATTERIES

6.22.1 Safety

The following points must be observed when working with batteries:

- good ventilation;
- protective clothing;
- supply of water available (running water preferable);
- first-aid equipment available, including eyewash;
- no smoking or naked lights permitted.

6.22.2 Lead-acid batteries

Incremental changes over the years have made the sealed and maintenance-free battery, now in common use, very reliable and long-lasting. This may not always appear to be the case to some end users, but note that quality is often related to the price the customer pays. Many bottom-of-the-range cheap batteries with a 12-month guarantee will last for 13 months (Figure 6.75).

The basic construction of a nominal 12 V lead-acid battery consists of six cells connected in series. Each cell producing approximately 2 V is housed in an individual compartment within a polypropylene or similar case. The active material is held in grids or baskets to form the positive and negative plates. Separators made from a microporous plastic insulate these plates from each other (Figure 6.76).

The grids, connecting strips and the battery posts are made from a lead alloy. For many years, this was lead antimony (PbSb), but this has now been largely replaced by lead calcium (PbCa). The newer materials cause less gassing of the electrolyte when the battery is fully charged. This has been one of the main reasons why sealed batteries became feasible as water loss is considerably reduced.

Modern batteries described as sealed do still have a small vent to stop the pressure build-up due to the

Figure 6.75 High-quality vehicle batteries.

Source: Bosch Media.

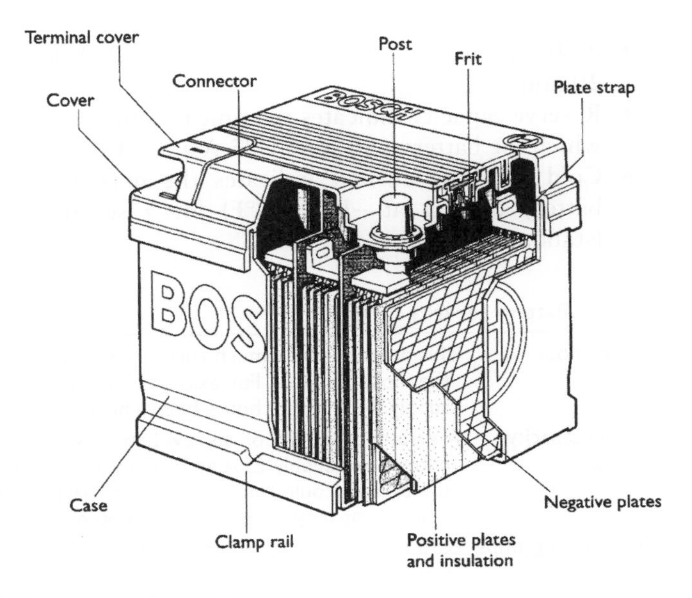

Figure 6.76 Vehicle battery components.

very small amount of gassing. A further requirement of sealed batteries is accurate control of charging voltage.

6.22.3 Battery rating

In simple terms, the characteristics or rating of a particular battery are determined by how much current it can produce and how long it can sustain this current. The rate at which a battery can produce current is determined by the speed of the chemical reaction. This in turn is determined by a number of factors:

- surface area of the plates;
- temperature;
- electrolyte strength;
- current demanded.

The actual current supplied therefore determines the overall capacity of a battery. The rating of a battery has to specify the current output and the time (Table 6.8).

Cold cranking amps (CCA) capacity rating methods do vary to some extent: British standards, DIN standards and SAE standards are the three main examples:

- BS 60 seconds
- DIN 30 seconds
- SAE 30 seconds

In summary, the capacity of a battery is the amount of electrical energy that can be obtained from it over a set time. It is usually given in ampere-hours, reserve capacity (RC) and cold cranking amps.

KEY FACT

The capacity of a battery is the amount of electrical energy that can be obtained from it over a set time.

- A 40 Ah battery means it should give 2 A for 20 hours.
- Reserve capacity indicates the time in minutes for which the battery will supply 25 A at 25°C.
- Cold cranking current indicates the maximum battery current at –18°C (0°F) for a set time (standards vary).

A battery for normal light vehicle use may be rated as follows: 44 Ah, 60 RC and 170 A CCA (BS). A 'heavy-duty' battery will have the same Ah rating as its 'standard duty' counterpart, but it will have a higher CCA and RC.

6.23 DIAGNOSTICS – BATTERIES

6.23.1 Servicing batteries

In use a battery requires very little attention other than the following when necessary:

- Corrosion should be cleaned from terminals using hot water.
- Terminals should be smeared with petroleum jelly or Vaseline *not* ordinary grease.
- Battery tops should be clean and dry.
- If not sealed, cells should be topped up with distilled water 3 mm above the plates (not very common now).
- Battery should be securely clamped in position.

6.23.2 Maintenance-free

By far the majority of batteries now available are classed as 'maintenance-free'. This implies that little attention is required during the life of the battery. Earlier batteries and some heavier types do, however, still require the electrolyte level to be checked and topped up periodically. Battery posts are still a little prone to corrosion and hence the usual service of cleaning with hot water if appropriate and the application of petroleum jelly or proprietary terminal grease is still recommended. Ensuring that the battery case and in particular the top remains clean will help to reduce the rate of self-discharge.

The state of charge of a battery is still very important, and in general it is not advisable to allow the state of charge to fall below 70% for long periods as the sulphate on the plates can harden, making recharging difficult. If a battery is to be stored for a long period (more than a few weeks), then it must be recharged every so often to prevent it from becoming sulphated.

Table 6.8 Battery capacity ratings

Ampere-hour capacity	This describes how much current the battery is able to supply for either 10 or 20 hours. The 20-hour figure is the most common. For example, a battery quoted as being 44 Ah (Ampere-hour) will be able, if fully charged, to supply 2.2 A for 20 hours before being completely discharged (cell voltage above 1.75 V).
Reserve capacity	A system used now on all new batteries is reserve capacity. This is quoted as a time in minutes for which the battery will supply 25 A at 25°C to a final voltage of 1.75 V per cell. This is used to give an indication of how long the battery could run the car if the charging system was not working. Typically, a 44 Ah battery will have a reserve capacity of approximately 60 minutes.
Cold cranking amps	Batteries are given a rating to indicate performance at high current output and at low temperature. A typical value of 170 A means that the battery will supply this current for one minute at a temperature of –18°C at which point the cell voltage will fall to 1.4 V (BS).

Recommendations vary, but a recharge every six weeks is a reasonable suggestion.

6.23.3 Charging

The recharging recommendations of battery manufacturers vary slightly. The following methods, however, are reasonably compatible and should not cause any problems. The efficiency of a battery is not 100%. Therefore, the recharging process must 'put back' the same Ah capacity as was used on discharge plus a bit more to allow for efficiency losses. It is therefore clear that the main question about charging is not how much, but at what rate.

The traditional recommendation was that the battery should be charged at a tenth of its Ah capacity for approximately 10 hours or less. This is based on the assumption that the Ah capacity is quoted at the 20-hour rate, as a tenth of this figure will make allowance for the charge factor. This figure is still valid but as Ah capacity is not always used nowadays, a different method of deciding the rate is necessary. One way is to set a rate at a sixteenth of the reserve capacity, again for up to 10 hours. The final suggestion is to set a charge rate at one-fortieth of the cold start performance figure, also for up to 10 hours. Clearly, if a battery is already half charged, half the time is required to recharge to full capacity.

KEY FACT

The ideal charge rate is determined as:
1/10 of the Ah capacity.
1/16 of the RC.
1/40 of the CCA.

The above-suggested charge rates are to be recommended as the best way to prolong battery life. They do all, however, imply a constant current charging source. A constant voltage charging system is often the best way to charge a battery. This implies that the charger, an alternator on a car for example, is held at a constant level and the state of charge in the battery will determine how much current will flow. This is often the fastest way to recharge a flat battery. If a constant voltage of less than 14.4 V is used, then it is not possible to cause excessive gassing and this method is particularly appropriate for sealed batteries.

Boost charging is a popular technique often applied in many workshops. It is not recommended as the best method but, if correctly administered and not repeated too often, it is suitable for most batteries. The key to fast or boost charging is that the battery temperature should not exceed 43°C. With sealed batteries, it is particularly important not to let the battery gas excessively in order to prevent the build-up of pressure. A rate of about five times the 'normal' charge setting will bring the battery to 70–80% of its full capacity within approximately one hour. Table 6.9 summarises the charging techniques for a lead-acid battery.

Table 6.9 Charging methods

Charging method	*Notes*
Constant voltage	Constant voltage will recharge any battery in seven hours or less without any risk of overcharging (14.4 V maximum)
Constant current	Ideal charge rate can be estimated as 1/10 of Ah capacity, 1/16 of reserve capacity or 1/40 of cold start current (charge time of 10–12 hours or pro rata original state)
Boost charging	At no more than five times the ideal rate, a battery can be brought up to approximately 70% of charge in about one hour
Smart charging	Let the charger do all the calculations and all the work

6.23.3.1 Smart chargers

Nowadays, there are a number of 'Smart' or 'Intelligent' battery chargers that are able to determine the ideal rate from the battery voltage and the current it will accept. Some also have features such as a 'recond' mode, which allows you to correct the acid stratification that often occurs in deeply discharged batteries – particularly leisure batteries. Some key features of a charger produced by a company called CTEK are as follows (Figure 6.77):

- Safe: No sparks and cannot harm vehicle electrics, so no need to disconnect the battery.
- Suitable for all types of 12 V lead-acid batteries up to 150 Ah.
- Connect and forget – can be left connected for months – ideal for vehicles used occasionally.
- Analysis mode to check if battery can hold charge.
- 10 day float maintenance for maximum charge level.
- 'Recond' mode – special programme to revive deeply discharged batteries.
- Supply mode – can be used as a 12 V power source to protect electrical settings.

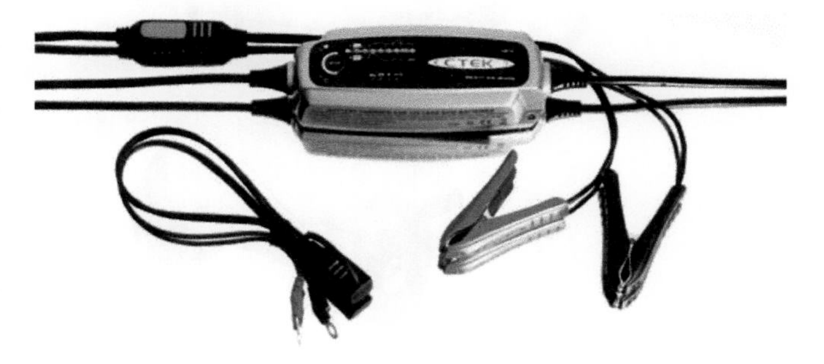

Figure 6.77 Smart battery charger.

Source: www.ctek.com.

Table 6.10 Battery faults

Symptom or fault	*Likely causes*					
Low state of charge	Charging system fault	Unwanted drain on battery	Electrolyte diluted	Incorrect battery for application		
Low capacity	Low state of charge	Corroded terminals	Impurities in the electrolyte	Sulphated	Old age – active material fallen from the plates	
Excessive gassing and temperature	Overcharging	Positioned too near exhaust component				
Short circuit cell	Damaged plates and insulators	Build-up of active material in sediment trap				
Open circuit cell	Broken connecting strap	Excessive sulphation	Very low electrolyte			
Service life shorter than expected	Excessive temperature	Battery has too low a capacity	Vibration excessive	Contaminated electrolyte	Long periods of not being used	Overcharging

6.23.4 Battery faults

Any electrical device can suffer from two main faults: these are either open circuit or short circuit. A battery is no exception but can also suffer from other problems such as low charge or low capacity. Often a problem, apparently with a vehicle battery, can be traced to another part of the vehicle such as the charging system. Table 6.10 lists all the common problems encountered with lead-acid batteries, together with typical causes (Figure 6.78).

Figure 6.78 Battery charger and engine starter.

Source: Bosch Media.

Most of the problems listed previously will require the battery to be replaced. In the case of sulphation, it is sometimes possible to bring the battery back to life with a very long low-current charge. A fortieth of the Ah capacity or about a two-hundredth of the cold start performance for approximately 50 hours is an appropriate rate. Some smart chargers are very good for this.

6.23.5 Testing batteries

For testing the state of charge of a non-sealed type of battery, it was traditional to use a hydrometer. The hydrometer is a syringe that draws electrolyte from a cell and a float, which will float at a particular depth in the electrolyte according to its density. The relative density or specific gravity is then read from the graduated scale on the float. A fully charged cell should show 1.280, when half charged 1.200 and if discharged 1.120.

Most vehicles are now fitted with maintenance free batteries and a hydrometer cannot be used to find the state of charge. This can, however, be determined from the voltage of the battery, as given in Table 6.11. An accurate voltmeter is required for this test – note the misleading surface charge shown in Figure 6.79.

KEY FACT

Accuracy: A good tip to reduce surface charge is, after switching off the engine, turn on the headlights for a few minutes, then turn them off, wait another few minutes – then take the reading.

Table 6.11 Battery voltages

Battery volts at 20°C	*State of charge*
12.0 V	Discharged (20% or less)
12.3 V	Half charged (50%)
12.7 V	Charged (100%)

Figure 6.79 Checking battery voltage. In this case the engine had just been switched off, so the reading shows a 'surface charge'.

The preferred way now to test a battery is with a dedicated battery tester. The example shown in Figure 6.80 will also quickly check the starter and charging systems as part of the process. Results can be printed out for the customer. At the end of the test process there are a number of different possible outcomes:

Battery test results:

1. good battery;
2. good but needs recharging;
3. replace;
4. bad cell, replace;
5. charge and retest.

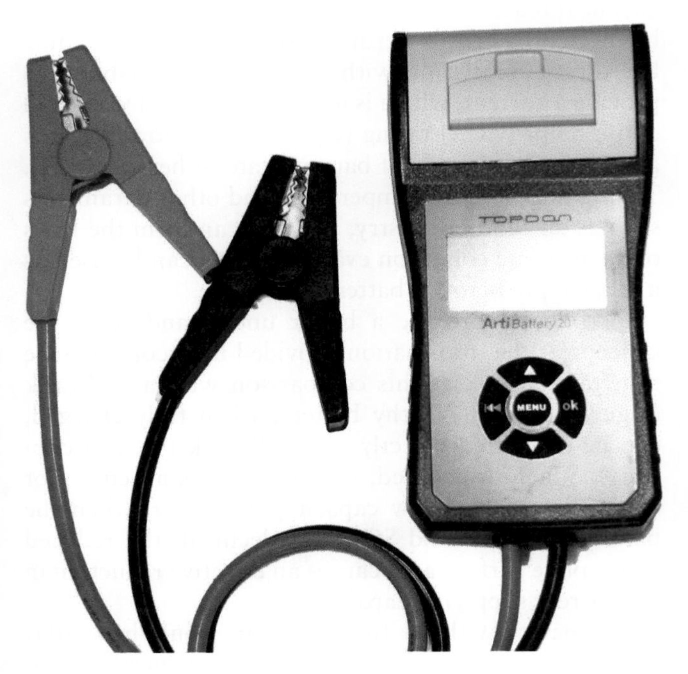

Figure 6.80 Battery, charging and starting tester.

Cranking test results:

1. cranking voltage and cranking time;
2. cranking voltage less than 9.6 V, the cranking system is abnormal;
3. cranking voltage higher than 9.6 V, the cranking system is OK.

Charging test results:

1. charging voltage normal;
2. charging voltage low;
3. charging voltage high;
4. no voltage output;
5. diode damaged.

Further checks are necessary to track down specific faults with the starting or charging system.

6.23.6 Battery diagnostics

Modern vehicles have sophisticated energy requirements and electronic consumers that need a stable, clean voltage supply. Workshops are seeing obscure faults with electronic systems, including fault code errors, brought on by failing batteries. Traditionally, a failing battery would manifest itself by having insufficient power to crank the engine and start the vehicle. This was often more apparent in winter, when cold starts need more torque to overcome the friction of a cold engine, with thicker, cold lubricating oil. However, with modern vehicles, a failing battery is likely to produce a fault, of an unrelated nature, before this 'non-start' symptom occurs. Battery technology has progressed in line with the vehicle systems, but a different method of establishing the serviceability was needed.

There are two traditional methods of checking a wet, lead-acid, vehicle battery. The first is state of charge (SOC), which can be determined via measuring the specific gravity (SG) of the electrolyte in each cell with a hydrometer (there is also a less accurate option, to measure the battery terminal voltage). Assuming the battery is reasonably well charged (>75%), then a performance test, indicating state-of-health (SOH), could be executed via a discharge test. This test is performed using a high rate discharge tester (Figure 6.81), with the appropriate load according to the battery capacity, and it would indicate the battery capability to supply a large current (as would be required under starting conditions). From these measurements, an experienced technician could make a judgement on the battery fitness for purpose.

Figure 6.81 A high-rate discharge tester can indicate battery state-of-health, but still relies on the skill and judgement of the technician. In addition, there are several health and safety-related issues to this approach! Note the tester in the picture is a fixed load, and not really suitable for the battery shown.

Source: www.autoelex.co.uk.

Figure 6.82 Intelligent, digital battery testers are much safer and more appropriate for testing modern battery technology.

Source: www.autoelex.co.uk.

There are several reasons that the methods mentioned previously are no longer applicable:

- Many modern batteries have no access to the cell electrolyte, thus hydrometer readings are simply not possible. Although the battery may have a built-in hydrometer, this is of limited use; it's just a general indicator.
- In order to execute a high-rate discharge test, the battery has to be disconnected from the vehicle – this can present time-consuming problems for the technician, e.g. lost radio codes, ECU memory loss, etc.
- There are health and safety issues – wet batteries contain acid and generate volatile gases. High rate load tests can create sparks and heat. All are potential problems in a safety-conscious workshop.
- The measurements still rely on the knowledge and experience of the technician to make a judgement on the battery SOH. This is subjective and could be the source of inaccurate diagnoses.

Along with the progress in battery and vehicle technology, technology developments have provided alternative methods for testing. Battery testers now use a completely different approach to evaluating the battery condition, providing an objective measurement of battery condition and capability, along with a more accurate SOH assessment. These testers are intelligent units with menu-guided test procedures (Figure 6.82). The technique itself is known as conductance measurement.

The conductance test is a completely different method of establishing the battery condition and performance and is ideally suited for modern vehicle battery test applications. It can also be applied to older, wet lead-acid batteries. The conductance tester applies an AC voltage, of known frequency and amplitude, across the battery terminals and monitors the subsequent current that flows with respect to phase shift and ripple (Figure 6.83). The AC voltage is superimposed on the battery's DC voltage and acts as brief charge and discharge pulses. This information is utilised to calculate the impedance (measure of opposition to alternating current) of the battery, and from this the conductance value can be established (impedance and conductance have a reciprocal relationship).

A conductance measurement provides a measure of the plate surface area, and this determines how much chemical reaction or power the battery can generate. It has been proven by experiment that a conductance value has direct correlation with the battery's capability to provide a current, which is normally specified via a cold crank amps (CCA) rating (Figure 6.84). However, it is also a good indicator of battery state of health (SOH). Taking into account temperature and other parameters such as age and chemistry, this test can form the basis of an accurate condition evaluation and can be used as a reliable predictor of battery end of life.

In order to provide a better understanding of the concept of the information provided by a conductance test, take a look at this comparison with a fuel tank (Figure 6.85). A healthy battery, when fully charged, can be compared directly to a full tank (i.e. full capacity). When discharged, it's the same as an empty or low fuel tank (i.e. low capacity). However, when the battery has aged and SOH has declined, the reduced active plate surface area causes an effective reduction in the current supplying capability.

Comparing with the fuel tank, this would be similar to damage to the fuel tank which has reduced its volume (e.g. a large dent in the tank). So, even if you fill

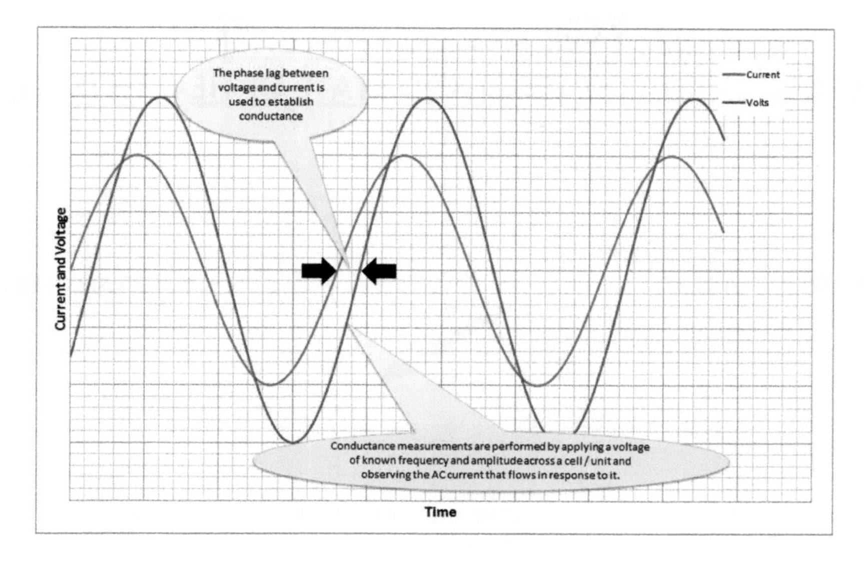

Figure 6.83 AC voltage across the DC battery allows a measurement of the battery's conducting capability to be made.

Source: www.autoelex.co.uk.

the tank and the gauge shows full, the actual capacity of the tank is reduced.

The conductance test is an accurate, repeatable method. The test can be applied to a battery connected in the vehicle. The test method is much safer for the operator and the vehicle. The result from the test is much more objective and factual. The conductance test can be applied to many modern battery systems. This is important as technology is currently changing and adapting to new demands and load profiles generated by the latest methods used to reduce vehicle tailpipe emissions. For example, stop/start and energy recovery systems.

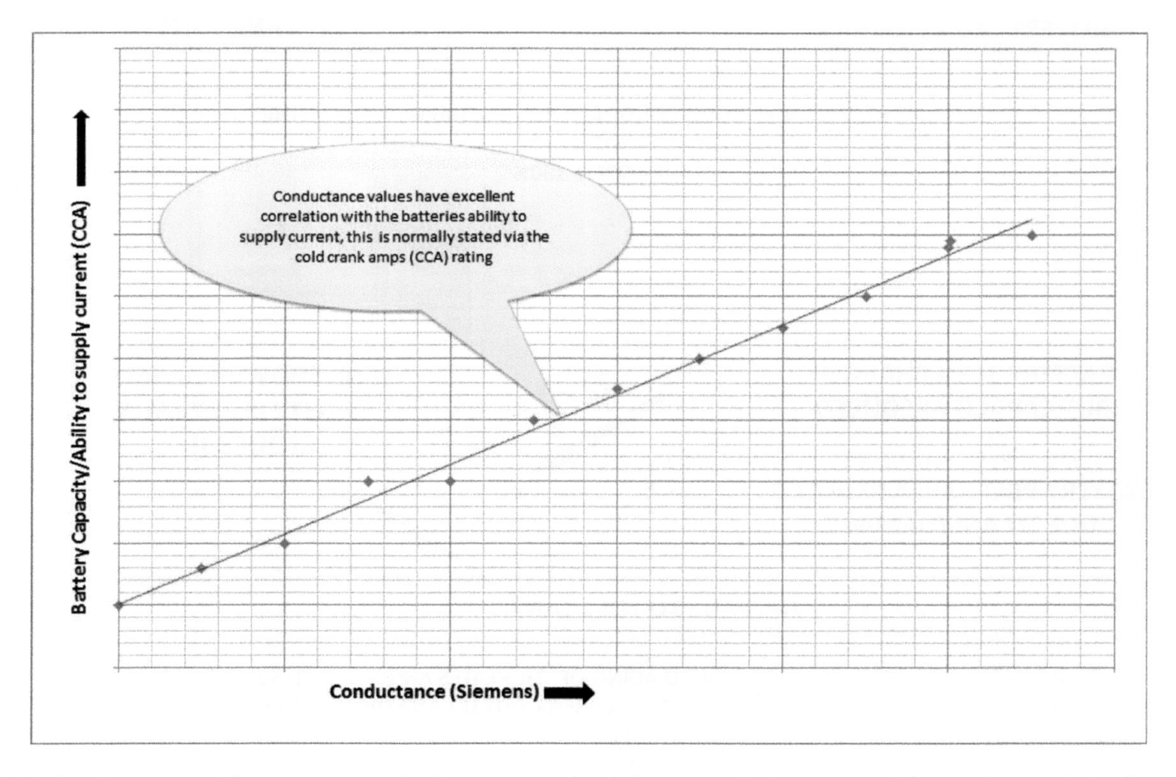

Figure 6.84 Conductance and battery capacity (with respect to the ability to supply large current) have a direct relationship.

Source: www.autoelex.co.uk.

FUEL TANK ANALOGY FOR BATTERY
STATE OF CHARGE (SOC) AND STATE OF HEALTH (SOH)

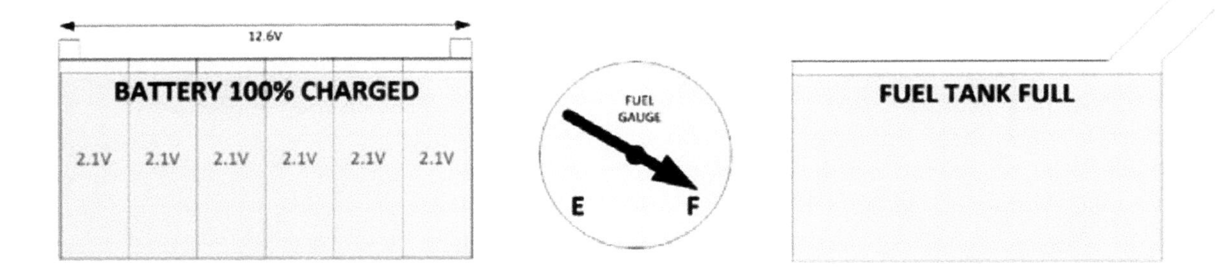

BATTERY IS IN GOOD CONDITION AND FULLY CHARGED, THEREFORE
SOC = 100%
SOH = 100%

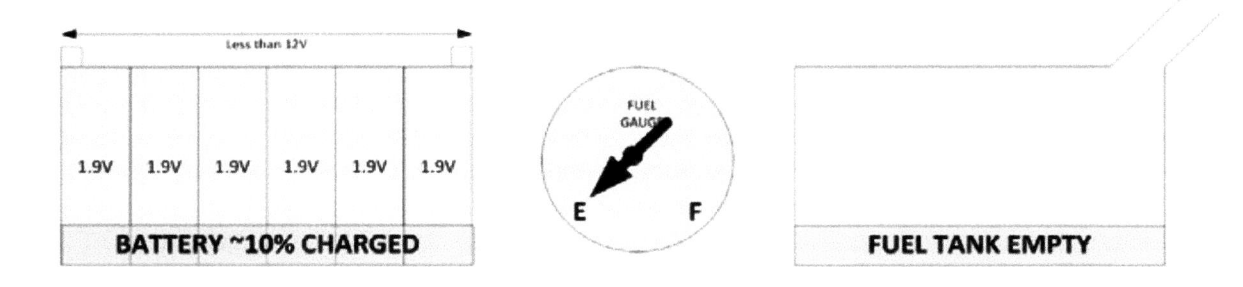

BATTERY IS IN GOOD CONDITION BUT DISCHARGED, THEREFORE
SOC = ~10%
SOH = 100%

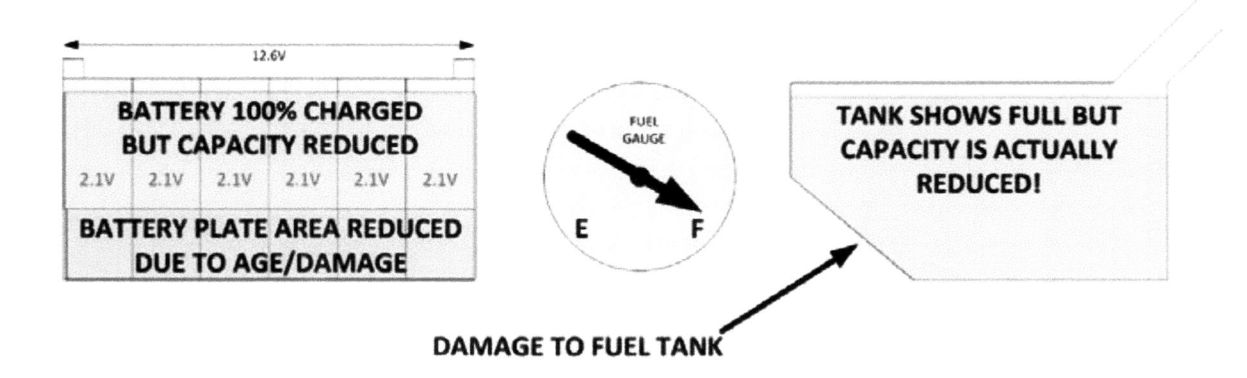

BATTERY IS IN POOR CONDITION DUE TO AGING OF THE PLATES AND REDUCED SURFACE AREA, THIS
REDUCES THE BATTERIES CAPABILITY TO SUPPLY CURRENT, BUT IT CAN STILL BE FULLY CHARGED, THEREFORE
SOC = 100%
SOH = less than 80%

Figure 6.85 Battery SOC and SOH, for illustration, compared to a fuel tank.

6.24 STARTING

6.24.1 Starter circuit

In comparison with most other circuits on the modern vehicle, the starter circuit is relatively simple (Figure 6.86). The problem to overcome, however, is that of volt drop in the main supply wires. A spring-loaded key switch usually operates the starter; the same switch also controls the ignition and accessories. The supply from the key switch, via a relay in many cases, causes the starter solenoid to operate and this in turn, by a set of contacts, controls the heavy current. In some cases, an extra terminal on the starter solenoid provides an output when cranking, usually used to bypass a dropping resistor on the ignition or fuel pump circuits. The problem of volt drop in the main supply circuit is due to the high current required by the starter, particularly under adverse starting conditions such as very low temperatures.

A typical cranking current for a light vehicle engine is in the order of 150 A, but this may peak in excess of 500 A to provide the initial stalled torque. It is generally accepted that a maximum volt drop of only 0.5 V should be allowed between the battery and starter when operating. An Ohm's law calculation indicates that the maximum allowed circuit resistance is 2.5 mΩ, when using a 12 V supply. This is a worst-case situation and lower resistance values are used in most applications. The choice of suitable conductors is therefore very important.

KEY FACT

A typical cranking current for a light vehicle engine is approximately 150 A, but this may peak in excess of 500 A to provide the initial stalled torque.

Many starting systems are now electronically controlled by the engine management and the security systems.

6.24.2 Pre-engaged starters

Pre-engaged starters are fitted to every modern vehicle (Figure 6.87). They provide a positive engagement with the ring gear, as full power is not applied until the pinion is fully in mesh.

KEY FACT

Pre-engaged starters provide a positive engagement with the ring gear, as full power is not applied until the pinion is fully in mesh.

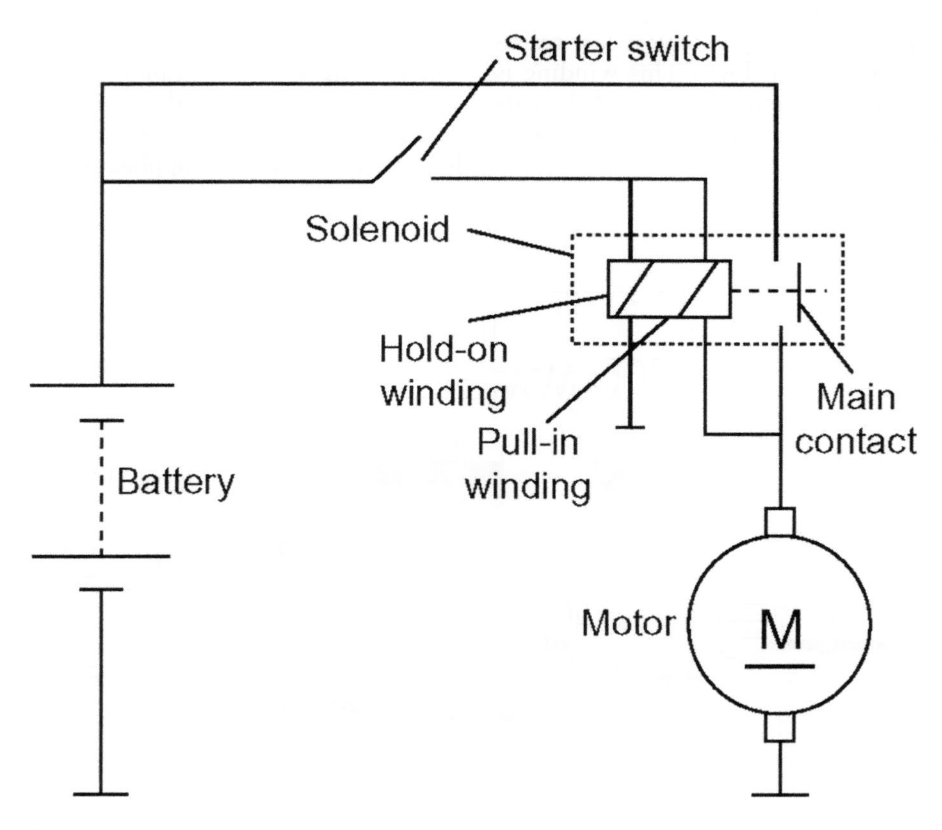

Figure 6.86 Starter circuit.

Figure 6.87 Intermediate transmission starter.

Source: Bosch Media.

Figure 6.88 shows the circuit associated with operating this type of pre-engaged starter. The basic operation of the pre-engaged starter is as follows. When the key switch is operated, a supply is made to terminal 50 on the solenoid. This causes two windings to be energised, the hold-on winding and the pull-in winding. Note that the pull-in winding is of very low resistance and hence a high current flows. This winding is connected in series with the motor circuit and the current flowing will allow the motor to rotate slowly to facilitate engagement.

At the same time the magnetism created in the solenoid attracts the plunger and via an operating lever pushes the pinion into mesh with the flywheel ring gear. When the pinion is fully in mesh, the plunger at the end of its travel causes a heavy-duty set of copper contacts to close. These contacts now supply full battery power to the main circuit of the starter motor. When the main contacts are closed, the pull-in winding is effectively switched off due to equal voltage supply on both ends. The hold-on winding holds the plunger in position as long as the solenoid is supplied from the key switch.

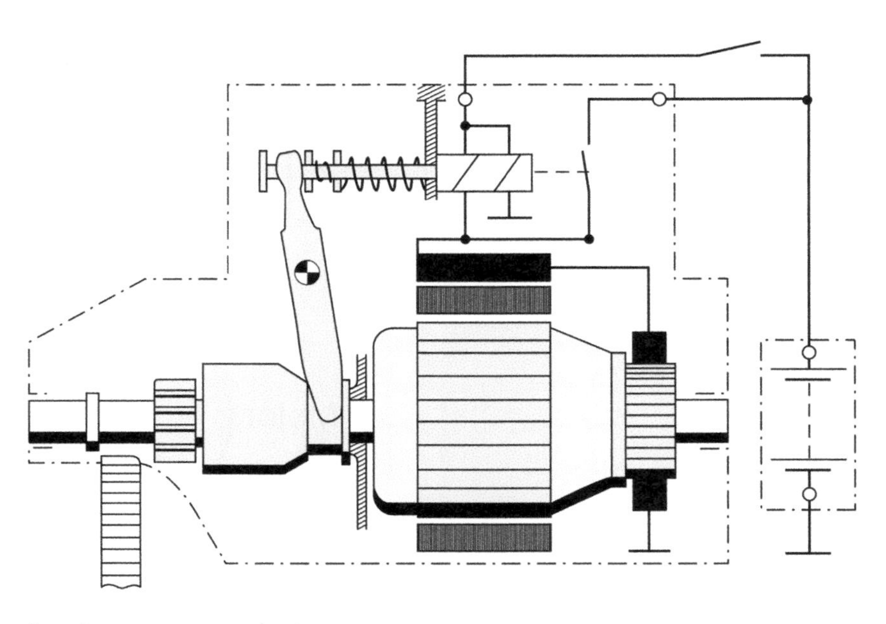

Figure 6.88 Starter circuit and engagement mechanism.

When the engine starts and the key is released, the main supply is removed and the plunger and pinion return to their rest positions under spring tension. A lost motion spring located on the plunger ensures that the main contacts open before the pinion is retracted from mesh.

During engagement if the teeth of the pinion hit the teeth of the flywheel (tooth to tooth abutment), the main contacts are allowed to close due to the engagement spring being compressed. This allows the motor to rotate under power and the pinion will slip into mesh.

The torque developed by the starter is passed through a one-way clutch to the ring gear. The purpose of this free-wheeling device is to prevent the starter being driven at excessively high speed if the pinion is held in mesh after the engine has started. The clutch consists of a driving and driven member with several rollers in between the two. The rollers are spring-loaded and either wedge-lock the two members together by being compressed against the springs, or free wheel in the opposite direction.

Starter motors used to have heavy-duty series windings, but most now use permanent magnets for the field. The advantages are less weight and smaller size. The reduction in weight provides a contribution towards reducing fuel consumption.

For applications with a higher power requirement, permanent magnet motors with intermediate epicyclic transmission are used (Figure 6.89). The sun gear is on the armature shaft and the planet carrier drives the pinion. The ring gear or annulus remains stationary and also acts as an intermediate bearing. This gear arrangement gives a reduction ratio of about 5:1.

This allows the armature to rotate at a higher and more efficient speed while still providing the torque, due to the gear reduction. Permanent magnet starters with intermediate transmission are available with power outputs of about 1.7 kW, suitable for spark ignition engines up to about 5 litres or compression ignition engines up to about 1.6 litres. This form of permanent magnet motors can give a weight saving of up to 40%.

KEY FACT

For applications with a higher power requirement, permanent magnet starter motors have an intermediate transmission.

6.24.3 Keyless starting system

In the Ford diagram shown in Figure 6.90, the powertrain control module (PCM) allows the engine to start only when the passive anti-theft system (PATS) reads a key which transmits a valid code. On a key-free vehicle, the passive key is recognised by the key-free module and if the key is valid the permission to start is issued directly. On vehicles with a manual transmission, it is necessary to depress the clutch pedal; on those with automatic transmission, the brake pedal must be pressed. On a key-free system, the key-free module switches on the control voltage for the starter relay.

KEY FACT

On a key-free system, the key-free module switches on the control voltage for the starter relay.

The PCM switches the ground in the control circuit of the starter relay which then connects power through to the starter solenoid. As soon as the speed of the engine has reached 750 rpm or the maximum permitted start time of 30 seconds has been exceeded, the PCM switches off the starter relay and therefore the starter motor. This protects the starter. If the engine does not turn or turns only slowly, the starting process is aborted by the PCM.

KEY FACT

Faults on key-free systems can be diagnosed using standard equipment, but a suitable scanner is almost essential.

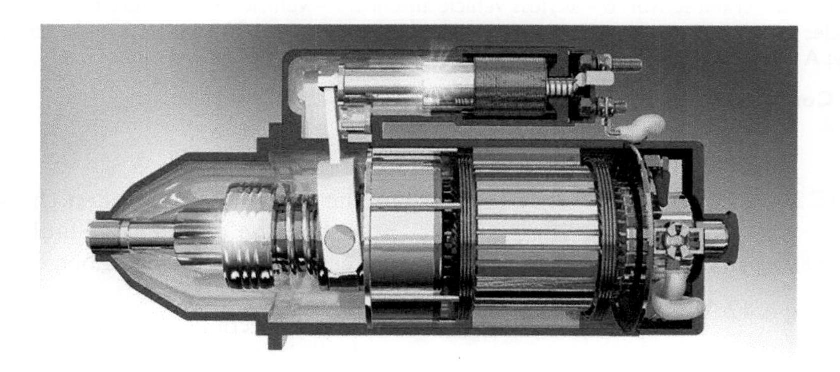

Figure 6.89 Permanent magnet fields are used in this starter motor.

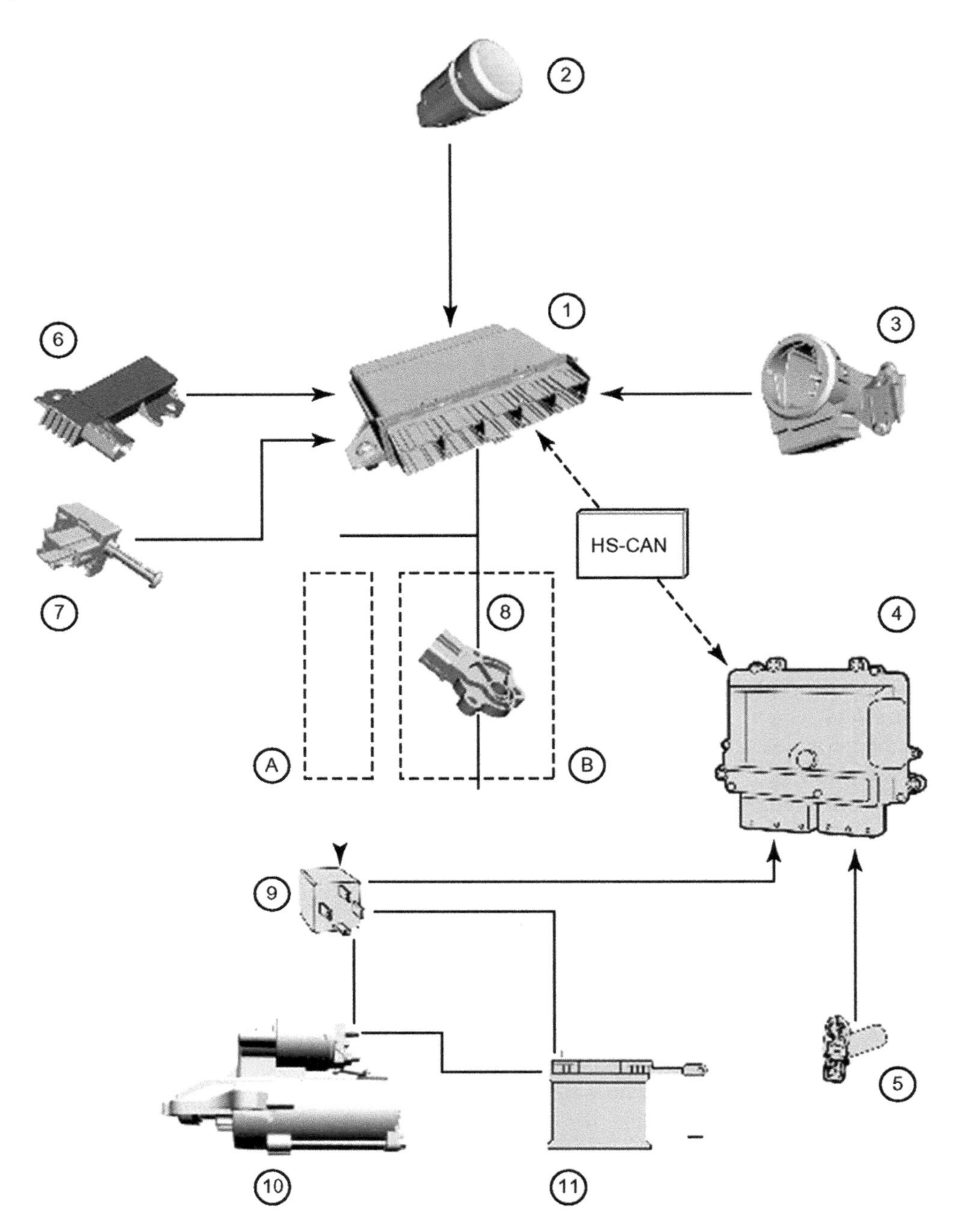

Figure 6.90 Keyless starting system: 1 – keyless vehicle module; 2 – start/stop button; 3 – electronic steering lock; 4 – powertrain control module; 5 – crank sensor; 6 – keyless vehicle antenna; 7 – vehicles with manual transmission: clutch pedal position switch/vehicles with automatic transmission: stoplamp switch; 8 – the TR sensor; 9 – starter relay; 10 – starter motor; 11 – battery; A – manual transmission (sensor not used); B – automatic transmission.

Source: Ford Motor Company.

6.25 DIAGNOSTICS – STARTING

6.25.1 Circuit testing procedure

The process of checking a 12 V starting system operation is shown in Figure 6.91.

The idea of these tests is to see if the circuit is supplying all the available voltage at the battery to the starter. If it is, then the starter is at fault, if not, then the circuit is at fault. The numbered voltmeters relate to the number of the test in the above list (Figure 6.92).

Note that connections to the starter should be made to the link between the solenoid contacts and the motor, not to the main supply terminal (Figure 6.93).

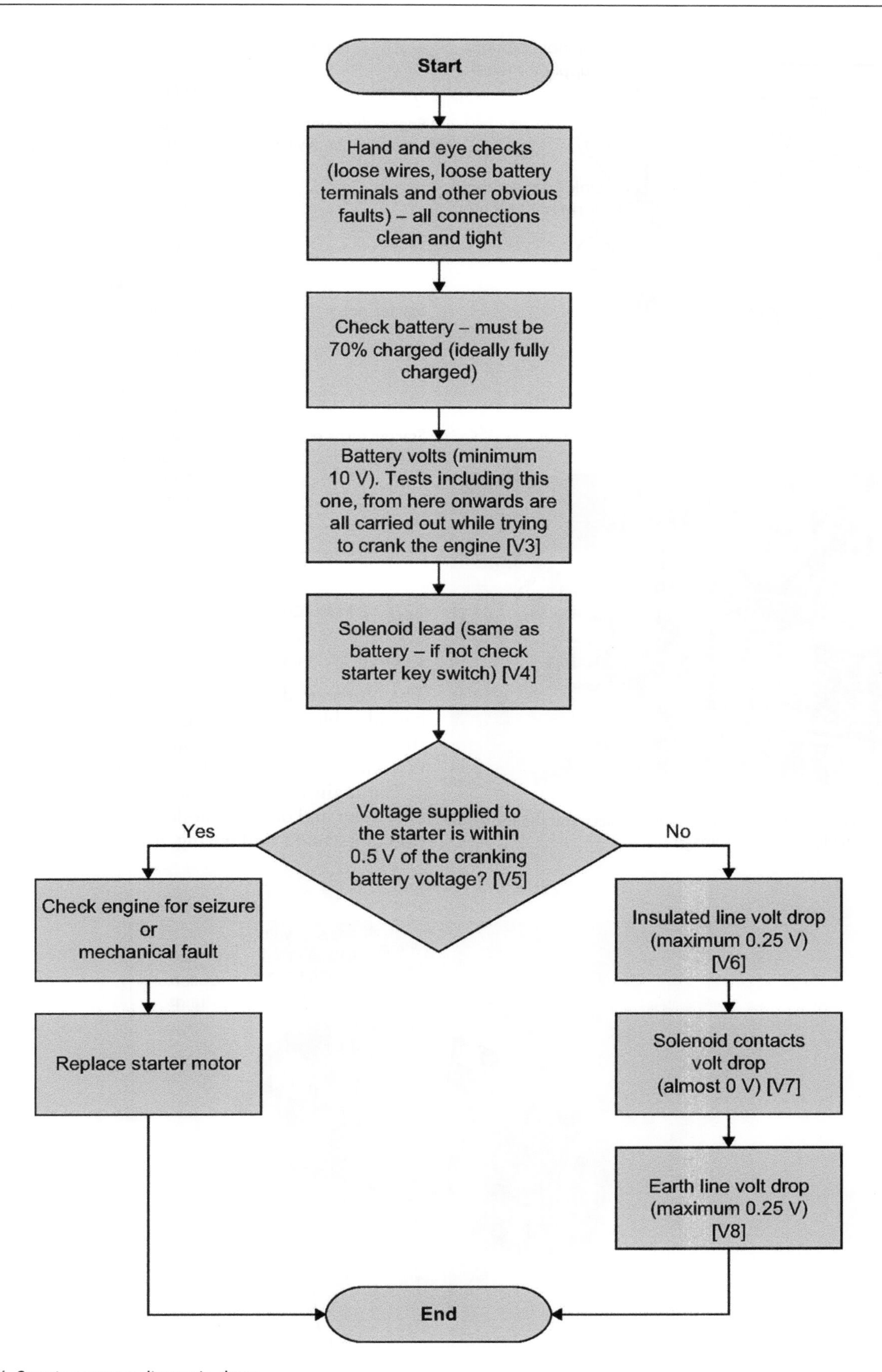

Figure 6.91 Starting system diagnosis chart.

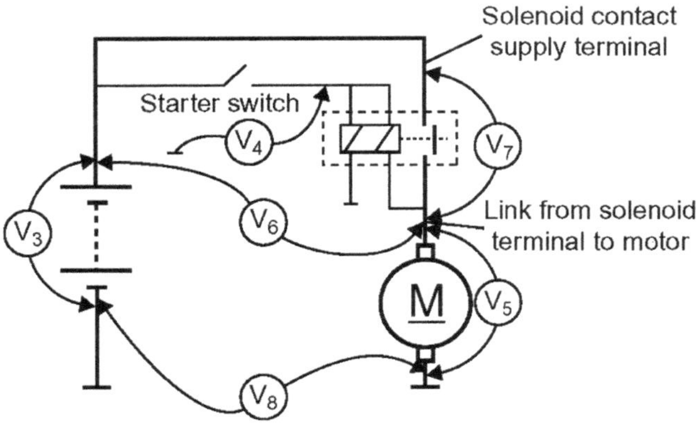

Figure 6.92 Starter circuit testing.

Figure 6.93 The link from the solenoid contacts and the motor is highlighted in this image.

6.25.2 Starting fault diagnosis table

Symptom	*Possible fault*
Engine does not rotate when trying to start	Battery connection loose or corroded
	Battery discharged or faulty
	Broken loose or disconnected wiring in the starter circuit
	Defective starter switch or automatic gearbox inhibitor switch
	Starter pinion or flywheel ring gear loose
	Earth strap broken. Loose or corroded
Starter noisy	Starter pinion or flywheel ring gear loose
	Starter mounting bolts loose
	Starter worn (bearings, etc.)
	Discharged battery (starter may jump in and out)
Starter turns engine slowly	Discharged battery (slow rotation)
	Battery terminals loose or corroded
	Earth strap or starter supply loose or disconnected
	High resistance in supply or earth circuit
	Internal starter fault

6.26 CHARGING

6.26.1 Introduction

The 'current' demands made by modern vehicles are considerable. The charging system must be able to meet these demands under all operating conditions and still fast charge the battery (Figure 6.94).

The main component of the charging system is the alternator and on most modern vehicles, with the exception of its associated wiring, it is the only component in

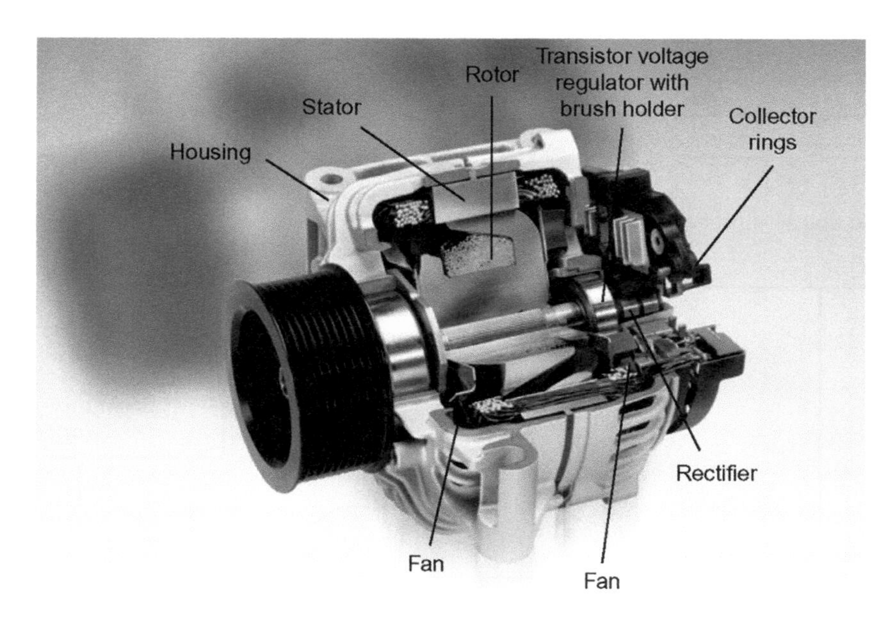

Figure 6.94 Alternator components.

the charging system. The alternator generates AC but must produce DC at its output terminal, as only DC can be used to charge the battery and run electronic circuits. The output of the alternator must be a constant voltage regardless of engine speed and current load.

KEY FACT

An alternator generates AC but must produce DC at its output terminal, as only DC can be used to charge the battery and run electronic circuits.

The charging system must meet the following criteria (when the engine is running):

- supply the current demands made by some or all loads;
- supply whatever charge current the battery demands;
- operate at idle speed;
- provide constant voltage under all conditions;
- have an efficient power to weight ratio;
- be reliable, quiet and resist contamination;
- be low maintenance;
- provide indication of correct operation.

6.26.2 Basic principles

When the alternator voltage is less than the battery (engine slow or not running for example), the direction of current flow is from the battery to the vehicle loads. The alternator diodes prevent current flowing into the alternator. When the alternator output is greater than the battery voltage, current will flow from the alternator to the vehicle loads and the battery.

It is clear, therefore, that the alternator output voltage must be above battery voltage at all times when the engine is running. The actual voltage used is critical and depends on a number of factors.

The main consideration for charging voltage is the battery terminal voltage when fully charged. If the charging system voltage is set to this value, then there can be no risk of overcharging the battery. This is known as the constant voltage charging technique. The figure of 14.2 ± 0.2 V is the accepted charging voltage for a 12 V system. Commercial vehicles generally employ two batteries in series at a nominal voltage of 24 V; therefore, the accepted charge voltage would be doubled. These voltages are used as the standard input for all vehicle loads. For the purpose of clarity, the text will just consider a 12 V system.

The other areas for consideration when determining charging voltage are any expected volt drops in the charging circuit wiring and the operating temperature of the system and battery. The voltage drops must be kept to a minimum, but it is important to note that the terminal voltage of the alternator may be slightly above that supplied to the battery.

6.26.3 Rectification of AC to DC

In order to full-wave rectify the output of a three-phase machine, six diodes are needed. These are connected in the form of a bridge, which consists of three positive diodes and three negative diodes. The output produced by this configuration is shown compared to the three phase signals (Figure 6.95).

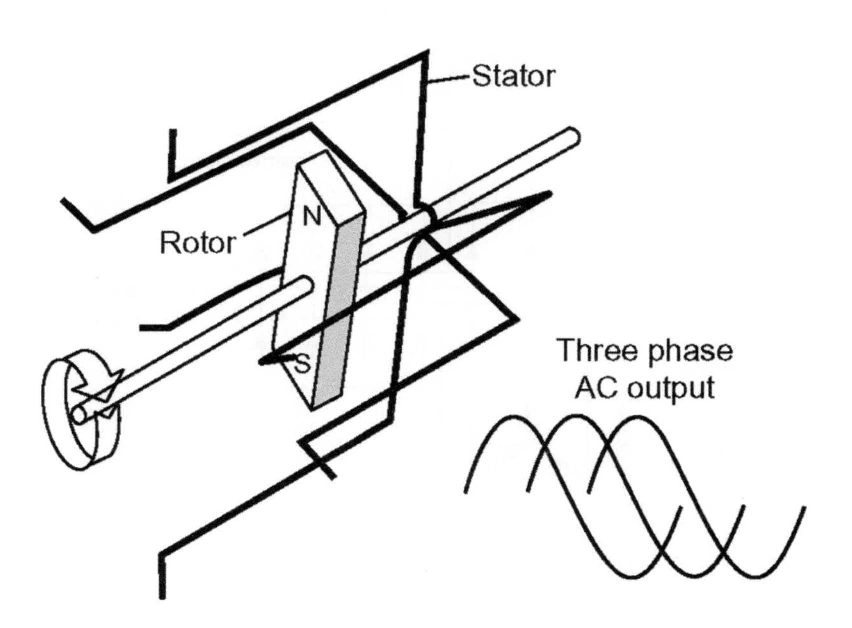

Figure 6.95 Alternator principle.

KEY FACT

In order to full-wave rectify the output of a three-phase machine, six diodes are needed.

Three positive field diodes are usually included in a rectifier pack. These are often smaller than the main diodes and are only used to supply a small current back to the field windings in the rotor. The extra diodes are known as the auxiliary, field or excitation diodes.

When a star wound stator is used, the addition of the voltages at the neutral point of the star is in theory 0 V. In practice, however, due to slight inaccuracies in the construction of the stator and rotor, a potential develops at this point. By employing two extra diodes, one positive and one negative connected to the star point, the energy can be collected. This can increase the power output of an alternator by up to 15%.

Figure 6.96 shows the full circuit of an alternator using an eight-diode main rectifier and three field diodes. The voltage regulator, which forms the starting point for the next section, is also shown in this diagram. The warning light in an alternator circuit, in addition to its function in warning of charging faults, also acts to supply the initial excitation to the field windings. An alternator will not always self-excite as the residual magnetism in the fields is not usually enough to produce a voltage which will overcome the 0.6 or 0.7 V needed to forward bias the rectifier diodes. A typical wattage for the warning light bulb is 2 W. Many manufacturers also connect a resistor in parallel with the bulb to assist in excitation and allow operation if the bulb blows. The charge warning light bulb is extinguished when the alternator produces an output from the field diodes, as this causes both sides of the bulb to take on the same voltage (a potential difference across the bulb of 0 V).

6.26.4 Regulation of output voltage

To prevent the vehicle battery from being overcharged, the regulated system voltage should be kept below the gassing voltage of the lead-acid battery. A figure of 14.2 ± 0.2 V is used for all 12 V charging systems. Accurate voltage control is vital with the ever-increasing use of electronic systems. It has also enabled the wider use of sealed batteries, as the possibility of overcharging is minimal.

KEY FACT

A figure of 14.2 ± 0.2 V is used for all 12 V charging systems – except a few that have a smart charging system – check data.

Voltage regulation is a difficult task on a vehicle alternator because of the constantly changing engine speed and loads on the alternator. The output of an alternator without regulation would rise linearly in proportion with engine speed. Alternator output is also proportional to magnetic field strength and this in turn is proportional to the field current. It is the task of the regulator to control this field current in response to alternator output voltage. The abrupt switching of the field current does not cause abrupt

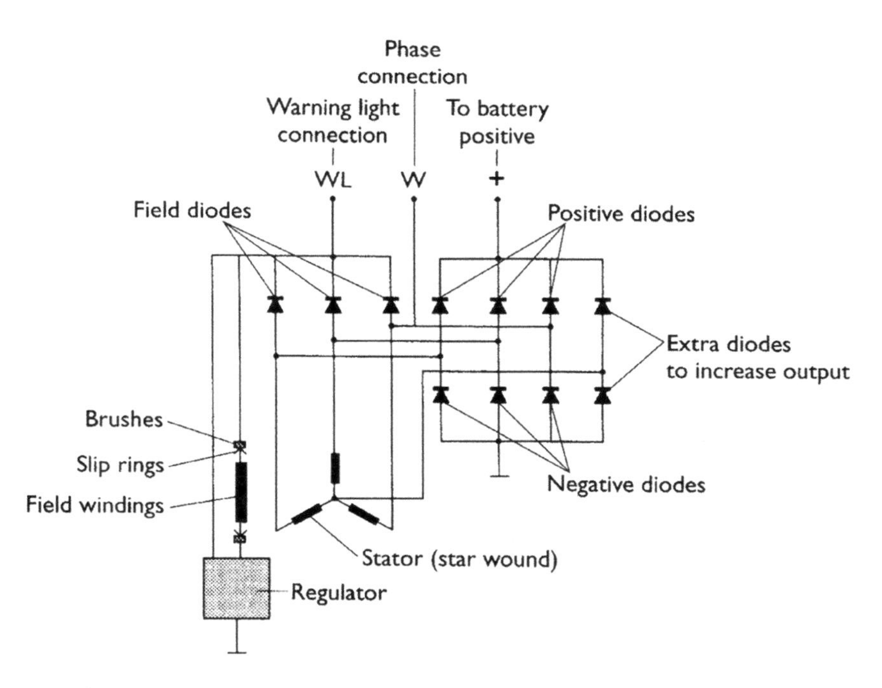

Figure 6.96 Alternator internal circuit.

changes in output voltage due to the very high inductance of the field (rotor) windings. The whole switching process also only takes a few milliseconds.

Regulators can be mechanical or electronic, the latter now almost universal on modern cars. The mechanical type uses a winding connected across the output of the alternator. The magnetism produced in this winding is proportional to output voltage. A set of normally closed contacts is attached to an armature, which is held in position by a spring. The supply to the field windings is via these contacts. When the output voltage rises beyond a pre-set level, say 14 V, the magnetism in the regulator winding will overcome spring tension and open the contacts. This switches off the field correct and causes alternator output to fall. As output falls below a pre-set level, the spring will close the regulator contacts again and so the process continues.

The problem with mechanical regulators is the wear on the contacts and other moving parts. This has been overcome with the use of electronic regulators which, due to more accurate tolerances and much faster switching, are far superior, producing a more stable output. Owing to the compactness and vibration resistance of electronic regulators, they are now fitted almost universally on the alternator reducing the number of connecting cables required.

The key to electronic voltage regulation is the Zener diode. This diode can be constructed to break down and conduct in the reverse direction at a precise level. This is used as the sensing element in an electronic regulator (Figure 6.97).

Electronic regulators can be made to sense either the battery voltage or the machine voltage (alternator) or a combination of the two. Most systems in use at present tend to be machine sensed, as this offers some protection against overvoltage in the event of the alternator being driven with the battery disconnected.

KEY FACT

Electronic regulators sense either the battery voltage or the machine voltage (alternator) or a combination of the two.

Overvoltage protection is required in some applications to prevent damage to electronic components. When an alternator is connected to a vehicle battery system, voltage, even in the event of regulator failure, will not often exceed approximately 20 V due to the low resistance and swamping effect of the battery. If an alternator is run with the battery disconnected (which is not recommended), a heavy-duty Zener diode connected across the output will offer some protection as, if the system voltage exceeds its breakdown figure, it will conduct and cause the system voltage to be kept within reasonable limits. This device is often referred to as a surge protection diode.

Figure 6.97 Voltage regulator and brush box.

6.26.5 Basic charging circuits

On many applications, the charging circuit is one of the simplest on the vehicle. The main output is connected to the battery via suitable size cable (or in some cases two cables to increase reliability and flexibility). The warning light is connected to an ignition supply on one side and to the alternator terminal at the other. A wire may also be connected to the phase terminal if it is utilised. Note that the output of the alternator is often connected to the starter main supply simply for convenience of wiring. If the wires are kept as short as possible, this will reduce voltage drop in the circuit. The volt drop across the main supply wire when the alternator is producing full output current should be less than 0.5 V (Figure 6.98).

Some systems have an extra wire from the alternator to 'sense' battery voltage directly. An ignition feed may also be found and this is often used to ensure instant excitement of the field windings. A number of vehicles link a wire from the engine management ECU to the alternator. This is used to send a signal to increase engine idle speed if the battery is low on charge.

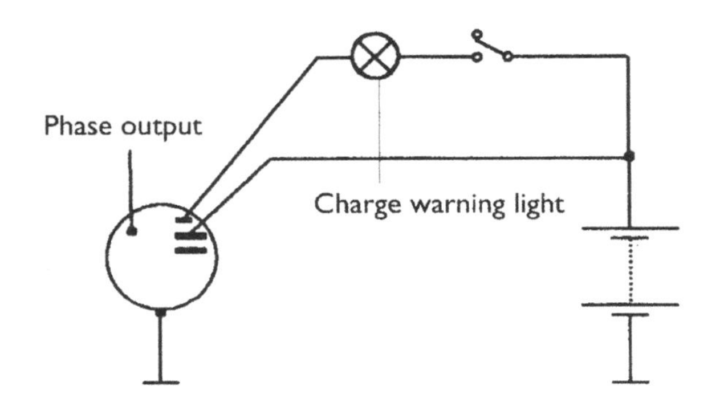

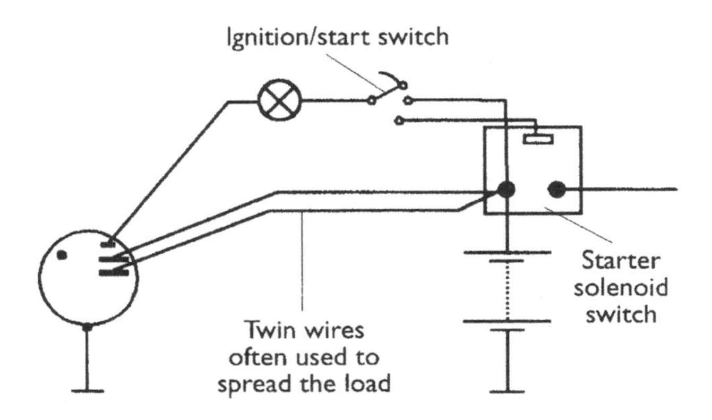

Figure 6.98 Example charging circuits.

6.26.6 Smart charging

The standard (simplified) method of testing a charging system is to measure battery voltage with the engine off and then again when running at about 3000 rpm with the lights on. If the voltage increases to about 14.2 V, then the system is almost certainly OK.

However, smart charging alternators are different. The alternator (Figure 6.99) is controlled by the engine management ECU, which monitors certain parameters

Figure 6.99 Valeo alternator.

Source: Valeo Media.

such as the engine temperature, battery temperature and electrical demand. If the alternator does not receive a signal from the PCM, the battery light is illuminated on the vehicle. This can sometimes be misdiagnosed as an alternator failure. Many smart charging systems are designed to be used only with a silver calcium battery not the lead acid type. This is because the voltages used may damage a lead acid battery and give incorrect readings.

6.27 DIAGNOSTICS – CHARGING

6.27.1 Testing procedure

After connecting a voltmeter across the battery and an ammeter in series with the alternator output wire(s), the process of checking the charging system operation is as shown in Figure 6.100.

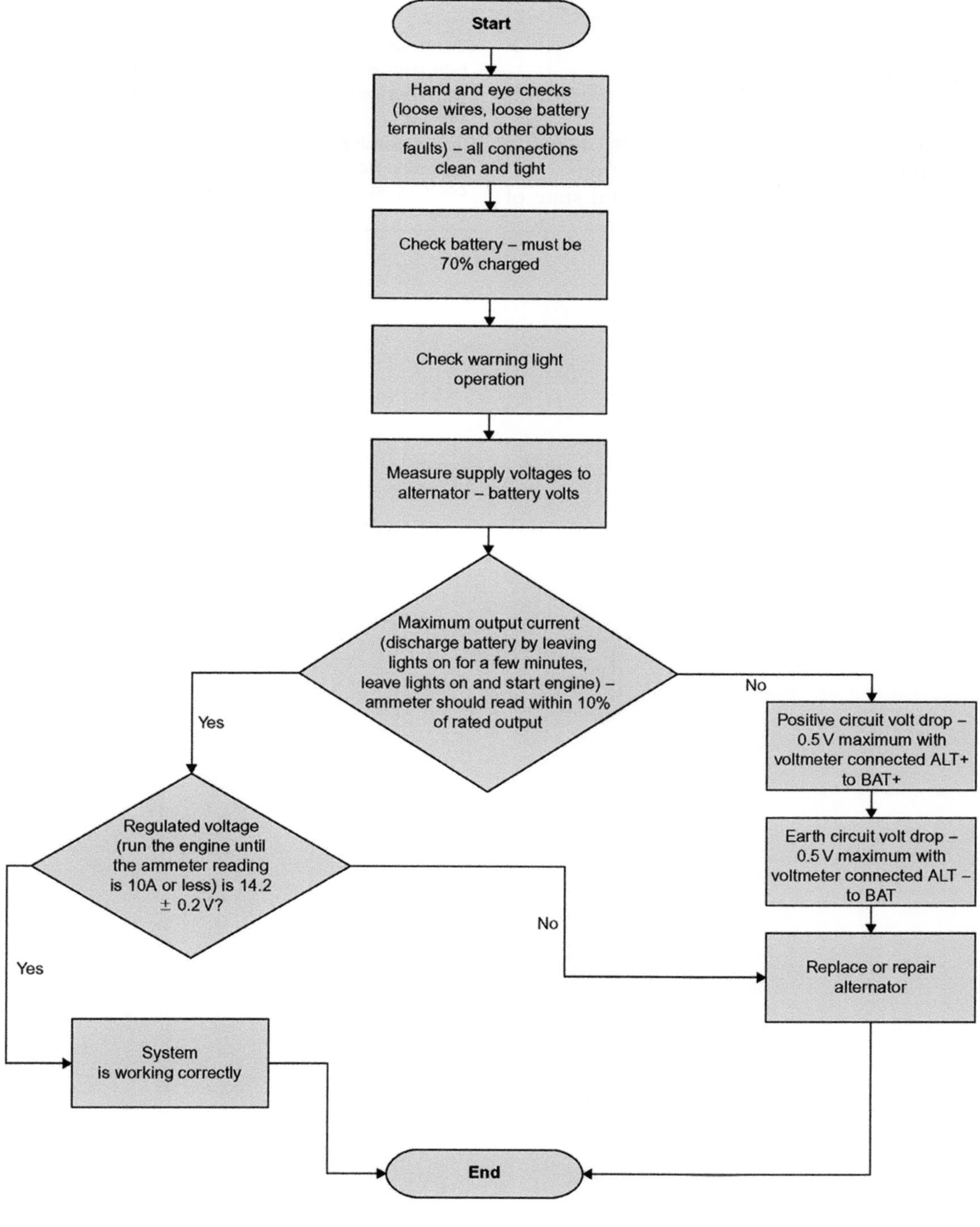

Figure 6.100 Charging system diagnosis chart.

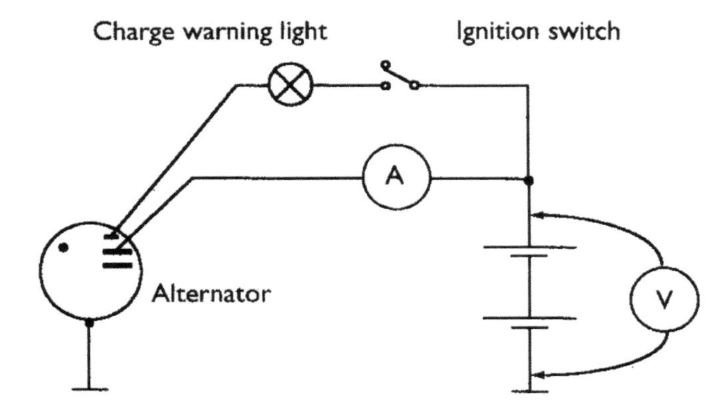

Figure 6.101 Alternator circuit testing.

If the alternator is found to be defective, then a quality replacement unit is the normal recommendation.

Repairs are possible but only if the general state of the alternator is good (Figure 6.101).

6.27.2 Charging fault diagnosis table

Symptom	*Possible fault*
Battery loses charge	Defective battery Slipping alternator drive belt Battery terminals loose or corroded Alternator internal fault (diode open circuit, brushes worn or regulator fault, etc.) Open circuit in alternator wiring, either main supply, ignition or sensing wires if fitted Short circuit component causing battery drain even when all switches are off High resistance in the main charging circuit
Charge warning light stays on when engine is running	Slipping or broken alternator drive belt Alternator internal fault (diode open circuit, brushes worn or regulator fault, etc.) Loose or broken wiring/connections
Charge warning light does not come on at any time	Alternator internal fault (brushes worn, open circuit or regulator fault, etc.) Blown warning light bulb Open circuit in warning light circuit

Chapter 7

Chassis systems

7.1 BRAKES

7.1.1 Introduction

The main braking system of a car works by hydraulics (Figure 7.1). This means that when the driver presses the brake pedal, liquid pressure forces pistons to apply brakes on each wheel. A handbrake system, usually operated by a lever and cables, is used for parking. Most handbrakes operate on the rear wheels.

Two types of light vehicle brakes are used. Disc brakes were traditionally used on the front wheels of cars but now are used on all four wheels of most modern vehicles. Braking pressure forces brake pads against both sides of a steel disc. Drum brakes are fitted on the rear wheels of some cars and on all wheels of older vehicles. Braking pressure forces brake shoes to expand outwards into contact with a drum. The important part of brake pads and shoes is a friction lining that grips well and withstands wear.

7.1.2 Principle of hydraulic braking

A complete system includes a master cylinder operating several wheel cylinders. The system is designed to give the power amplification needed for braking the particular vehicle. On any vehicle, when braking, a lot of the weight is transferred to the front wheels. Most braking effort is therefore designed to work on the front brakes. Some cars have special hydraulic valves to limit rear wheel braking. This reduces the chance of the rear wheels locking and skidding.

The main benefits of hydraulic brakes are as follows:

- almost immediate reaction to pedal pressure (no free play as with mechanical linkages);
- automatic even pressure distribution (fluid pressure effectively remains the same in all parts of the system);
- increase in force (liquid lever).

Caution and regular servicing is required to ensure the following:

- no air must be allowed in the hydraulic circuits (air compresses and would not transfer the force);
- correct adjustment must be maintained between shoe linings to drums and pads to discs (otherwise the pedal movement would be too large);
- lining materials must be free from contamination (such as oil, grease or brake fluid).

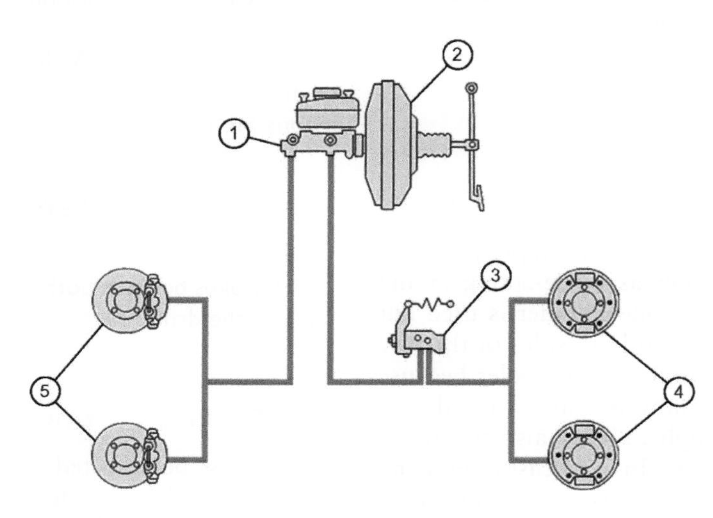

Figure 7.1 Brake system: 1 – master cylinder; 2 – brake servo or booster; 3 – pressure regulator; 4 – brake shoes; 5 – brake discs (rotors) and pads.

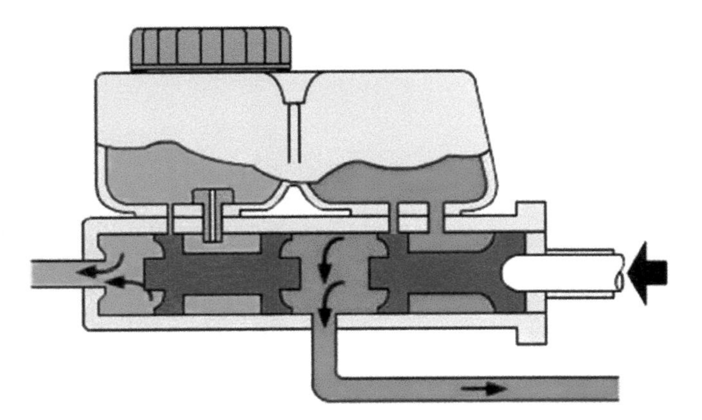

Figure 7.2 Master cylinder operation.

Figure 7.3 Brake calliper and ventilated disc (rotor).

A separate mechanical system is a good safety feature. Most vehicles have the mechanical handbrake working on the rear wheels but a few have it working on the front – take care.

KEY FACT

Most braking systems have a separate mechanical brake which is a good safety feature.

Note the importance of flexible connections to allow for suspension and steering movement. These flexible pipes are made of high-quality rubber and are covered in layers of strong mesh to prevent expansion when under pressure.

Extra safety is built into braking systems by using a double-acting master cylinder (Figure 7.2). This is often described as tandem and can be thought of as two cylinders in one housing. The pressure from the pedal acts on both cylinders but fluid cannot pass from one to the other. Each cylinder is then connected to a complete circuit. This can be by a number of methods:

- diagonal split;
- separate front and rear;
- duplicated front.

7.1.3 Disc and drum brake systems

Figure 7.3 shows a typical disc brake, calliper pads and disc. The type shown is known as single-acting sliding calliper. This is because only one cylinder is used but pads are still pressed equally on both sides of the disc by the sliding action. Disc brakes keep cooler because they are in the air stream and only part of the disc is heated as the brakes are applied. They also throw off water better than drum brakes. In most cases, servicing is minimal. Disc brakes are self-adjusting, and replacing pads is usually a simple task. In the type shown, just one bolt has to be removed to hinge the calliper upwards.

KEY FACT

Disc brakes are self-adjusting.

Disc brakes provide for good braking and are less prone to brake fade than drum brakes. This is because they are more exposed and can get rid of heat more easily. Brake fade occurs when the brakes become so hot that they cannot transfer energy any more and stop working. This type of problem can happen, for example, after keeping the car brakes on for a long time when travelling down a long steep hill. This is why a lower gear should be used to employ the engine as a brake. It is clearly important to use good-quality pads and linings because inferior materials can fail if overheated.

Drum brakes operate by shoes being forced onto the inside of the drum (Figure 7.4). Shoes can be moved by double- or single-acting cylinders. The most common layout is to use one double-acting cylinder and brake shoes on each rear wheel of the vehicle, and disc brakes on the front wheels. A double-acting cylinder simply means that as fluid pressure acts through a centre inlet, pistons are forced out of both ends.

KEY FACT

Drum brakes are more affected by wet and heat than disc brakes because both water and heat are trapped inside the drum.

7.1.4 Brake adjustments

Brakes must be adjusted so that the minimum movement of the pedal starts to apply the brakes. The adjustment in question is the gap between the pads and disc and/or the shoes and drum.

Figure 7.4 Drum brakes showing the shoes and wheel cylinder.

Disc brakes are self-adjusting because as pressure is released it moves the pads just away from the disc. Drum brakes are different because the shoes are moved away from the drum to a set position by a pull off spring. The set position is adjustable and this can be done in a number of ways.

- Self-adjusting drum brakes are almost universal now. On light vehicles, a common type uses an offset ratchet which clicks to a wider position if the shoes move beyond a certain amount when operated. Modern cars frequently have a self-adjusting handbrake.
- Screwdriver adjustment through a hole in the wheel and drum is also used. This is often a type of nut on a threaded bar which pushes the shoes out as it is screwed along the thread. This method can also have an automatic adjuster fitted.
- An adjustment screw on the back plate is now quite an old method in which a screw or square head protruding from the back plate moves the shoes by a snail cam.

The adjustment procedure stated by the manufacturer must be followed. As a guide, though, most recommend tightening the adjuster until the wheels lock and then moving it back until the wheel is just released. You must ensure that the brakes are not rubbing as this would build up heat and wear the friction material very quickly. As an aid to fault diagnosis, the effects of incorrect adjustment are as follows:

- reduced braking efficiency;
- unbalanced braking;
- excessive pedal travel.

KEY FACT

You must ensure that the brakes are not rubbing as this would build up heat and wear the friction material very quickly.

7.1.5 Servo-assisted braking

Servo systems are designed to give little assistance for light braking but increase the assistance as pedal pressure is increased. A common servo system uses low pressure (vacuum) from the manifold on one side, and the higher atmospheric pressure on the other side of a diaphragm. The low pressure is taken via a non-return safety valve from the engine inlet manifold. This pressure difference causes a force, which is made to act on the master cylinder (Figure 7.5).

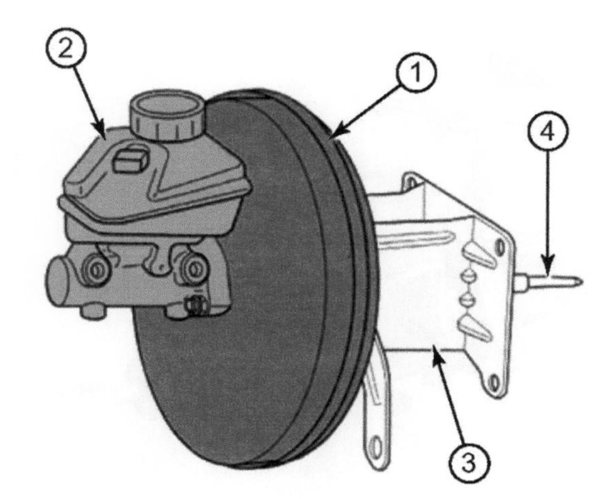

Figure 7.5 Servo unit: 1 – brake servo (booster); 2 – master cylinder and fluid reservoir; 3 – bracket; 4 – diaphragm rod connection to brake pedal.

KEY FACT

Servo systems are designed to give little assistance for light braking but increase the assistance as pedal pressure is increased.

Hydraulic power brakes use the pressure from an engine-driven pump. This pump will often be the same one used to supply the power-assisted steering. Pressure from the pump is made to act on a plunger in line with the normal master cylinder. As the driver applies force to the pedal, a servo valve opens in proportion to the force applied by the driver. The hydraulic assisting force is therefore also proportional. This maintains the important 'driver feel'.

A hydraulic accumulator (a reservoir for fluid under pressure) is incorporated into many systems. This is because the pressure supplied by the pump varies with engine speed. The pressure in the accumulator is kept between set pressures in the region of 70 bar.

Warning: If you have to disconnect any components from the braking system on a vehicle fitted with an accumulator, you must follow the manufacturer's recommendations on releasing the pressure first.

Vehicles without an engine need some other form of servo assistance and many manufacturers now use the iBooster (Figure 7.6) or similar technology.

The control principle behind the iBooster is similar to that of vacuum brake boosters: in vacuum brake systems, a valve controls the air supply to provide a boost to the force applied from the driver's foot. With the iBooster, the actuation of the brake pedal is detected via an integrated differential travel sensor and this information is sent to the control unit. The control unit determines the control signals for the electric motor, while a three-stage gear unit converts the torque of the motor into the necessary boost power. The power supplied by the booster is converted into hydraulic pressure in a standard master brake cylinder.

Figure 7.6 iBooster.

Source: Bosch Media.

7.1.6 Brake fluid

Always use new and approved brake fluid when topping up or renewing the system. The manufacturer's recommendations must always be followed. Brake fluid is hygroscopic, which means that over a period of time it absorbs water. This increases the risk of the fluid boiling due to the heat from the brakes. Pockets of steam in the system would not allow full braking pressure to be applied. Many manufacturers recommend that the fluid should be changed at regular intervals – in some cases once per year or every 30 000 km.

SAFETY FIRST

Make sure the correct grade of fluid is used.

7.2 DIAGNOSTICS – BRAKES

7.2.1 Systematic testing

If the reported fault is the handbrake not holding, proceed as follows:

1. Confirm the fault by trying to pull away with the handbrake on.
2. Check the foot brake operation. If correct, this suggests the brake shoes and drums (or pads and discs) are likely to be in good order.
3. Consider this: Do you need to remove the wheels and drums or could it be a cable fault?
4. Check cable operation by using an assistant in the car while you observe.
5. Renew the cable if seized.
6. Check handbrake operation and all associated systems.

7.2.2 Test equipment

7.2.2.1 Brake fluid tester

Because brake fluid can absorb a small amount of water, it must be renewed or tested regularly. It becomes dangerous if the water turns into steam inside the cylinders or pipes, causing the brakes to become ineffective. The tester measures the moisture content of the fluid.

7.2.2.2 Brake roller test

This is the type of test carried out as part of the annual safety test. The front or rear wheels are driven into a pair of rollers. The rollers drive each wheel of the car

Figure 7.7 Gauges on a rolling road brake tester.

and as the brakes are applied the braking force affects the rotation. A measure of braking efficiency can then be worked out (Figure 7.7).

7.2.3 Dial gauge

A dial gauge, sometimes called a clock gauge or a dial test indicator (DTI), is used to check the brake disc for run out. The symptoms of this would often be vibration or pulsation when braking. Manufacturers recommend maximum run out figures. In some cases, the disc can be re-ground but if in any doubt it is often safer and more cost effective to fit new discs. This would also be done in pairs (Figure 7.8).

SAFETY FIRST

Note: You should always refer to the manufacturer's instructions appropriate to the equipment you are using.

7.2.4 Test results

Some of the information you may have to get from other sources such as data books or a workshop manual is listed in Table 7.1.

Figure 7.8 Checking for brake disc run out with a dial gauge.

Table 7.1 Tests and information required

Test carried out	*Information required*
Brake roller test	Required braking efficiency: 50% for first-line brakes, 25% for second-line brakes and 16% for the parking brake. On modern vehicles, half of the main system is the second line (dual-line brakes). Old vehicles had to use the parking brake as the second line, therefore it had to work at 25%
Brake fluid condition	Manufacturers specify maximum moisture content

7.2.5 Brakes fault diagnosis table 1

Symptom	*Possible faults*	*Suggested action*
Excessive pedal travel	Incorrect adjustment	Adjust it! But check condition as well
Poor performance when stopping	Pad and/or shoe linings worn	Renew
	Seized calliper or wheel cylinders	Renew or free off, if possible, and safe
	Contaminated linings	Renew (both sides)
Car pulls to one side when braking	Seized calliper or wheel cylinder on one side	Overhaul or renew if piston or cylinder is worn
	Contaminated linings on one side	Renew (both sides)
Spongy pedal	Air in the hydraulic system	Bleed system and then check for leaks
	Master cylinder seals failing	Overhaul or renew
Pedal travels to the floor when pressed	Fluid reservoir empty	Refill, bleed system and check for leaks
	Failed seals in master cylinder	Overhaul or renew
	Leak from a pipe or union	Replace or repair as required
Brakes overheating	Shoe return springs broken	Renew (both sides)
	Callipers or wheel cylinders sticking	Free off or renew, if in any doubt
Brake judder	Linings worn	Renew
	Drums out of round	Renew
	Discs have excessive run out	Renew
Squeaking	Badly worn linings	Renew
	Dirt in brake drums	Clean out with proper cleaner
	Anti-squeal shims missing at rear of pads	Replace and smear with copper grease

7.2.6 Brakes fault diagnosis table 2

Symptom	*Possible cause*
Brake fade	Incorrect linings Badly lined shoes Distorted shoes Overloaded vehicle Excessive braking
Spongy pedal	Air in system Badly lined shoes Shoes distorted or incorrectly set Faulty drums Weak master cylinder mounting
Long pedal	Discs running out pushing pads back Distorted damping shims Misplaced dust covers Drum brakes need adjustment Fluid leak Fluid contamination Worn or swollen seals in master cylinder Blocked filler cap vent
Brakes binding	Brakes or handbrake maladjusted No clearance at master cylinder push rod Seals swollen Seized pistons Shoe springs weak or broken Servo faulty
Hard pedal – poor braking	Incorrect linings Glazed linings Linings wet, greasy or not bedded correctly Servo unit inoperative Seized calliper pistons Worn dampers causing wheel bounce
Brakes pulling	Seized pistons Variation in linings Unsuitable tyres or pressures Loose brakes Greasy linings Faulty drums, suspension or steering
Fall in fluid level	Worn disc pads External leak Leak in servo unit
Disc brake squeal – pad rattle	Worn retaining pins Worn discs No pad damping shims or springs
Uneven or excessive pad wear	Disc corroded or badly scored Incorrect friction material
Brake judder	Excessive disc or drum run out Calliper mounting bolts loose Worn suspension or steering components

7.2.7 Brake hydraulic faults

Brake hose clamps will assist in diagnosing hydraulic faults and enable a fault to be located quickly. Proceed as follows:

1. Clamp all hydraulic flexible hoses and check the pedal.
2. Remove the clamps one at a time and check the pedal again (each time).
3. The location of air in the system or the faulty part of the system will now be apparent.

7.2.8 Electronic brake system bleeding

On many braking systems it is necessary to use a scan tool as part of a brake diagnosis process. It may also be necessary to use the scanner to bleed the hydraulic systems. Bi-directional functions performed with any scan tool use the vehicle's software to activate tasks within the module. After performing a brake repair, a typical method of bleeding the system will be (always refer to manufacturer's information):

1. Use the manufacturer's recommended brake fluid
2. Keep reservoir full during the bleeding process
3. Make sure the EBC is disabled while using the scan tool
4. Put the transmission in Park (or manual in neutral) and apply the parking brake
5. Connect the scan tool to the DLC with the ignition off
6. Turn the ignition on and start the scan tool
7. Go to the appropriate scan tool menus (for example):

   - Diagnosis
   - OBD
   - ABS/TRAC/VSC
   - ECB Utility

Select the appropriate option from the menu, for example it may be that just the front or rear brakes have been installed or repaired. Bleed the brakes in the order listed in the manufacturer's information.

7.3 ANTILOCK BRAKES

7.3.1 Introduction

The reason for the development of the antilock brake system (ABS) is very simple. Under braking conditions, if one or more of the vehicle wheels lock (begins to skid), then this has a number of consequences:

- braking distance increases;
- steering control is lost;
- tyre wear is abnormal.

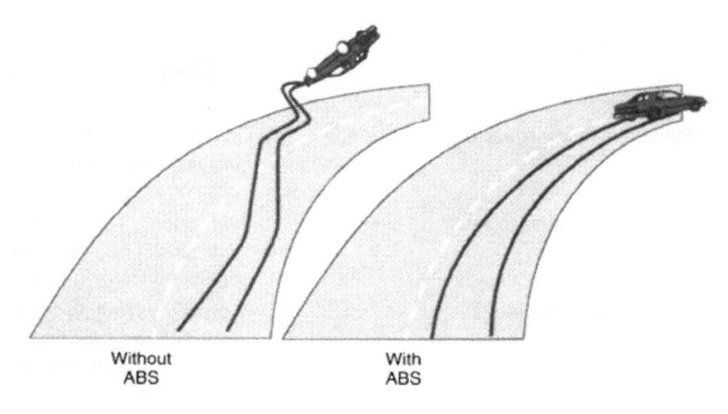

Figure 7.9 Advantages of ABS.

The obvious consequence is that an accident is far more likely to occur. The maximum deceleration of a vehicle is achieved when maximum energy conversion is taking place in the brake system. This is the conversion of kinetic energy to heat energy at the discs and brake drums. The potential for this conversion process when a tyre skids, even on a dry road, is far less. A good driver can pump the brakes on and off to prevent locking but electronic control can achieve even better results.

ABS is becoming more common on lower price vehicles, which should be a contribution to safety. It is important to remember, however, that for normal use, the system is not intended to allow faster driving and shorter braking distances. It should be viewed as operating in an emergency only. Figure 7.9 shows how ABS can help to maintain steering control even under very heavy braking conditions.

KEY FACT

The maximum deceleration of a vehicle is achieved when maximum energy conversion is taking place in the brake system – not between the tyres and road.

7.3.2 Requirements of ABS

A good way of considering the operation of a complicated system is to ask, 'What must the system be able to do?' In other words, 'What are the requirements?' These can be considered for ABS under the following headings.

A number of different types of antilock brake systems are in use, but all operate to achieve the requirements as set out in the following table.

Fail safe system	In the event of the ABS failing, conventional brakes must still operate to their full potential. In addition, a warning must be given to the driver. This is normally in the form of a simple warning light.
Manoeuvrability must be maintained	Good steering and road holding must continue when the ABS is operating. This is arguably the key issue as being able to swerve round a hazard while still braking hard is often the best course of action.
Immediate response must be available	Even over a short distance, the system must react such as to make use of the best grip on the road. The response must be appropriate whether the driver applies the brakes gently or slams them on hard.
Operational influences	Normal driving and manoeuvring should produce no reaction on the brake pedal. The stability and steering must be retained under all road conditions. The system must also adapt to braking hysteresis when the brakes are applied, released and then re-applied. Even if the wheels on one side are on dry tarmac and the other side on ice, the yaw (rotation about the vertical axis of the vehicle) of the vehicle must be kept to a minimum and only increase slowly to allow the driver to compensate.
Controlled wheels	In its basic form, at least one wheel on each side of the vehicle should be controlled on a separate circuit. It is now general for all four wheels to be controlled on passenger vehicles.
Speed range of operation	The system must operate under all speed conditions down to walking pace. At this very slow speed, even when the wheels lock, the vehicle will come to rest very quickly. If the wheels did not lock, then in theory the vehicle would never stop.
Other operating conditions	The system must be able to recognise aquaplaning and react accordingly. It must also still operate on an uneven road surface. The one area still not perfected is braking from slow speed on snow. The ABS will actually increase stopping distance in snow but steering will be maintained. This is considered to be a suitable trade-off.

7.3.3 General system description

As with other systems, ABS can be considered as a central control unit with a series of inputs and outputs. An ABS is represented by the closed loop system block diagram shown in Figure 7.10. The most important of the inputs are the wheel speed sensors and the main output is some form of brake system pressure control.

The task of the control unit is to compare signals from each wheel sensor to measure the acceleration or deceleration of an individual wheel. From this data and pre-programmed look-up tables, brake pressure to one or more of the wheels can be regulated. Brake pressure can be reduced, held constant or allowed to increase. The maximum pressure is determined by the driver's pressure on the brake pedal.

From the wheel speed sensors, the electronic control unit (ECU) calculates the components given in Table 7.2.

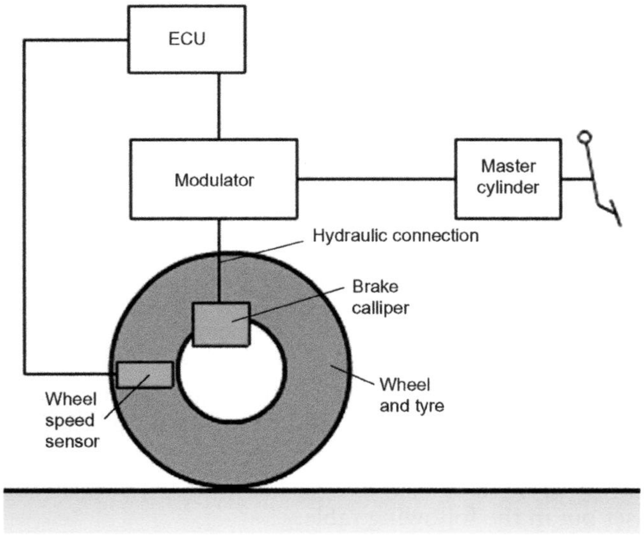

Figure 7.10 ABS closed loop block diagram.

7.3.4 ABS components

There are a few variations between manufacturers involving a number of different components. For the majority of systems, however, there are three main components.

KEY FACT

As with many other systems, ABS can be considered as a central control unit with a series of inputs and outputs.

7.3.4.1 Wheel speed sensors

Most of these devices are simple inductance sensors and work in conjunction with a toothed wheel. They consist of a permanent magnet and a soft iron rod around which is wound a coil of wire. As the toothed wheel rotates, the changes in inductance of the magnetic circuit generates a signal, the frequency and voltage of which are

Table 7.2 ABS calculations

Vehicle reference speed	Determined from the combination of two diagonal wheel sensor signals. After the start of braking, the ECU uses this value as its reference.
Wheel acceleration or deceleration	This is a live measurement that is constantly changing.
Brake slip	Although this cannot be measured directly, a value can be calculated from the vehicle reference speed. This figure is then used to determine when/if ABS should take control of the brake pressure.
Vehicle deceleration	During brake pressure control, the ECU uses the vehicle reference speed as the starting point and decreases it in a linear manner. The rate of decrease is determined by the evaluation of all signals received from the wheel sensors. Driven and non-driven wheels on the vehicle must be treated in different ways as they behave differently when braking. A logical combination of wheel deceleration/acceleration and slip are used as the controlled variable. The actual strategy used for ABS control varies with the operating conditions.

proportional to wheel speed. The frequency is the signal used by the ECU. The coil resistance is in the order of 800–1000 Ω. Coaxial cable is used to prevent interference affecting the signal. Some systems now use Hall effect sensors.

7.3.4.2 Electronic control unit

The function of the ECU is to take in information from the wheel sensors and calculate the best course of action for the hydraulic modulator.

The heart of a modern ECU consists of two microprocessors such as the Motorola 68HC11, which run the same programme independently of each other. This ensures greater security against any fault which could adversely affect braking performance, because the operation of each processor should be identical. If a fault is detected, the ABS disconnects itself and operates a warning light. Both processors have non-volatile memory into which fault codes can be written for later service and diagnostic access. The ECU also has suitable input signal processing stages and output or driver stages for actuator control.

SAFETY FIRST

Note: ABS problems may require specialist attention – but don't be afraid to check the basics. An important note, however, is that some systems require special equipment to reinitialise the ECU if it has been disconnected.

The ECU performs a self-test after the ignition is switched on. A failure will result in disconnection of the system. The following list forms the self-test procedure:

- current supply;
- exterior and interior interfaces;
- transmission of data;
- communication between the two microprocessors;
- operation of valves and relays;
- operation of fault memory control;
- reading and writing functions of the internal memory.

All this takes less than 300 ms.

7.3.4.3 Hydraulic modulator

A hydraulic modulator has three operating positions:

1. pressure build-up brake line open to the pump;
2. pressure holding brake line closed;
3. pressure release brake line open to the reservoir.

The valves are controlled by electrical solenoids, which have a low inductance so they react very quickly. The motor only runs when ABS is activated. Figure 7.11 shows an ABS hydraulic modulator with integrated ECU.

Figure 7.11 Hydraulic modulators.

Source: Bosch Media.

7.4 DIAGNOSTICS – ANTILOCK BRAKES

7.4.1 Systematic testing procedure

If the reported fault is the ABS warning light staying on, proceed as given in Figure 7.12 (a scanner may be needed to reset the light).

7.4.2 Antilock brakes fault diagnosis table

Symptom	*Possible cause*
ABS not working and/or warning light on	Wheel sensor or associated wiring open circuit/high resistance Wheel sensor air gap incorrect Power supply/earth to ECU low or not present Connections to modulator open circuit No supply/earth connection to pump motor Modulator windings open circuit or high resistance
Warning light comes on intermittently	Wheel sensor air gap incorrect Wheel sensor air gap contaminated Loose wire connection

7.4.3 Bleeding antilock brakes

Special procedures may be required to bleed the hydraulic system when ABS is fitted. Refer to appropriate data for the particular vehicle. This process often requires the use of a diagnostic tool.

7.5 TRACTION CONTROL

7.5.1 Introduction

The steerability of a vehicle is not only lost when the wheels lock up on braking, the same effect arises if the wheels spin when driving off under severe acceleration. Electronic traction control has been developed as a supplement to ABS. This control system prevents the wheels from spinning when moving off or when accelerating sharply while on the move. In this way, an individual wheel which is spinning is braked in a controlled manner. If both or all of the wheels are spinning, the drive torque is reduced by means of an engine control function. Traction control has become known as traction control system (TCS), anti-slip regulation (ASR) or just traction control (TCR).

Traction control is not normally available as an independent system but in combination with ABS. This is because many of the components required are the same as for the ABS. Traction control only requires a change in logic control in the ECU and a few extra control elements such as control of the throttle. Figure 7.13 shows a block diagram of a traction control system. Note the links with ABS and the engine control system.

Traction control will intervene to

- maintain stability;
- reduce yawing moment reactions;
- provide optimum propulsion at all speeds;
- reduce driver workload.

An automatic control system can intervene in many cases more quickly and precisely than the driver of the vehicle. This allows stability to be maintained at a time when the driver might not have been able to cope with the situation.

7.5.2 Control functions

Control of tractive force can be by a number of methods.

Throttle control	This can be via an actuator, which can move the throttle cable, or if the vehicle employs a drive by wire accelerator, then control will be in conjunction with the engine management ECU. This throttle control will be independent of the drivers throttle pedal position. This method alone is relatively slow to control engine torque.
Ignition control	If ignition is retarded, the engine torque can be reduced by up to 50% in a very short space of time. The timing is adjusted by a set ramp from the ignition map value.
Braking effect	If the spinning wheel is restricted by brake pressure, the reduction in torque at the affected wheel is very fast. Maximum brake pressure is not used to ensure that passenger comfort is maintained.

KEY FACT

Traction control is usually combined with ABS.

7.5.3 System operation

The description that follows is for a vehicle with an electronic accelerator (drive by wire). A simple sensor determines the position of the accelerator and, taking into account other variables such as engine temperature and speed for example, the throttle is set at the optimum position by a servomotor. When accelerating, the increase in engine torque leads to an increase in driving torque at the wheels. To achieve optimum acceleration, the maximum possible driving torque must be transferred to the road. If driving torque exceeds that which can be transferred then wheel slip will occur on at least one wheel. The result of this is that the vehicle becomes unstable.

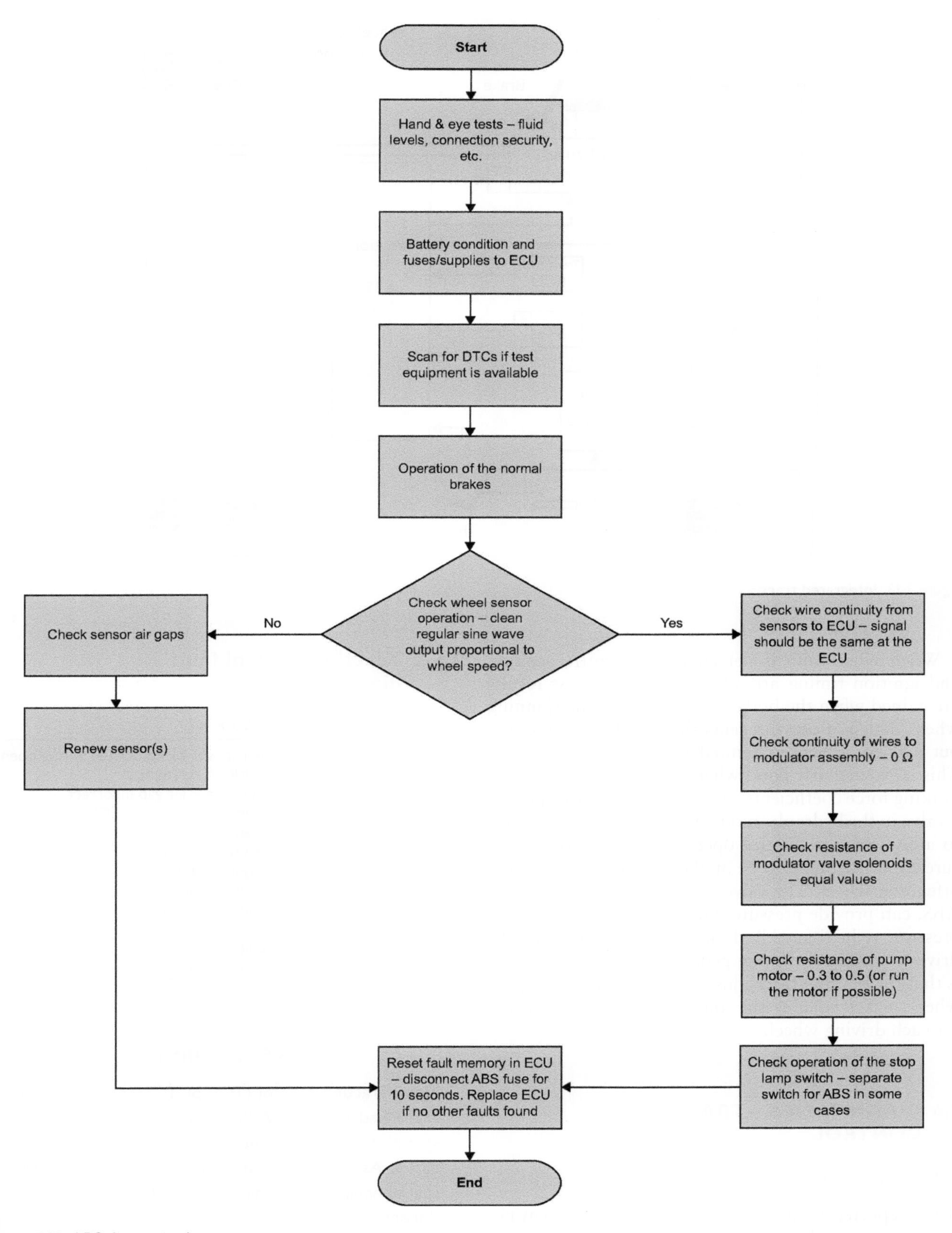

Figure 7.12 ABS diagnosis chart.

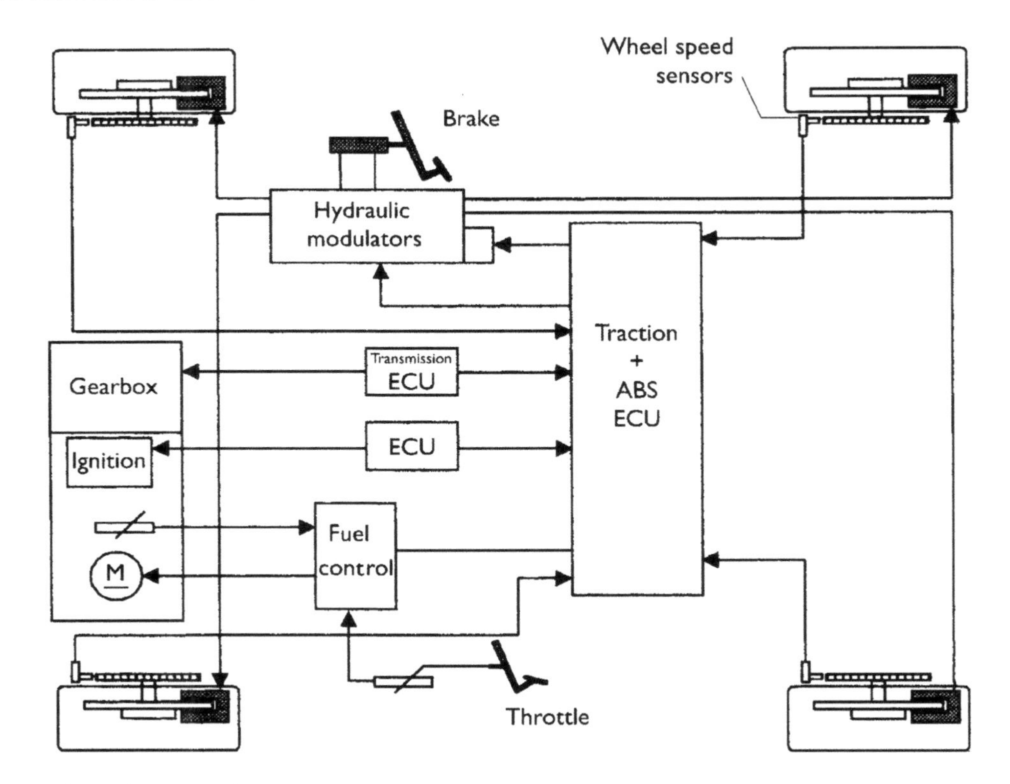

Figure 7.13 Integrated traction control system.

When wheel spin is detected, the throttle position and ignition timing are adjusted, but the best results are gained when the brakes are applied to the spinning wheel. This not only prevents the wheel from spinning but acts to provide a limited slip differential action. This is particularly good when on a road with varying braking force coefficients. When the brakes are applied, a valve in the hydraulic modulator assembly moves over to allow traction control operation. This allows pressure from the pump to be applied to the brakes on the offending wheel. The valves, in the same way as with ABS, can provide pressure build-up, pressure hold and pressure reduction. All these take place without the driver touching the brake pedal. The summary of this is that the braking force must be applied to the slipping wheel so as to equalise the combined braking coefficient for each driving wheel.

7.6 DIAGNOSTICS – TRACTION CONTROL

7.6.1 Systematic testing

If the reported fault is the traction control system not working, proceed as given in Figure 7.14.

7.6.2 Traction control fault diagnosis table

Symptom	*Possible cause*
Traction control inoperative	Wheel sensor or associated wiring open circuit/high resistance Wheel sensor air gap incorrect Power supply/earth to ECU low or not present Switch open circuit ABS system fault Throttle actuator inoperative or open circuit connections Communication link between ECUs open circuit ECU needs to be initialised

SAFETY FIRST

Note: Traction control (TCR or TCS or ASC) is usually linked with the ABS and problems may require specialist attention – but don't be afraid to check the basics. As with ABS, note that some systems require special equipment to reinitialise the ECU if it has been disconnected.

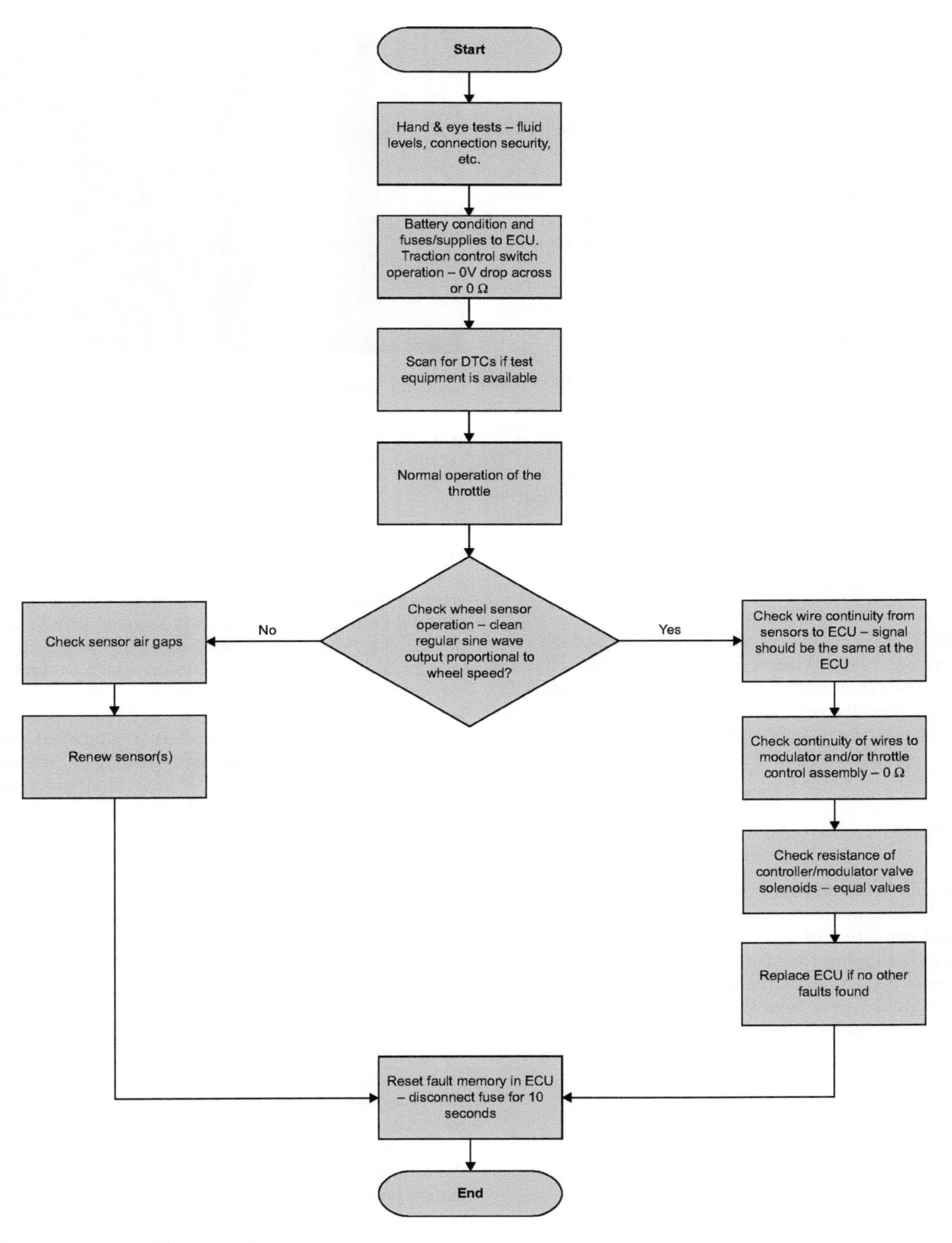

Figure 7.14 Traction control diagnosis chart.

7.7 STEERING AND TYRES

The tyre performs two basic functions:

1. It acts as the primary suspension, cushioning the vehicle from the effects of a rough surface.
2. It provides frictional contact with the road surface. This allows the driving wheels to move the vehicle. The tyres also allow the front wheels to steer and the brakes to slow or stop the vehicle.

The tyre is a flexible casing which contains air. Tyres are manufactured from reinforced synthetic rubber. The tyre is made of an inner layer of fabric plies which are wrapped around bead wires at the inner edges. The bead wires hold the tyre in position on the wheel rim. The fabric plies are coated with rubber, which is moulded to form the side walls and the tread of the tyre. Behind the tread is a reinforcing band, usually made of steel, rayon or glass fibre. Modern tyres are mostly tubeless, so they have a thin layer of rubber coating the inside to act as a seal.

7.7.1 Construction of a tubeless radial tyre

The wheel is made with a leak-proof rim and the valve is rubber mounted into a hole formed in the well of the rim. The tyre is made with an accurate bead, which fits tightly onto the rim. A thin rubber layer in the interior of the tyre makes an airtight seal.

KEY FACT

Modern tyres are almost always tubeless.

The plies of a radial tyre pass from bead to bead at 90° to the circumference, or radially. There is a rigid belt band consisting of several layers of textile or metallic threads running round the tyre under the tread. Steel wire is often used in the construction of radial tyres. The radial tyre is flexible but retains high strength. It has good road holding and cornering power. In addition, radial tyres are economical due to their low 'rolling resistance' (Figure 7.15).

KEY FACT

The plies of a radial tyre pass from bead to bead at 90° to the circumference, or radially.

A major advantage of a radial tyre is its greatly improved grip even on wet roads. This is because the rigid belt band holds the tread flat on the road surface, when cornering. The rigid belt band also helps with the escape of water from under the tyre.

Figure 7.15 Tyres.

7.7.2 Steering box and rack

Steering boxes contain a spiral gear known as a worm gear, or something similar, which rotates with the steering column. One form of design has a nut wrapped round the spiral and is therefore known as a worm and nut-steering box. The grooves can be filled with recirculating ball bearings, which reduce backlash or slack in the system and also reduce friction, making steering lighter. On vehicles with independent front suspension, an idler unit is needed together with a number of links and several joints. The basic weakness of the steering box system is in the number of swivelling joints and connections. If there is just slight wear at a number of points, the steering will not feel, or be, positive.

KEY FACT

The steering rack is now used on almost all light vehicles because it is simple in design, long lasting and makes optimum use of available space.

The steering rack is now used almost without exception on light vehicles. This is because it is simple in design and very long lasting. The wheels turn on two large swivel joints. Another ball joint (often called a track rod end) is fitted on each swivel arm.

A further ball joint to the ends of the rack connects the track rods. The rack is inside a lubricated tube and gaiters protect the inner ball joints. The pinion meshes with the teeth of the rack, and as it is turned by the steering wheel, the rack is made to move back and forth, turning the front wheels on their swivel ball joints. On many vehicles now, the steering rack is augmented with hydraulic power assistance (Figure 7.16).

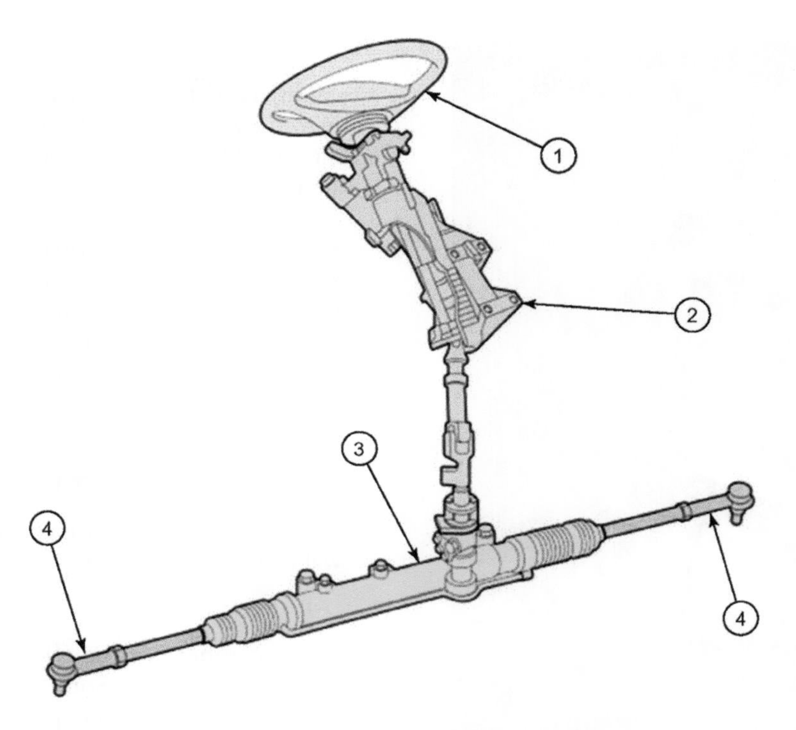

Figure 7.16 Steering system: 1 – steering wheel; 2 – column; 3 – rack; 4 – track rod ends.

7.7.3 Power-assisted steering

Rack and pinion steering requires more turning effort than a steering box, although this is not too noticeable with smaller vehicles. However, most cars, in particular heavier ones with larger engines or with wider tyres, which scrub more, often benefit from power steering.

Many vehicles use a belt-driven hydraulic pump to supply fluid under pressure for use in the system. Inside the rack and pinion housing is a hydraulic valve, which is operated as the pinion is turned for steering. The valve controls the flow of oil into a cylinder, which has a piston connected to the rack. This assists with the steering effort quite considerably.

A well-designed system will retain 'feel' of road conditions for the driver to control the car. Steering a slow-moving heavier vehicle when there is little room can be tiring or even impossible for some drivers. This is where power steering has its best advantage. Many modern systems are able to make the power steering progressive. This means that as the speed of the vehicle increases, the assistance provided by the power steering reduces. This maintains better driver feel.

Many modern systems use electric power steering. This employs a powerful electric motor as part of the steering linkage. There are two main types of electric power steering:

- replacing the conventional hydraulic pump with an electric motor while the ram remains much the same;
- a drive motor, which directly assists with the steering and which has no hydraulic components (the most common method now).

KEY FACT

Many modern systems use electric power steering.

The second system uses a small electric motor acting directly on the steering via an epicyclic gear train. This completely replaces the hydraulic pump and servo cylinder. It also eliminates the fuel penalty of the conventional pump and greatly simplifies the drive arrangements. Engine stall when the power steering is operated at idle speed is also eliminated.

An optical torque sensor is used to measure driver effort on the steering wheel. The sensor works by measuring light from an LED which is shining through holes that are aligned in discs at either end of a 50 mm torsion bar fitted into the steering column. Other methods are also used, but the optical is most common (Figure 7.17).

7.7.4 Steering characteristics

The steering characteristics of a vehicle, or in other words, the way in which it reacts when cornering, can be described by one of three headings:

1. oversteer;
2. understeer;
3. neutral.

Figure 7.17 Electric power steering.

Source: ZF.

Oversteer occurs when the rear of the vehicle tends to swing outward more than the front during cornering. This is because the slip angle on the rear axle is significantly greater than the front axle. This causes the vehicle to travel in a tighter circle, hence the term oversteer. If the steering angle is not reduced, the vehicle will break away and all control will be lost. Turning the steering towards the opposite lock will reduce the front slip angle.

Understeer occurs when the front of the vehicle tends to swing outward more than the rear during cornering. This is because the slip angle on the rear axle is significantly smaller than the front axle. This causes the vehicle to travel in a greater circle, hence the term understeer. If the steering angle is not increased, the vehicle will be carried out of the corner and all control will be lost. Turning the steering further into the bend will increase the front slip angle. Front-engined vehicles tend to understeer because the centre of gravity is situated in front of the vehicle centre. The outward centrifugal force therefore has a greater effect on the front wheels than on the rear.

Neutral steering occurs when the centre of gravity is at the vehicle centre and the front and rear slip angles are equal. The cornering forces are therefore uniformly spread. Note, however, that understeer or oversteer can still occur if the cornering conditions change.

7.7.5 Camber

On many cars, the front wheels are not mounted vertically to the road surface. Often they are tilted outwards at the top. This is called positive camber (Figure 7.18), and has the following effects:

- easier steering, less turning effort required;
- less wear on the steering linkages;
- less stress on main components;
- smaller scrub radius, which reduces the effect of wheel forces on the steering.

KEY FACT

Typical value for camber is approximately 0.5° (values will vary so check specs).

Negative camber has the effect of giving good cornering force (Figure 7.19). Some cars have rear wheels with negative camber. With independent suspension systems, wheels can change their camber from positive through neutral to negative as the suspension is compressed. This varies, however, with the design and position of the suspension hinge points.

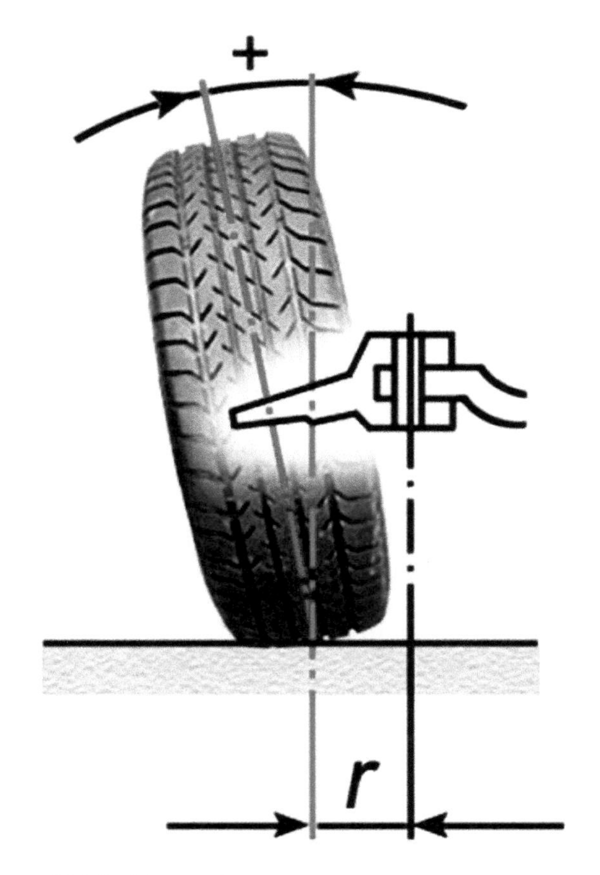

Figure 7.18 Positive camber.

Figure 7.19 Negative camber.

7.7.6 Castor

The front wheels tend to straighten themselves out after cornering. This is due to a castor action. Supermarket trolley wheels automatically run straight when pushed because the axle on which they rotate is behind the swivel mounting. Vehicle wheels get the same result by leaning the swivel pin mountings back so that the wheel axle is moved slightly behind the line of the swivel axis. The further the axle is behind the swivel, the stronger will be the straightening effect. The main effects of a positive castor angle (Figure 7.20) are

- self-centring action;
- help in determining the steering torque when cornering.

Negative castor is used on some front wheel drive vehicles to reduce the return forces when cornering (Figure 7.21). Note that a combination of steering geometry angles is used to achieve the desired effect. This means that in some cases the swivel axis produces the desired self-centre action so the castor angle may need to be negative to reduce the return forces on corners.

KEY FACT

Typical castor value is approximately 2–4° (values will vary so check specs).

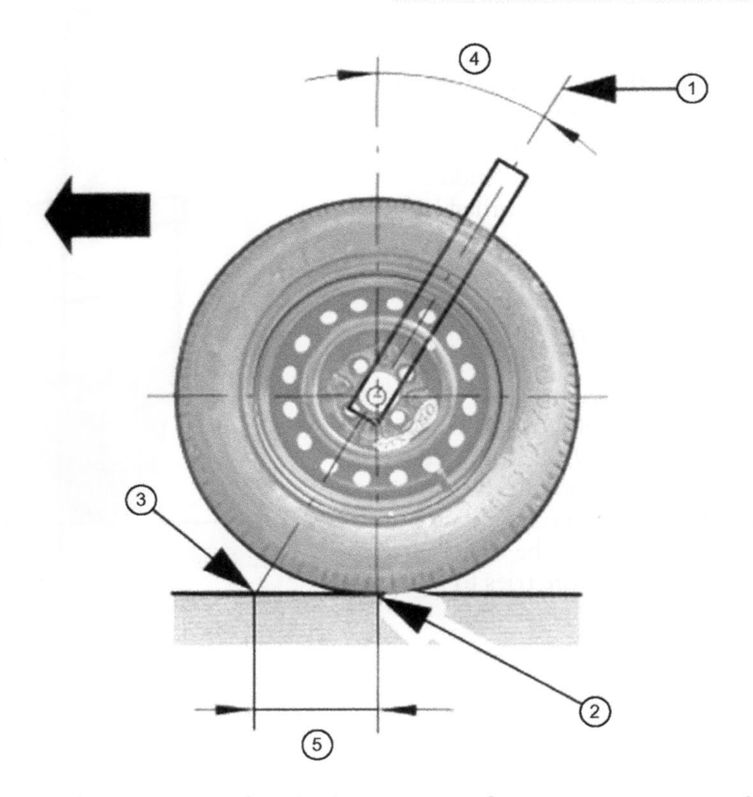

Figure 7.20 Castor angle – Positive: 1 – steering axis; 2 – wheel contact point; 3 – positive castor point of intersection of steering axis with the road surface; 4 – castor angle; 5 – castor distance.

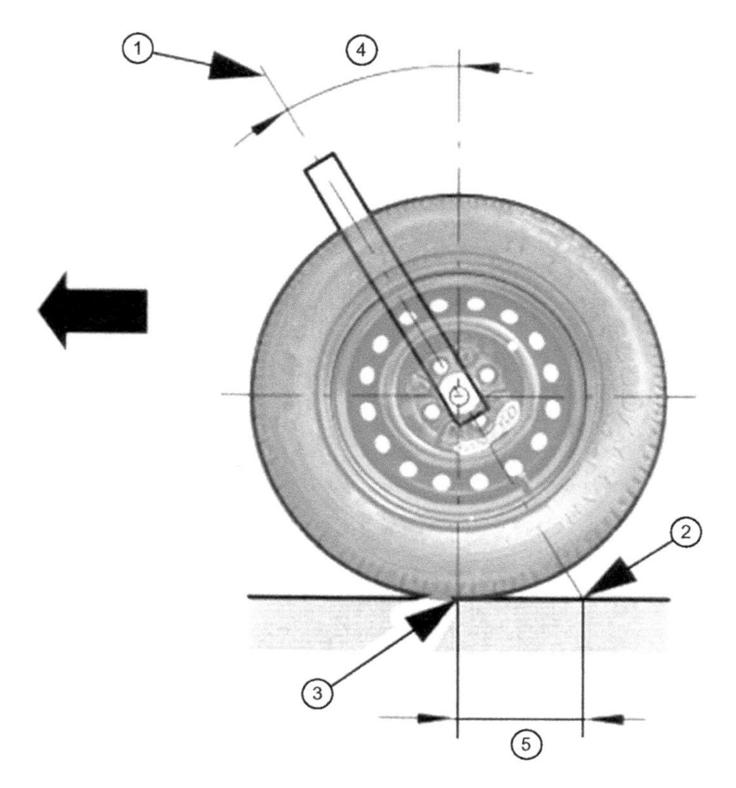

Figure 7.21 Castor angle – Negative: 1 – steering axis; 2 – wheel contact point; 3 – negative castor point of intersection of steering axis with the road surface; 4 – castor angle; 5 – castor distance.

7.7.7 Swivel axis inclination

The swivel axis is also known as the steering axis. Swivel axis inclination (Figure 7.22) means the angle compared to vertical made by the two swivel joints when viewed from the front or rear. On a strut-type suspension system, the angle is broadly that made by the strut. This angle always leans towards the middle of the vehicle. The swivel axis inclination (also called king pin inclination) is mainly for

- producing a self-centre action;
- improving steering control on corners;
- giving a lighter steering action.

Scrub radius, wheel camber and swivel axis inclination all have an effect on one another. The swivel axis inclination mainly affects the self-centring action, also known as the aligning torque. Because of the axis inclination, the vehicle is raised slightly at the front as the wheels are turned. The weight of the vehicle therefore tries to force the wheels back into the straight-ahead position.

DEFINITION

SAI: Swivel axis inclination.

KPI: King pin inclination.

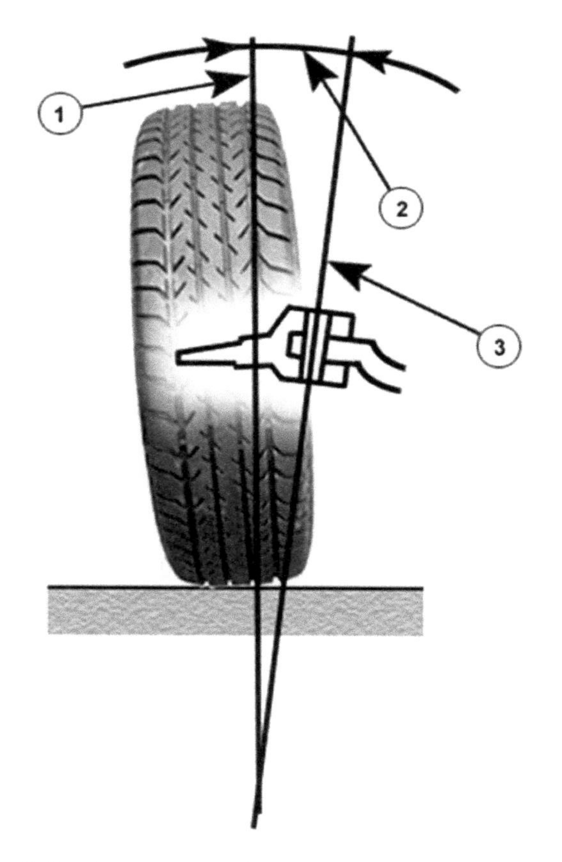

Figure 7.22 King pin or swivel axis inclination: 1 – perpendicular; 2 – swivel angle; 3 – steering axis.

7.7.8 Tracking

As a front wheel drive car drives forward, the tyres pull on the road surface, taking up the small amount of free play in the mountings and joints. For this reason, the tracking is often set toe-out so that the wheels point straight ahead when the vehicle is moving. Rear wheel drive tends to make the opposite happen because it pushes against the front wheels. The front wheels are therefore set toe-in. When the car moves, the front wheels are pushed out taking up the slack in the joints, so the wheels again end up straight ahead. The amount of toe-in or toe-out is very small, normally not exceeding 5 mm (the difference in the distance between the front and rear of the front wheels). Correctly set tracking ensures true rolling of the wheels and therefore reduced tyre wear. Figure 7.23 shows wheels set toe-in and toe-out.

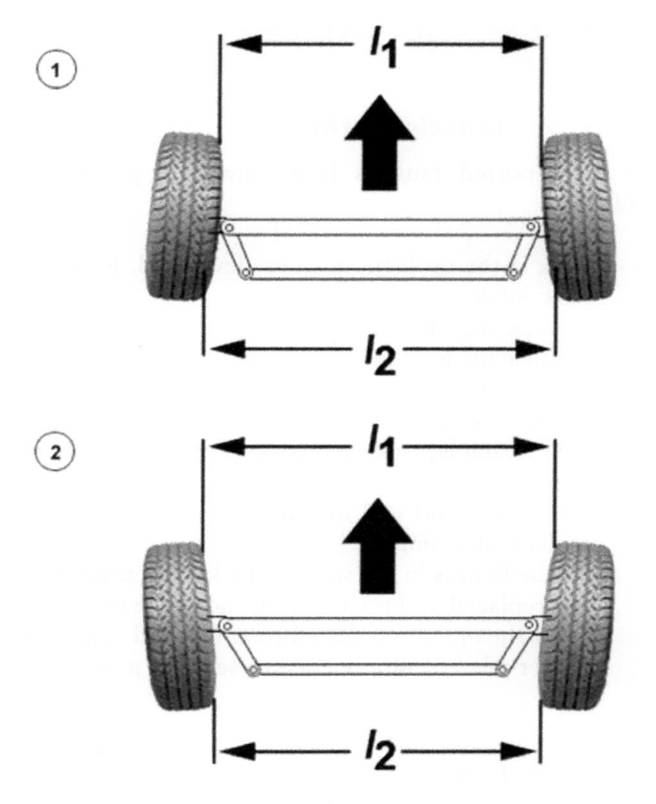

Figure 7.23 Tracking: 1 – toe-in; 2 – toe-out.

KEY FACT

Typical KPI/SAI value is approximately 7–9° (values will vary so check specs).

7.7.9 Scrub radius

The scrub radius is the distance between the contact point of the steering axis with the road and the wheel centre contact point. The purpose of designing in a scrub radius is to reduce the steering force and to prevent steering shimmy. It also helps to stabilise the straight-ahead position. It is possible to design the steering with a negative, positive or zero scrub radius as described in Table 7.3.

From the information given, you will realise that decisions about steering geometry are not clear-cut. One change may have a particular advantage in one area but a disadvantage in another. To assist with fault diagnosis, a good understanding of steering geometry is essential.

KEY FACT

As a front wheel drive car drives forward, the tyres pull on the road surface, forcing the toe-out of the wheels inwards – resulting in them being parallel (but check data).

Table 7.3 Scrub radius

Scrub radius	*Description*	*Properties*
Negative	The contact point of the steering axis hits the road between the wheel centre and the outer edge of the wheel	Braking forces produce a torque which tends to make the wheel turn inwards. The result of this is that the wheel with the greatest braking force is turned in with greater torque. This steers the vehicle away from the side with the heaviest braking producing a built-in counter steer action which has a stabilising effect.
Positive	The contact point of the steering axis hits the road between the wheel centre and the inner edge of the wheel	A positive scrub radius makes turning the steering easier. However, braking forces produce a torque, which tends to make the wheel turn outwards. The result of this is that the wheel with the greatest braking force is turned out with greater torque. Under different road conditions, this can have the effect of producing an unwanted steering angle.
Zero	The contact point of the steering axis hits the road at the same place as the wheel centre	This makes the steering heavy when the vehicle is at rest because the wheel cannot roll at the steering angle. However, no separate turning torque about the steering axis is created.

7.8 DIAGNOSTICS – STEERING AND TYRES

7.8.1 Systematic testing

If the reported fault is heavy steering, proceed as follows:

1. Ask if the problem has just developed. Road test to confirm.
2. Check the obvious such as tyre pressures. Is the vehicle loaded to excess? Check geometry?
3. Assuming tyre pressure and condition is as it should be, we must move on to further tests.
4. For example, jack up and support the front of the car. Operate the steering lock to lock. Disconnect one track rod end and move the wheel on that side, and so on.
5. If the fault is in the steering rack, then this should be replaced and the tracking should be set.
6. Test the operation with a road test and inspect all other related components for security and safety.

7.8.2 Test equipment

7.8.2.1 Tyre pressure gauge and tread depth gauge

These are often underrated pieces of test equipment. Correctly inflated tyres make the vehicle handle better, stop better and use less fuel. The correct depth of tread means the vehicle will be significantly safer to drive, particularly in wet conditions.

7.8.2.2 Tracking gauges

The toe-in and toe-out of a vehicle's front wheels are very important. Many types of tracking gauges are available. One of the most commonly used is a frame placed against each wheel with a mirror on one side and a moveable viewer on the other (Figure 7.24). The viewer is moved until marks are lined up and the tracking can then be measured.

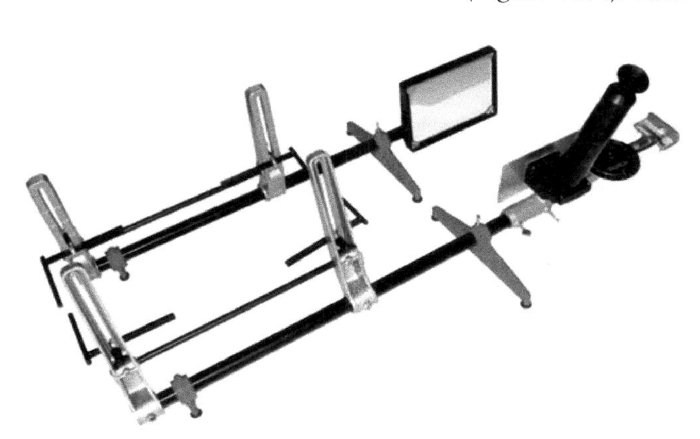

Figure 7.24 Basic mirror-type tracking gauges.

Figure 7.25 Wheel balancer.

KEY FACT

The scrub radius is the distance between the contact point of the steering axis with the road and the wheel centre contact point.

7.8.2.3 Wheel balancer

This is a large fixed piece of equipment in most cases. The wheel is removed from the car, fixed onto the machine and spun at high speed. Sensors in the tester measure the balance of the wheel. The tester then indicates the amount of weight which should be added to a certain position. The weight is added by fitting small lead weights (Figure 7.25).

7.8.3 Four-wheel alignment

Standard front-wheel alignment is simply a way of making sure the wheels are operating parallel with one another and that the tyres meet the road at the correct angle. Four-wheel alignment makes sure that the rear wheels follow the front wheels in a parallel path.

Different manufacturers set different specifications for the angles created between the suspension, steering, wheels and the frame of the vehicle. When these angles are correct, the vehicle is properly aligned.

SAFETY FIRST

Note: You should always refer to the manufacturer's instructions appropriate to the equipment you are using.

Figure 7.26 Laser alignment systems give digital readouts.

The main reasons for correct alignment are to ensure that the vehicle achieves

- minimum rolling friction;
- maximum tyre mileage;
- stability on the road;
- steering control for the driver.

Diagnosing incorrect alignments is usually just a matter of examining the

- tyres for unusual wear;
- wheels for damage;
- steering wheel for position.

In addition, a road test is usually necessary to check that the vehicle is not pulling to one side, wandering or weaving. Four basic wheel settings or angles determine whether a vehicle is properly aligned (Figures 7.26 and 7.27).

- Camber is the inward or outward tilt of a wheel compared to a vertical line. If the camber is out of adjustment, it will cause tyre wear on one side of the tyre's tread.
- Castor is the degree that the car's steering axis is tilted forward or backward from the vertical as viewed from the side of the car. If the castor is out of adjustment, it can cause problems with selfcentring and wander. Castor has little effect on tyre wear.
- Toe refers to the directions in which two wheels point relative to each other. Incorrect toe causes rapid tyre wear to both tyres equally. Toe is the most common adjustment and it is always adjustable on the front wheels and is adjustable on the rear wheels of some cars.

Figure 7.27 Scale to show angle of wheels.

7.8.4 Test results

Some of the information you may have to get from other sources such as data books or a workshop manual is listed in Table 7.4.

7.8.5 Tyres fault diagnosis table

The following table lists some of the faults which can occur if tyres and/or the vehicle are not maintained correctly. Figure 7.28 shows the same.

Symptom	*Possible cause/fault*
Wear on both outer edges of the tread	Under inflation
Wear in the centre of the tread all round the tyre	Over inflation
Wear just on one side of the tread	Incorrect camber
Feathering	Tracking not set correctly
Bald patches	Unbalanced wheels or unusual driving technique!

7.8.6 Tyre inflation pressures

The pressure at which the tyres should be set is determined by a number of factors such as

- load to be carried;
- number of plies;
- operating conditions;
- section of the tyre.

Tyre pressures must be set at the values recommended by the manufacturers. Pressure will vary according to the temperature of the tyre – this is affected by operating conditions. Tyre pressure should always be adjusted when the tyre is cold and be checked at regular intervals.

SAFETY FIRST

Tyre pressures must always be set at the values recommended by the manufacturer.

Table 7.4 Tests and information required

Test carried out	*Information required*
Tracking	The data for tracking will be given as either an angle or a distance measurement. Ensure you use the appropriate data for your type of test equipment. The distance will be a figure such as 3 mm toe-in, and if as an angle such as 50' toe-in (50' means 50 minutes). The angle of 1° is split into 60 minutes, so in this case the angle is 50/60 or 5/6 of a degree.
Pressures	A simple measurement which should be in bars. You will find, however, many places still use PSI (pounds per square inch). As in all other cases, only compare like with like.
Tread depth	The minimum measurement (e.g. 1.6 mm over 75% of the tread area but please note the current local legal requirements).

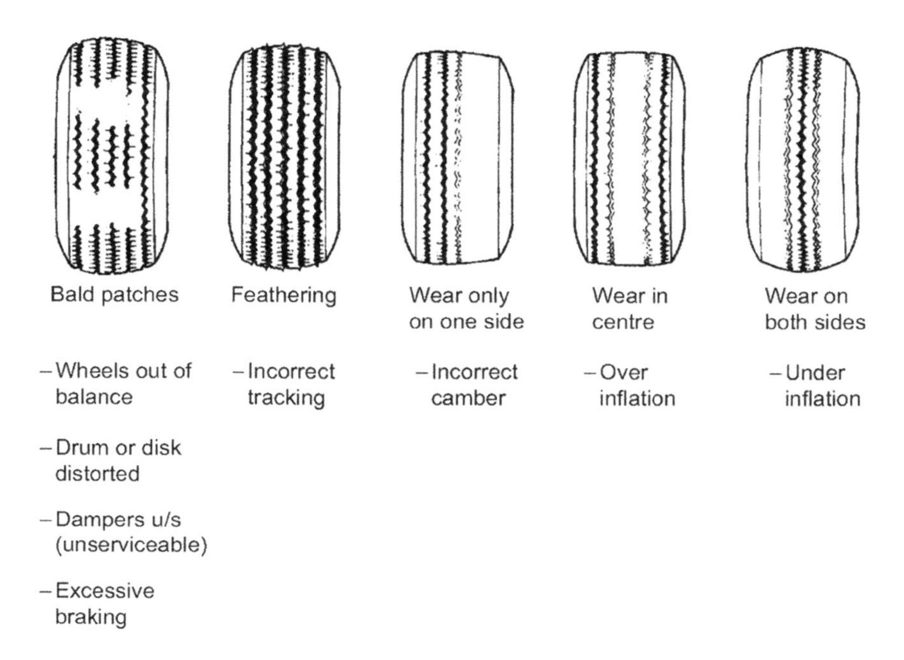

Figure 7.28 Tread wear patterns are a useful diagnostic aid.

7.8.7 Steering fault diagnosis

Symptom	*Possible faults*	*Suggested action*
Excessive free play at steering wheel	Play between the rack and pinion or in the steering box	Renew in most cases but adjustment may be possible
	Ball joints or tie-rod joints worn	Renew
	Column coupling loose or bushes worn	Secure or renew
Vehicle wanders, hard to keep in a straight line	As above	As above
	Alignment incorrect	Adjust to recommended setting
	Incorrect tyre pressure or mix of tyre types is not suitable	Adjust pressures or replace tyres as required
	Worn wheel bearings	Renew
Stiff steering	Wheel alignment incorrect	Adjust to recommended setting
	Tyre pressures too low	Adjust pressures
	Ball joints or rack seizing	Renew
Wheel wobble	Wheels out of balance	Balance or renew
	Wear in suspension linkages	Renew
	Alignment incorrect	Adjust to recommended setting
Under steer or over steer	Tyre pressures incorrect	Adjust pressures
	Dangerous mix of tyre types	Replace tyres as required
	Excessive free play in suspension or steering system	Renew components as required

7.8.8 Steering, wheels and tyres fault diagnosis

Symptom	*Possible cause*
Wandering or instability	Incorrect wheel alignment Worn steering joints Wheels out of balance Wheel nuts or bolts loose
Wheel wobble	Front or rear Wheels out of balance Damaged or distorted wheels/tyres Worn steering joints
Pulling to one side	Defective tyre Excessively worn components Incorrect wheel alignment
Excessive tyre wear	Incorrect wheel alignment Worn steering joints Wheels out of balance Incorrect inflation pressures Driving style! Worn dampers
Excessive free-play	Worn track rod end or swivel joints Steering column bushes worn Steering column universal joint worn
Stiff steering	Lack of steering gear lubrication Seized track rod end joint or suspension swivel joint Incorrect wheel alignment Damage to steering components

7.8.9 ADAS and alignment

Advanced driver assistance systems (ADAS) have been developed to automate/adapt/enhance vehicle systems for safety and better driving. Safety features are designed to avoid collisions and accidents by offering technologies that alert the driver to potential problems, or to avoid collisions by implementing safeguards and taking over control of the vehicle. Adaptive features may automate lighting, provide adaptive cruise control, automate braking, incorporate GPS traffic warnings, connect to smartphones, alert drivers to other cars or dangers, keep the car in the correct lane or show what is in blind spots.

DEFINITION

ADAS: Advanced driver assistance systems.

To avoid being locked out of future service and repair work, workshops must understand these developments as ADAS technology is an increasingly crucial component in vehicle design.

As one of the original equipment (OE) manufacturers that have pioneered ADAS technology, Hella is ideally placed to help the service and repair sector meet the significant challenges it faces (Figures 7.29 and 7.30).

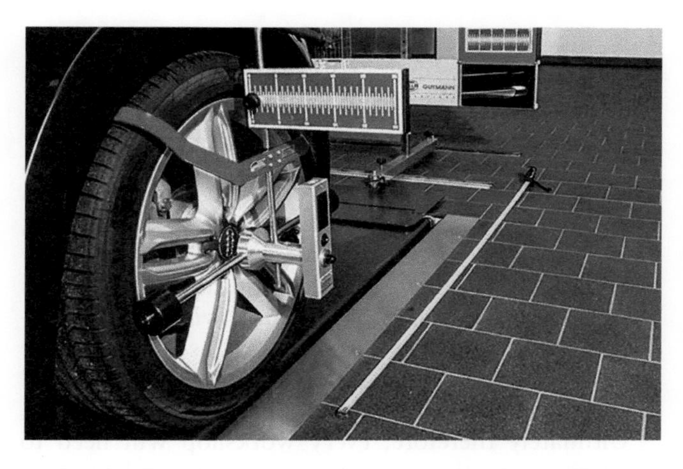

Figure 7.29 Camera System Calibration (CSC) tool allows precise recalibration.

Source: Hella-Gutmann Solutions.

A perfect example of ADAS technology is adaptive lighting, such as Hella's Matrix LED system. Through camera-controlled information processes, this technology provides 'intelligent' intense, but glare-free lighting, which, the company says, has already helped to reduce road accidents at night by 30%.

Since 2016, any new vehicle is required to have a minimum of two items of ADAS technology to achieve the current 'industry standard' 5-star Euro NCAP safety rating. This minimum requirement includes autonomous emergency braking (AEB) and lane departure warning (LDW), both of which are complex systems that require special equipment to diagnose any faults and to recalibrate.

By its very nature, ADAS technology, which also includes hazard recognition, blind spot detection and fatigue warning, is of fundamental importance to driver and pedestrian safety and therefore needs to function precisely, efficiently and reliably. These systems and the

Figure 7.30 Wheel alignment is part of the system.

Source: Hella Gutmann Solutions.

associated actions they perform, are all controlled by sensors, cameras and complex software, which means that incorrect inputs from any of these elements can have catastrophic consequences. Those doing service and repair work on these vehicles, therefore, have extra responsibility.

ADAS technology will have the biggest impact on the body shop and windscreen replacement sectors, where accident damage and glass replacement logically require the system to be reset. Traditional workshops concentrating on mechanical repair will also be confronted by it because even something as commonplace as an adjustment to the vehicle's steering geometry will require the system's recalibration.

Ultimately therefore, every workshop will need to have access to the necessary equipment and training to be able to correctly reset the ADAS components following an accident, windscreen replacement or geometry-related repair. In fact, the vehicle manufacturers now specify that recalibration of ADAS must take place after work has been performed, otherwise the workshop could potentially face liability.

KEY FACT

Workshops will need to have access to the necessary equipment and training to be able to correctly reset the ADAS components.

Naturally, there could be serious repercussions if recalibration is not undertaken or is carried out incorrectly, with the potential deactivation of systems and corresponding ADAS function. There is also the possibility that the motorist could pursue legal action against the workshop if the calibration is not carried out properly.

SAFETY FIRST

There could be serious repercussions if ADAS recalibration is not undertaken or is carried out incorrectly.

7.9 SUSPENSION

7.9.1 Introduction

The purpose of a suspension system can best be summarised by the following requirements:

- cushion the car, passengers and load from road surface irregularities;
- resist the effects of steering, braking and acceleration, even on hills and when loads are carried;
- keep tyres in contact with the road at all times;
- work in conjunction with the tyres and seat springs to give an acceptable ride at all speeds.

The above list is difficult to achieve completely, so some sort of compromise has to be reached. Because of this, many different methods have been tried, and many are still in use. Keep these four main requirements in mind and it will help you understand why some systems are constructed in different ways.

A vehicle needs a suspension system to cushion and damp out road shocks so providing comfort to the passengers and preventing damage to the load and vehicle components. A spring between the wheel and the vehicle body allows the wheel to follow the road surface. The tyre plays an important role in absorbing small road shocks. It is often described as the primary form of suspension. The vehicle body is supported by springs located between the body and the wheel axles. Together with the damper, these components are referred to as the suspension system.

As a wheel hits a bump in the road, it is moved upwards with quite some force. An unsprung wheel is affected only by gravity, which will try to return the wheel to the road surface, but most of the energy will be transferred to the body. When a spring is used between the wheel and the vehicle body, most of the energy in the bouncing wheel is stored in the spring and not passed to the vehicle body. The vehicle body will now only move upwards through a very small distance compared to the movement of the wheel.

7.9.2 Suspension system layouts

On older types of vehicle, a beam axle was used to support two stub axles. Beam axles are now rarely used in car suspension systems, although many commercial vehicles use them because of their greater strength and constant ground clearance (Figure 7.31).

The need for a better suspension system came from the demand for improved ride quality and improved handling. Independent front suspension (IFS) was developed

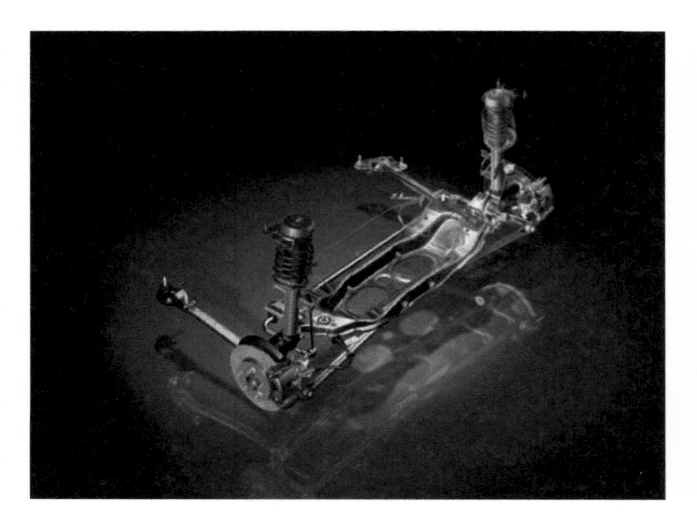

Figure 7.31 Rear suspension layout.

Source: Ford Media.

Figure 7.32 Front wishbone suspension system.

Figure 7.33 Front suspension struts.

Source: Jaguar Media.

to meet this need (Figure 7.32). The main advantages of independent front suspension are as follows:

- When one wheel is lifted or drops, it does not affect the opposite wheel.
- The unsprung mass is lower, therefore the road wheel stays in better contact with the road.
- Problems with changing steering geometry are reduced.
- There is more space for the engine at the front.
- Softer springing with larger wheel movement is possible.

KEY FACT

A suspension system reduces road shocks so providing comfort to the passengers and preventing damage to the load and vehicle components.

There are a number of basic suspension systems in common use. Figure 7.33 shows a front suspension layout on a Jaguar.

7.9.3 Front axle suspensions

As with most design aspects of the vehicle, compromise often has to be reached between performance, body styling and cost. Table 7.5 compares the common front axle suspension systems.

Table 7.5 Front axle suspension

Name	*Description*	*Advantages*	*Disadvantages*
Double transverse arms	Independently suspended wheels located by two arms perpendicular to direction of travel. The arms support stub axles.	Low bonnet line. Only slight changes of track and camber with suspension movements.	A large number of pivot points is required. High production costs.
Transverse arms with leaf spring	A transverse arm and a leaf spring locate the wheel.	The spring can act as an anti-roll bar, hence low cost.	Harsh response when lightly loaded. Major changes of camber as vehicle is loaded.
Transverse arm with McPherson strut	A combination of the spring, damper, wheel hub, steering arm and axle joints in one unit.	Only slight changes in track and camber with suspension movement. Forces on the joints are reduced because of the long strut.	The body must be strengthened around the upper mounting. A low bonnet line is difficult to achieve.
Double trailing arms	Two trailing arms support the stub axle. These can act on torsion bars often formed as a single assembly.	No change in castor, camber or track with suspension movement Can be assembled and adjusted off the vehicle.	Lots of space is required at the front of the vehicle. Expensive to produce. Acceleration and braking cause pitching movements which in turn changes the wheel base.

Table 7.6 Rear axle suspension

Name	*Description*	*Advantages*	*Disadvantages*
Rigid axle with leaf springs	The final drive, differential and axle shafts are all one unit	Rear track remains constant, reducing tyre wear Good directional stability because no camber change causes body roll on corners Low cost Strong design for load carrying	High unsprung mass The interaction of the wheels causes lateral movement, reducing tyre adhesion when the suspension is compressed on one side
Rigid axle with A-bracket	Solid axle with coil springs and a central joint supports the axle on the body	Rear of the vehicle pulls down on braking which stabilises the vehicle	High cost Large unsprung mass
Rigid axle with compression/tension struts	Coil springs provide the springing and the axle is located by struts	Suspension extension is reduced when braking or accelerating The springs are isolated from these forces	High loads on the welded joints High weight overall Large unsprung mass
Torsion beam trailing arm axle	Two links are used, connected by a 'U' section that has low torsional stiffness but high resistance to bending	Track and camber does not change Low unsprung mass Simple to produce Space saving	Torsion bar springing on this system can be more expensive than coil springs
Torsion beam axle with Panhard rod	Two links are welded to an axle tube or 'U' section and lateral forces are taken by a Panhard rod	Track and camber do not change Simple flexible joints to the bodywork	Torsion bar springing on this system can be more expensive than coil springs
Trailing arms	The pivot axis of the trailing arms is at 90° to the direction of vehicle travel	When braking, the rear of the vehicle pulls down, giving stable handling Track and camber do not change Space saving	Slight change of wheel base when the suspension is compressed
Semi-trailing arms – fixed-length drive shafts	The trailing arms are pivoted at an angle to the direction of travel Only one universal joint (UJ) is required because the radius of the suspension arm is the same as the drive shaft when the suspension is compressed	Only very small dive when braking Lower cost than when variable length shafts are used	Sharp changes in track when the suspension is compressed resulting in tyre wear Slight tendency to oversteer
Semi-trailing arms – variable length drive shafts	The final drive assembly is mounted to the body and two UJs are used on each shaft	The two arms are independent of each other Only slight track changes	Large camber changes High cost because of the drive shafts and joints

7.9.4 Rear axle suspensions

Table 7.6 compares the common rear axle suspension systems.

7.9.5 Anti-roll bar

The main purpose of an anti-roll bar is to reduce body roll on corners. The anti-roll bar can be thought of as a torsion bar. The centre is pivoted on the body and each end bends to make connection with the suspension/wheel assembly. When the suspension is compressed on both sides, the anti-roll bar has no effect because it pivots on its mountings (Figure 7.34).

As the suspension is compressed on just one side, a twisting force is exerted on the anti-roll bar. The anti-roll bar is now under torsional load. Part of this load is transmitted to the opposite wheel, pulling it upwards. This reduces the amount of body roll on corners. The disadvantages are that some of the 'independence' is lost and the overall ride is harsher. Anti-roll bars can be fitted to both front and rear axles.

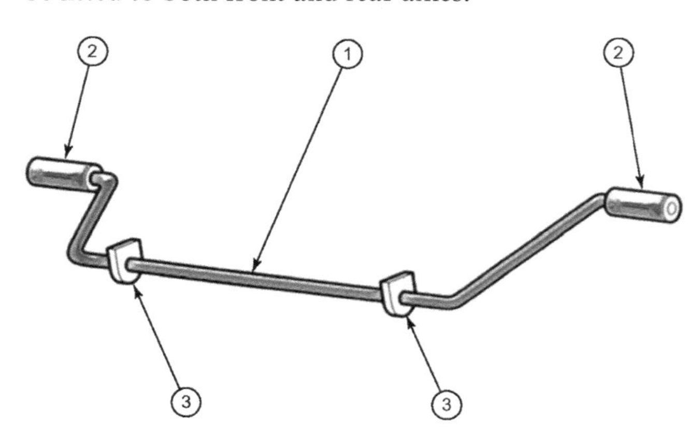

Figure 7.34 Anti-roll bar: 1 – torsion bar; 2 – pivots on suspension lower arms; 3 – fixings to vehicle body.

7.9.6 Springs

The requirements of the springs can be summarised as follows:

- absorb road shocks from uneven surfaces;
- control ground clearance and ride height;
- ensure good tyre adhesion;
- support the weight of the vehicle;
- transmit gravity forces to the wheels.

There are a number of different types of spring in use on modern light vehicles. Table 7.7 lists these together with their main features.

7.9.7 Dampers

The functions of a damper can be summarised as follows (Figure 7.35):

- ensure directional stability;
- ensure good contact between the tyres and the road;
- prevent build-up of vertical movements;
- reduce oscillations;
- reduce wear on tyres and chassis components.

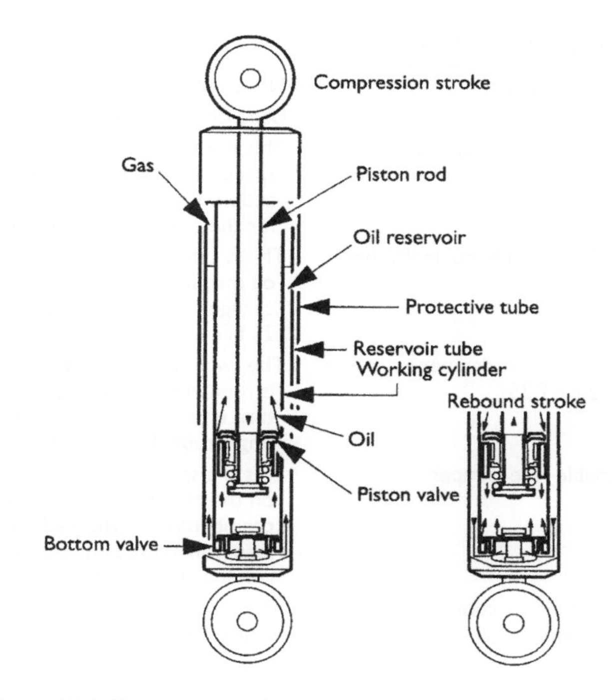

Figure 7.35 Twin-tube gas damper.

Table 7.7 Spring features

Name	*Comments*	*Characteristics*
Coil springs	The most common spring currently in use on light vehicles. The coil spring is a torsion bar wound into a spiral.	Can be progressive if the diameter of the spring is tapered conically Cannot transmit lateral or longitudinal forces, hence the need for links or arms Little internal damping Little or no maintenance High travel
Leaf springs	These springs can be single or multiple leaf. They are most often mounted longitudinally. Nowadays they are only used on commercial vehicles.	Can transmit longitudinal and lateral forces Short travel High internal damping High load capacity Maintenance may be required Low height but high weight
Torsion bar springs	A torsion bar is a spring where twisting loads are applied to a steel bar. They can be round or square section, solid or hollow. Their surface must be finished accurately to eliminate pressure points, which may cause cracking and fatigue failure. They can be fitted longitudinally or laterally.	Maintenance free but can be adjusted Transmit longitudinal and lateral forces Limited self-damping Linear rate Low weight May have limited fatigue life
Rubber springs	Nowadays rubber springs are only used as a supplement to other forms of springs. They are, however, popular on trailers and caravans.	Progressive rate Transmit longitudinal and lateral forces Short travel Low weight and low cost Their springing and damping properties can change with temperature
Air springs	Air springs can be thought of as being like a balloon or football on which the car is supported. The system involves compressors and air tanks. They are not normally used on light vehicles.	Expensive Good-quality ride Electronic control can be used Progressive spring rate High production cost
Hydro-pneumatic springs	A hydro-pneumatic spring is a gas spring with hydraulic force transmission. Nitrogen is usually used as the gas. The damper can be built in as part of the hydraulic system. The springs can be hydraulically connected together to reduce pitch or roll. Ride height control can be achieved by pumping oil into or out of the working chamber.	Progressive rate Ride height control Damping built-in Pressurised oil supply is required Expensive and complicated

Table 7.8 Types of damper

Friction damper	Not used on cars today, but you will find this system used as part of caravan or trailer stabilisers.
Lever-type damper	Used on earlier vehicles, the lever operates a piston which forces oil into a chamber.
Twin-tube telescopic damper	This is a commonly used type of damper; it consists of two tubes. An outer tube forms a reservoir space and contains the oil displaced from an inner tube. Oil is forced through a valve by the action of a piston as the damper moves up or down. The reservoir space is essential to make up for the changes in volume as the piston rod moves in and out.
Single-tube telescopic damper	This is often referred to as a gas damper. However, the damping action is still achieved by forcing oil through a restriction. The gas space behind a separator piston is to compensate for the changes in cylinder volume caused as the piston rod moves. It is at a pressure of approximately 25 bar.
Twin-tube gas damper (Figure 7.35)	The twin-tube gas damper is an improvement on the well-used twin-tube system. The gas cushion is used in this case to prevent oil foaming. The gas pressure on the oil prevents foaming, which in turn ensures constant operation under all operating conditions. Gas pressure is lower than for a single-tube damper at approximately 5 bar.
Variable rate damper	This is a special variation of the twin-tube gas damper. The damping characteristics vary depending on the load on the vehicle. Bypass grooves are machined in the upper half of the working chamber. With light loads, the damper works in this area with a soft damping effect. When the load is increased, the piston moves lower down the working chamber away from the grooves resulting in full damping effect.
Electronically controlled dampers	These are dampers where the damping rate can be controlled by solenoid valves inside the units. With suitable electronic control, the characteristics can be changed within milliseconds to react to driving and/or load conditions.

There are a number of different types of damper. These are listed in Table 7.8.

KEY FACT

Strictly speaking, a damper damps out spring oscillations, but as it also absorbs shock it is also called a shock absorber.

7.10 DIAGNOSTICS – SUSPENSION

7.10.1 Systematic testing

If the reported fault is poor handling, proceed as follows:

1. Road test to confirm the fault.
2. With the vehicle on a lift, inspect obvious items like tyres and dampers.
3. Consider if the problem is suspension related or in the steering, for example. You may have decided this from road testing.
4. Inspect all the components of the system you suspect; for example, dampers for correct operation and suspension bushes for condition and security. Let's assume the fault was one front damper not operating to the required standard.
5. Renew both of the dampers at the front to ensure balanced performance.
6. Road test again and check for correct operation of the suspension and other systems.

7.10.2 Test equipment

7.10.2.1 Damper tester

The operating principle of a damper tester is shown in Figure 7.36, which indicates that the damper is not operating correctly in this case. This is a device that will draw a graph to show the response of the dampers. It may be useful for providing paper evidence of the operating condition but a physical examination is normally adequate.

SAFETY FIRST

Note: You should always refer to the manufacturer's instructions appropriate to the equipment you are using.

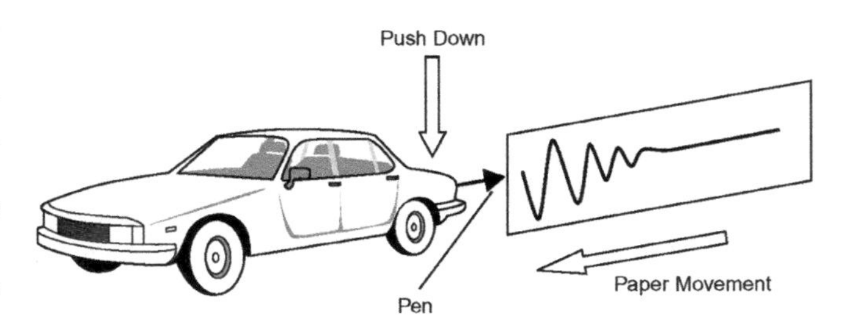

Figure 7.36 Representation of a damper (shock absorber) test – the symptoms suggest a faulty damper.

Table 7.9 Tests and information required

Test carried out	*Information required*
Damper operation	The vehicle body should move down as you press on it, bounce back just past the start point and then return to the rest position
Suspension bush condition	Simple levering, if appropriate, should not show excessive movement, cracks or separation of rubber bushes
Trim height	This is available from data books as a measurement from say the wheel centre to a point on the car wing above

7.10.3 Test results

Some of the information you may have to get from other sources such as data books or a workshop manual is listed in Table 7.9.

7.10.4 Suspension fault diagnosis table 1

Symptom	*Possible faults*	*Suggested action*
Excessive pitch or roll when driving	Dampers worn	Replace in pairs
Car sits lopsided	Broken spring Leak if hydraulic suspension	Replace in pairs Rectify by replacing unit or fitting new pipes
Knocking noises	Excessive free play in a suspension joint	Renew
Excessive tyre wear	Steering/suspension geometry incorrect (may be due to accident damage)	Check and adjust or replace any 'bent' or out of true components

7.10.5 Suspension fault diagnosis table 2

Symptom	*Possible cause*
Excessive pitching	Defective dampers Broken or weak spring Worn or damaged anti-roll bar mountings
Wandering or instability	Broken or weak spring Worn suspension joints Defective dampers
Wheel wobble	Worn suspension joints
Pulling to one side	Worn suspension joints Accident damage to suspension alignment
Excessive tyre wear	Worn suspension joints Accident damage to suspension alignment Incorrect trim height (particularly hydrolastic systems)

7.11 ACTIVE SUSPENSION

7.11.1 Active suspension operation

A traditional or a conventional suspension system, consisting of springs and dampers, is passive. In other words, once it has been installed in the car, its characteristics do not change (Figure 7.37).

The main advantage of a conventional suspension system is its predictability. Over time, the driver will become familiar with a car's suspension and understand its capabilities and limitations. The disadvantage is that the system has no way of compensating for situations beyond its original design (Figure 7.38).

An active suspension system (also known as computerised ride control) has the ability to adjust itself continuously. It monitors and adjusts its characteristics to suit the current road conditions. As with all electronic control systems, sensors supply information to an ECU which in turn outputs to actuators. By changing its characteristics in response to changing road conditions, active suspension offers improved handling, comfort, responsiveness and safety (Figure 7.39).

KEY FACT

An active suspension system has the ability to adjust itself continuously.

Active suspension systems consist of the following components:

- electronic control unit;
- adjustable dampers and springs;
- sensors at each wheel and throughout the car;
- levelling compressor (some systems).

Components vary between manufacturers, but the principles are the same.

Active suspension works by constantly sensing changes in the road surface and feeding that information to the ECU, which in turn controls the suspension springs and dampers. These components then act upon the system to modify the overall suspension characteristics by adjusting damper stiffness, ride height (in some cases) and spring rate.

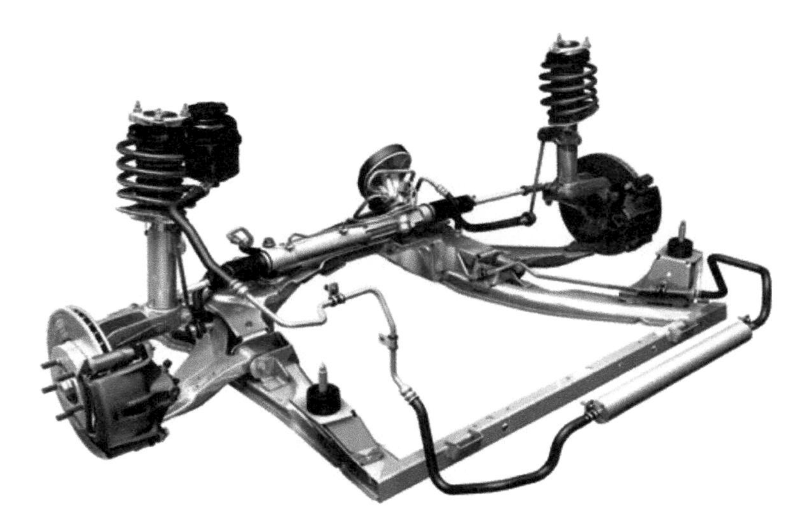

Figure 7.37 Jaguar suspension system.

Source: Jaguar Media.

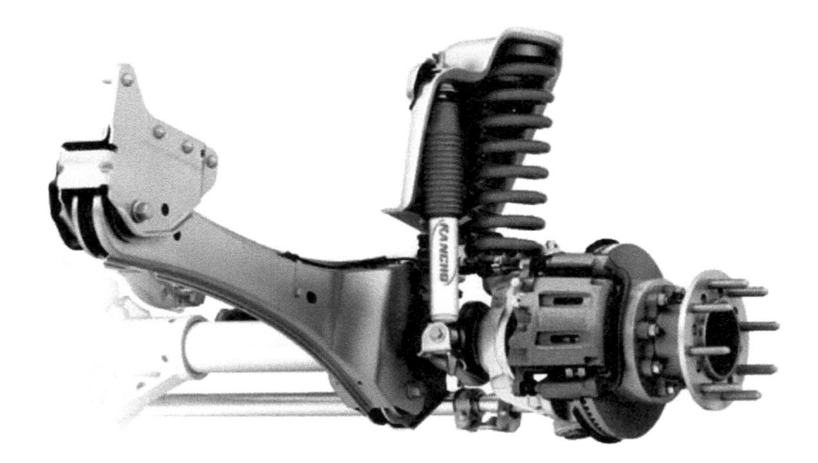

Figure 7.38 Suspension system.

Source: Ford Media.

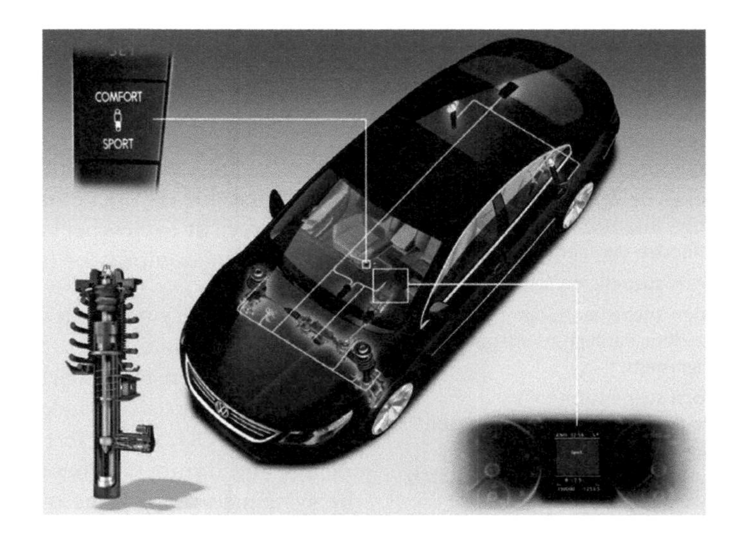

Figure 7.39 Active suspension also allows adjustments; in this case, between sport.

Figure 7.40 Potholes.

Assume that a car with conventional suspension is cruising down the road and then, after turning left, hits a series of potholes on the right-hand side, each one larger than the next (Figure 7.40). This would present a serious challenge to a conventional suspension system because the increasing size of the holes could set up an oscillation loop and bottom out the system. An active system would react very differently.

Sensors send information to the ECU about yaw and lateral acceleration. Other sensors measure excessive vertical travel, particularly in the right-front region of the car, and a steering angle sensor provides information on steering position.

The ECU analyses this information in approximately 10 ms. It then sends a signal to the right-front spring to stiffen up. A similar signal is sent to the right-rear spring, but this will not be stiffened as much. The rigidity of the suspension dampers on the right-hand side of the vehicle is therefore increased. Because of these actions, the vehicle will drive through the corner, with little impact on driveability and comfort.

One of the latest types of sensor is produced by Bosch. The sensor simultaneously monitors three of a vehicle's movement axes – two acceleration or inclination axes (*ax*, *ay*) and one axis of rotation (Ω*z*).

Previously, at least two separate sensors were required for this. The integration of the sensors for lateral acceleration and yaw rate reduces space requirements in the vehicle and the assembly work for the complete system (Figure 7.41).

Figure 7.41 Integrated sensor.

Source: Bosch Media.

There are a number of ways of controlling the suspension. However, in most cases it is done by controlling the oil restriction in the damper. On some systems, ride height is controlled by opening a valve and supplying pressurised fluid from an engine-driven compressor.

Figure 7.42 Suspension strut and actuator connection.

Source: Delphi Media.

Other systems use special fluid in the dampers that reacts to a magnetic field, which is applied from a simple electromagnetic coil. The case study of a Delphi system in the next section looks at this method in detail (Figure 7.42).

The improvements in ride comfort are considerable, which is why active suspension technology is becoming more popular. In simple terms, sensors provide the input to a control system that in turn actuates the suspension dampers in a way that improves stability and comfort.

7.11.2 Delphi MagneRide case study

MagneRide was the industry's first semi-active suspension technology that employs no electromechanical valves and small moving parts. The MagneRide magneto-rheological (MR) fluid-based system consists of MR fluid-based single-tube struts, shock absorbers (dampers), a sensor set and an on-board controller (Figure 7.43).

MR fluid is a suspension of magnetically soft particles such as iron microspheres in a synthetic hydrocarbon base fluid. When MR fluid is in the 'off' state, it is not magnetised, and the particles exhibit a random pattern. But in the 'on' or magnetised state, the applied magnetic field aligns the metal particles into fibrous structures, changing the fluid rheology to a near plastic state (Figure 7.44).

DEFINITION

Rheology: The study of friction between liquids.

By controlling the current to an electromagnetic coil inside the piston of the damper, the MR fluid's shear strength is changed, varying the resistance to fluid flow. Fine-tuning of the magnetic current allows for any state between the low forces of 'off' to the high forces of 'on' to be achieved in the damper. The result is continuously variable real-time damping (Figure 7.45).

The layout in Figure 7.46 shows the inputs and outputs of the MagneRide system. Note the connections with the ESP system and how the information is shared over the controller area network (CAN).

The MagneRide system, produced by Delphi, uses a special fluid in the dampers. The properties of this fluid are changed by a magnetic field. This allows for very close control of the damping characteristics and a significant improvement in ride comfort and quality.

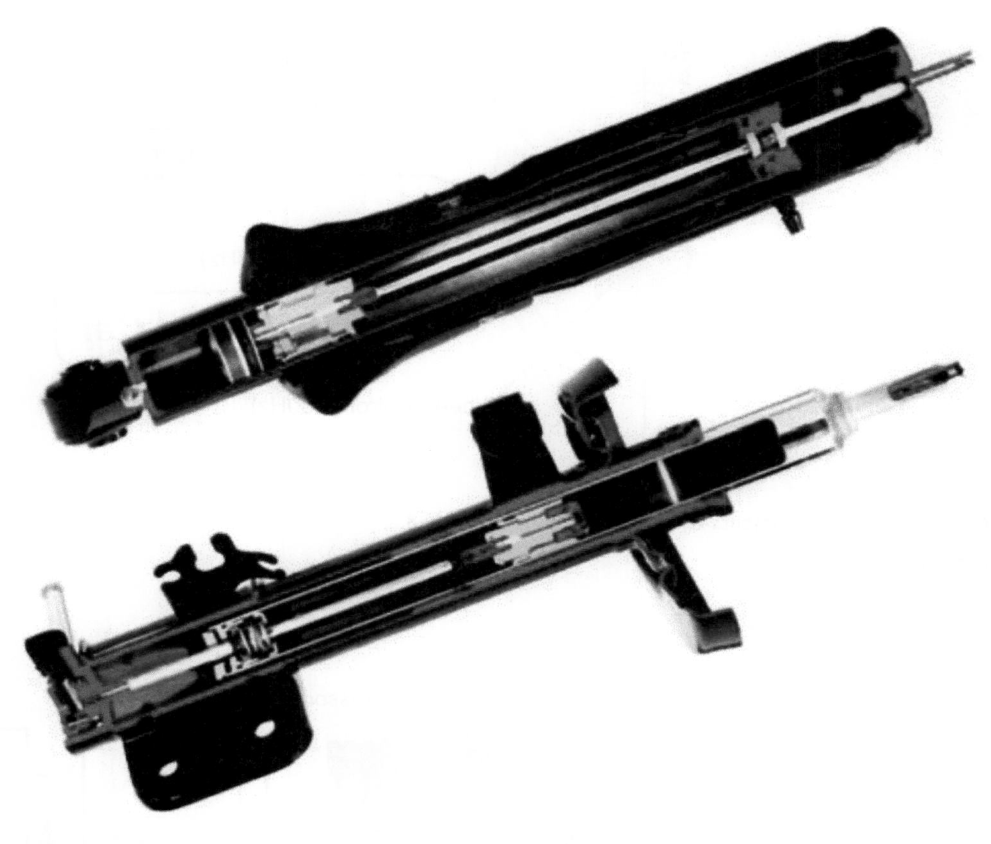

Figure 7.43 MagneRide suspension components.

Source: Delphi Media.

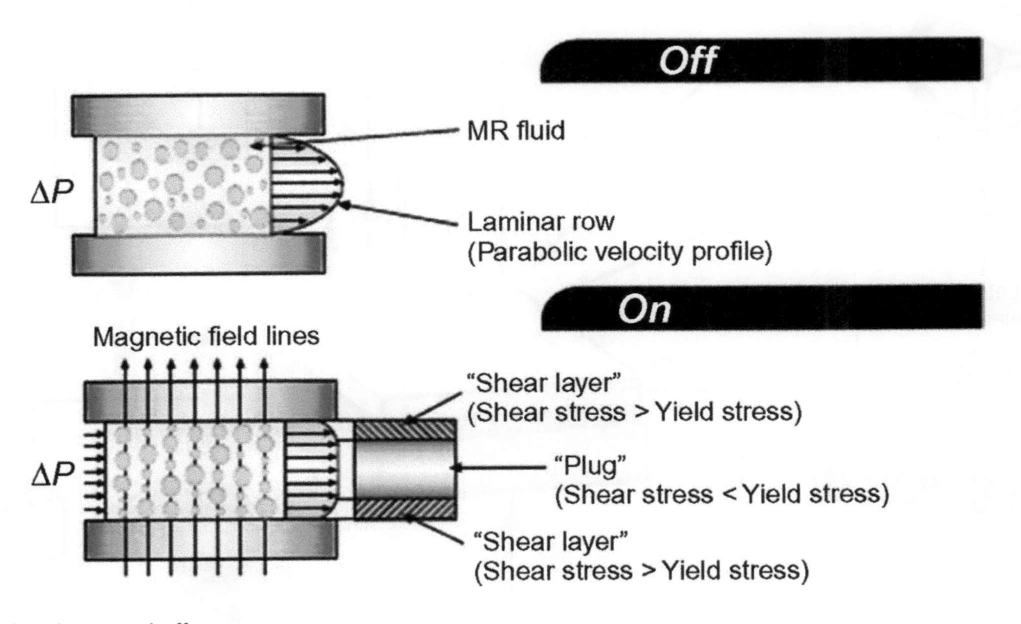

Figure 7.44 Fluid in the on and off states.

Source: Delphi Media.

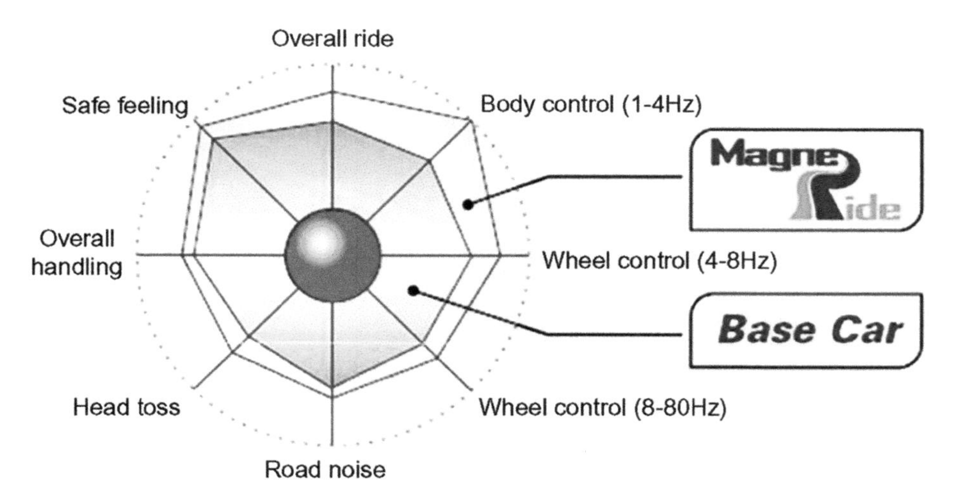

Figure 7.45 Representation of improvements when suspension is controlled.

Source: Delphi Media.

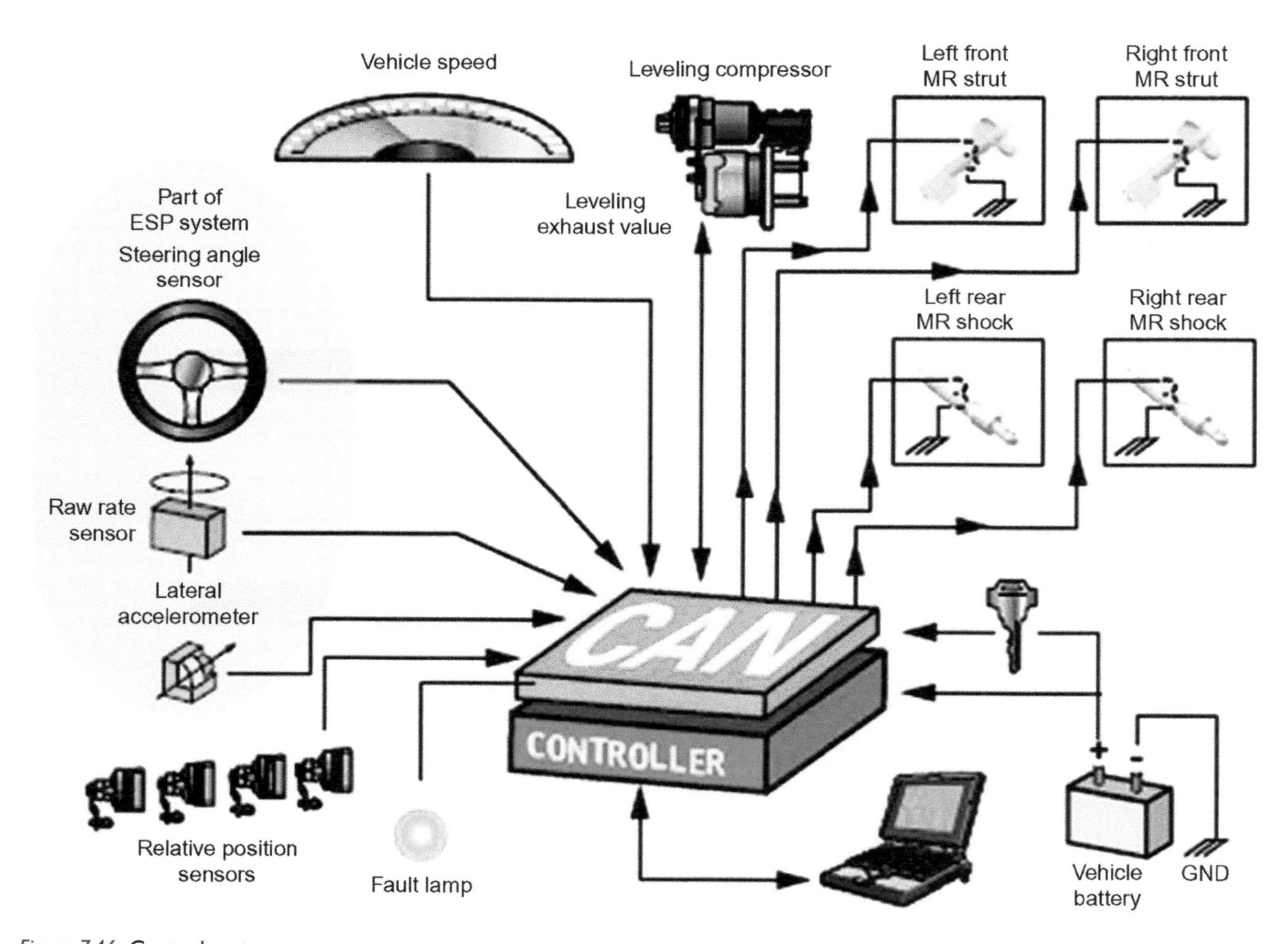

Figure 7.46 Control system.

Source: Delphi Media.

7.12 DIAGNOSTICS – ACTIVE SUSPENSION

7.12.1 Systematic testing

Even for an active system it may be useful to treat it like a conventional system at first. If the reported fault is poor handling and/or an MIL and code indicates a suspension problem, proceed as follows:

1. Confirm the fault and scan for codes.
2. With the vehicle on a lift, inspect obvious items like tyres and dampers for leaks.
3. Inspect all the components of the system you suspect and look for leaks.
4. Check mountings for damage and sensor wiring.
5. Check sensors and actuators (see Section 7.12.2 and Chapter 4).

7.12.2 Back to the black box

Active suspension systems now revolve around an ECU, and the ECU can be considered to be a 'black box'; in other words, we know what it should do but the exact details of how it does it are less important.

KEY FACT

Most vehicle systems involve an ECU.

Treating the ECU as a 'black box' allows us to ignore its complexity. The theory is that if all the sensors and associated wiring to the 'black box' are OK, all the output actuators and their wiring are OK and the supply/earth (ground) connections are OK, then the fault must be in the 'black box'. Most ECUs are very reliable, however, and it is far more likely that the fault will be found in the inputs or outputs (Figure 7.47).

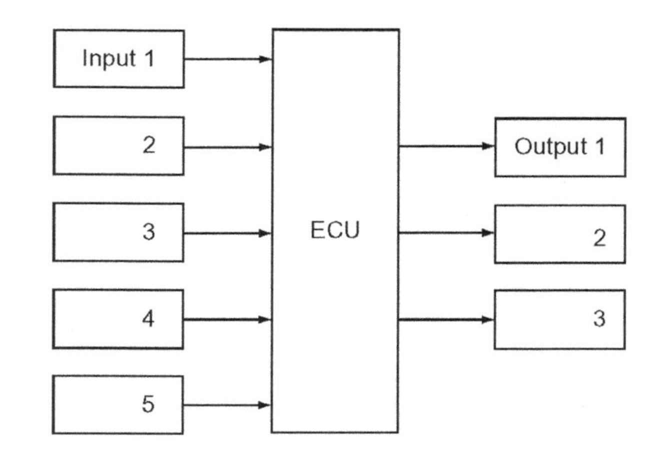

Figure 7.47 System block diagram.

Normal faultfinding or testing techniques can be applied to the sensors and actuators. For example, if the system uses four movement sensors, then an easy test is to measure their output. Even if the correct value is not known, it would be very unlikely for all four to be wrong at the same time so a comparison can be made. If the same reading is obtained on the end of the sensor wires at the ECU, then almost all of the 'inputs' have been tested with just a few readings. The same technique will often work with 'outputs'.

KEY FACT

If the readings of all similar items connected to an ECU are the same, then it is reasonable to assume the figure is almost certainly correct.

Don't forget that no matter how complex the electronics in an ECU, they will not work without a good power supply and an earth.

Chapter 8

Electrical systems

8.1 ELECTRONIC COMPONENTS AND CIRCUITS

8.1.1 Introduction

This section describing the principles and applications of various electronic components and circuits is not intended to explain their detailed operation. Overall, an understanding of basic electronic principles will help to show how electronic systems operate. These can range from a simple interior light delay unit to the most complicated engine management system. Testing individual electronic components is a useful diagnostic procedure.

8.1.2 Components

The symbols for the electronic components discussed in this section are shown in Figure 8.1.

Resistors are probably the most widely used component in electronic circuits. Two factors must be considered when choosing a suitable resistor: the ohms value and the power rating. Resistors are used to limit current flow and provide fixed voltage drops. Most resistors used in electronic circuits are made from small carbon rods; the size of the rod determines the resistance. Carbon resistors have a negative temperature coefficient (NTC) and this must be considered for some applications. Thin-film resistors have more stable temperature properties and are constructed by depositing a layer of carbon onto an insulated former such as glass. The resistance value can be manufactured very accurately by spiral grooves cut into the carbon film. For higher power applications, resistors are usually wire wound. Variable forms of most resistors are available. The resistance of a circuit is its opposition to current flow.

KEY FACT

Resistors are used to limit current flow and provide fixed voltage drops.

A capacitor is a device for storing an electric charge. In its simple form, it consists of two plates separated by an insulating material. One plate can have excess electrons compared to the other. On vehicles, its main uses are for reducing arcing across contacts and for radio interference suppression circuits as well as in electronic control units (ECUs). Capacitors are described as two plates separated by a dielectric. The area of the plates, the distance between them and the nature of the dielectric determine the value of capacitance. Metal foil sheets insulated by a type of paper are often used to construct capacitors. The sheets are rolled up together inside a tin can. To achieve higher values of capacitance it is necessary to reduce the distance between the plates in order to keep the overall size of the device manageable. This is achieved by immersing one plate in an electrolyte to deposit a layer of oxide typically 10^{-4} mm thick, thus ensuring a higher capacitance value. The problem, however, is that this now makes the device polarity conscious and only able to withstand low voltages.

KEY FACT

A capacitor is a device for storing an electric charge.

Diodes are often described as one-way valves, and for most applications, this is an acceptable description. A diode is a PN junction allowing electron flow from the N-type material to the P-type material. The materials are usually constructed from doped silicon. Diodes are not perfect devices and a voltage of approximately 0.6 V is required to switch the diode on in its forward biased direction.

KEY FACT

Diodes are often described as one-way valves.

Zener diodes are very similar in operation with the exception that they are designed to break down and conduct in the reverse direction at a predetermined voltage. They can be thought of as a type of pressure relief valve.

Transistors are the devices that have allowed the development of today's complex and small electronic systems. The transistor is used as either a solid state

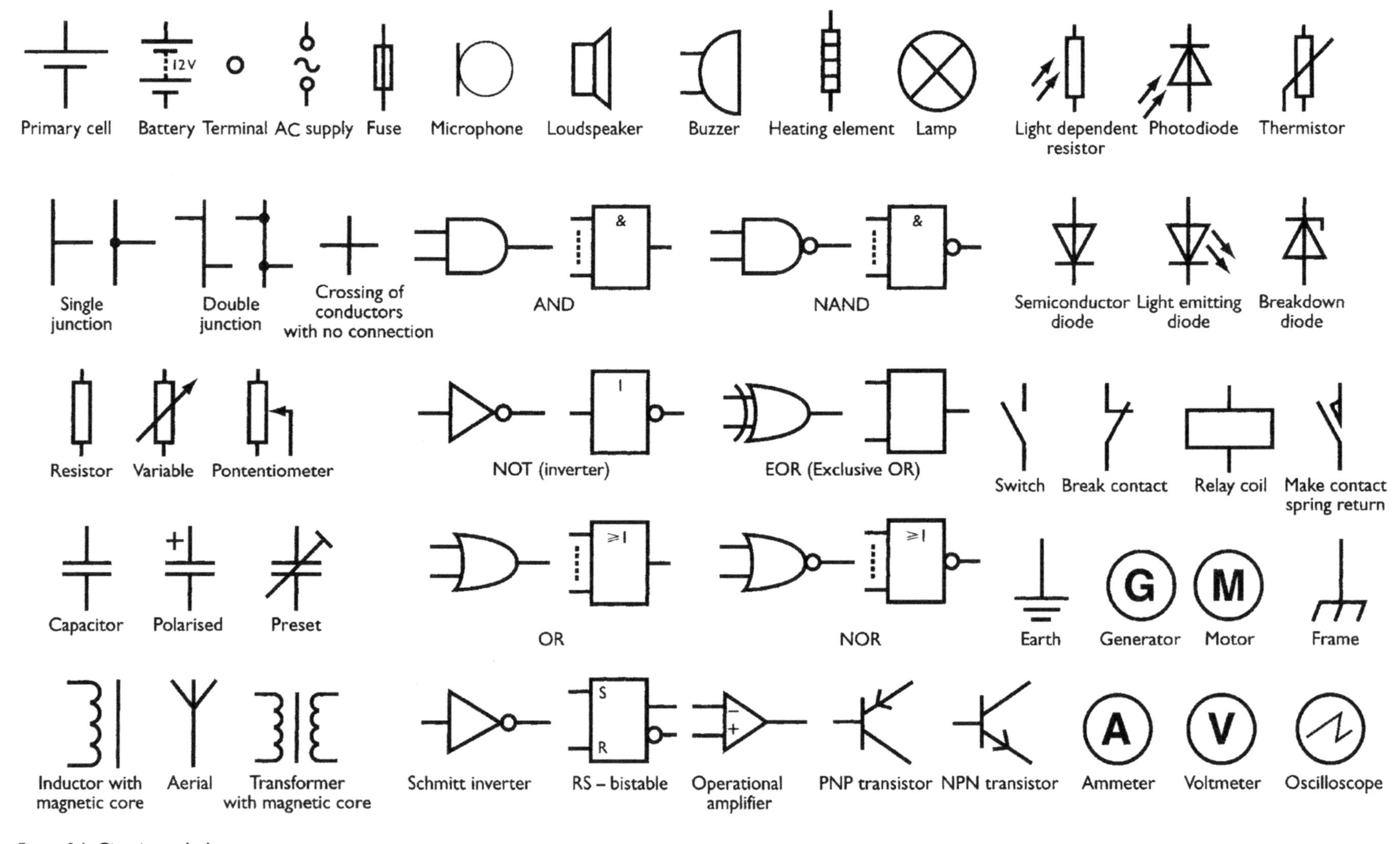

Figure 8.1 Circuit symbols.

switch or as an amplifier. They are constructed from the same P- and N-type semiconductor materials as the diodes and can be made in either NPN or PNP format. The three terminals are known as the base, collector and emitter. When the base is supplied with the correct bias, the circuit between the collector and emitter will conduct. The base current can be in the order of 50–200 times less than the emitter current. The ratio of the current flowing through the base compared to the current through the emitter is an indication of the amplification factor of the device.

KEY FACT

The transistor is used as either a solid state switch or as an amplifier.

A Darlington pair is a simple combination of two transistors which will give a high current gain, typically several thousand. The transistors are usually mounted on a heat sink and overall the device will have three terminals marked as a single transistor – base, collector and emitter. The input impedance of this type of circuit is in the order of 1 MΩ; hence, it will not load any previous part of a circuit connected to its input. The Darlington pair configuration is used for many switching applications. A common use of a Darlington pair is for the switching of coil primary current in the ignition circuit.

Another type of transistor is the field effect transistor (FET). This device has higher input resistance than the bipolar type described above. They are constructed in their basic form as n-channel or p-channel devices. The three terminals are known as the gate, source and drain. The voltage on the gate terminal controls the conductance of the circuit between the drain and the source.

Inductors are most often used as part of an oscillator or amplifier circuit. The basic construction of an inductor is a coil of wire wound on a former. It is the magnetic effect of the changes in current flow which gives this device the properties of inductance. Inductance is a difficult property to control, particularly as the inductance value increases. This is due to magnetic coupling with other devices. Iron cores are used to increase the inductance value.

This also allows for adjustable devices by moving the position of the core. Inductors, particularly of higher values, are often known as chokes and may be used in DC circuits to smooth the voltage.

8.1.3 Integrated circuits

Integrated circuits (ICs) are constructed on a single slice of silicon often known as a substrate. Combinations of some of the components mentioned previously can be used to carry out various tasks such as switching, amplifying and logic functions. The components required for these circuits can be made directly onto the slice of silicon. The great advantage of this is not just the size of the ICs but the speed at which they can be made to work due to the short distances between components. Switching speed in excess of 1 MHz is typical.

Figure 8.2 Silicon wafer used in the construction of integrated circuits.

KEY FACT

Integrated circuits or ICs are constructed on a single slice of silicon often known as a substrate.

The range and type of integrated circuits now available is so extensive that a chip is available for almost any application. The integration level of chips is now exceeding VLSI (very large scale integration). This means that there can be more than 100 000 active elements on one chip. Development in this area is moving so fast that often the science of electronics is now concerned mostly with choosing the correct combination of chips, and discrete components are only used as final switching or power output stages. Figure 8.2 shows a highly magnified view of a typical IC.

8.1.4 Digital circuits

With some practical problems, it is possible to express the outcome as a simple yes/no or true/false answer.

Let's take a simple example: if the answer to the first or the second question is 'yes', then switch on the brake warning light; if both answers are 'no' then switch it off.

Table 8.1 Electronic component test methods

Component	*Test method*
Resistor	Measure the resistance value with an ohmmeter and compare this to the value written or colour coded on the component.
Capacitor	A capacitor can be difficult to test without specialist equipment, but try this: Charge the capacitor up to 12 V and connect it to a digital voltmeter. As most digital meters have an internal resistance of approximately 10 MΩ, calculate the expected discharge time (T = 5CR) and see if the device complies. A capacitor from a contact breaker ignition system should take approximately five seconds to discharge in this way.
Inductor	An inductor is a coil of wire, so a resistance check is the best method to test for continuity.
Diode	Many multimeters have a diode test function. If so, the device should read open circuit in one direction and approximately 0.4–0.6 V in the other direction. This is its switch on voltage. If no meter is available with this function, then wire the diode to a battery via a small bulb; it should light with the diode one way and not the other.
LED	LEDs can be tested by connecting them to a 1.5 V battery. Note the polarity though; the longest leg or the flat side of the case is negative.
Transistor (bipolar)	Some multimeters even have transistor testing connections but, if not available, the transistor can be connected into a simple circuit as in Figure 8.4 and voltage tests carried out as shown. This also illustrates a method of testing electronic circuits in general. It is fair to point out that without specific data it is difficult for the non-specialist to test unfamiliar circuit boards. It's always worth checking for obvious breaks and dry joints though.
Digital components	A logic probe can be used. This is a device with a very high internal resistance, so it does not affect the circuit under test. Two different coloured lights are used; one glows for a 'logic 1' and the other for 'logic 0'. Specific data is required in most cases but basic tests can be carried out.

1. Is the handbrake on?
2. Is the level in the brake fluid reservoir low?

In this case, we need the output of an electrical circuit to be 'on' when either one or both of the inputs to the circuit are 'on'. The inputs will be via simple switches on the handbrake and in the brake reservoir. The digital device required to carry out the above task is an OR gate. An OR gate for use on this system would have two inputs (a and b) and one output (c). Only when 'a' OR 'b' is supplied, will 'c' produce a voltage.

Once a problem can be described in logic states then a suitable digital or logic circuit can also determine the answer to the problem. Simple circuits can also be constructed to hold the logic state of their last input; these are in effect simple forms of 'memory'. By combining vast quantities of these basic digital building blocks, circuits can be constructed to carry out the most complex tasks in a fraction of a second. Because of IC technology, it is now possible to create hundreds of thousands if not millions of these basic circuits on one chip. This has given rise to the modern electronic control systems used for vehicle applications as well as all the countless other uses for a computer.

In electronic circuits, true/false values are assigned voltage values. In one system, known as TTL (transistor-transistor-logic), true or logic '1' is represented by a voltage of 3.5 V and false or logic '0' by 0 V.

8.1.5 Electronic component testing

Individual electronic components can be tested in a number of ways but a digital multimeter is normally the favourite option. Table 8.1 suggests some methods of testing components removed from the circuit (Figures 8.3 and 8.4).

KEY FACT

In electronic circuits, true/false values are assigned voltage values.

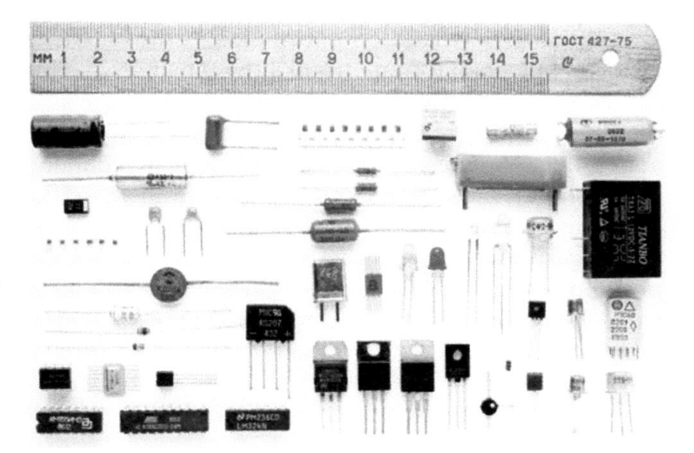

Figure 8.3 Electronic components.

Source: Kae, Wikimedia Commons.

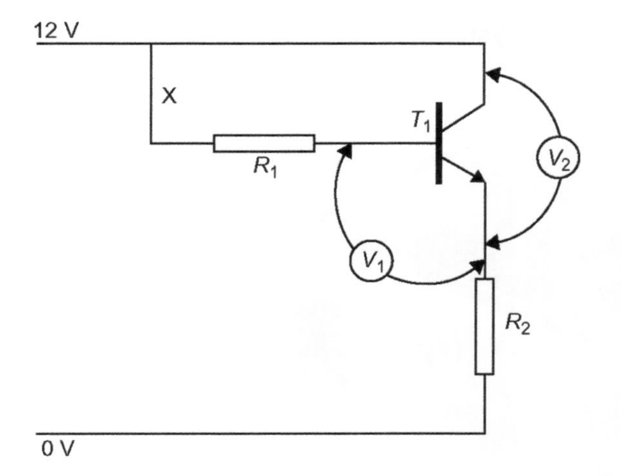

Figure 8.4 Transistor testing: Use resistors $R_1 = R_2$ of approximately 1 kΩ, when connected as shown, V_1 should read 0.6–0.7 V and V_2 approximately 1 V. Disconnect wire X, V_1 should now read 0 V and V_2 12 V.

8.2 MULTIPLEXING

8.2.1 Overview

The number of vehicle components which are networked has considerably increased the requirements for the vehicle control systems to communicate with one another. The CAN (controller area network) developed by Bosch is today's communication standard in passenger cars. However, there are a number of other systems.

DEFINITION

CAN: Controller area network.

Multiplexing is a process of combining several messages for transmission over the same signal path. The signal path is called the data bus. The data bus is basically just a couple of wires connecting the control units together. A data bus consists of a communication or signal wire and a ground return, serving all multiplex system nodes. The term 'node' is given to any sub-assembly of a multiplex system (such as a control unit) that communicates on the data bus.

On some vehicles, early multiplex systems used three control units (Figure 8.5). These were the door control unit, the driver's side control unit and the passenger's side control unit. These three units replaced the following:

- integrated unit;
- interlock control unit;
- door lock control unit;
- illumination light control;
- power window control unit;
- security alarm control unit.

When a switch is operated, a coded digital signal is generated and communicated, according to its priority, via the data bus. All control units receive the signal but only the control unit for which the signal is intended will activate the desired response.

Only one signal can be sent on the bus at any one time. Therefore, each signal has an identifier that is unique throughout the network. The identifier defines not only the content but also the priority of the message. Some systems make changes or adjustments to their operation much faster than other systems. Therefore, when two signals are sent at the same time, it is the system which

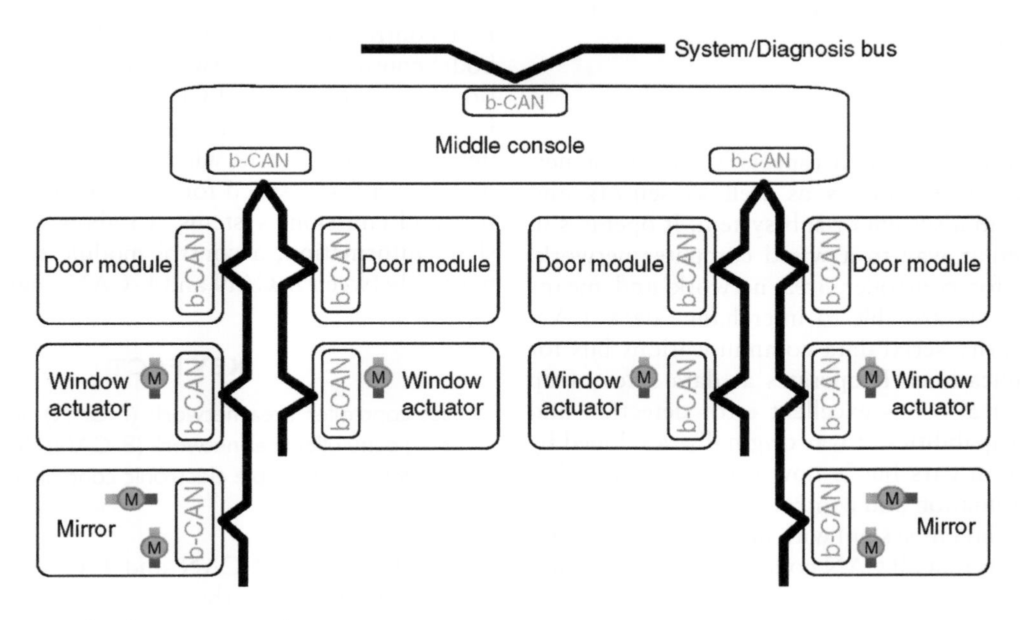

Figure 8.5 Sub-system for doors on an earlier system.

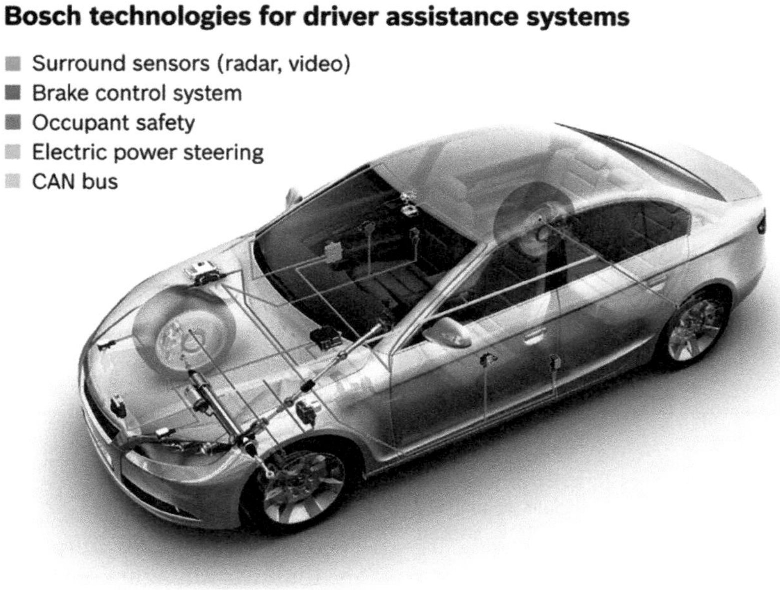

Bosch technologies for driver assistance systems

- Surround sensors (radar, video)
- Brake control system
- Occupant safety
- Electric power steering
- CAN bus

Figure 8.6 A data bus connects all networked components.

requires the message most urgently whose signal takes priority (Figure 8.6).

A multiplex control system has the advantage of self-diagnosis. This allows quick and easy troubleshooting and verification using diagnostic trouble codes (DTCs).

Many vehicles contain over a kilometre of wiring to supply all their electrical components. Luxury models may contain considerably more because of elaborate drivers' aids. The use of multiplexing means that considerably less wiring is used in a vehicle along with fewer multi-plugs and connectors, etc. An additional advantage of multiplexing is that existing systems can be upgraded or added to without modification to the original system.

8.2.2 Controller area network (CAN)

CAN is a serial bus system especially suited for networking 'intelligent' devices as well as sensors and actuators within a system or sub-system. It operates in a broadly similar way to a wired computer network. CAN stands for controller area network and means that control units are able to interchange data. CAN is a high-integrity serial data communications bus for real-time applications. It operates at data rates of up to 1 Mbit/s. It also has excellent error detection and confinement capabilities. It was originally developed by Bosch for use in cars but is now used in many other industrial automation and control applications.

CAN is a serial bus system with multi-master capabilities. This means that all CAN nodes are able to transmit data and several CAN nodes can request use of the bus simultaneously. In CAN networks, there is no addressing of subscribers or stations, like on a computer network, but instead, prioritised messages are transmitted. A transmitter sends a message to all CAN nodes (broadcasting). Each node decides on the basis of the identifier received whether it should process the message or not. The identifier also determines the priority that the message enjoys in competition for bus access (Figure 8.7).

KEY FACT

CAN is a serial bus system with multi-master capabilities.

Fast controller area network (F-CAN) and basic (or body) controller area network (B-CAN) share information between multiple ECUs. B-CAN communication is transmitted at a slower speed for convenience related items such as electric windows. F-CAN information moves at a faster speed for real-time functions such as fuel and emissions systems. To allow both systems to share information, a control module translates information between B-CAN and F-CAN (Figure 8.8).

KEY FACT

Fast controller area network (F-CAN) and basic (or body) controller area network (B-CAN) share information between multiple electronic control units (ECUs).

The ECUs on the B-CAN and F-CAN transmit and receive information in the form of structured messages

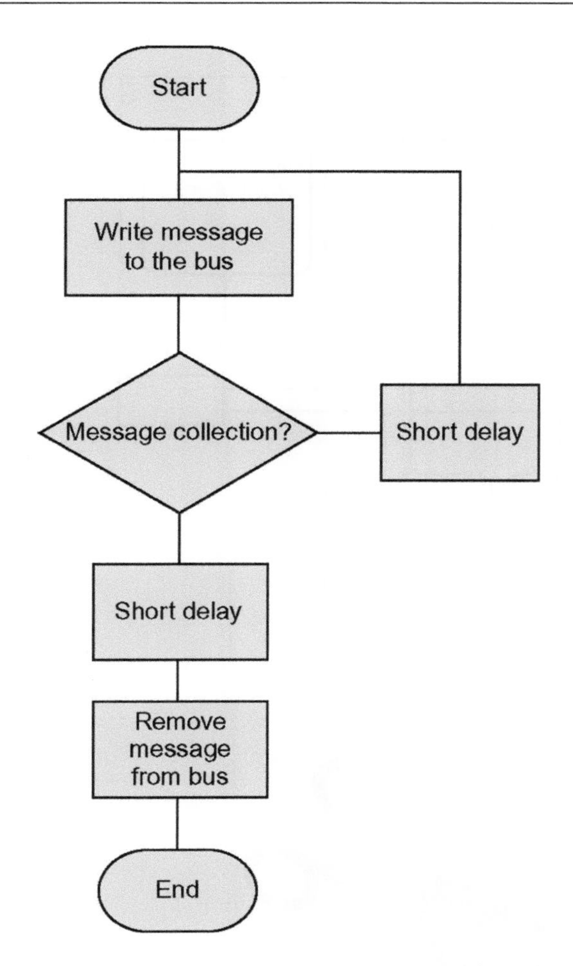

Figure 8.7 Much simplified CAN message protocol flow chart.

that may be received by several different ECUs on the network at one time. These messages are transmitted and received across a communication circuit that consists of a single wire that is shared by all the ECUs. However, as messages on the F-CAN network are typically of higher importance, a second wire is used for communication circuit integrity monitoring. This CAN-high and CAN-low circuit forms the CAN bus (Figure 8.9).

A multiplex control unit is usually combined with the under-dash fuse/relay box. It controls many of the vehicle systems related to body electrics and the B-CAN. It also carries out much of the remote switching of various hardwired and CAN-controlled systems.

One of the outstanding features of the CAN protocol is its high transmission reliability. The CAN controller registers a station's error and evaluates it statistically in order to take appropriate measures. These may extend to disconnecting the CAN node producing the errors (Figure 8.10).

Each CAN message can transmit from 0 to 8 bytes of user information. Longer messages can be sent by using segmentation, which means slicing a longer message into smaller parts. The maximum transmission rate is specified as 1 Mbit/s. This value applies to networks up to 40 m which is more than enough for normal cars and trucks.

CAN is a serial bus system designed for networking ECUs as well as sensors and actuators.

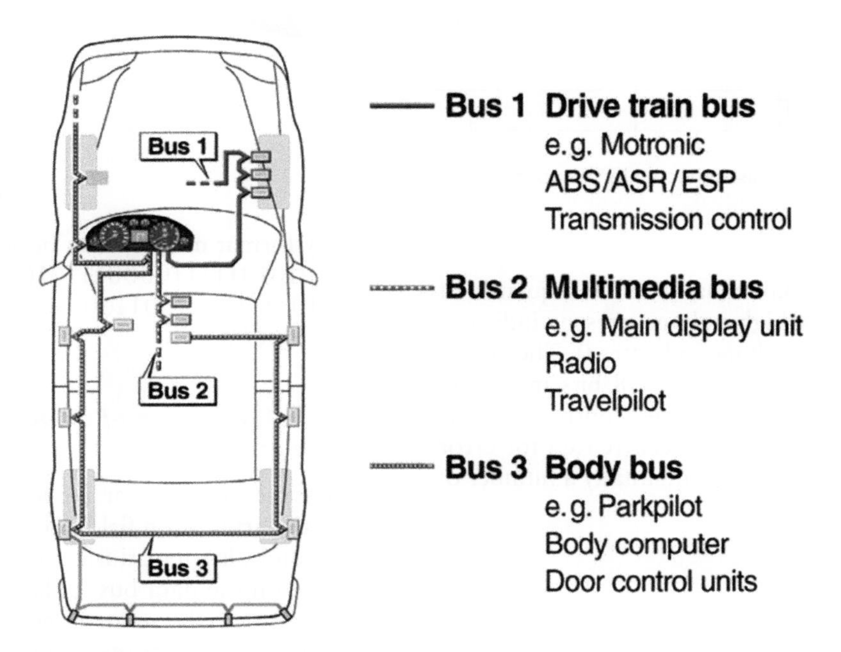

Figure 8.8 Three different speed buses in use.

Source: Bosch Media.

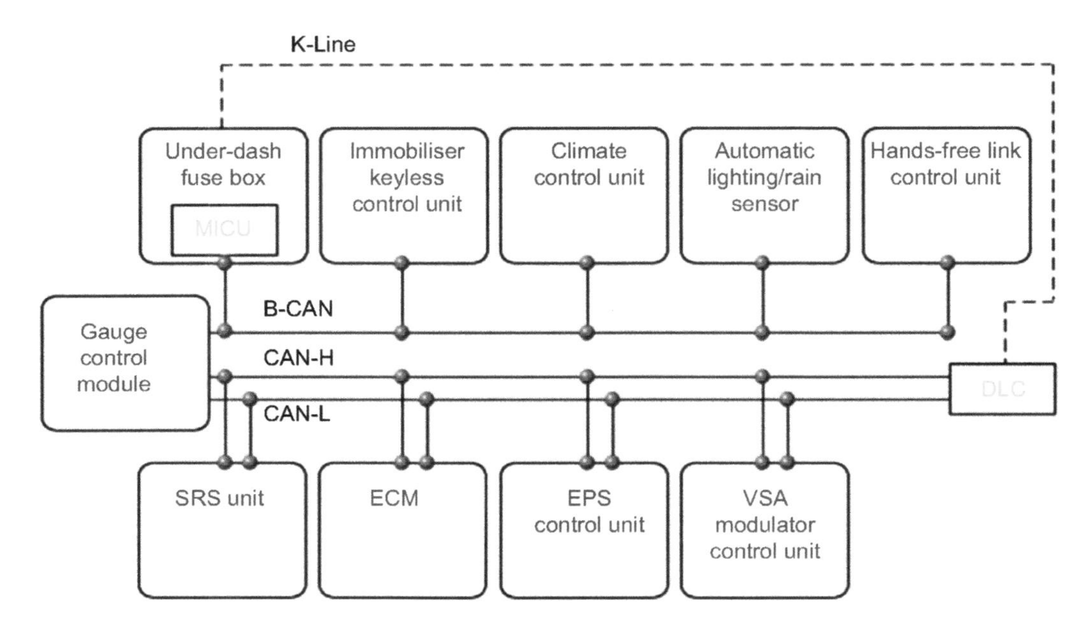

Figure 8.9 F-CAN uses CAN-H (high) and CAN-L (low) wires.

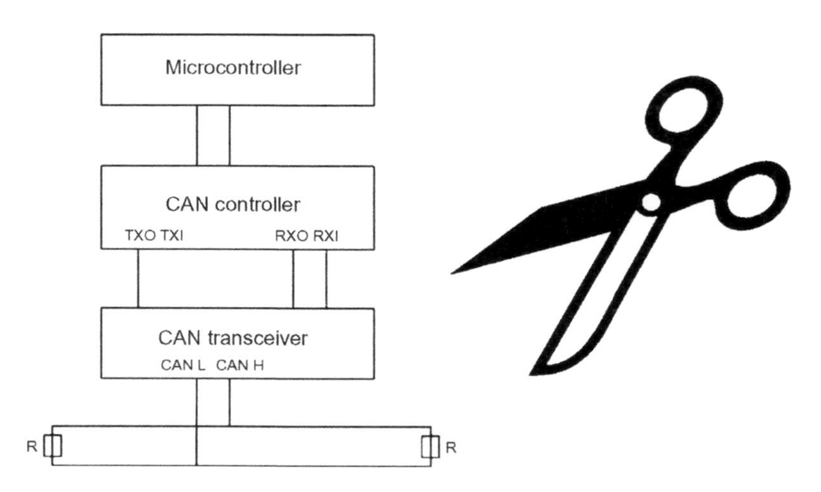

Figure 8.10 CAN nodes can be disconnected by the control program.

8.2.3 CAN data signal

The CAN message signal consists of a sequence of binary digits or bits. A high voltage present indicates the value 1, a low or no voltage indicates 0. The actual message can vary between 44 and 108 bits in length. This is made up of a start bit, name, control bits, the data itself, a cyclic redundancy check (CRC) for error detection, a confirmation signal and finally a number of stop bits (Figure 8.11).

KEY FACT

The CAN message signal consists of a sequence of binary digits or bits.

A binary format message can be something like 10001 01010001010111110000011101011110101010101000 1111101011110011001100000111111010101000011 11111111000000001

8.2.3.1 *3-452 Section of an actual electrical signal*

The message identifier or name portion of the signal (part of the arbitration field) identifies the message destination and also its priority. As the transmitter puts a message on the data bus it also reads the name back from the bus. If the name is not the same as the one it sent, then another transmitter must be in operation, which has a higher priority. If this is the case, it will stop transmission of its own message. This is very important in the case of motor vehicle data transmission.

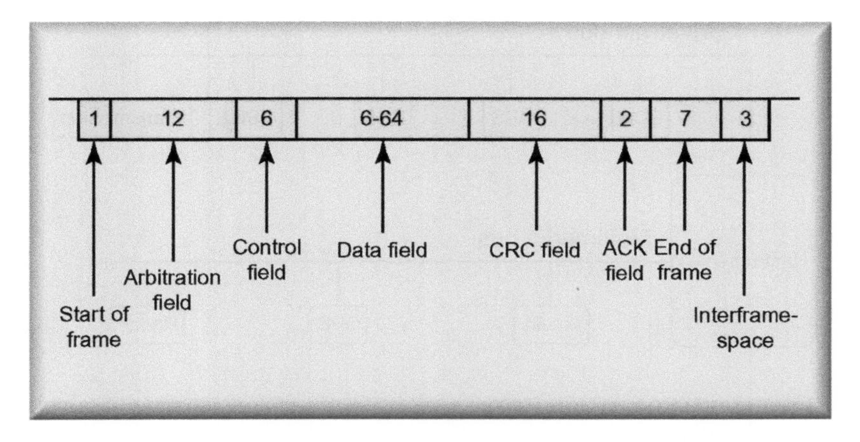

Figure 8.11 Message format (the three spaces are not part of the message).

Errors in a message are recognised by what is known as a CRC. This is an error detection scheme in which all the bits in a block of data are divided by a predetermined binary number. A check character, known to the transmitter and receiver, is determined by the remainder. If an error is recognised, the message on the bus is destroyed. This in turn is recognised by the transmitter, which then sends the message again. This technique, when combined with additional tests, makes it possible to discover all faulty messages.

The CRC field is part of the overall message. (The basic idea behind CRCs is to treat the message string as a single binary word M, and divide it by a keyword k that is known to both the transmitter and the receiver. The remainder r left after dividing M by k constitutes the 'check word' for the given message. The transmitter sends both the message string M and the check word r, and the receiver can then check the data by repeating the calculation, dividing M by the keyword k, and verifying that the remainder is r.)

Because each node in effect monitors its own output, interrupts disturbed transmissions and acknowledges correct transmissions, faulty stations can be recognised and uncoupled (electronically) from the bus. This prevents other transmissions from being disturbed.

DEFINITION

The cyclic redundancy check (CRC) field is part of the overall message. (The basic idea behind CRCs is to treat the message string as a single binary word M, and divide it by a keyword k that is known to both the transmitter and the receiver. The remainder r left after dividing M by k constitutes the 'check word' for the given message. The transmitter sends both the message string M and the check word r, and the receiver can then check the data by repeating the calculation, dividing M by the keyword k, and verifying that the remainder is r.)

8.2.4 Local interconnect network (LIN)

A local interconnect network (LIN) is a serial bus system especially suited for networking 'intelligent' devices, sensors and actuators within a sub-system. It is a concept for low-cost automotive networks, which complements existing automotive multiplex networks such as CAN.

KEY FACT

LIN is a concept for low-cost automotive networks, which complements existing automotive multiplex networks such as CAN.

LIN enables the implementation of a hierarchical vehicle network. This allows further quality enhancement and cost reduction of vehicles (Figure 8.12).

The LIN standard includes the specification of the transmission protocol, the transmission medium, the interface between development tools, and the interfaces for software programming. LIN guarantees the interoperability of network nodes from the viewpoint of hardware and software, and predictable electromagnetic compatibility (EMC) behaviour (Figure 8.13).

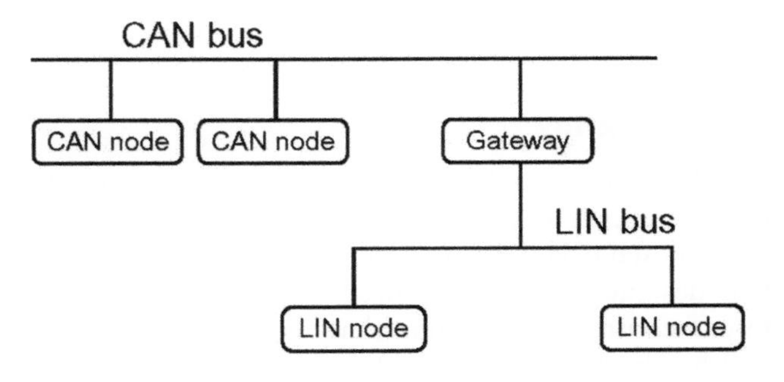

Figure 8.12 Structure using CAN and LIN.

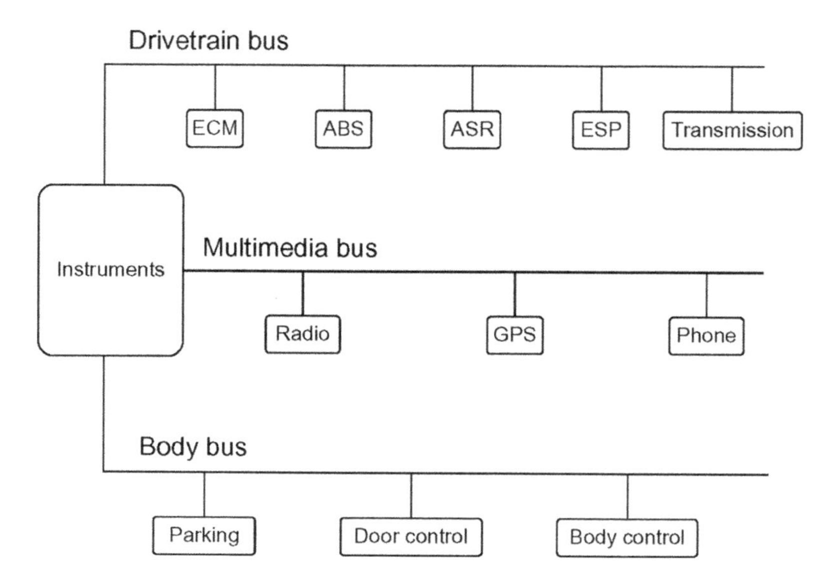

Figure 8.13 Standards allow communication between different systems.

DEFINITION

EMC: Electromagnetic compatibility. EMC requirements stipulate that a device shall not cause interference within itself or in other devices, or be susceptible to interference from other devices.

LIN is a time-triggered single-master, multiple-slave network concept. It is based on common interface hardware, which makes it a low-cost solution. Additional attributes of LIN are

- multicast reception with self-synchronisation;
- selectable length of message frames;
- data checksum security and error detection;
- single-wire implementation;
- speed up to 20 kbit/s.

LIN provides a cost-efficient bus communication where the bandwidth and versatility of CAN are not required. It is used for non-critical systems.

8.2.5 FlexRay

FlexRay is a fast and fault-tolerant bus system for automotive use. It was developed using the experience of well-known original equipment manufactures (OEMs). It is designed to meet the needs of current and future in-car control applications that require a high bandwidth. The bit rate for FlexRay can be programmed to values up to 10 Mbit/s (Figure 8.14).

Figure 8.14 FlexRay logo.

KEY FACT

FlexRay can cope with the requirements of X-by-wire systems.

The data exchange between the control devices, sensors and actuators in automobiles is mainly carried out via CAN systems. However, the introduction of X-by-wire systems has resulted in increased requirements. This is especially so with regard to error tolerance and speed of message transmission. FlexRay meets these requirements by message transmission in fixed time slots and by fault-tolerant and redundant message transmission on two channels (Figure 8.15).

The physical layer means the hardware, that is, the actual components and wires. FlexRay works on the principle of time division multiple access (TDMA). This means that components or messages have fixed time

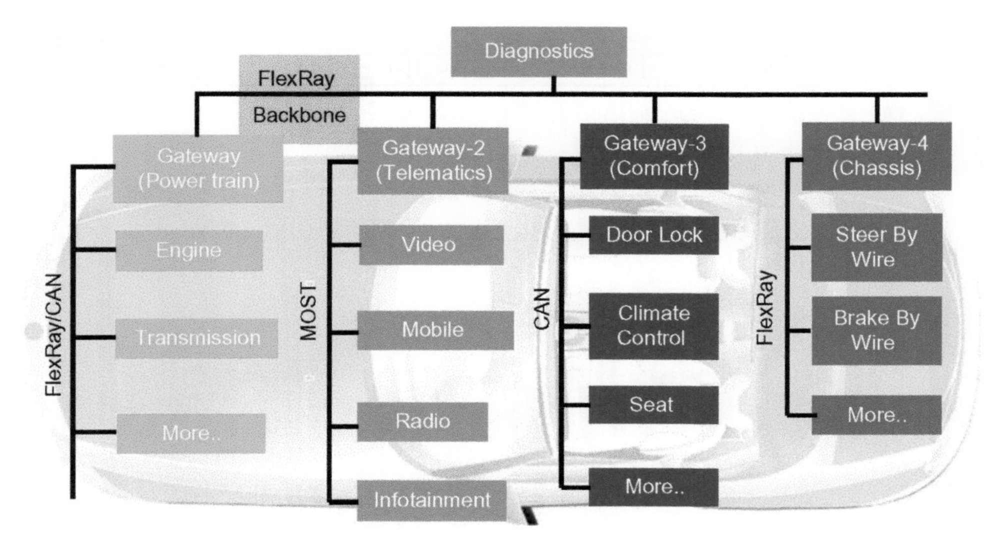

Figure 8.15 FlexRay backbone.

slots in which they have exclusive access to the data bus. These time slots are repeated in a cycle and are just a few milliseconds long.

FlexRay communicates via two physically separated lines with a data rate of up to 10 Mbit/s on each. The two lines are mainly used for redundant and therefore fault-tolerant message transmission, but they can also transmit different messages.

KEY FACT

FlexRay communicates via two physically separated lines with a data rate of up to 10 Mbit/s on each.

FlexRay is a fast and fault-tolerant bus system that was developed to meet the needs of high bandwidth applications such as X-by-wire systems. Error tolerance and speed of message transmission in these systems is essential (Figure 8.16).

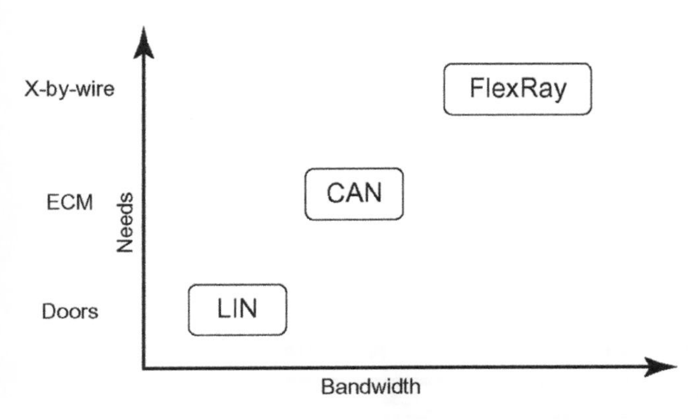

Figure 8.16 Comparing requirements and data rates of the three systems.

8.3 DIAGNOSTICS – MULTIPLEXING

The integrity of the signal on the CAN can be checked in two ways. The first way is to examine the signal on a dual-channel scope connected to the CAN-high and CAN-low lines (Figure 8.17).

In this display, it is possible to verify that:

- data is being continuously exchanged along the CAN bus;
- the voltage levels are correct;
- a signal is present on both CAN lines.

CAN uses a differential signal, so the signal on one line should be a coincident mirror image of the data on the other line. The usual reasons for examining the CAN signals is where a CAN fault has been indicated by OBD, or to check the CAN connection to a suspected faulty CAN node. Manufacturers' data should be referred to for precise waveform parameters.

KEY FACT

CAN uses a differential signal, so the signal on one line should be a coincident mirror image of the data on the other line.

The CAN data in Figure 8.18 is captured on a much faster timebase and allows the individual state changes to be examined. This enables the mirror image nature of the signals and the coincidence of the edges to be verified.

The signals are equal and opposite and they are of the same amplitude (voltage). The edges are clean and coincident with each other. This shows that the vehicle data

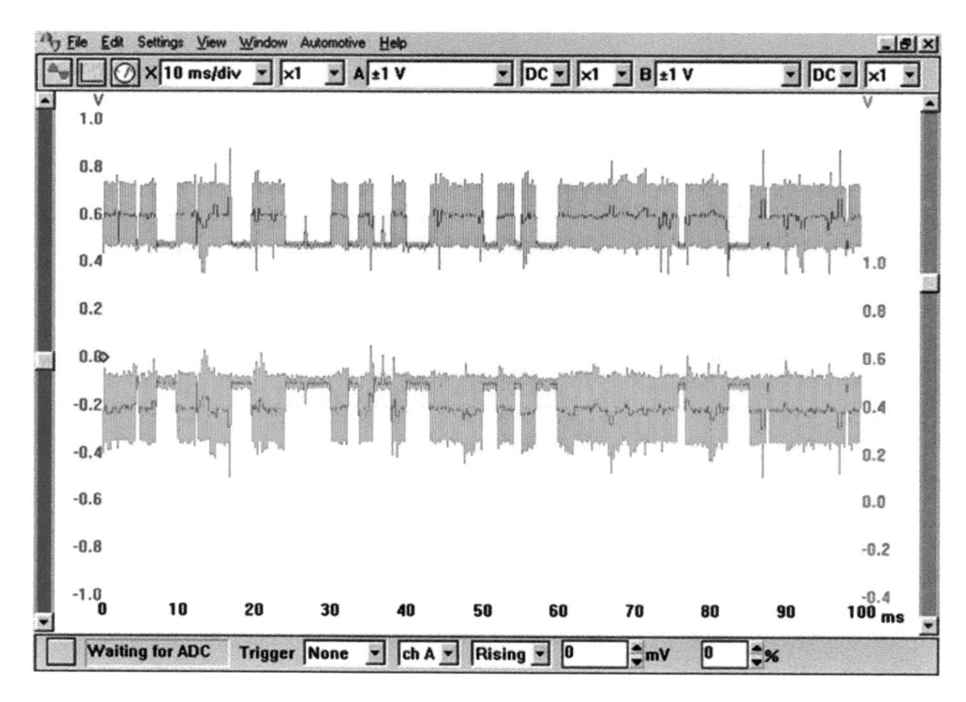

Figure 8.17 CAN signals.

bus (CAN bus) is enabling communication between the nodes and the CAN controller unit. This test effectively verifies the integrity of the bus at this point in the network. If a particular node is not responding correctly, the fault is likely to be the node itself. The rest of the bus should work correctly.

It is usually recommended to check the condition of the signals present at the connector of each of the ECUs on the network. The data at each node will always be the same on the same bus. Remember that much of the data on the bus is safety critical, so do *not* use insulation piercing probes!

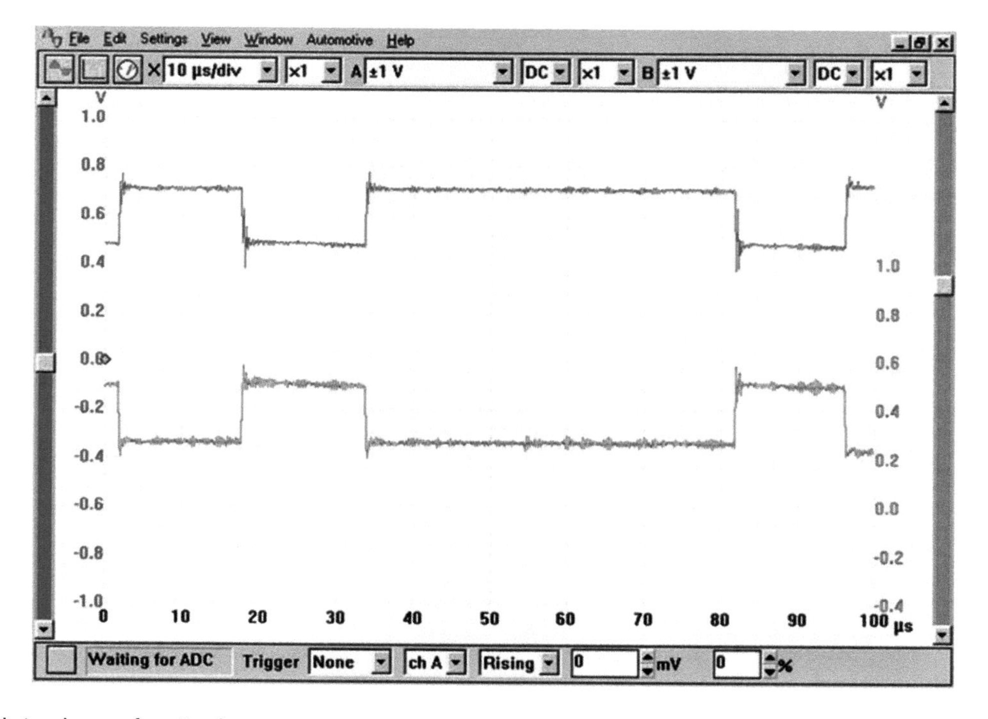

Figure 8.18 CAN signals on a fast timebase.

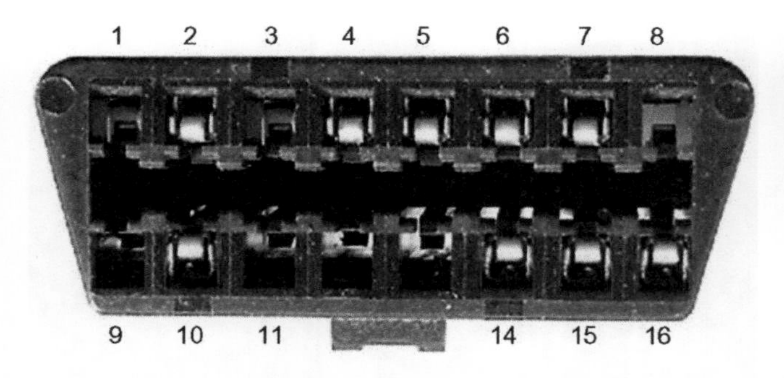

Figure 8.19 16 pin data link connector.

PicoTech have produced the CAN test box. This gives easy access to the 16 pins of the diagnostic connector that is fitted to all modern vehicles. Depending on the configuration of the vehicle, this may allow you to check power, ground and CAN bus signal quality (Figure 8.19).

Connector design and location is dictated by an industry wide (OBD2) standard. Vehicle manufacturers can use the empty DLC terminals for whatever they would like. However, the DLC of every vehicle is required to provide pins 4 and 5 and 16 as defined below. Further, after the CAN protocol was fully implemented in the 2008 model year, all vehicles must use pins 6 and 14 as defined below:

- Terminal 2 – SAE J1850 10.4 kbits/s (kbps) variable pulse width serial data (GM Class-2) or SAE J1850 41.6 kbps pulse width modulation serial data high line (Ford)
- Terminal 4 – Scan tool chassis ground
- Terminal 5 – Common signal ground for serial data lines (Logic Low)
- Terminal 6 – ISO 11898/15765/SAE J2284 CAN serial data high line
- Terminal 7 – ISO 9141 K serial data line or ISO 14230 (Keyword 2000) serial data line (DaimlerChrysler/Honda/Toyota)
- Terminal 10 – SAE J1850 41.6 kbps pulse width modulation serial data low line (Ford)
- Terminal 14 – ISO 11898/15765/SAE J2284 CAN serial data low line
- Terminal 15 – ISO 9141 L serial data line or ISO 14230 (Keyword 2000) serial data line (DaimlerChrysler/Honda)
- Terminal 16 – Scan tool power (unswitched battery positive voltage)

With the test leads supplied, a PicoScope automotive scope, or any other suitable scope, may be connected to the CAN test box. This allows the monitoring of any signals present, such as CAN high and CAN low.

The CAN test box has a 2.5 m cable so that work can be carried out at a convenient location away from the diagnostic connector. An additional pass-through connector allows a scan tool to be connected at the same time as a scope. Its 4 mm sockets are backlit by LEDs to show the state of each pin on the connector. The CAN test box is powered by the diagnostic connector, so batteries or a mains adaptor are not needed (Figure 8.20).

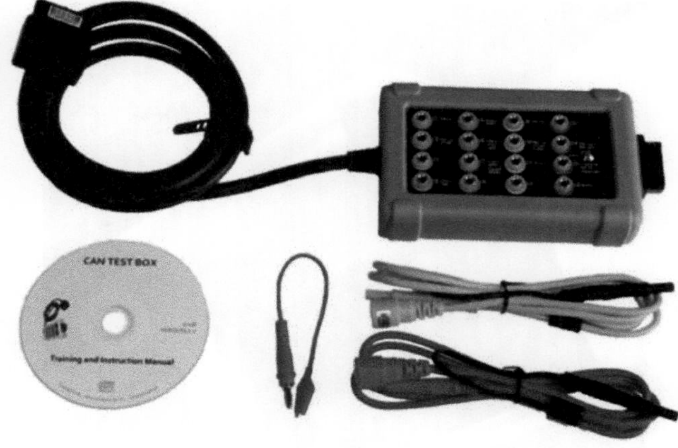

Figure 8.20 CAN test box.

Source: PicoTech.

In the two CAN scope patterns shown previously, the second is on a timebase 1000 times faster than the first so that more details of the signal are shown. The connection for one of the traces is to pin 6 and the other to pin 14.

The second way of checking the CAN signals is to use a suitable reader or scanner.

The KTS 200 controller diagnostic tester from Bosch is a good example of an OBD/CAN reader as it offers a wide range of features (Figure 8.21). It reads diagnostic codes and CAN data. The device can be used both as a full controller diagnostic tester, complete with a testing scope, and for straightforward servicing work on vehicles.

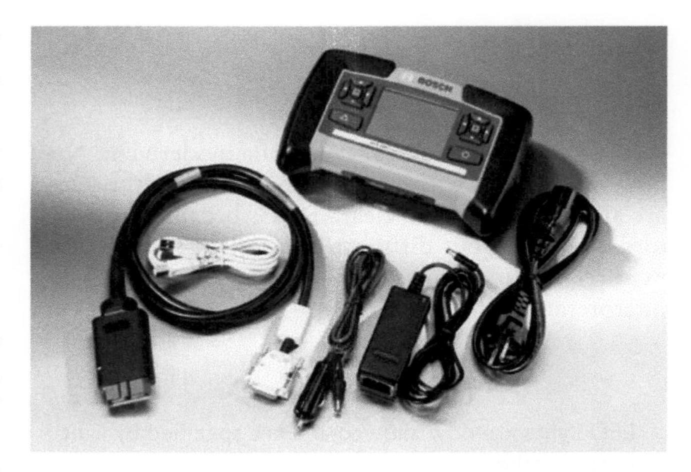

Figure 8.21 KTS 200 kit.

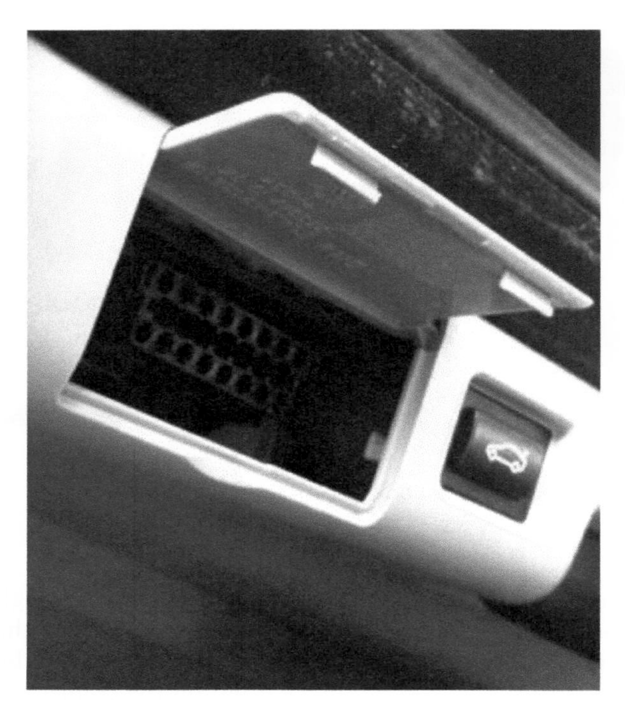

Figure 8.22 OBD connector on a BMW.

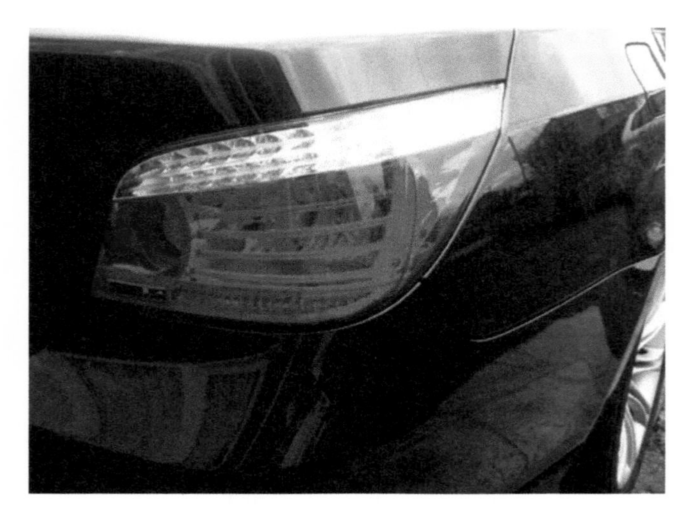

Figure 8.23 BMW rear lights.

It is powered via the diagnostic cable, the cigarette lighter cable or a power pack.

OBD has been in use for some time in its different formats. However, the CAN protocol is a popular standard and is making significant in roads into the market. Since 2008, all vehicles sold in the European Union and United States are required to have implemented CAN. This should finally eliminate the ambiguity of the several existing signalling protocols (Figure 8.22).

KEY FACT

Since 2008, all vehicles sold in the European Union and United States are required to have implemented CAN.

8.4 LIGHTING

8.4.1 External lights

Figure 8.23 shows the rear lights of a modern car. Note how in common with many manufacturers, the lenses are almost smooth and clear. This is because the reflectors now carry out diffusion of the light. Regulations exist relating to external lights. Table 8.2 is a simplified interpretation of current rules.

KEY FACT

LED lights are now allowed and are specified by light output rather than wattage.

8.4.2 Lighting circuits

Figure 8.24 shows a simplified lighting circuit. While this representation helps to demonstrate the way in which a lighting circuit operates, it is not now used in this simple form. The circuit does, however, help to show in a simple way how various lights in and around the vehicle operate with respect to each other.

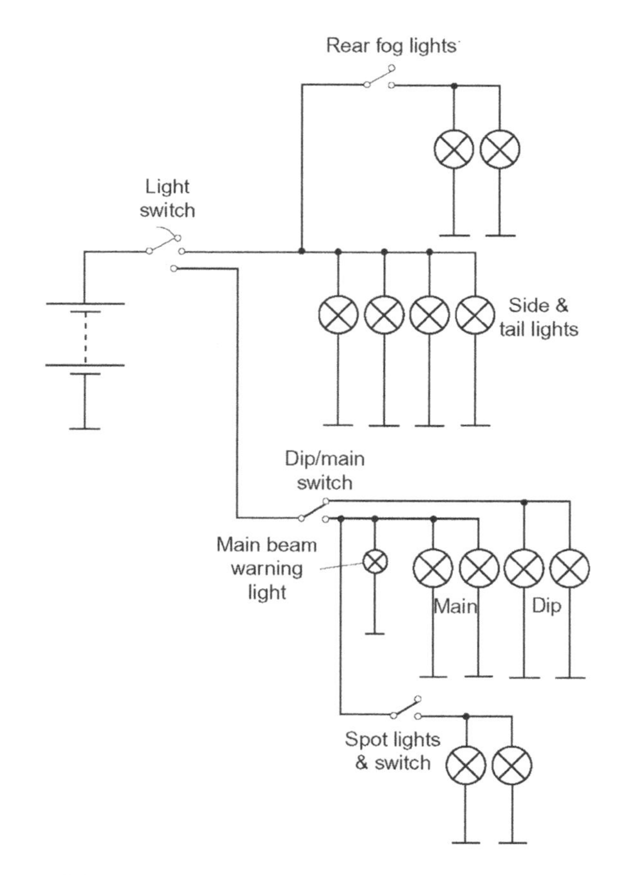

Figure 8.24 Simplified lighting circuit.

Table 8.2 Lighting features

Sidelights	A vehicle must have two sidelights each with wattage of less than 7 W. Most vehicles have the sidelight incorporated as part of the headlight assembly.
Rear lights	Again two must be fitted each with wattage not less than 5 W. Lights used in Europe must be 'E' marked and show a diffused light. Position must be within 400 mm from the vehicle edge and over 500 mm apart and between 350 and 1500 mm above the ground.
Brake lights	Two lights often combined with the rear lights. They must be between 15 and 36 W each, with diffused light, and must operate when any form of first-line brake is applied. Brake lights must be between 350 and 1500 mm above the ground and at least 500 mm apart in a symmetrical position. High-level brake lights are now allowed, and if fitted, must operate with the primary brake lights.
Reverse lights	No more than two lights may be fitted with a maximum wattage each of 24 W. The light must not dazzle and either be switched automatically from the gearbox or with a switch incorporating a warning light. Safety reversing 'beepers' are now often fitted in conjunction with this circuit, particularly on larger vehicles.
Day running lights	Volvo use day running lights as these are in fact required in Sweden and Finland. These lights come on with the ignition and must only work in conjunction with the rear lights. Their function is to indicate that the vehicle is moving or about to move. They switch off when parking or headlights are selected.
Rear fog lights	One or two may be fitted, but if only one, it must be on the offside or centre line of the vehicle. They must be between 250 and 1000 mm above the ground and over 100 mm from any brake light. The wattage is normally 21 W and they must only operate when either the side lights, headlights or front fog lights are in use.
Front spot and fog lights	If front spot lights are fitted (auxiliary driving lights), they must be between 500 and 1200 mm above the ground and more than 400 mm from the side of the vehicle. If the lights are non-dipping, then they must only operate when the headlights are on main beam. Front fog lamps are fitted below 500 mm from the ground and may only be used in fog or falling snow. Spot lamps are designed to produce a long beam of light to illuminate the road in the distance. Fog lights are designed to produce a sharp cut-off line such as to illuminate the road just in front of the vehicle but without reflecting back or causing glare.

For example, fog lights can be wired to work only when the side lights are on. Another example is how the headlights cannot be operated without the side lights first being switched on.

Dim dip headlights were an attempt to stop drivers just using side lights in semi-dark or poor visibility conditions. The circuit is such that when side lights and ignition are on together, then the headlights will come on automatically at about one-sixth of normal power.

Dim dip lights are achieved in one of two ways. The first uses a simple resistor in series with the headlight bulb and the second is to use a 'chopper' module which switches the power to the headlights on and off rapidly. In either case, the 'dimmer' is bypassed when the driver selects normal headlights. The most cost-effective method is using a resistor, but this has the problem that the resistor (approximately 1 Ω) gets quite hot and hence has to be positioned appropriately. Figure 8.25 shows a typical vehicle lighting circuit.

SAFETY FIRST

Note: If there is any doubt as to the visibility or conditions, switch on dipped headlights. If your vehicle is in good order, it will not discharge the battery.

8.4.3 Gas discharge lighting

Xenon gas discharge headlamps (GDL) are now fitted to some vehicles. They have the potential to provide more effective illumination and new design possibilities for the front of a vehicle. The conflict between aerodynamic styling and suitable lighting positions is an economy/safety trade-off, which is undesirable. The new headlamps make a significant contribution towards improving this situation because they can be relatively small. The GDL system consists of three main components:

- Bulb – this operates in a very different way from conventional incandescent bulbs. A much higher voltage is needed.
- Ballast system – this contains an ignition and control unit and converts the electrical system voltage into the operating voltage required by the lamp. It controls the ignition stage and run up as well as regulating during continuous use and finally monitors operation as a safety aspect.
- Headlamp – the design of the headlamp is broadly similar to conventional units. However, in order to meet the limits set for dazzle, a more accurate finish is needed and hence more production costs are involved.

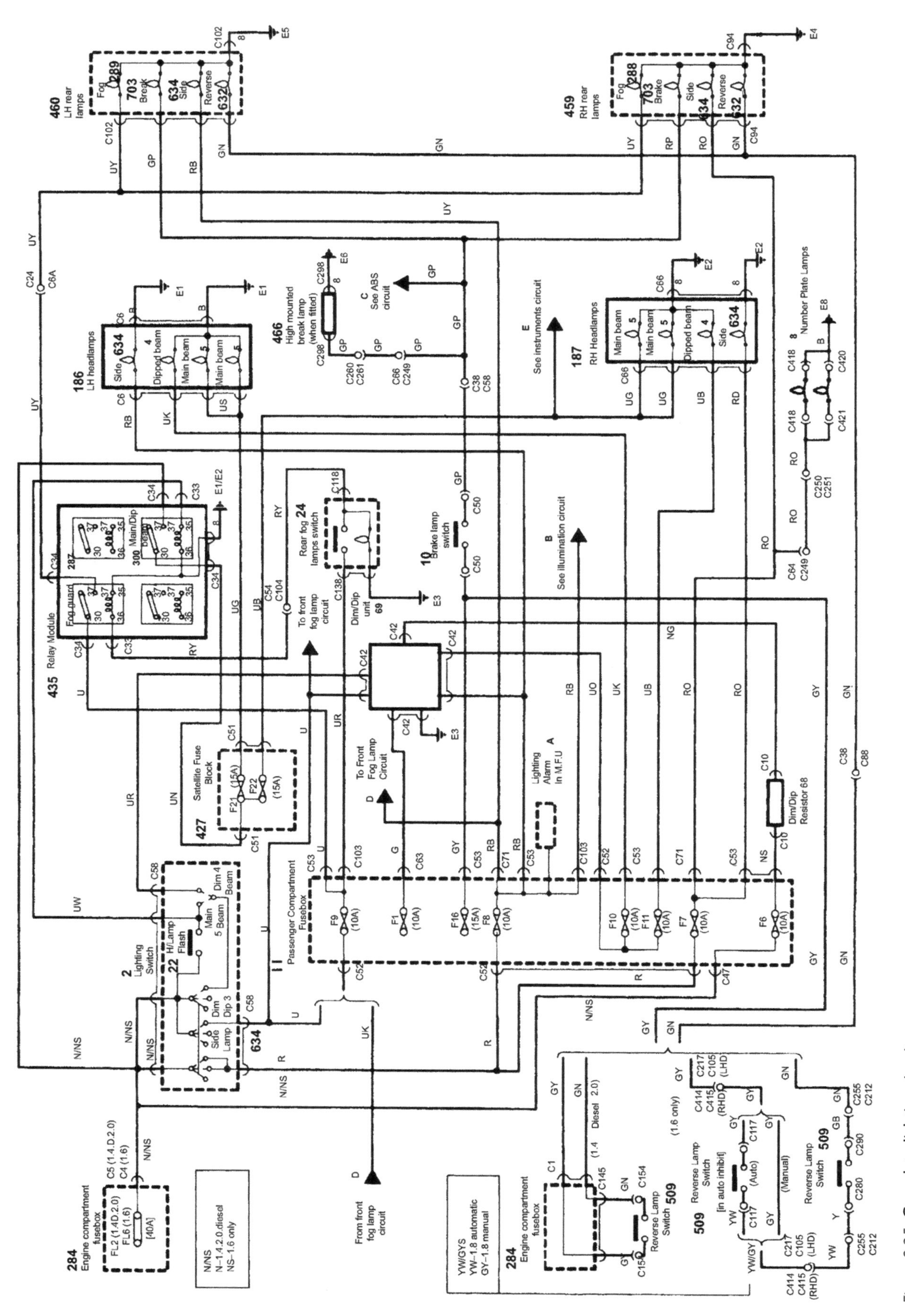

Figure 8.25 Complete lighting circuit.

Figure 8.26 Xenon lighting.

8.4.4 LED lighting

The advantages of LED lighting are clear, the greatest being reliability. LEDs have a typical rated life of over 50 000 hours compared to just a few thousand for incandescent lamps. The environment in which vehicle lights have to survive is hostile to say the least. Extreme variations in temperature and humidity as well as serious shocks and vibration have to be endured.

LEDs are more expensive than bulbs but the potential savings in design costs due to sealed units being used and greater freedom of design could outweigh the extra expense. A further advantage is that they turn on quicker than ordinary bulbs. This time is approximately the difference between 130 ms for the LEDs and 200 ms for bulbs. If this is related to a vehicle brake light at motorway speeds, then the increased reaction time equates to about a car length. This is potentially a major contribution to road safety. LEDs as high-level brake lights are becoming popular because of the shock resistance, which will allow them to be mounted on the boot lid (Figures 8.26 and 8.27).

Figure 8.27 Adaptive lighting using LEDs.

KEY FACT

LEDs have a typical rated life of over 50000 hours compared to just a few thousand for conventional lamps, which must make them incandescent.

Heavy vehicle side marker lights are an area of use where LEDs have proved popular. Many lighting manufacturers are already producing lights for the aftermarket. Being able to use sealed units will greatly increase the life expectancy. Side indicator repeaters are a similar issue due to the harsh environmental conditions.

8.5 DIAGNOSTICS – LIGHTING

8.5.1 Testing procedure

The process of checking a lighting system circuit is broadly presented in Figure 8.28.

Figure 8.29 shows a simplified dim dip lighting circuit with meters connected for testing. A simple principle to keep in mind is that the circuit should be able to supply all the available battery voltage to the consumers (bulbs, etc.). A loss of 5% may be acceptable.

KEY FACT

A circuit should be able to supply a minimum of 95% of the available battery voltage to the consumers (bulbs, etc.).

With all the switches in the 'on' position appropriate to where the meters are connected, the following readings should be obtained:

- V1 12.6 V (if less, check battery condition);
- V2 0–0.2 V (if more, the ignition switch contacts have a high resistance);
- V3 0–0.2 V (if more, the dim dip relay contacts have a high resistance);
- V4 0–0.2 V (if more, there is a high resistance in the circuit between the output of the light switch and the junction for the tail lights);
- V7 12–12.6 V if on normal dip or approximately 6 V if on dim dip (if less, then there is a high resistance in the circuit – check other readings, etc., to narrow down the fault).

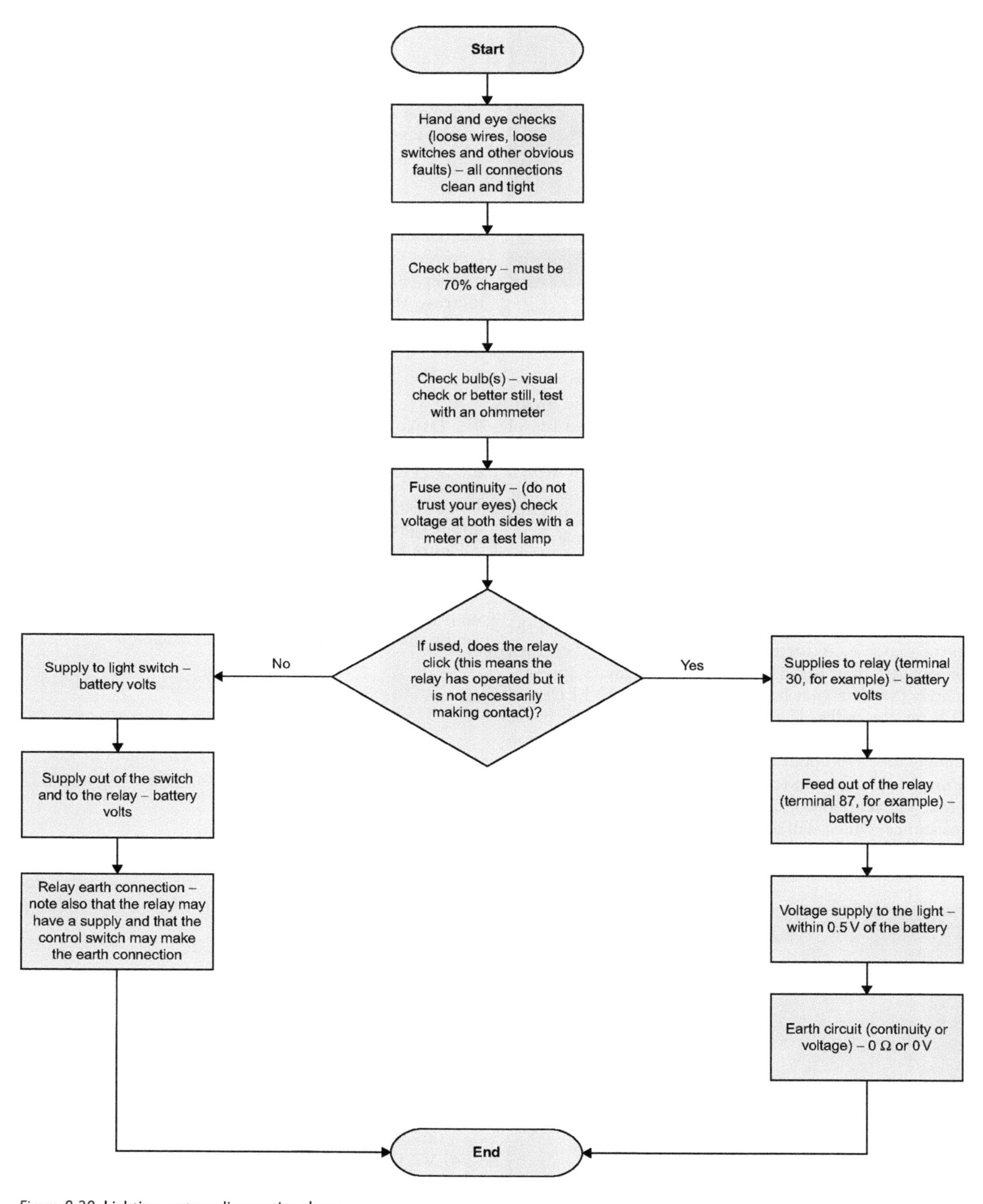

Figure 8.28 Lighting system diagnostics chart.

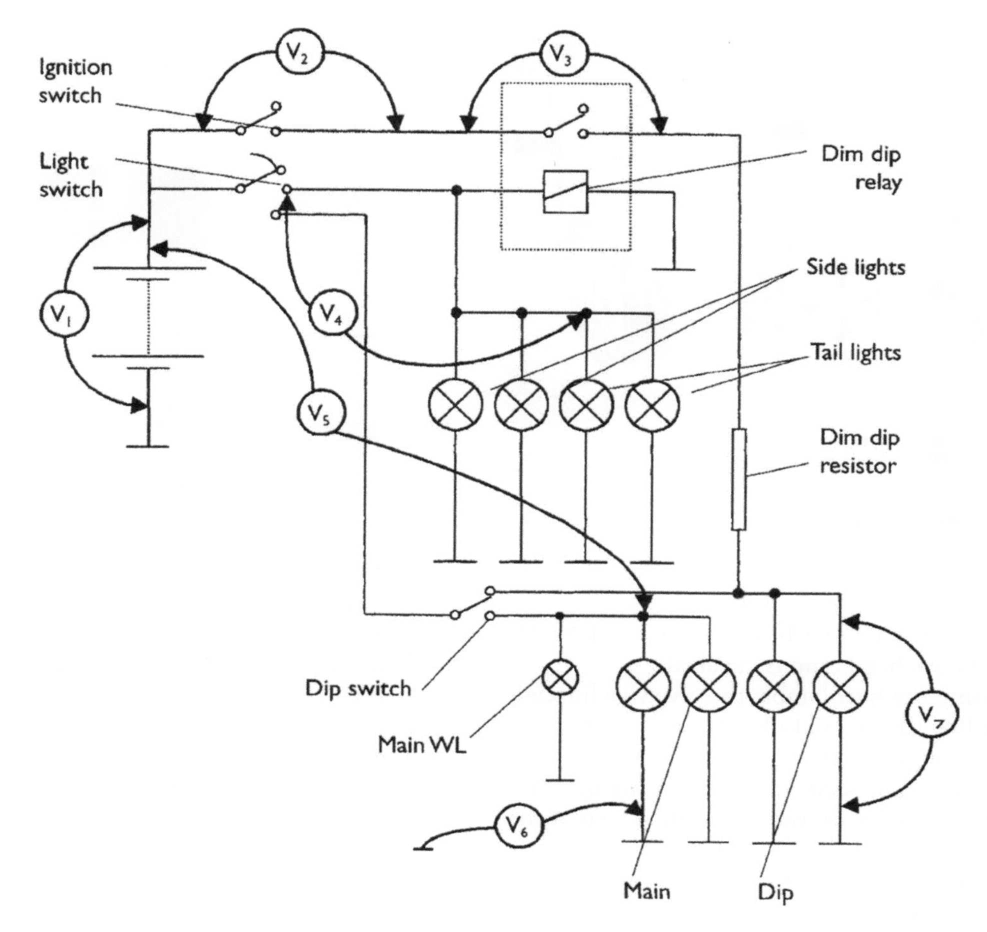

Figure 8.29 Lighting circuit under test.

8.5.2 Lighting fault diagnosis table

Symptom	*Possible fault*
Lights dim	High resistance in the circuit Low alternator output Discoloured lenses or reflectors
Headlights out of adjustment	Suspension fault Loose fittings Damage to body panels Adjustment incorrect
Lights do not work	Bulbs blown Fuse blown Loose or broken wiring/ connections/fuse Relay not working Corrosion in light units Switch not making contact

8.5.3 Headlight beam setting

Many types of beam-setting equipment are available and most work on the same principle. The method is the same as using an aiming board but is more convenient and accurate due to easier working and because less room is required.

Move the beam setter into position in front of the headlamp to be checked, and align the beam setter box with the middle of the headlamp (Figures 8.30 and 8.31). It must not be more than 3 cm out of line horizontally or vertically. The distance between the front edge of beam setter box and the headlamp should be

Figure 8.30 Headlights.

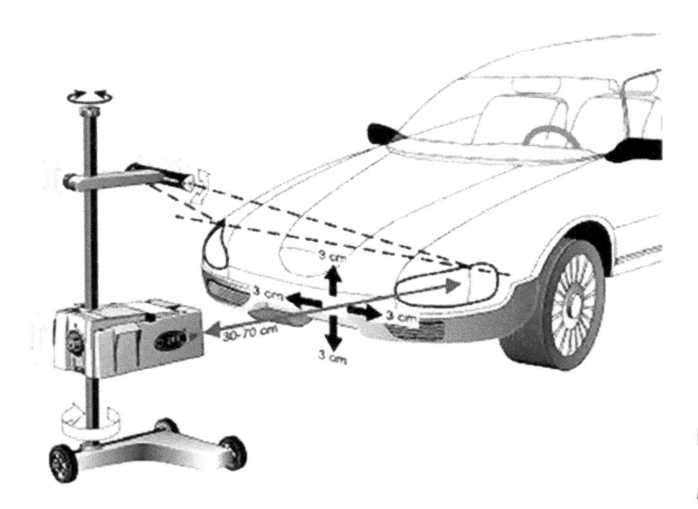

Figure 8.31 Headlamp alignment.

Source: Hella.

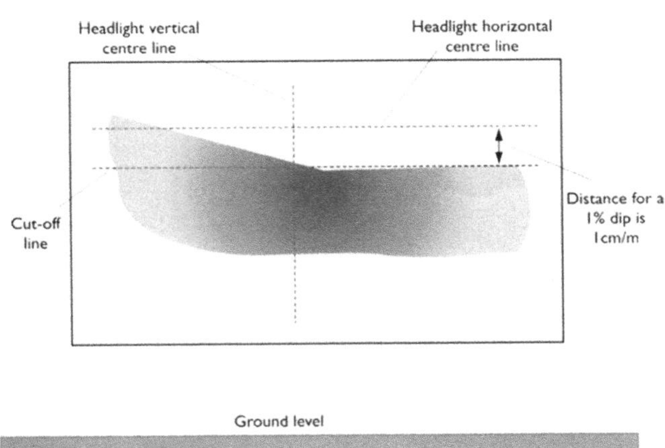

Figure 8.32 Asymmetric dip beam pattern.

between 30 and 70 cm. The beam setter must be re-adjusted before each headlamp is checked.

When adjusting the headlamps, the given inclination for the cut-off line (from a data book, etc.) must be set on the alignment equipment.

The beam is now adjusted until the cut-off line and break off are in the correct position on the screen of the aligner.

To set the headlights of a car using an aiming board, the following procedure should be adopted.

1. Park the car on level ground square onto a vertical aiming board at a distance of 10 m, if possible. The car should be unladen except for the driver.
2. Mark out the aiming board as shown in Figure 8.32.
3. Bounce the suspension to ensure it is level.
4. With the lights set on dip beam, adjust the cut-off line to the horizontal mark, which will in most cases be 1 cm below the height of the headlight centre for every 1 m the car is away from the board. The break-off point should be adjusted to the centre line of each light in turn.

Note: If the required dip is 1% then 1 cm per 1 m. If 1.2% is required, then 1.2 cm per 1 m, etc. Always check data for actual settings.

In the UK, the required alignment is part of the MOT test. The requirements are summarised in Figure 8.33.

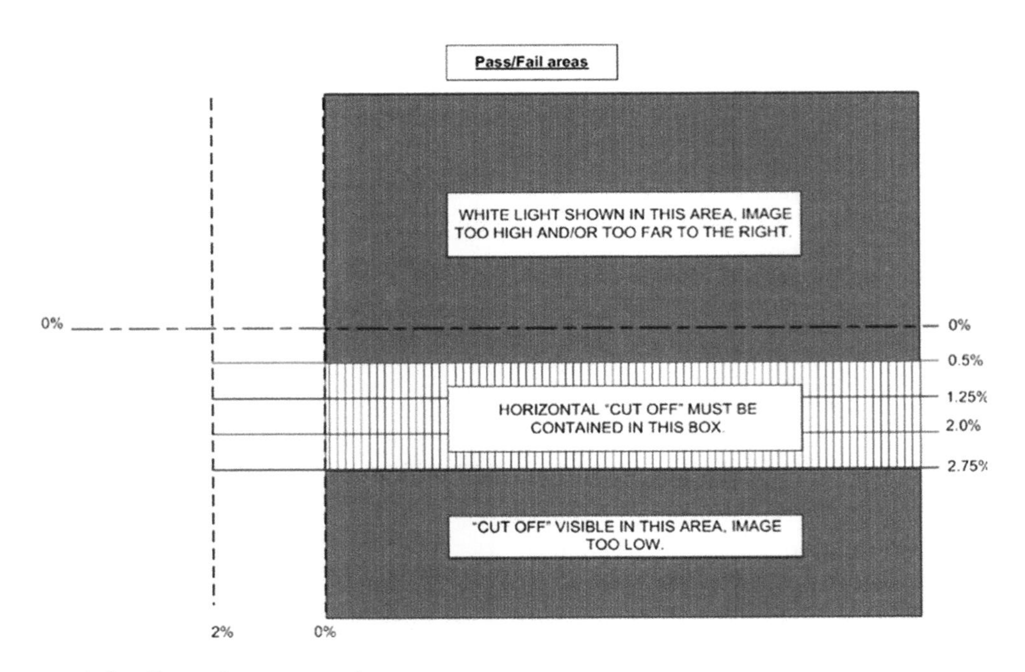

Figure 8.33 Asymmetric headlamp alignment requirements.

8.6 AUXILIARIES

8.6.1 Wiper motors and linkages

Most wiper linkages consist of series or parallel mechanisms. Some older types use a flexible rack and wheel boxes similar to the operating mechanism of many sunroofs. One of the main considerations for the design of a wiper linkage is the point at which the blades must reverse, because of the high forces on the motor and linkage at this time. If the reverse point is set so that the linkage is at its maximum force transmission angle, then the reverse action of the blades puts less strain on the system. This also ensures smoother operation (Figure 8.34).

KEY FACT

Most wiper linkages consist of series or parallel mechanisms.

Most if not all wiper motors now in use are permanent magnet motors. The drive is taken via a worm gear to increase torque and reduce speed. Three brushes may be used to allow two-speed operation. The normal speed operates through two brushes placed in the usual positions opposite to each other. For a fast speed, the third brush is placed closer to the earth brush. This reduces the number of armature windings between them, which reduces resistance, hence increasing current and therefore speed. Typical specifications for wiper motor speed and hence wipe frequency are 45 rpm at normal speed and 65 rpm at fast speed. The motor must be able to overcome the starting friction of each blade at a minimum speed of 5 rpm (Figure 8.35).

The wiper motor or the associated circuit often has some kind of short circuit protection (Figure 8.36). This is to protect the motor in the event of stalling, if frozen to the screen, for example. A thermal trip of some type is often used or a current sensing circuit in the wiper ECU if fitted.

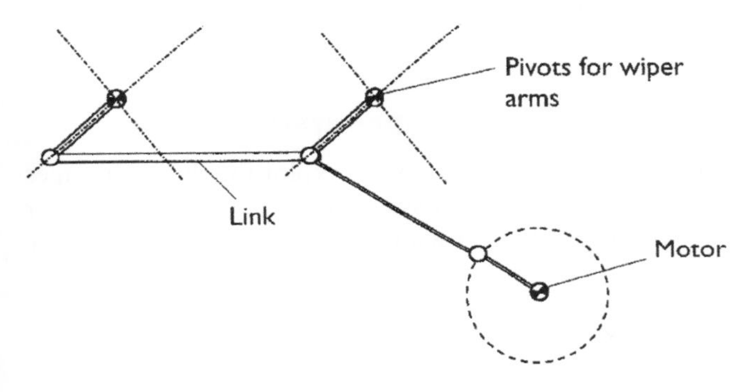

Figure 8.34 Wiper linkage.

Figure 8.36 Wiper motor.

Figure 8.35 Wiper motor using three brushes for two-speed operation.

The windscreen washer system usually consists of a simple DC permanent magnet motor driving a centrifugal water pump. The water, preferably with a cleaning additive, is directed onto an appropriate part of the screen by two or more jets. A non-return valve is often fitted in the line to the jets to prevent water siphoning back to the reservoir. This also allows 'instant' operation when the washer button is pressed. The washer circuit is normally linked in to the wiper circuit such that when the washers are operated the wipers start automatically and will continue for several more sweeps after the washers have stopped.

8.6.2 Wiper circuits

Figure 8.37 shows a circuit for fast, slow and intermittent wiper control. The switches are shown in the off position and the motor is stopped and is in its park position. Note that the two main brushes of the motor are connected together via the limit switch, delay unit contacts and the wiper switch. This causes regenerative braking because of the current generated by the motor due to its momentum after the power is switched off. Being connected to a very low resistance loads up the 'generator' and it stops instantly when the park limit switch closes.

When either the delay contacts or the main switch contacts are operated, the motor will run at slow speed. When fast speed is selected, the third brush on the motor is used. On switching off, the motor will continue to run until the park limit switch changes over to the position shown. This switch is only in the position shown when the blades are in the parked position.

Many vehicles use a system with more enhanced facilities. This is regulated by what may be known as a central control unit (CCU), a multifunction unit (MFU) or a general electronic module (GEM). These units often control other systems as well as the wipers, thus allowing reduced wiring bulk under the dash area. Electric windows, headlights and heated rear window, to name just a few, are now often controlled by a central unit (Figure 8.38).

KEY FACT

A central control unit (CCU), a multi-function unit (MFU) or a general electronic module (GEM) is now often used to control a range of auxiliary components.

Using electronic control, a CCU allows the following facilities for the wipers:

- front and rear wash/wipe;
- intermittent wipe;
- time delay set by the driver;
- reverse gear selection rear wipe operation;
- rear wash/wipe with 'dribble wipe' (an extra wipe several seconds after washing);
- stall protection.

8.6.3 Two-motor wiper system

More and more carmakers are exploiting the advantages of the two-motor wiper systems where each of the wiper arms is driven by its own electric motor. The advantage

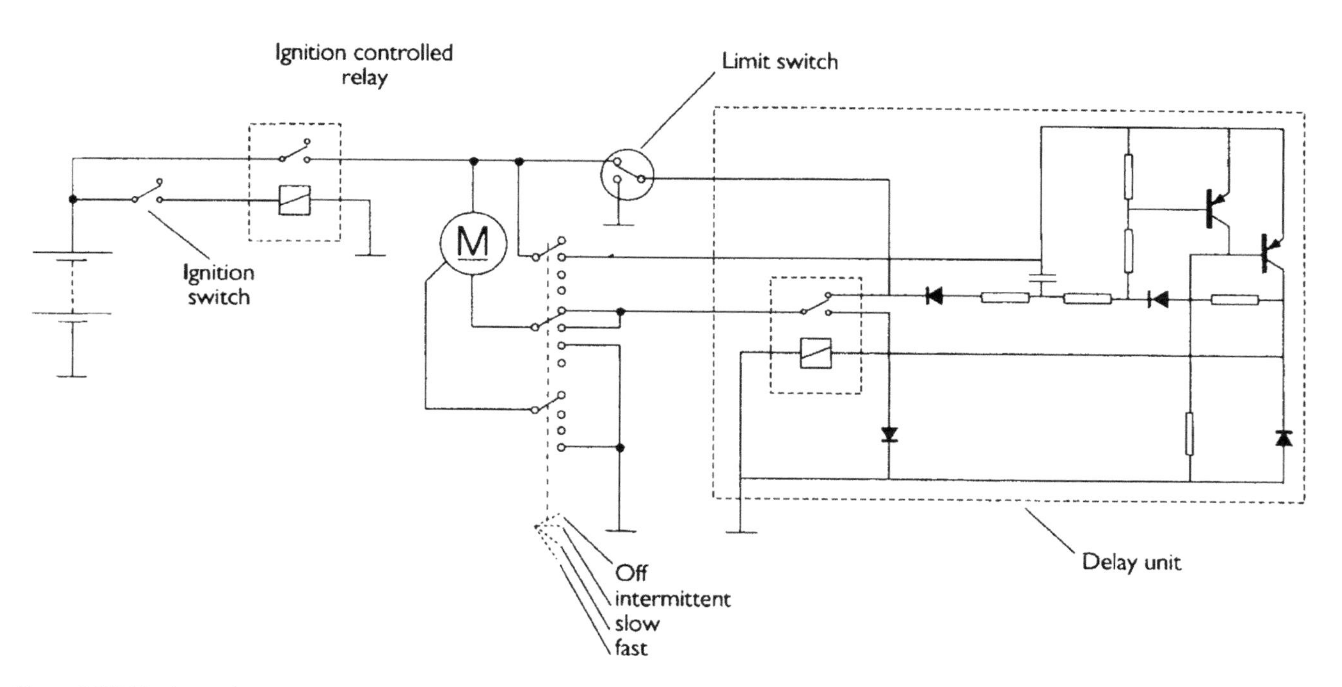

Figure 8.37 Traditional wiper circuit.

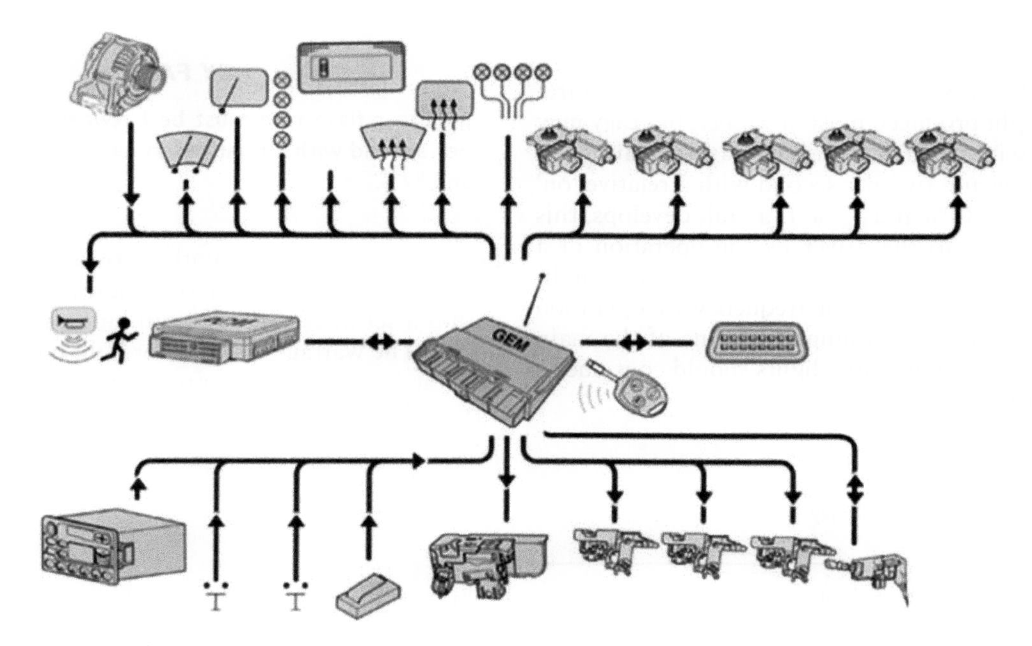

Figure 8.38 GEM and components.

Source: Ford Motor Company.

is the largest possible wiped area, yet they are compact in construction. This system is fitted, for example, in the new Ford Galaxy, as well as in the recently launched Mercedes-Benz S-Class.

The two-motor wiper system synchronises its two drives entirely electronically (Figure 8.39). Integrated sensors continuously monitor the precise position of the wiper arms. This allows the change in direction to be individually determined; the change can, therefore, always take place very close to the A-pillar, which provides the widest possible field of view under all conditions. When the wiper is switched off, the wiper arms and blades can disappear completely under the engine bonnet/hood. This improves aerodynamics and reduces the risk of injury to pedestrians and cyclists in the event of an accident. It is also possible for the wiper equipment to work fully automatically when combined with the rain and light sensors.

Figure 8.39 Two wiper motors must be synchronised.

Source: Bosch Media.

KEY FACT

The two-motor wiper system synchronises its two drives electronically.

At present, approximately 5% of all the cars manufactured in Europe are fitted with a two-motor wiper system. This proportion is expected to double over the next five years.

8.6.4 Headlight wipers and washers

There are two ways in which headlights are cleaned, first by high-pressure jets and second by small wiper blades with low-pressure water supply. The second method is in fact much the same as windscreen cleaning but on a smaller scale. The high-pressure system tends to be favoured but can suffer in very cold conditions due to the fluid freezing. It is expected that the wash system should be capable of approximately 50 operations before refilling of the reservoir is necessary. Headlight cleaners are often combined with the windscreen washers. They operate each time the windscreen washers are activated, if the headlights are also switched on.

A retractable nozzle for headlight cleaners is often used. When the water pressure is pumped to the nozzle, it is pushed from its retracted position, flush with the bodywork. When the washing is completed, the jet is then retracted back into the housing.

8.6.5 Indicators and hazard lights

Direction indicators have a number of statutory requirements. The light produced must be amber, but they may be grouped with other lamps. The flashing rate must be between one and two per second with a relative 'on' time of between 30% and 57%. If a fault develops, this must be apparent to the driver by the operation of a warning light on the dashboard. The fault can be indicated by a distinct change in frequency of operation or the warning light remaining on. If one of the main bulbs fails then the remaining lights should continue to flash perceptibly.

KEY FACT

Indicator flash rate must be between one and two per second with a relative 'on' time of between 30% and 57%.

Legislation as to the mounting position of the exterior lamps exists such that the rear indicator lights must be within a set distance of the tail lights and within a set height. The wattage of indicator light bulbs is normally 21 W at 6, 12 or 24 V as appropriate. Figure 8.40 shows a typical indicator and hazard circuit.

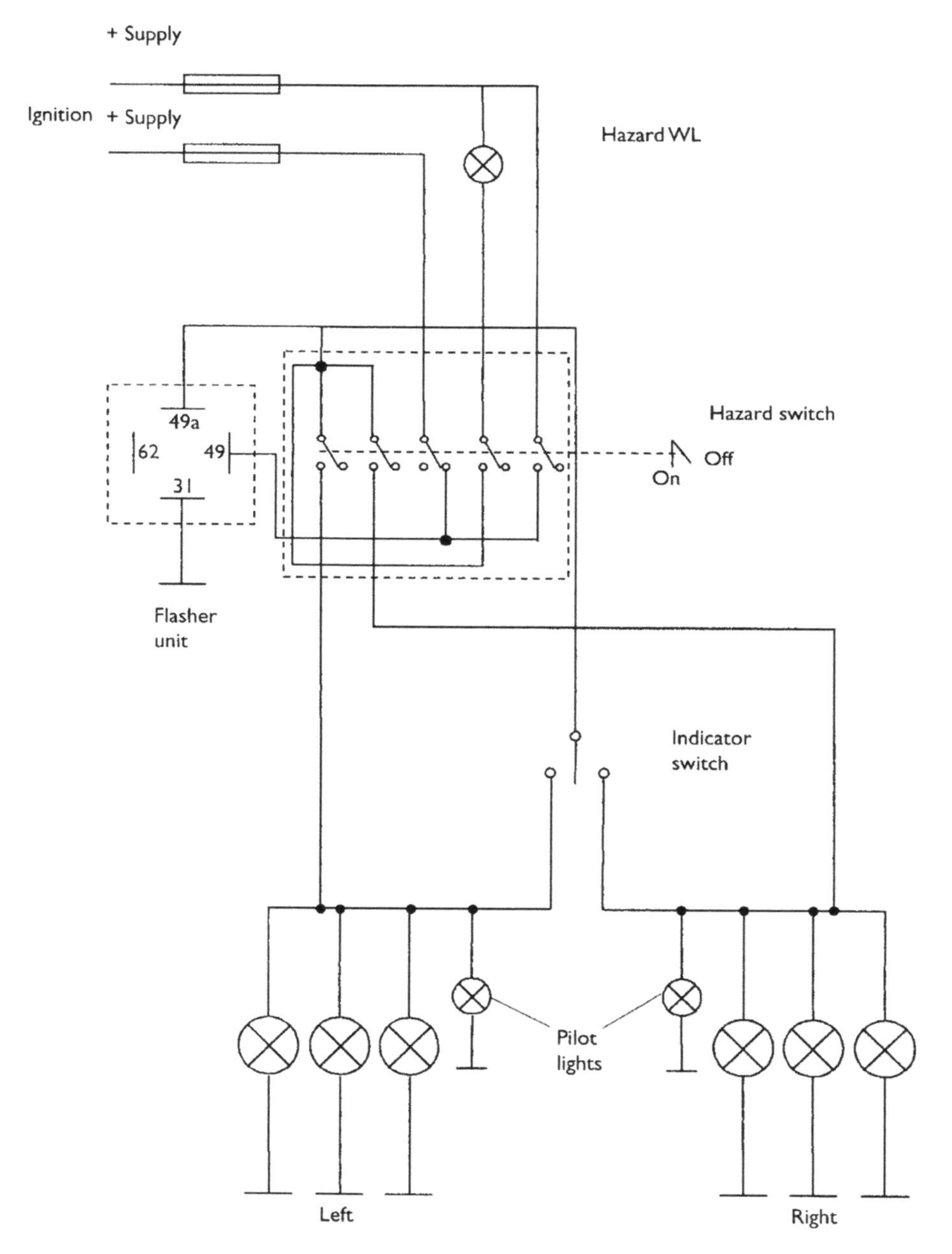

Figure 8.40 Indicator and hazard circuit.

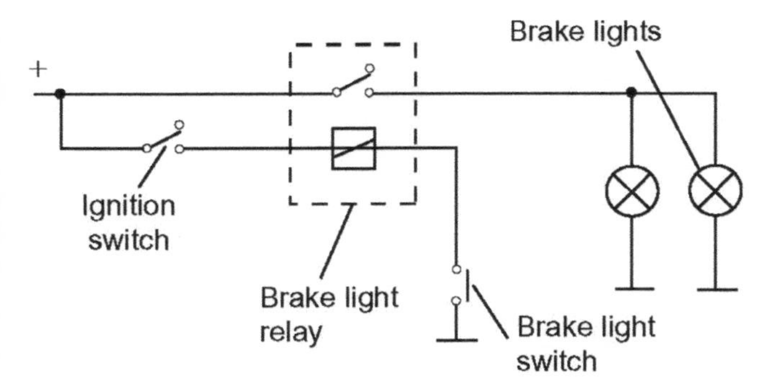

Figure 8.41 Simple relay operated circuit used for brake lights (stop lights).

Flasher units are rated by the number of bulbs they are capable of operating. When towing a trailer or caravan, the unit must be able to operate at a higher wattage. Most units use a relay for the actual switching as this is not susceptible to voltage spikes and also provides an audible signal.

KEY FACT

Flasher units are rated by the number of bulbs they are capable of operating.

8.6.6 Brake lights

Most brake light circuits incorporate a relay to switch the lights, which is in turn operated by a spring-loaded switch on the brake pedal. Links from this circuit to cruise control may be found. This is to cause the cruise control to switch off as the brakes are operated (Figure 8.41).

8.6.7 Electric horns

Regulations in most countries state that the horn (or audible warning device) should produce a uniform sound. This makes sirens and melody type fanfare horns illegal. Most horns draw a large current so are switched by a suitable relay.

The standard horn operates by simple electromagnetic switching. Current flow causes an armature, to which is attached a tone disc, to be attracted towards a stop. This opens a set of contacts which disconnects the current allowing the armature and disc to return under spring tension. The whole process keeps repeating when the horn switch is on. The frequency of movement and hence the fundamental tone is arranged to lie between 1.8 and 3.5 kHz. This note gives good penetration through traffic noise. Twin horn systems, which have a high- and low-tone horn, are often used. This produces a more pleasing sound but is still very audible in both town and higher speed conditions (Figure 8.42).

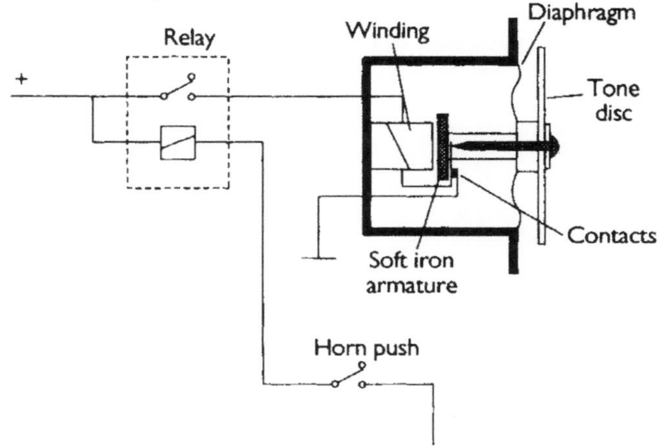

Figure 8.42 Typical horn together with its associated circuit.

KEY FACT

The standard horn operates by simple electromagnetic switching.

8.6.8 Engine cooling fan motors

Most engine cooling fan motors (radiator cooling) are simple PM types. The fans used often have the blades placed asymmetrically (balanced but not in a regular pattern) to reduce noise when operating (Figure 8.43).

DEFINITION

PM: Permanent magnet.

When twin cooling fans and motors are fitted, they can be run in series or parallel. This is often the case

Figure 8.43 Engine cooling fan in position.

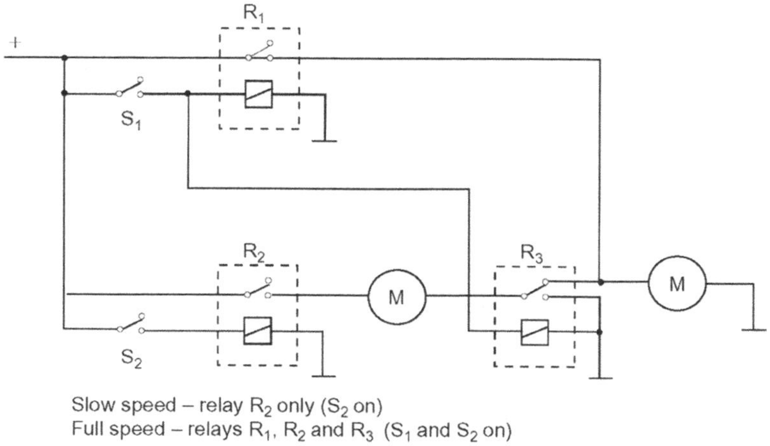

Slow speed – relay R_2 only (S_2 on)
Full speed – relays R_1, R_2 and R_3 (S_1 and S_2 on)

Figure 8.44 Two-speed, twin cooling fan circuit.

Figure 8.45 Remember to check the obvious first – like this blown fuse.

when air conditioning is used as the condenser is usually placed in front of the radiator and extra cooling air speed may be needed. A circuit for series or parallel operation of cooling fans is shown in Figure 8.44.

KEY FACT

A simple way to check for a blown fuse (Figure 8.45) is to use a test lamp on each side.

8.7 CENTRAL CONTROL

8.7.1 Overview

For many years, the trend with automotive electrical systems has been towards some sort of networked central control. This makes sense because many systems can share one source of information and, with the proper equipment, diagnostics can be made easier. Also, centralization allows facilities to be linked and improved. For example, networking and centralization of control units makes it easier to have a system where the engine will not start if a door is open, or selection of reverse gear can operate the rear wiper when the fronts are switched on.

KEY FACT

Centralization systems allow facilities to be linked and improved.

The basic central control system can be simplified in a way that is represented by Figure 8.46.

The most common usage of central control is for body systems such as lighting, wipers, doors, seats and windows. In some cases, these systems are controlled

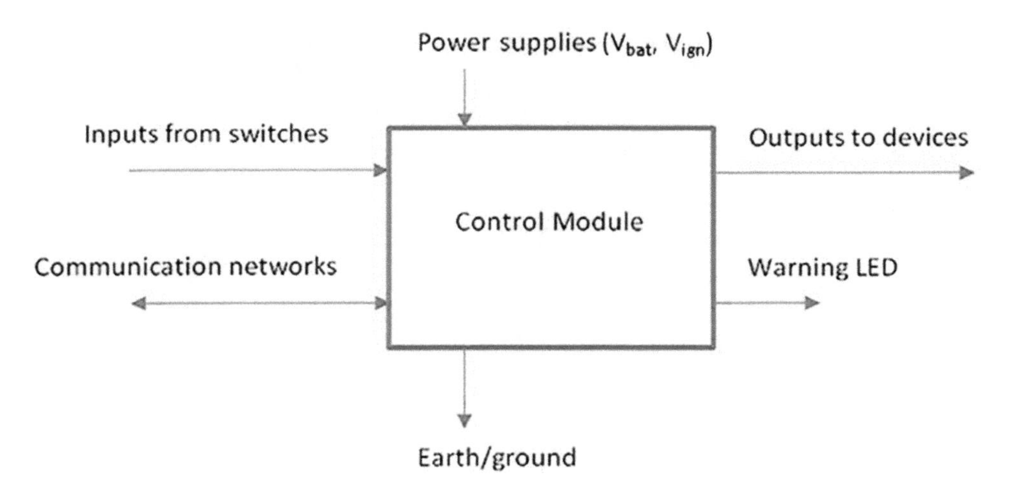

Figure 8.46 Central control system (simplified).

by slave units via a communication network, in other cases, one unit controls everything. In almost all cases, this central unit is networked to other ECUs.

KEY FACT

The most common usage of central control is for body systems such as lighting, wipers, doors, seats and windows.

Some central control modules connect via normal wires to switches that supply normal voltage on/off signals, others use switches that communicate on the CAN or LIN networks. The outputs from the module are sent via relays or solid-state switches on standard wires.

Manufacturers have different names for these systems and the control units, but most have a similar function. Four example names are:

- body control module (BCM);
- general electronic module (GEM);
- central control unit (CCU);
- central control module (CCM).

8.7.2 System and circuit diagrams

Figures 8.47 and 8.48 show a full circuit diagram (in two halves) that has been adapted from materials supplied by Ford Motor Company. The circuit shows a general electronic module (GEM) and how it is used to control the wipers (in this case). Note that the multi-function wiper switch contains a series of switch contacts that all connect directly back to the GEM. Also note the CAN connection to the module.

Figure 8.49 is from a Ford vehicle with adaptive front lighting. The light switch in this case has a supply (30) and an earth/ground (31) connection, but that all commands to operate the lights are sent via the LIN bus. The GEM supplies outputs to operate the lights and a separate module is used in this case for the adaptive features of the lights.

8.7.3 Control units

A central body control module (BCM) is the primary hub that maintains body functions, such as:

- internal and external lighting;
- security and access control;
- comfort features for doors and seats;
- other convenience controls.

A single board computer (SBC) combines voltage regulation with a CAN or LIN physical interface in a single package. H-bridge drivers and a series of high-side switches drive high-current loads and replace relays. The gateway serves as the information bridge between various in-car communication networks, including Ethernet, FlexRay, CAN, LIN and MOST protocols. It also serves as the car's central diagnostic interface.

KEY FACT

H-bridge drivers and high-side switches drive high-current loads and replace relays.

8.8 DIAGNOSTICS – AUXILIARY

8.8.1 Testing procedure

The process of checking an auxiliary system circuit is broadly as presented in Figure 8.50.

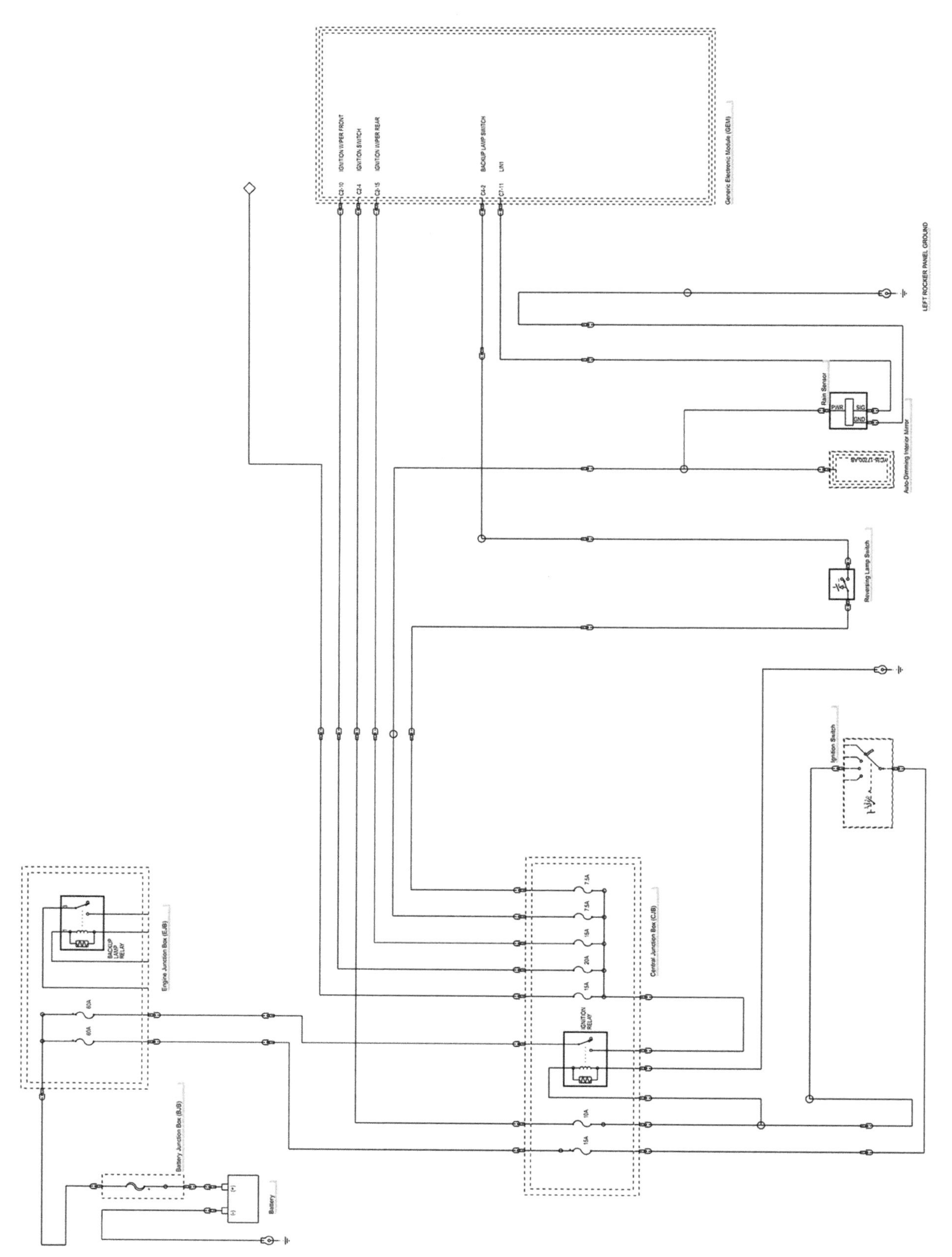

Figure 8.47 Part 1 of a wiper circuit using a central control module.

Source: Ford Motor Company.

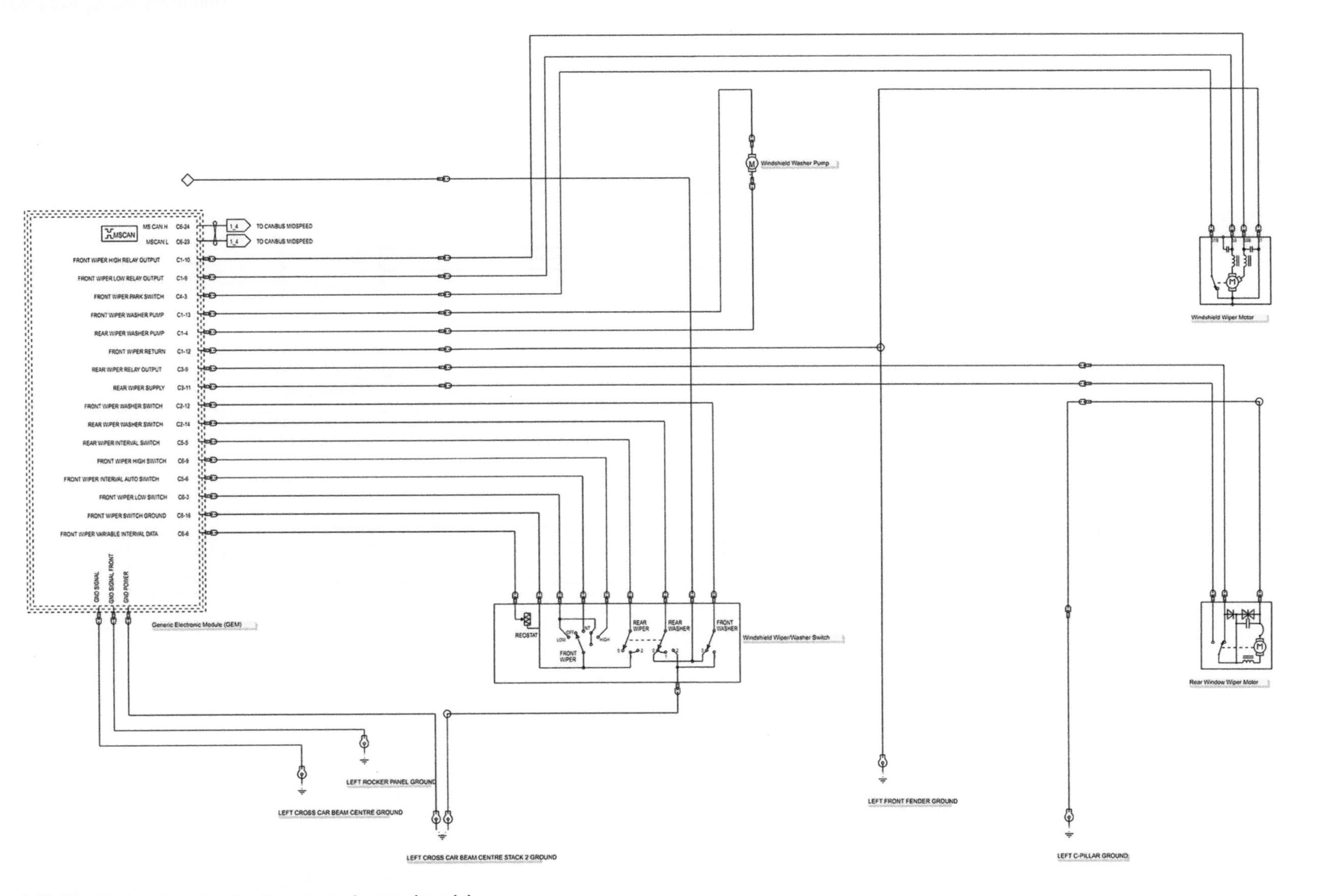

Figure 8.48 Part 2 of a wiper circuit using a central control module.

Source: Ford Motor Company.

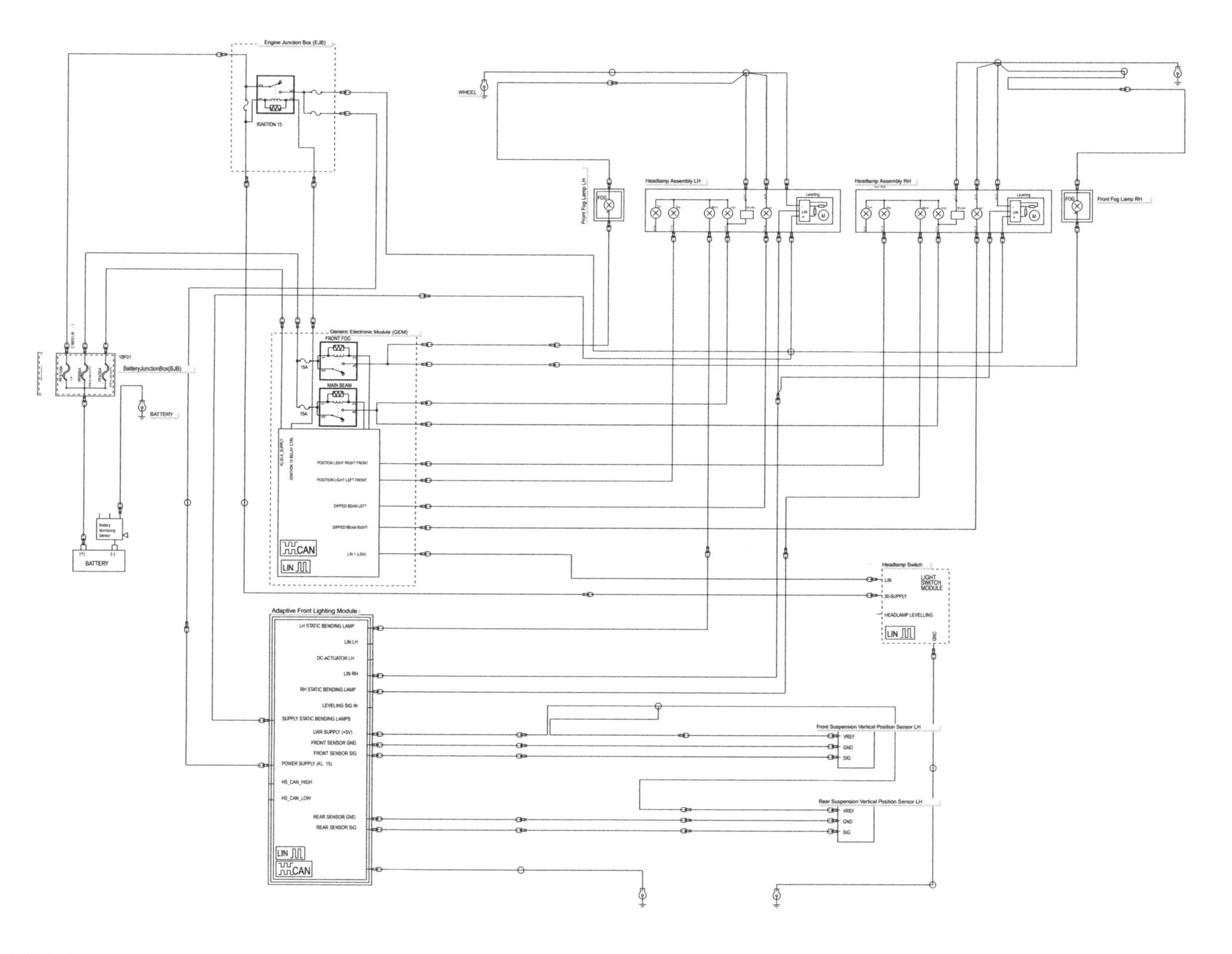

Figure 8.49 Lighting circuit using a general electronic module.
Source: Ford Motor Company.

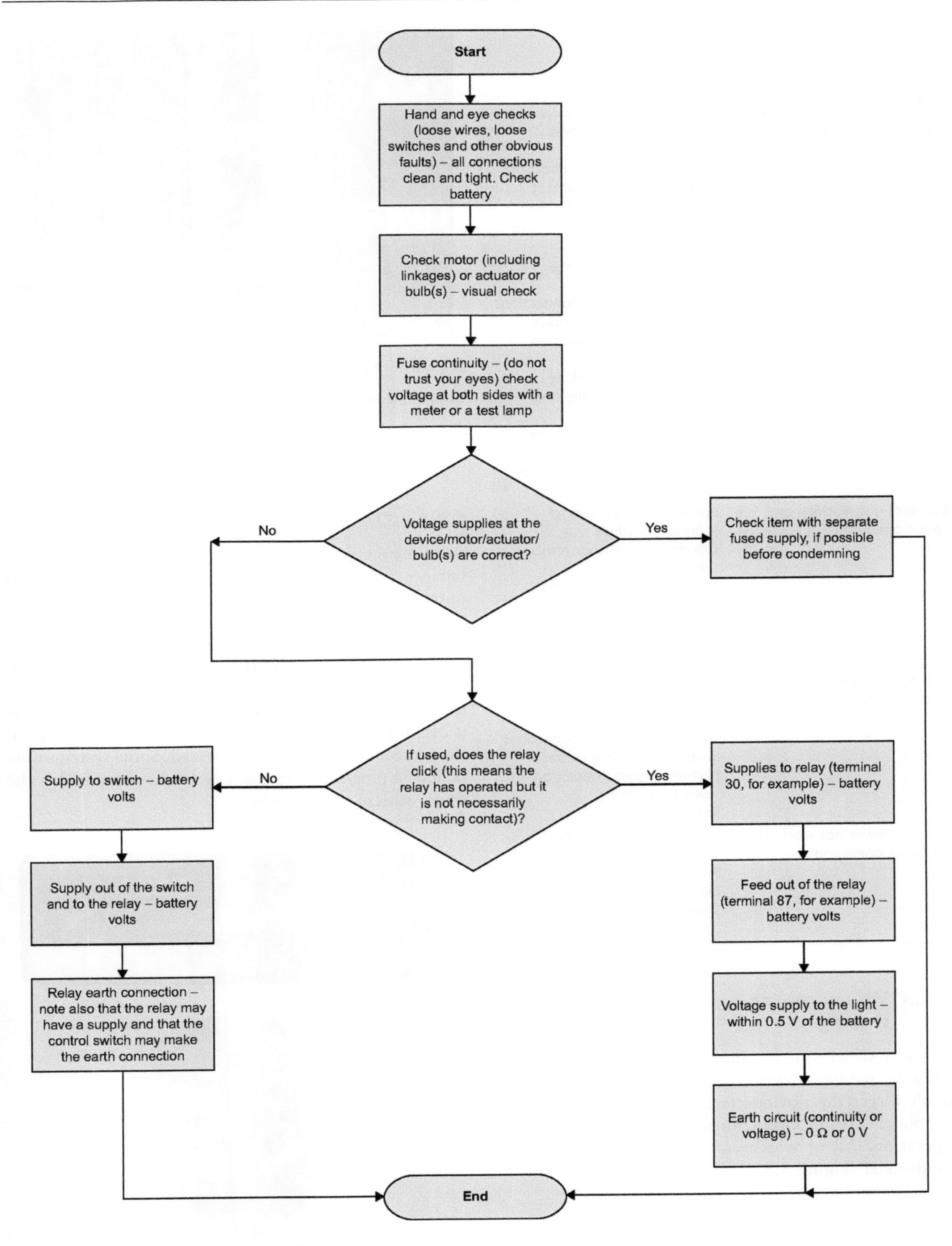

Figure 8.50 Auxiliary systems diagnosis chart.

8.8.2 Auxiliaries fault diagnosis table

Symptom	*Possible fault*
Horn not working or poor sound quality	Loose or broken wiring/ connections/fuse Corrosion in horn connections Switch not making contact High-resistance contact on switch or wiring Relay not working
Wipers not working or poor operation	Loose or broken wiring/ connections/fuse Corrosion in wiper connections Switch not making contact High-resistance contact on switch or wiring Relay/timer not working Motor brushes or slip ring connections worn Limit switch contacts open circuit or high resistance Blades and/or arm springs in poor condition
Washers not working or poor operation	Loose or broken wiring/ connections/fuse Corrosion in washer motor connections Switch not making contact Pump motor poor or not working Blocked pipes or jets Incorrect fluid additive used
Indicators not working or incorrect operating speed	Bulb(s) blown Loose or broken wiring/ connections/fuse Corrosion in horn connections Switch not making contact High-resistance contact on switch or wiring Relay not working
Heater blower not working or poor operation	Loose or broken wiring/ connections/fuse Switch not making contact Motor brushes worn Speed selection resistors open circuit

8.8.3 Wiper motor and circuit testing

Modern wiper systems may need the assistance of a suitable scanner when diagnosing faults. However, don't forget the obvious tests such as correct voltage supplies, earth/ground connections and correct switch operation. All of which can be tested using a simple multimeter (Figure 8.51).

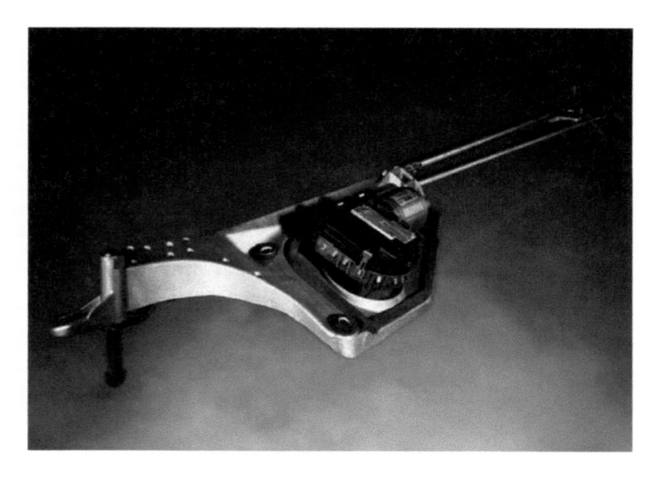

Figure 8.51 Reversible wiper motor and drive linkage.

Source: Bosch Media.

8.9 IN-CAR ENTERTAINMENT (ICE), SECURITY AND COMMUNICATIONS

8.9.1 ICE

Controls on most sets will include volume, treble, bass, balance and fade. A digital display will provide a visual output of operating condition (Figure 8.52). This is also linked into the vehicle lighting to prevent glare at night. Track selection and programming for one or several compact discs is possible. An MP3 input is now often provided.

Many in-car entertainment (ICE) systems are coded to deter theft. The code is activated if the main supply is disconnected and will not allow the set to work until the correct code has been re-entered. Some systems now include a plug in electronic 'key card', which makes the set worthless when removed.

Figure 8.52 ICE display and sub-woofer.

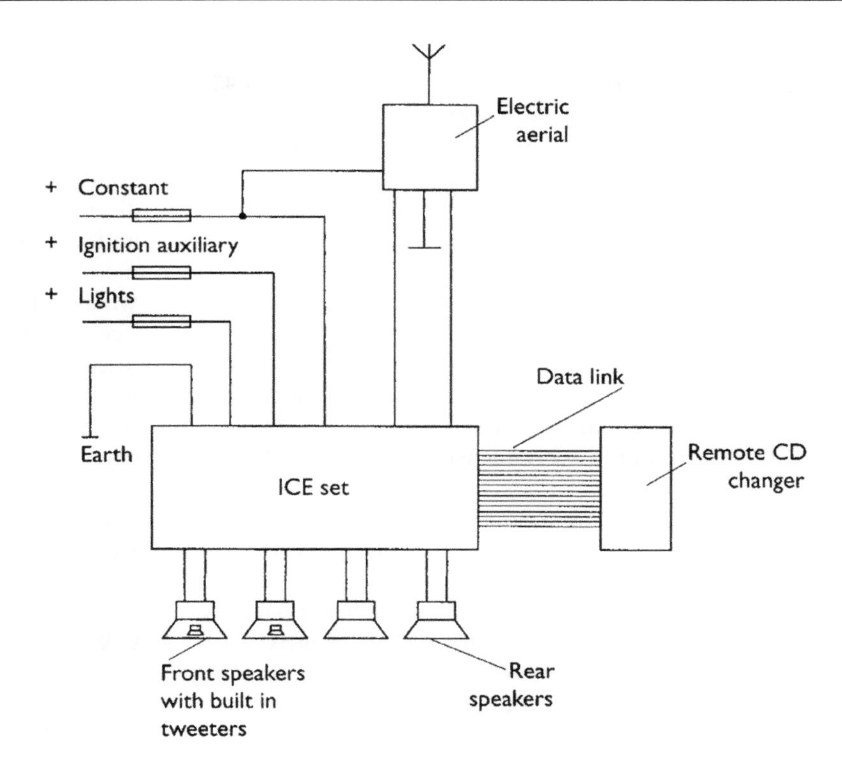

Figure 8.53 ICE circuit.

DEFINITION

MP3: An audio file format, based on MPEG (Moving Picture Expert Group) technology. It creates very small files suitable for streaming or downloading over the Internet.

Good ICE systems include at least six speakers, two larger speakers in the rear parcel shelf to produce good low-frequency reproduction, two front-door speakers for mid-range and two front-door tweeters for high-frequency notes (Figure 8.53). Speakers are a very important part of a sound system. No matter how good the receiver or CD player is, the sound quality will be reduced if inferior speakers are used. Equally, if the speakers are of a lower power output rating than the set, distortion will result at best and damage to the speakers at worst. Speakers fall generally into the following categories:

- tweeters high-frequency reproduction;
- mid-range frequency reproduction (treble);
- woofers low-frequency reproduction (bass);
- sub-woofers very low frequency reproduction.

The radio data system (RDS) has become a standard on many radio sets. It is an extra inaudible digital signal which is sent with FM broadcasts in a similar way to how teletext is sent with TV signals. RDS provides information so a receiver can appear to act intelligently. The possibilities available when RDS is used are as follows.

- The station name can be displayed in place of the frequency.
- There can be automatic tuning to the best available signal for the chosen radio station. For example, in the United Kingdom, a journey from the south of England to Scotland would mean the radio would have to be retuned up to 10 times. RDS will do this without the driver even knowing.
- Traffic information broadcasts can be identified and a setting made so that whatever you are listening to at the time can be interrupted.

DEFINITION

RDS: Radio data system. Traffic information system on FM. RDS shows station name display and delivers traffic bulletins; it also locks onto the best possible frequency for a station in a particular part of the country.

The radio broadcast data system (RBDS) is an extension of RDS which has been in use in Europe since 1984. The system allows the broadcaster to transmit text information at the rate of approximately 1200 bits/s. The information is transmitted on a 57 kHz suppressed sub-carrier as part of the FM MPX signal.

RBDS was developed for the North American market by the National Radio Systems Committee (NRSC), a joint committee composed of the Electronic Industries Association (EIA) and the National Association of Broadcasters (NAB). The applications for the transmission of text to the vehicle are interesting and include

- song title and artist;
- traffic, accident and road hazard information;
- stock information;
- weather.

In emergency situations, the audio system can be enabled to interrupt the cassette, CD or normal radio broadcast to alert the user.

8.9.2 Security systems

Car and alarm manufacturers are constantly fighting to improve security. Building the alarm system as an integral part of the vehicle electronics has made significant improvements. Even so, retrofit systems can still be very effective. Three main types of intruder alarm are used:

- switch operated on all entry points;
- battery voltage sensed;
- volumetric sensing.

There are three main ways to disable the vehicle:

- ignition circuit cut off;
- starter circuit cut off;
- engine ECU code lock.

Most alarm systems are made for 12 V, negative earth vehicles. They have electronic sirens and give an audible signal when arming and disarming. They are all triggered when the car door opens and will automatically reset after a period of time, often one or two minutes. The alarms are triggered instantly when entry point is breached. Most systems are two pieces, with separate control unit and siren; most will have the control unit in the passenger compartment and the siren under the bonnet.

Most systems now come with remote 'keys' that use small button-type batteries and may have an LED that shows when the signal is being sent. They operate with one vehicle only. Intrusion sensors such as car movement and volumetric sensing can be adjusted for sensitivity.

When operating with flashing lights, most systems draw approximately 5 A. Without flashing lights (siren only), the current drawn is less than 1 A. The sirens produce a sound level of approximately 95 dB, when measured 2 m in front of the vehicle.

The system, as is usual, can be considered as a series of inputs and outputs. This is particularly useful for diagnosing faults. Most factory-fitted alarms are combined with the central door locking system. This allows the facility mentioned in a previous section known as lazy lock. Pressing the button on the remote unit, as well as setting the alarm, closes the windows and sunroof and locks the doors (Figure 8.54).

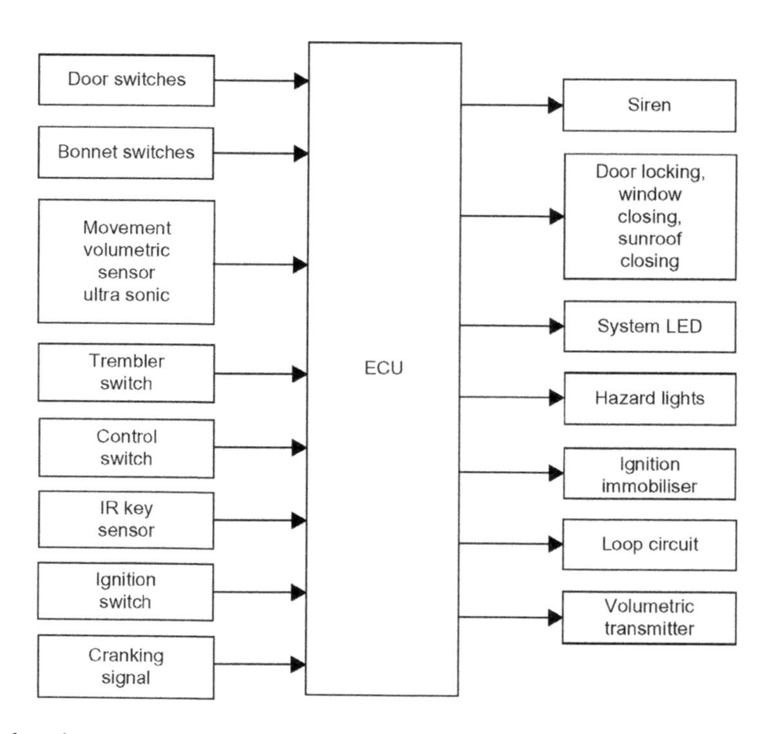

Figure 8.54 Block diagram of an alarm system.

KEY FACT

Most factory-fitted alarms are combined with the central door locking system.

A security code in the engine ECU is a powerful deterrent. This can only be 'unlocked' to allow the engine to start when it receives a coded signal. Ford and other manufacturers use a special ignition key which is programmed with the required information. Even the correct 'cut' key will not start the engine.

Of course, nothing will stop the car being lifted onto a truck and driven away, but this technique will mean a new engine control ECU will be needed.

8.10 DIAGNOSTICS – ICE, SECURITY AND COMMUNICATION

8.10.1 Testing procedure

The process of checking an ICE system circuit is broadly as presented in Figure 8.55.

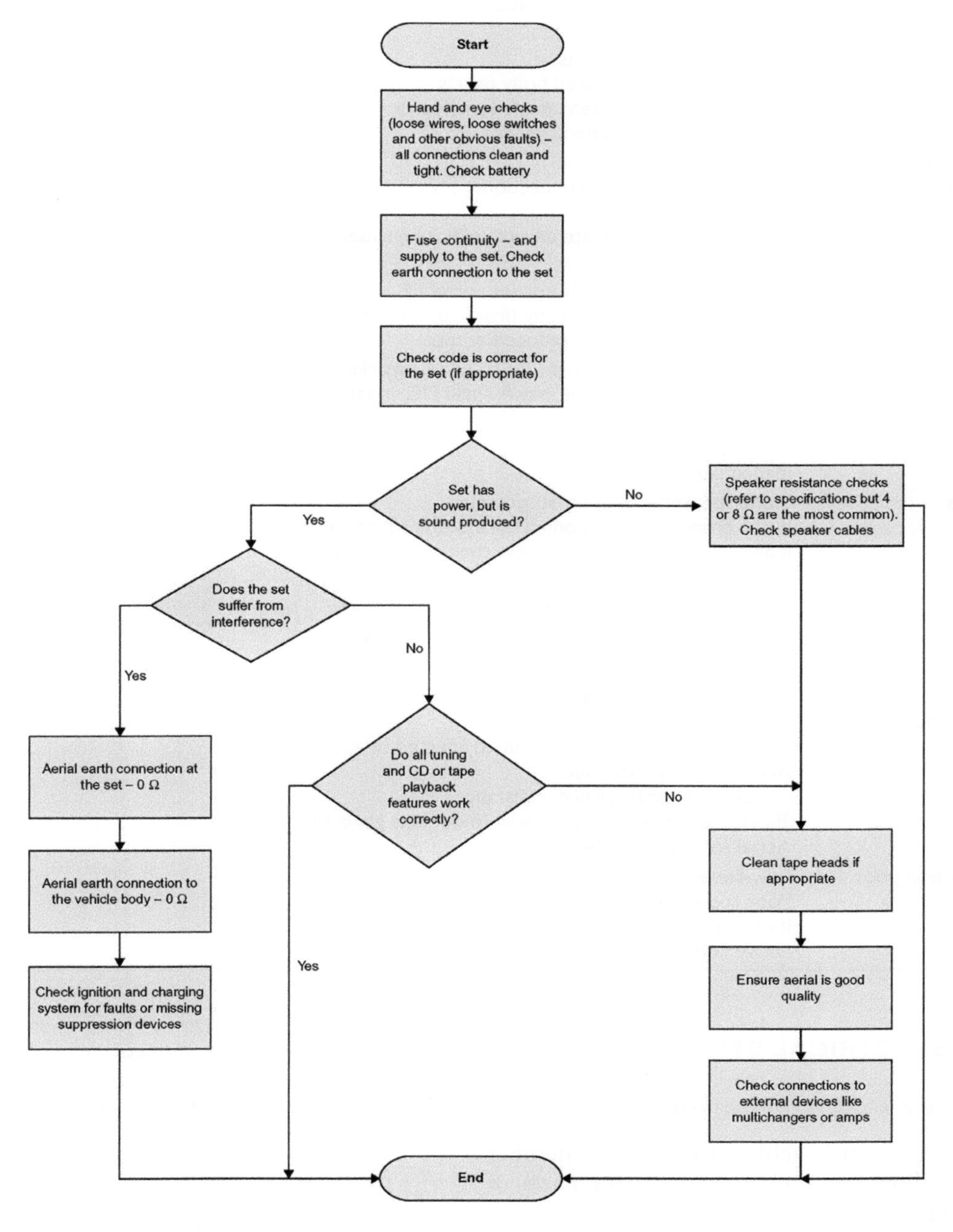

Figure 8.55 ICE system diagnosis chart.

8.10.2 ICE, security and communication system fault diagnosis table

Symptom	*Possible fault*
Alarm does not operate	Fuse blown Not set correctly Remote key battery discharged Open circuit connection to alarm unit ECU fault Receiver/transmitter fault Volumetric transmitter/receiver fault
Alarm goes off for no apparent reason	Drain on battery Loose connection Vibration/trembler/movement detection circuit set too sensitive Self-discharge in the battery Window left open allowing wind or even a bird or insect to cause interior movement Somebody really is trying to steal the car Loose connection
Radio interference	Tracking HT components Static build up on isolated body panels High-resistance or open circuit aerial earth Suppression device open circuit
ICE system does not produce sound	Set not switched on! Loose or open circuit connections Trapped wires Connections to separate unit (amplifier, equaliser, etc.) incorrect Fuse blown
Unbalanced sound	Fade or balance controls not set correctly Speakers not wired correctly (front right, front left, rear right, rear left, etc.) Speaker open circuit or reduced output
Phasing	Speaker polarity incorrect. This should be marked, but if not, use a small battery to check all speakers are connected the same way. A small DC voltage will move the speaker cone in one direction.
Speaker rattle	Insecure speaker(s) Trim not secure Inadequate baffles
Crackling noises	If one speaker – then try substitution If one channel – swap connections at the set to isolate the fault If all channels but only the radio then check interference Radio set circuit fault
Vibration	Incorrect or loose mounting
Hum	Speaker cables routed next to power supply wires Set fault
Distortion	Incorrect power rating speakers
Poor radio reception	Incorrect tuning 'Dark' spot/area. FM signals can be affected by tall buildings, etc. Aerial not fully extended Aerial earth loose or high resistance Tuner not trimmed to the aerial (older sets generally) Aerial sections not clean
Telephone reception poor	Low-battery power Poor reception area Interference from the vehicle Loose connections on hands-free circuit

8.11 BODY ELECTRICAL SYSTEMS

8.11.1 Electric seat adjustment

Adjustment of the seat is achieved by using a number of motors to allow positioning of different parts of the seat. A typical motor reverse circuit is shown in Figure 8.56.

When the switch is moved, one of the relays will operate and this changes the polarity of the supply to one side of the motor. Movement is often possible in the following ways:

- front to rear;
- cushion height rear;

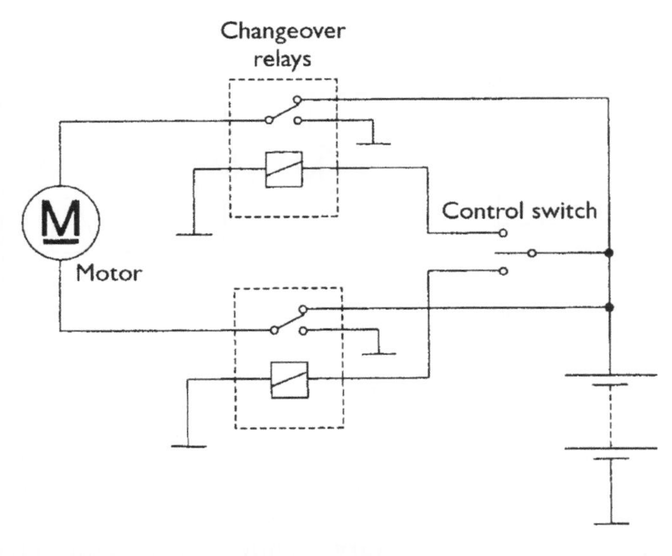

Figure 8.56 Motor reverse circuit using a centre-off changeover switch and two changeover relays.

- cushion height front;
- backrest tilt;
- headrest height;
- lumbar support.

When seat position is set, some vehicles have position memories to allow automatic repositioning, if the seat has been moved. This is often combined with electric mirror adjustment. Figure 8.57 shows how the circuit is constructed to allow position memory. As the seat is moved, a variable resistor, mechanically linked to the motor, is also moved. The resistance value provides feedback to an ECU.

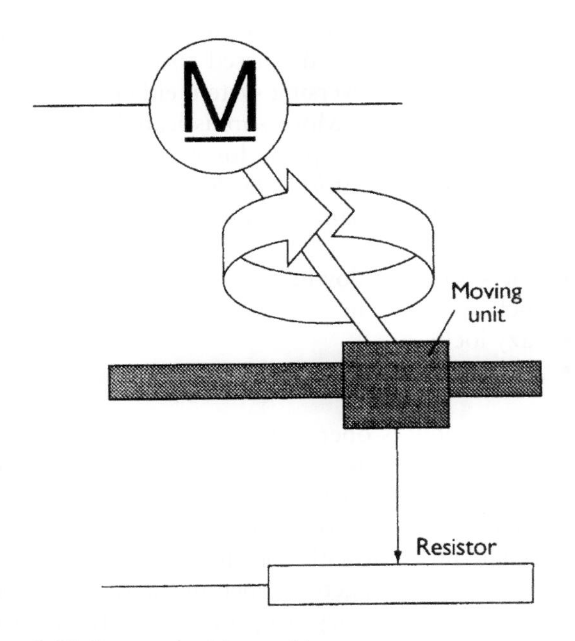

Figure 8.57 One method for position memory.

This can be 'remembered' in a number of ways; the best technique is to supply the resistor with a fixed voltage such that the output relative to the seat position is proportional to position. This voltage can then be 'analogue to digital' converted, which produces a simple 'number' to store in a digital memory. When the driver presses a memory recall switch, the motor relays are activated by the ECU until the number in memory and the number fed back from the seat are equal. This facility is often isolated when the engine is running to prevent the seat moving into a dangerous position as the car is being driven. Position of the seats can still be adjusted by operating the switches as normal.

8.11.2 Electric mirrors

Many vehicles have electrical adjustment of mirrors, particularly on the passenger side. The system used is much the same as has been discussed above in relation to seat movement. Two small motors are used to move the mirror vertically or horizontally. Many mirrors also contain a small heating element on the rear of the glass. This is operated for a few minutes when the ignition is first switched on and can also be linked to the heated rear window circuit. Figure 8.58 shows an electrically operated mirror circuit, which includes feedback resistors for positional memory.

8.11.3 Electric sunroof operation

The operation of an electric sunroof is similar to the motor reverse circuit discussed earlier in this chapter. However, further components and circuitry are needed to allow the roof to slide, tilt and stop in the closed position. The extra components used are a micro switch and a latching relay. A latching relay works in much the same way as a normal relay except that it locks into position each time it is energised. The mechanism used

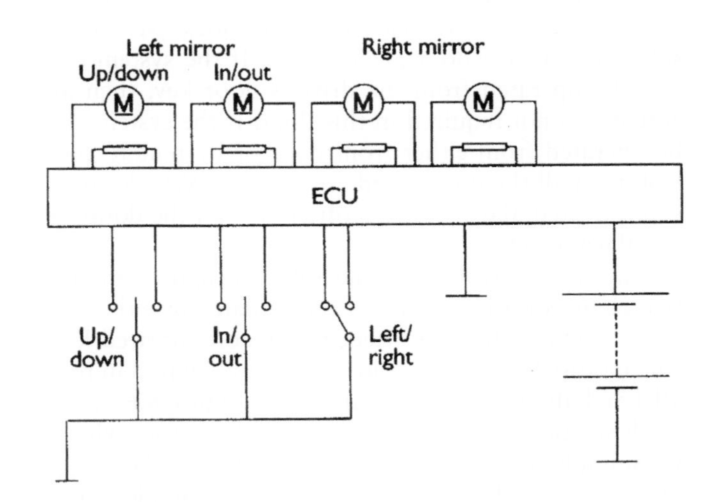

Figure 8.58 Mirror adjustment circuit.

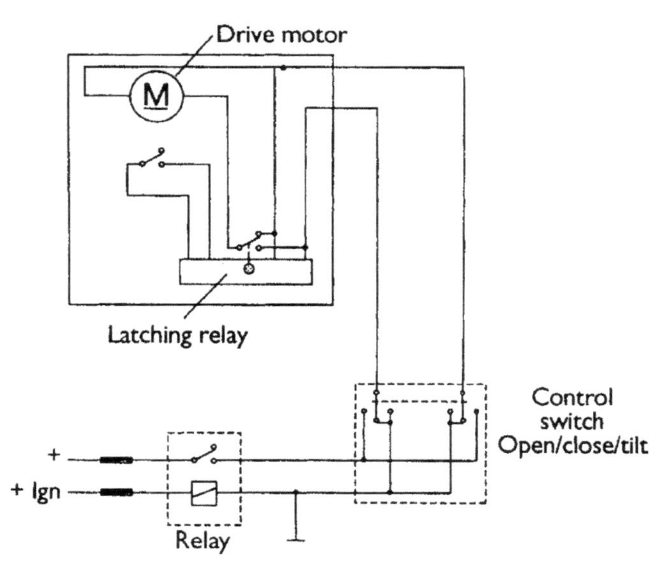

Figure 8.59 Sunroof circuit example.

to achieve this is much like that used in ball point pens that use a button on top (Figure 8.59).

The micro switch is mechanically positioned such as to operate when the roof is in its closed position. A rocker switch allows the driver to adjust the roof. The switch provides the supply to the motor to run it in the chosen direction. The roof will be caused to open or tilt. When the switch is operated to close the roof, the motor is run in the appropriate direction until the micro switch closes when the roof is in its closed position. This causes the latching relay to change over, which stops the motor. The control switch has now to be released. If the switch is pressed again, the latching relay will once more change over and the motor will be allowed to run.

8.11.4 Door locking circuit

When the key is turned in the driver's door lock, all the other doors on the vehicle should also lock. Motors or solenoids in each door achieve this. If the system can only be operated from the driver's door key, then an actuator is not required in this door. If the system can be operated from either front door or by remote control, then all the doors need an actuator. Vehicles with sophisticated alarm systems often lock all the doors as the alarm is set.

The main control unit in the following figure contains two changeover relays. These are actuated by either the door lock switch or, if fitted, the remote infrared key. The motors for each door lock are simply wired in parallel and all operate at the same time (Figure 8.60).

Most door actuators are now small motors which, via suitable gear reduction, operate a linear rod in either direction to lock or unlock the doors. A simple motor reverse circuit is used to achieve the required action.

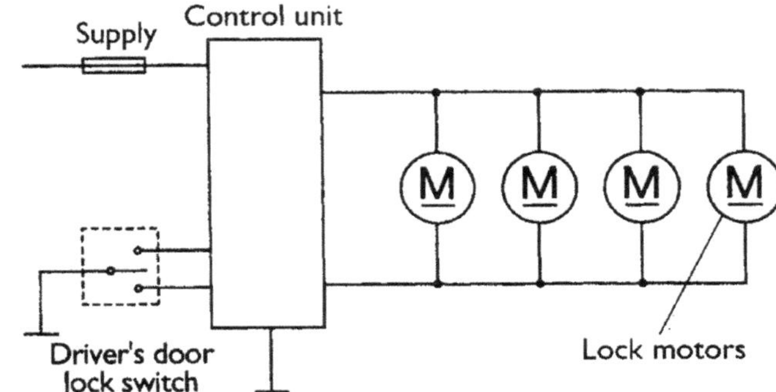

Figure 8.60 Door lock circuit.

Remote central door locking is controlled by a small hand-held transmitter and an infrared or RF sensor receiver unit as well as a decoder in the main control unit. This layout will vary slightly between different manufacturers. When the infrared key is operated by pressing a small switch, a complex code is transmitted. The number of codes used is well in excess of 50 000. The infrared sensor picks up this code and sends it in an electrical form to the main control unit. If the received code is correct, the relays are triggered and the door locks are either locked or unlocked. If an incorrect code is received on three consecutive occasions when attempting to unlock the doors, then the infrared system will switch itself off until the door is opened by the key. This will also reset the system and allow the correct code to again operate the locks. This technique prevents a scanning-type transmitter unit from being used to open the doors.

8.11.5 Electric window operation

The basic form of electric window operation is similar to many of the systems discussed so far in this chapter, that is a motor-reversing system either by relays or directly by a switch. More sophisticated systems are now becoming more popular for reasons of safety as well as improved comfort. The following features are now available from many manufacturers:

- one-shot up or down;
- inch up or down;
- lazy lock;
- back-off.

When a window is operated in one-shot or one-touch mode, the window is driven in the chosen direction until either the switch position is reversed, the motor stalls or the ECU receives a signal from the door lock circuit. The problem with one-shot operation is that if a child, for example, gets trapped in the window, there is a serious risk of injury. To prevent this, the back-off feature is used. An extra commutator is fitted to the

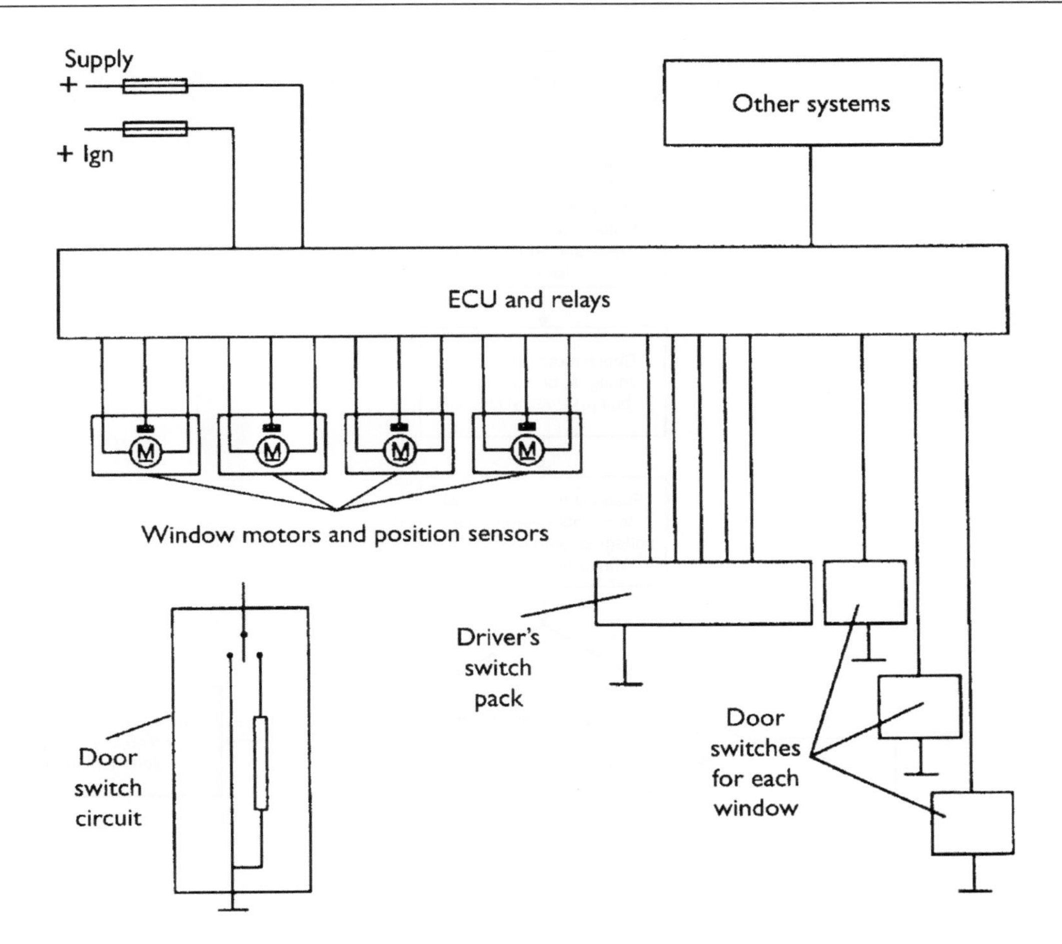

Figure 8.61 Electric window circuit.

motor armature and produces a signal via two brushes, proportional to the motor speed. If the rate of change of speed of the motor is detected as being below a certain threshold when closing, then the ECU will reverse the motor until the window is fully open. By counting the number of pulses received, the ECU can also determine the window position.

> **KEY FACT**
>
> To prevent children (or others) becoming trapped in an auto-close window, a back-off feature is used.

This is important, as the window must not reverse when it stalls in the closed position. In order for the ECU to know the window position, it must be initialised. This is often done simply by operating the motor to drive the window first fully open and then fully closed. If this is not done, then the one-shot close will not operate. On some systems, Hall effect sensors are used to detect motor speed. Other systems sense the current being drawn by the motor and use this as an indication of speed.

Lazy lock feature allows the car to be fully secured by one operation of a remote key. This is done by the link between the door lock ECU and the window and sunroof ECUs. A signal is supplied which causes all the windows to close in turn and then the sunroof, and finally locks the doors. The alarm will also be set, if required. The windows close in turn to prevent the excessive current demand which would occur if they all tried to operate at the same time.

A circuit for electric windows is shown in Figure 8.61. Note the connections to other systems such as door locking and the rear window isolation switch. This is commonly fitted to allow the driver to prevent rear window operation for child safety, for example.

8.12 DIAGNOSTICS – BODY ELECTRICAL

8.12.1 Testing procedure

The following procedure is very generic but, with a little adaptation, can be applied to any electrical system. Refer to manufacturer's recommendations if in any doubt. The process of checking any system circuit is broadly as presented in Figure 8.62.

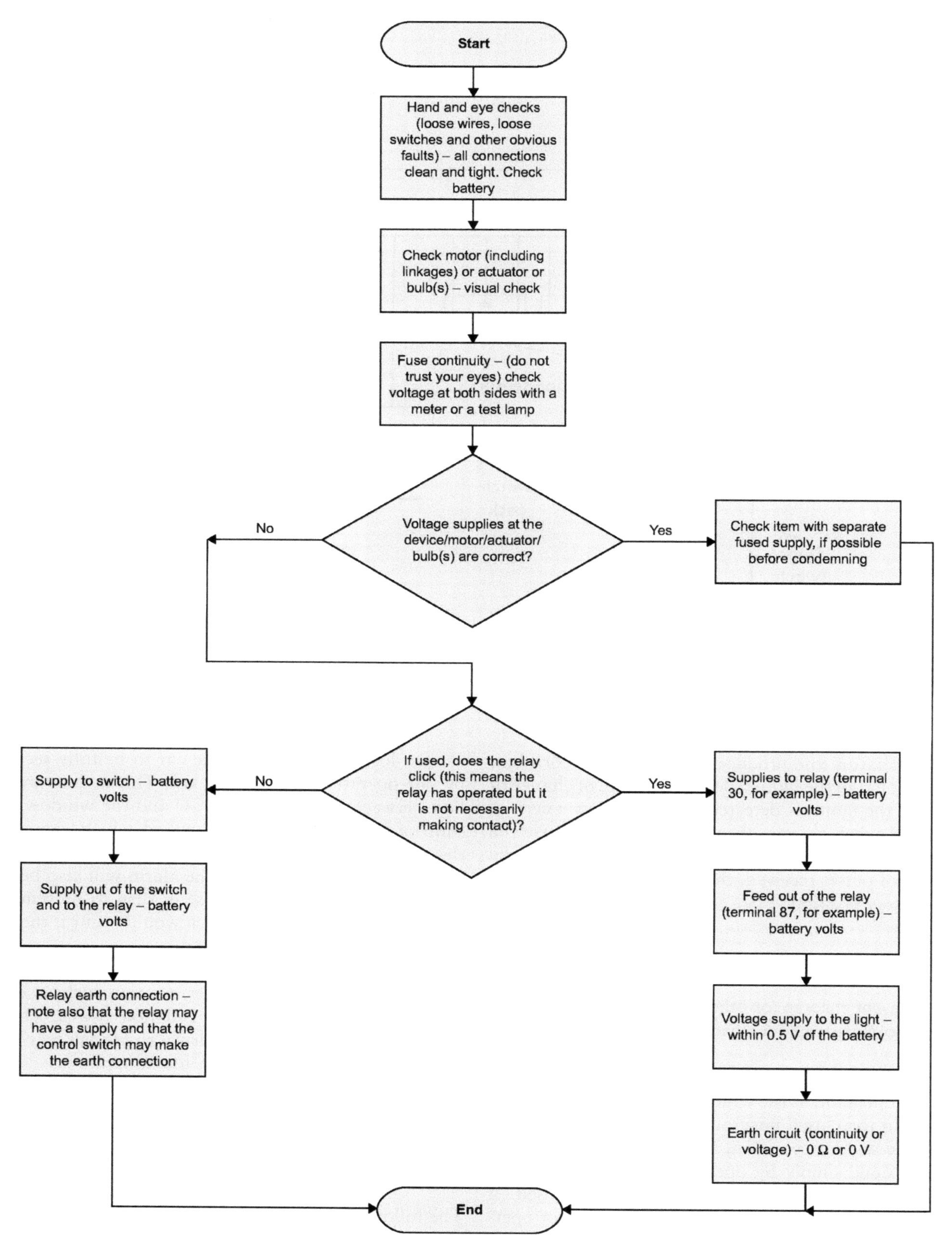

Figure 8.62 Auxiliary and body systems diagnosis chart.

8.12.2 Body electrical systems fault diagnosis table

Symptom	*Possible fault*
Electric units not operating	If ALL units not operating Open circuit in main supply Main fuse blown
Unit = window, door lock, mirror, etc.	Relay coil or contacts open circuit or high resistance If ONE unit is not operating Fuse blown Control switch open circuit Motor seized or open circuit Back-off safety circuit signal incorrect (windows)

8.12.3 Circuit systematic testing

The circuit shown in Figure 8.63 is for a power hood (meaning roof in this case) on a vehicle. The following faultfinding guide is an example of how to approach a problem with a system such as this in a logical manner.

If the power hood will not operate with the ignition switch at the correct position and the handbrake applied, proceed as follows:

1. Check fuses 6 and 13.
2. Check 12 V supply on N wire from fuse 6.
3. Check for 12 V on GS wire at power hood relay.

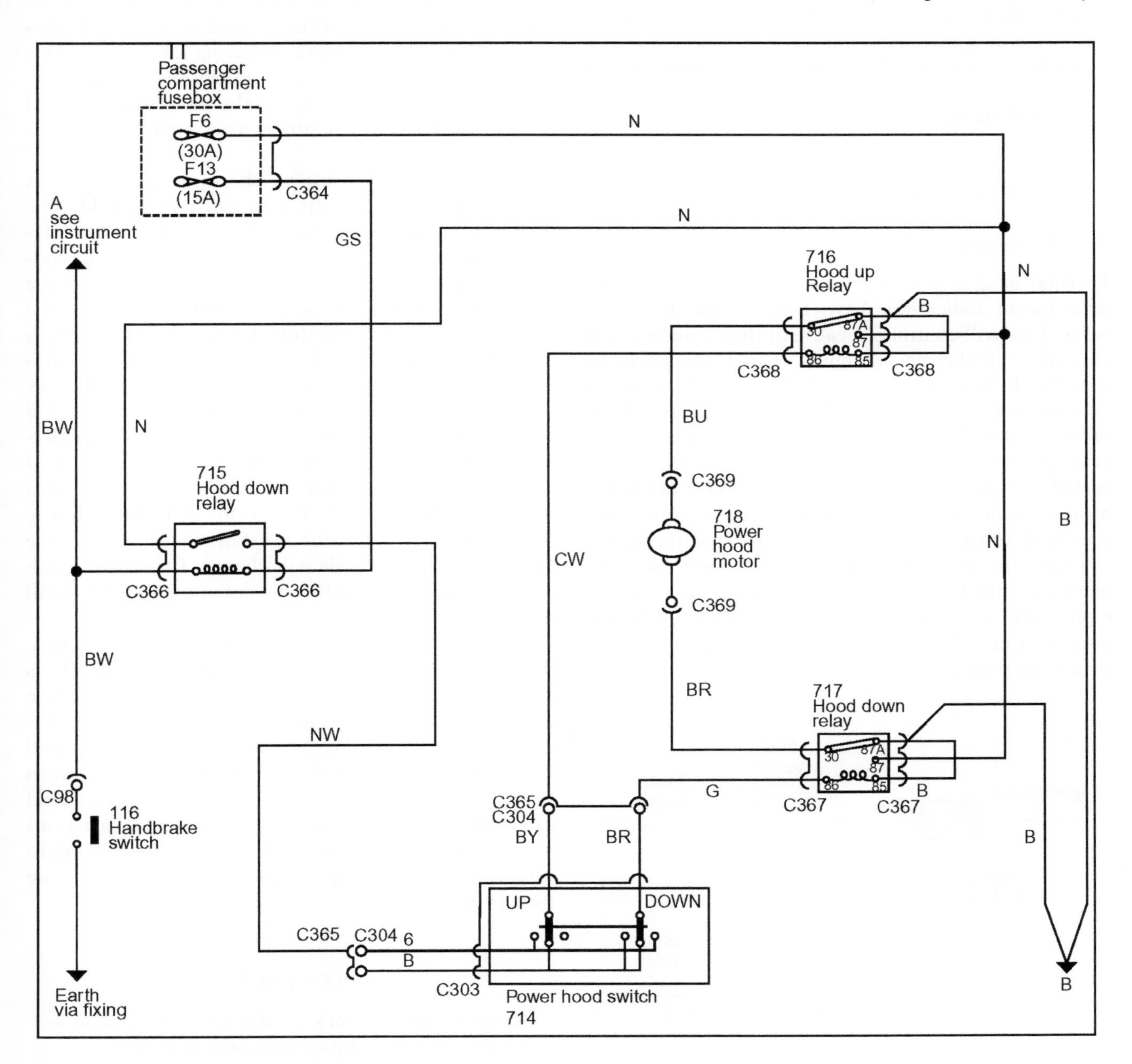

Figure 8.63 Power hood (roof) circuit.

4. Check continuity from power hood relay to earth on BW wire.
5. Check power hood relay.
6. Check for 12 V on NW wire at hood switch. Check for 12 V on N wire at hood up and down relays.
7. Check continuity from hood up and down relays to earth on B wire.
8. Check switch operation.
9. Check pump motor operation.

If the power hood will operate in one direction only, proceed as follows:

1. Check for 12 V on N wire at hood up or down relay as appropriate.
2. Check continuity from hood up or down relay to earth on B wire.
3. Check relay.

8.13 INSTRUMENTATION

8.13.1 Gauges

Thermal gauges, which are ideal for fuel and engine temperature indication, have been in use for many years. This will continue because of their simple design and inherent 'thermal' damping. The gauge works by utilising the heating effect of electricity and the widely adopted benefit of the bimetal strip. As a current flows through a simple heating coil wound on a bimetal strip, heat causes the strip to bend. The bimetal strip is connected to a pointer on a suitable scale. The amount of bend is proportional to the heat, which in turn is proportional to the current flowing. Provided the sensor can vary its resistance in proportion to the measurement (e.g. fuel level), the gauge will indicate a suitable representation as long as it has been calibrated for the particular task. Figure 8.64 shows a representation of a typical thermal gauge circuit.

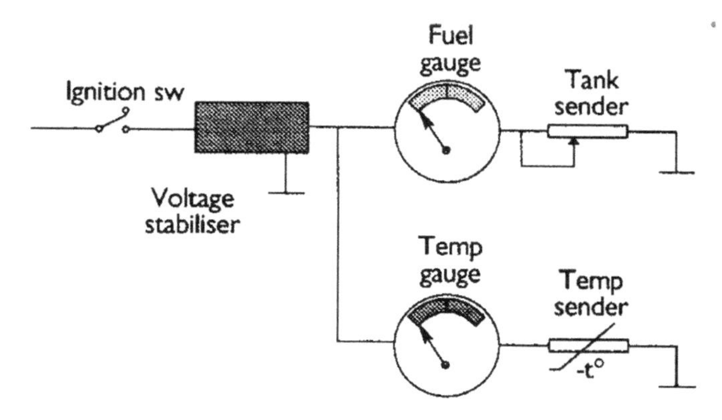

Figure 8.64 Simplified thermal gauge circuit.

Thermal-type gauges are used with a variable resistor and float in a fuel tank or with a thermistor in the engine water jacket. The resistance of the fuel tank sender can be made non-linear to counteract any non-linear response of the gauge. The sender resistance is at a maximum when the tank is empty.

KEY FACT

With a bimetal strip, the amount of bend is proportional to the heat, which in turn is proportional to the current flowing.

A constant voltage supply is required to prevent changes in the vehicle system voltage affecting the reading. This is because if system voltage increased, the current flowing would increase and hence the gauges would read higher. Most voltage stabilisers are simple Zener diode circuits.

Air-cored gauges work on the same principle as a compass needle lining up with a magnetic field. The needle of the display is attached to a very small permanent magnet. Three coils of wire are used and each produces a magnetic field. The magnet will line up with the resultant of the three fields. The current flowing and the number of turns (ampere-turns) determine the strength of the magnetic flux produced by each coil. As the number of turns remains constant, the current is the key factor. Figure 8.65 shows the principle of the air-cored gauge together with the circuit for use as a temperature indicator. The ballast resistor on the left is used to limit maximum current and the calibration resistor is used for calibration. The thermistor is the temperature sender. As the thermistor resistance is increased, the current in all three coils will change. Current through C will be increased but the current in coils A and B will decrease.

The air-cored gauge has a number of advantages. It has almost instant response, and as the needle is held in a magnetic field it will not move as the vehicle changes position. The gauge can be arranged to continue to register the last position even when switched off or, if a small 'pull off' magnet is used, it will return to its zero position. As a system voltage change would affect the current flowing in all three coils, variations are cancelled out negating the need for voltage stabilisation. Note that the operation is similar to the moving iron gauge.

KEY FACT

Air-cored gauges work on the same principle as a compass needle lining up with a magnetic field.

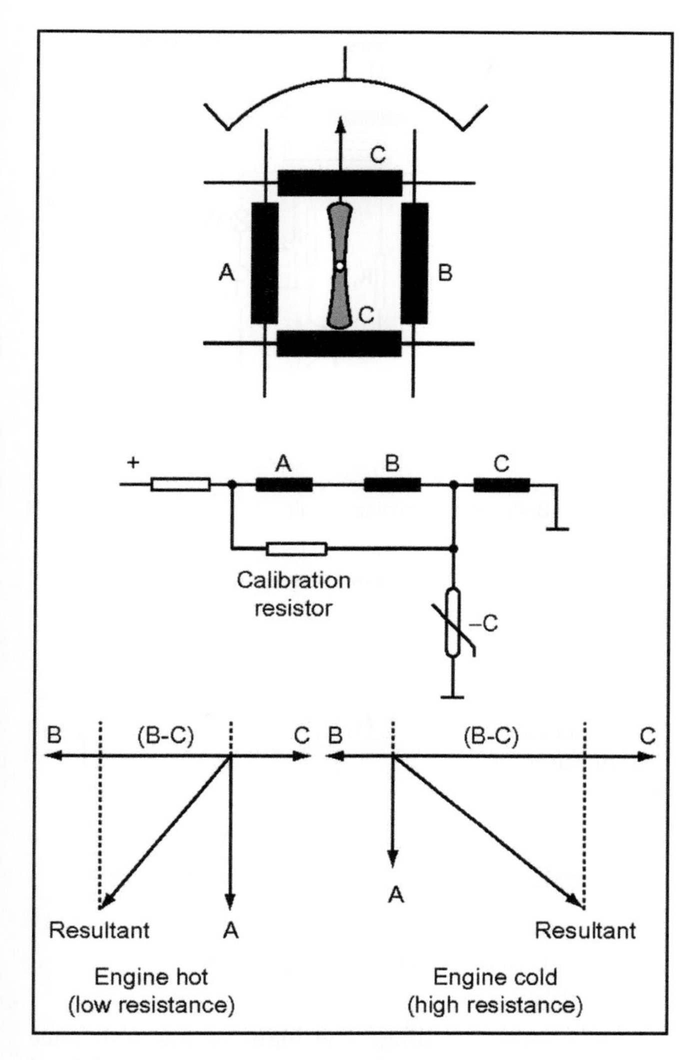

Figure 8.65 Principle of an air-cored gauge and the circuit used for engine temperature.

8.13.2 Digital instrumentation

The block diagram shown in Figure 8.66 is the representation of a digital instrumentation system. All signal conditioning and logic functions are carried out in the ECU. This will often form part of the dashboard assembly. Standard sensors provide information to the ECU, which in turn will drive suitable displays. The ECU contains a ROM (read only memory) section, which allows it to be programmed to a specific vehicle. The gauges used are as described in the above sections. Some of the extra functions available with this system are described briefly as follows:

- Low fuel warning light: Can be made to illuminate at a particular resistance reading from the fuel tank sender unit.
- High engine temperature warning light: Can be made to operate at a set resistance of the thermistor.

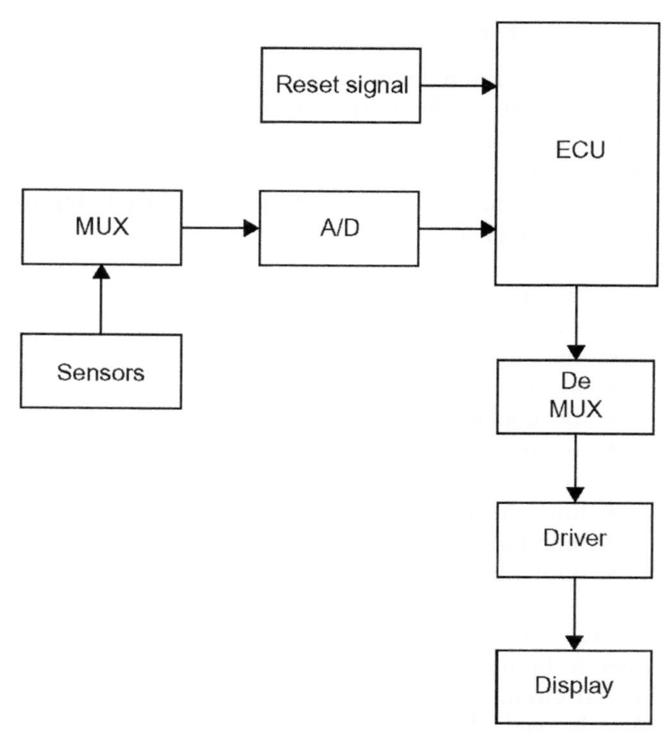

Figure 8.66 Digital instrumentation block diagram.

- Steady reading of the temperature gauge: To prevent the gauge fluctuating as the cooling system thermostat operates, the gauge can be made to read only at say five set figures. For example, if the input resistance varies from 240 to 200 Ω as the thermostat operates, the ECU will output just one reading corresponding to 'normal' on the gauge. If the resistance is much higher or lower, the gauge will read to one of the five higher or lower positions. This gives a low resolution but high readability for the driver.
- Oil pressure or other warning lights can be made to flash: This is more likely to catch the driver's attention.
- Service or inspection interval warning lights can be used: The warning lights are operated broadly as a function of time but, for example, the service interval is reduced if the engine experiences high speeds and/or high temperatures. Oil condition sensors are also used to help determine service intervals.
- Alternator warning light: Works as normal, but the same or an extra light can be made to operate if the output is reduced or if the drive belt slips. This is achieved by a wire from one phase of the alternator providing a pulsed signal, which is compared to a pulsed signal from the ignition. If the ratio of the pulses changed, this would indicate a slipping belt.

8.13.3 Vehicle condition monitoring

Vehicle condition monitoring (VCM) is a sort of enhancement to the normal instrumentation system. For example, a warning light added to a gauge as shown in Figure 8.67.

A system may include driver information relating to the following list:

- high engine temperature;
- low fuel;
- low brake fluid;
- worn brake pads;
- low coolant level;
- low oil level;
- low screen washer fluid;
- low outside temperature;
- bulb failure;
- doors, bonnet, hood or boot open warning.

A circuit is shown in Figure 8.68 that can be used to operate bulb failure warning lights for whatever particular circuit it is monitoring. The simple principle is that the reed relay is only operated when the bulb being monitored is drawing current. The fluid and temperature level monitoring systems work in a similar way to the systems described earlier, but in some cases the level of a fluid is by a float and switch.

Oil level can be monitored by measuring the resistance of a heated wire on the end of the dip stick. A small current is passed through the wire to heat it. How much of the wire is covered by oil will determine its temperature and therefore resistance.

Many of the circuits monitored use a dual-resistance system so that the circuit itself is also checked (Figure 8.69). In effect, it will produce one of three possible outputs: high-resistance, low-resistance or an out-of-range reading. The high- or low-resistance readings are used to indicate say correct fluid level and low fluid level. A figure outside these limits would indicate a circuit fault of either a short or open circuit connection.

Figure 8.67 Low fuel warning.

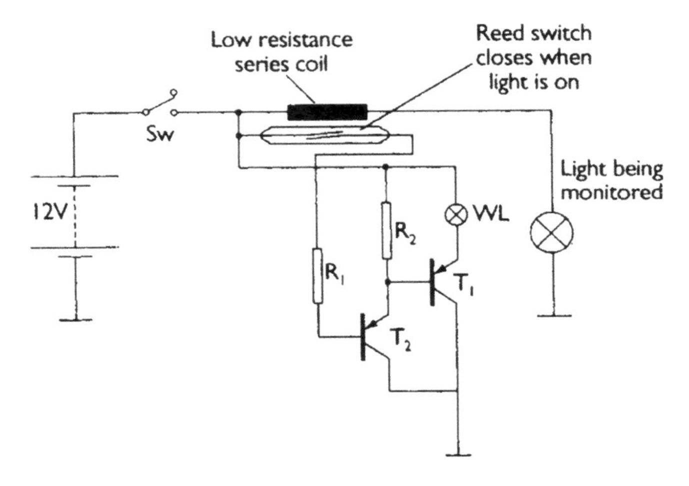

Figure 8.68 Bulb failure warning circuit.

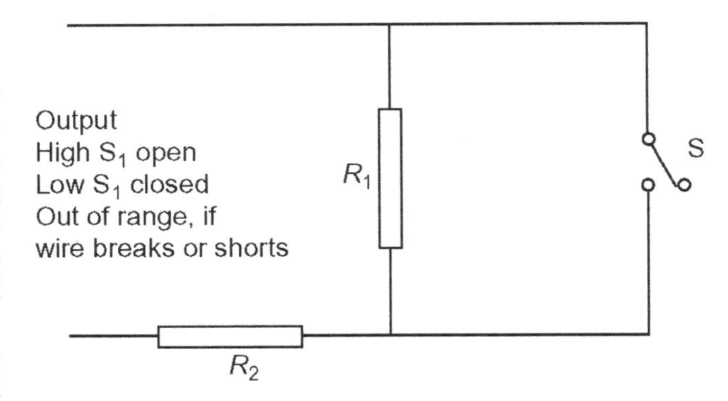

Figure 8.69 Dual-resistance self-testing system.

KEY FACT

Many of the circuits monitored use a dual-resistance system so that the circuit itself is also checked.

The display on earlier cars was often just a collection of LEDs or a backlit liquid crystal display (LCD). These were even arranged into suitable patterns and shapes such as to represent the circuit or system being monitored.

However, the current trend is to just present information as requested by the driver or to present only what is important at any particular time. Low outside temperature or ice warning is often a large snowflake, or a tyre pressure issue is notified using a symbol shown in Figure 8.70.

Figure 8.70 Low tyre pressure warning symbol.

8.13.4 Trip computer

The trip computer used on many top range vehicles is arguably an expensive novelty, but is popular nonetheless. The functions available on most systems are

- time and date;
- elapsed time or a stop watch;
- estimated time of arrival;
- average fuel consumption;
- range on remaining fuel;
- trip distance.

The above details can usually be displayed in imperial, US or metric units as required. Note that several systems use the same inputs and that several systems 'communicate' with each other. This makes the overall wiring very bulky – if not complicated.

8.13.5 Displays

If the junction of a diode is manufactured in a certain way, light will be emitted from the junction when a current is made to pass in the forward biased direction. This is an LED and will produce red, yellow or green light with slight changes in the manufacturing process. LEDs are used extensively as indicators on electronic equipment and in digital displays. They last for a very long time (50 000 hours) and draw only a small current (Figure 8.71).

LED displays are tending to be replaced with the liquid crystal type for automobile use, which can be backlit to make it easier to read in the daylight. However, LEDs are still popular for many applications. The actual display will normally consist of a number of LEDs arranged into a suitable pattern for the required output. This can range from the standard seven-segment display to show numbers, to a custom-designed speedometer display.

KEY FACT

LED displays are tending to be replaced with the liquid crystal type for automobile use, which can be backlit to make it easier to read in the daylight.

Liquid crystals are substances that do not melt directly from a solid to the liquid phase, but first pass through a para-crystalline stage in which the molecules are partially ordered. In this stage, a liquid crystal is a cloudy or translucent fluid but still has some of the optical properties of a solid crystal.

Mechanical stress, electric and magnetic fields, pressure and temperature can alter the molecular structure of liquid crystals. A liquid crystal also scatters light that shines on it. Because of these properties, liquid crystals are used to display letters and numbers on calculators, digital watches and automobile instrument displays. LCDs are also used for portable computer screens and even television screens. The LCD has many more areas of potential use and developments are ongoing. In particular, this type of display is now good enough to reproduce pictures and text on computer and TV screens.

Many displays now are high resolution and touch operated. This means that designers can create a wide range of options. Figure 8.72 is just one example.

Figure 8.71 Instrument display combining analogue and digital displays.

Figure 8.72 Touchscreen display.

Source: Honda Media.

8.14 DIAGNOSTICS – INSTRUMENTS

8.14.1 Testing procedure

The process of checking a thermal gauge fuel or temperature instrument system is broadly as presented in Figure 8.73.

8.14.2 Instrumentation fault diagnosis table

Symptom	*Possible fault*
Fuel and temperature gauges both read high or low	Voltage stabiliser
Gauges read full/hot or empty/cold all the time	Short/open circuit sensors Short or open circuit wiring
Instruments do not work	Loose or broken wiring/connections/fuse Inoperative instrument voltage stabiliser Sender units (sensor) faulty Gauge unit fault (not very common)

8.14.3 Black box technique for instrumentation

Instrumentation systems, like most others, now revolve around an ECU. The ECU is considered to be a 'black box'; in other words, we know what it should do, but how it does it is irrelevant. Figure 8.74 shows an instrumentation system where the instrument pack could be considered as a black box. Normal faultfinding or testing techniques can now be applied to the sensors and supply circuits.

SAFETY FIRST

Warning: The circuit supply must always be off when carrying out ohmmeter tests.

Remember also the 'sensor to ECU method' of testing described in Chapter 2. A resistance test carried out on a component such as the tank unit (lower right) would give a direct measure of its resistance. A second reading at the instrument pack between the GB and BP wires, if the same as the first, would confirm that the circuit is in good order.

8.15 HEATING, VENTILATION AND AIR CONDITIONING (HVAC)

8.15.1 Ventilation and heating

To allow fresh air from outside the vehicle to be circulated inside the cabin, a pressure difference must be created. This is achieved by using a plenum chamber. A plenum chamber by definition holds a gas (in this case air) at a pressure higher than the ambient pressure. The plenum chamber on a vehicle is usually situated just below the windscreen, behind the bonnet. When the vehicle is moving, the airflow over the vehicle will cause a higher pressure in this area. Suitable flaps and drains are utilised to prevent water entering the car through this opening.

DEFINITION

Plenum chamber: A pressurised housing containing a gas or fluid (typically air) at positive pressure (pressure higher than surroundings). One function of a plenum can be to equalise pressure for more even distribution, because of irregular supply or demand.

By means of distribution trunking, control flaps and suitable 'nozzles', the air can be directed as required. This system is enhanced with the addition of a variable speed blower motor. When extra air is forced into

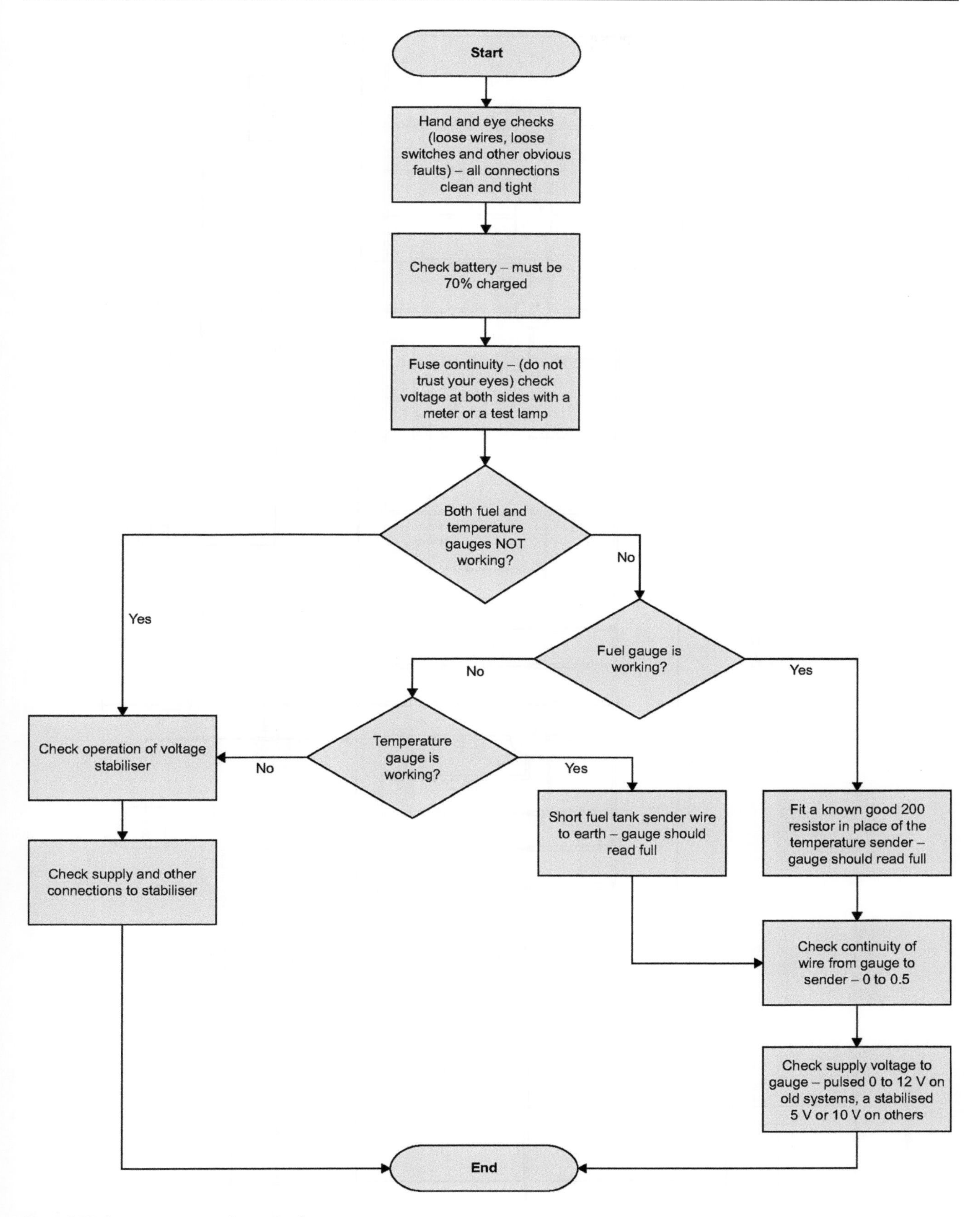

Figure 8.73 Instrumentation diagnosis chart.

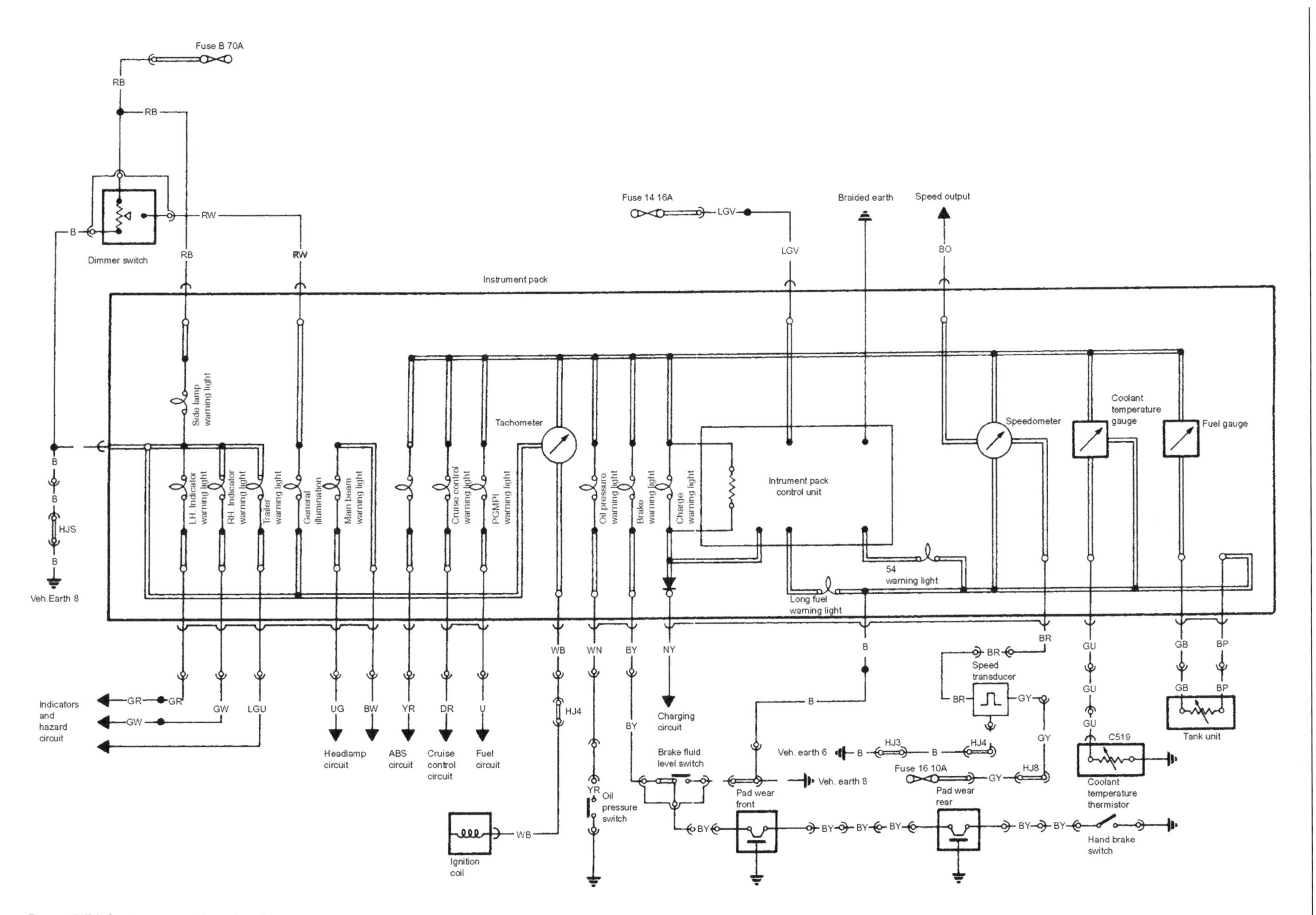

Figure 8.74 Instrumentation circuit.

a vehicle cabin, the interior pressure would increase if no outlet was available. Most passenger cars have the outlet grilles on each side of the vehicle above or near the rear quarter panels or doors.

8.15.2 Heating system – water-cooled engine

Heat from the engine is utilised to increase the temperature of the car interior. This is achieved by use of a heat exchanger, called the heater matrix. Because of the action of the thermostat in the engine cooling system, the water temperature remains broadly constant. This allows for the air being passed over the heater matrix to be heated by a set amount depending on the outside air temperature and the rate of airflow. A source of hot air is therefore available for heating the vehicle interior. However, some form of control is required over how much heat (if any) is required. The method used on most modern vehicles is the blending technique. This is simply a control flap, which determines how much of the air being passed into the vehicle is directed over the heater matrix. The main drawback of this system is the change in airflow with vehicle speed. Some systems use a valve to control the hot coolant flowing to the heater matrix (Figure 8.75).

DEFINITION

HVAC: Heating, ventilation and air conditioning.

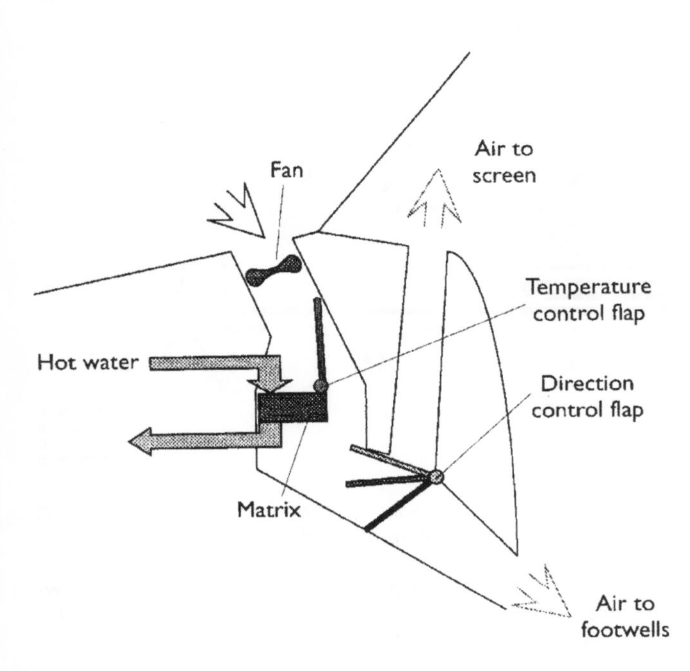

Figure 8.75 Heating and ventilation system.

By a suitable arrangement of flaps, it is possible to direct air of the chosen temperature to selected areas of the vehicle interior. In general, basic systems allow the warm air to be adjusted between the inside of the windscreen and the driver and passenger footwells. Most vehicles also have small vents directing warm air at the driver's and front passenger's side windows. Fresh cool air outlets with directional nozzles are also fitted.

One final facility, which is available on many vehicles, is the choice between fresh or recirculated air. The main reason for this is to decrease the time it takes to demist or defrost the vehicle windows, and simply to heat the car interior more quickly to a higher temperature. The other reason is that the outside air may not be very clean, for example, in heavy congested traffic.

8.15.3 Heater blower motors

The motors used to increase airflow are simple permanent magnet two brush motors. The blower fan is often the centrifugal type, and in many cases, the blades are positioned asymmetrically to reduce resonant noise. Varying the voltage supplied controls motor speed. This is achieved by using dropping resistors. The speed in some cases is made 'infinitely' variable, by the use of a variable resistor. In most cases, the motor is controlled to three or four set speeds.

Figure 8.76 shows a circuit diagram typical of a three-speed control system. The resistors are usually wire wound and are placed in the air stream to prevent overheating. These resistors will have low values in the region of 1 Ω or less (Figure 8.77).

8.15.4 Electronic heating control

Most vehicles that have electronic control of the heating system also include air conditioning, which is covered in the next section. However, a short description at this stage will help to lead into the more complex systems.

This system requires control of the blower motor, blend flap, direction flaps and the fresh or recirculated

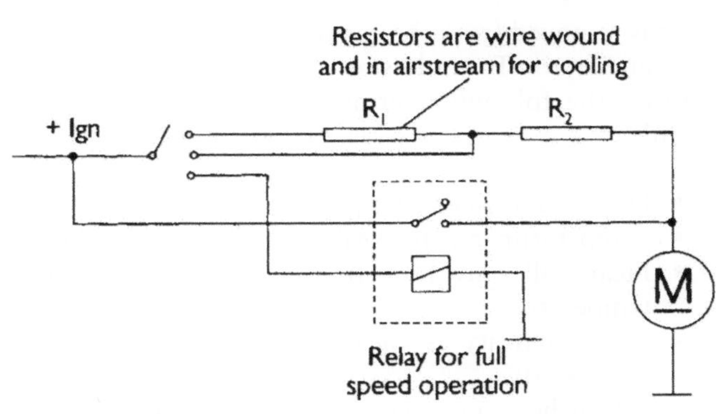

Figure 8.76 Three-speed motor control circuit.

Figure 8.77 Blower motor and fan.

air flap. The technique involves one or a number of temperature sensors suitably positioned in the vehicle interior to provide information for the ECU. The ECU responds to information received from these sensors and sets the controls to their optimum positions. The whole arrangement is in fact a simple closed loop feedback system with the air temperature closing the loop. The ECU has to compare the position of the temperature control switch with the information that is supplied by the sensors and either cool or heat the car interior as required.

8.15.5 Air conditioning introduction

A vehicle fitted with air conditioning allows the temperature of the cabin to be controlled to the ideal or most comfortable value determined by the ambient conditions. The system as a whole still utilises the standard heating and ventilation components, but with the important addition of an evaporator, which both cools and dehumidifies the air.

Air conditioning can be manually controlled or, as is not often the case, combined with some form of electronic control. The system as a whole can be thought of as a type of refrigerator or heat exchanger. Heat is removed from the car interior and dispersed to the outside air. To understand the principle of refrigeration, the following terms and definitions will be useful.

- Heat is a form of energy.
- Temperature means the degree of heat of an object.
- Heat will only flow from a higher to a lower temperature.
- Heat quantity is measured in ‘calories’ (more often kcal).
- 1 kcal heat quantity changes the temperature of 1 kg of liquid water by 1°C.
- Change of state is a term used to describe the changing of a solid to liquid, a liquid to a gas, a gas to a liquid or a liquid to a solid.
- Evaporation is used to describe the change of state from a liquid to a gas.
- Condensation is used to describe the change of state from gas to liquid.
- Latent heat describes the energy required to evaporate a liquid without changing its temperature (breaking of molecular bonds), or the amount of heat given off when a gas condenses back into a liquid without changing temperature (making of molecular bonds).

Latent heat in the change of state of a refrigerant is the key to air conditioning. A simple example of this is that if you put a liquid such as methylated spirits on your hand it feels cold. This is because it evaporates and the change of state (liquid to gas) uses heat from your body. This is why the process is often thought of as ‘unheating’ rather than cooling.

The refrigerant used in many air conditioning systems is known as R134A. This substance changes state from liquid to gas at –26.3°C. R134A is hydrofluorocarbon (HFC) based rather than chlorofluorocarbon (CFC), due to the problems with atmospheric ozone depletion associated with CFC-based refrigerants. A key to understanding refrigeration is to remember that low-pressure refrigerant will have low temperature, and high-pressure refrigerant will have a high temperature.

Figure 8.78 shows the basic principle of an air conditioning – or refrigeration – system. The basic components are the evaporator, condenser and pump or compressor. The evaporator is situated in the car; the condenser is outside the car, usually in the air stream; and the compressor is driven by the engine.

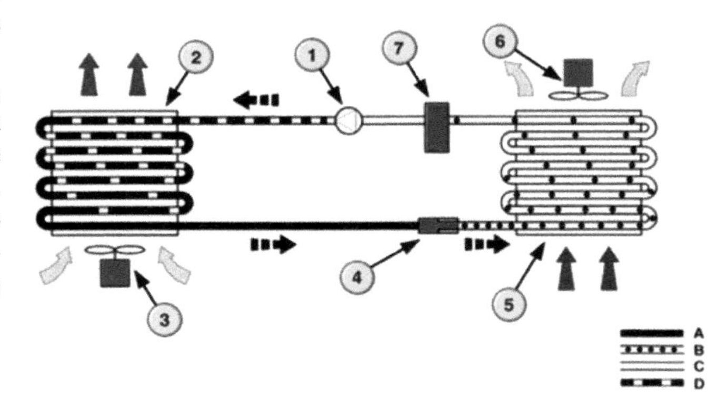

Figure 8.78 AC system layout: 1 – compressor; 2 – condenser; 3 – auxiliary fan (depending on model); 4 – fixed orifice tube; 5 – evaporator; 6 – heater/air conditioning blower; 7 – suction accumulator/drier; A – high-pressure warm liquid; B – low-pressure cool liquid; C – low-pressure, gaseous and cool; D – high-pressure, gaseous and hot.

Source: Ford Motor Company.

Figure 8.79 AC components.

KEY FACT

On an AC system, the evaporator is situated in the car; the condenser is outside the car, usually in the air stream; and the compressor is driven by the engine.

As the pump operates, it will cause the pressure on its intake side to fall, which will allow the refrigerant in the evaporator to evaporate and draw heat from the vehicle interior. The high pressure or output of the pump is connected to the condenser. The pressure causes the refrigerant to condense (in the condenser), thus giving off heat outside the vehicle as it changes state. Figure 8.79 shows some typical components of an air conditioning system.

8.15.6 Air conditioning overview

The operation of the system is a continuous cycle. The compressor pumps low pressure but heatladen vapour from the evaporator, compresses it and pumps it as a superheated vapour under high pressure to the condenser. The temperature of the refrigerant at this stage is much higher than the outside air temperature, hence it gives up its heat via the fins on the condense as it changes state back to a liquid.

This high-pressure liquid is then passed to the receiver drier where any vapour which has not yet turned back to a liquid is stored, and a desiccant bag removes any moisture (water) that is contaminating the refrigerant. The high-pressure liquid is now passed through the thermostatic expansion valve and is converted back to a low-pressure liquid as it passes through a restriction in the valve into the evaporator. This valve is the element of the system that controls the refrigerant flow and hence the amount of cooling provided. As the liquid changes state to a gas in the evaporator, it takes up heat from its surroundings, thus cooling or 'unheating' the air that is forced over the fins. The low-pressure vapour leaves the evaporator returning to the pump, thus completing the cycle.

If the temperature of the refrigerant increases beyond certain limits, condenser cooling fans can be switched in to supplement the ram air effect. A safety switch is fitted in the high-pressure side of most systems. It is often known as a high–low pressure switch, as it will switch off the compressor if the pressure is too high due to a component fault, or if the pressure is too low due to a leakage, thus protecting the compressor.

8.15.7 Automatic temperature control

Full temperature control systems provide a comfortable interior temperature in line with the passenger controlled input. The ECU has full control of fan speed, air distribution, air temperature, fresh or recirculated air and the air conditioning pump. These systems will soon be able to control automatic demist or defrost when reliable sensors are available. A single button currently will set the system to full defrost or demist.

A number of sensors are used to provide input to the ECU.

- Ambient temperature sensor mounted outside the vehicle to allow compensation for extreme temperature variation. This device is usually a thermistor.
- Solar light sensor mounted on the fascia panel. This device is a photodiode and allows a measurement of direct sunlight from which the ECU can determine whether to increase the air to the face vents.
- The in-car temperature sensors are simple thermistors but to allow for an accurate reading a small motor and fan can be used to take a sample of interior air and direct it over the sensing elements.
- A coolant temperature sensor is used to monitor the temperature of the coolant supplied to the heater matrix. This sensor is used to prevent operation of the system until coolant temperature is high enough to heat the vehicle interior.
- Driver input control switches.

KEY FACT

Control of the HVAC flaps can be either by solenoid controlled vacuum actuators or by small motors.

The ECU takes information from all of the above sources and will set the system in the most appropriate manner as determined by the software. Control of the flaps can be either by solenoid controlled vacuum actuators or by small motors. The main blower motor is controlled by a heavy-duty power transistor and is constantly variable. These systems are able to provide a comfortable interior temperature in exterior conditions ranging from –10 to +35°C even in extreme sunlight.

8.15.8 Seat heating

The concept of seat heating is very simple. A heating element is placed in the seat, together with an on-off switch and a control to regulate the heat. However, the design of these heaters is more complex than first appears. The heater must meet the following criteria:

- The heater must only supply the heat loss experienced by the person's body.
- Heat is to be supplied only at the major contact points.
- Leather and fabric seats require different systems due to their different thermal properties.
- Heating elements must fit the design of the seat.
- The elements must pass the same rigorous tests as the seat, such as squirm, jounce and bump tests.

In order for the passengers (including the driver) to be comfortable, rigorous tests have been carried out to find the optimum heat settings and the best position for the heating elements. Many tests are carried out on new designs, using a manikin with sensors attached, to measure the temperature and heat flow (Figure 8.80).

The cable used for most heating elements consists of multi-strand alloyed copper. This cable may be coated with tin or insulated as the application demands. The heating element is laminated and bonded between layers of polyurethane foam.

Figure 8.80 Heated seat.

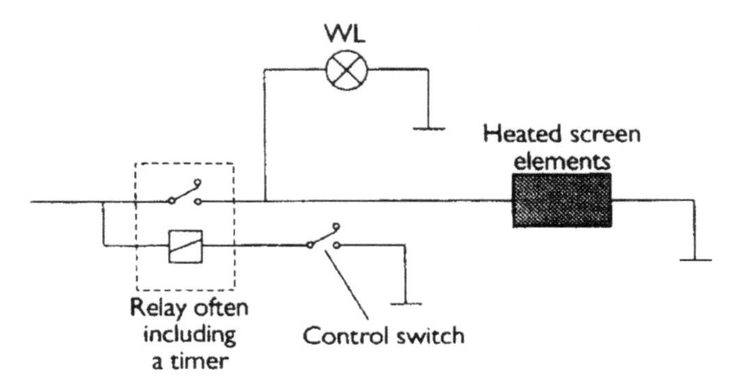

Figure 8.81 Screen heating circuit.

8.15.9 Screen heating

Heating of the rear screen involves a very simple circuit. The heating elements consist of a thin metallic strip bonded to the glass. When a current is passed through the elements, heat is generated and the window will defrost or demist. This circuit can draw high current, 10–15 A being typical. Because of this, the circuit will often contain a timer relay to prevent the heater being left on for too long. The timer will switch off after 10–15 minutes. The elements are usually positioned to defrost the main area of the screen and the rest position of the rear wiper blade if fitted (Figure 8.81).

Front windscreen heating is being introduced on many vehicles. This of course presents more problems than the rear screen, as vision must not be obscured. The technology, drawn from the aircraft industry, involves very thin wires cast in to the glass. As with the heated rear window, this device can consume a large current and is operated by timer relay.

8.15.10 PTC EV heaters

The heater on high voltage electric vehicle is effectively an electric fire. It produces heat by using PTC elements in an air flow or exchanges the heat into coolant. PTC or positive temperature coefficient elements increase in resistance as temperature increases, and because of this they are self-limiting with respect to current flow.

BorgWarner, for example, has technology that is designed to provide rapid cabin heating while making the most efficient possible use of energy in order to conserve battery power.

Unlike legacy vehicles, EVs don't generate a significant source of waste heat that can be used to heat the cabin (Figure 8.82). BorgWarner's high-voltage cabin heater relies on ceramic PTC components to warm the air stream coming from the blower. It self-regulates to ensure that high-power heating is available in cold temperatures when it is needed most. As temperatures rise and heating demand decreases, the energy level is automatically reduced. The heater offers up to 7 kW of power, provides dual-zone functionality for increased efficiency and boasts nearly silent operation.

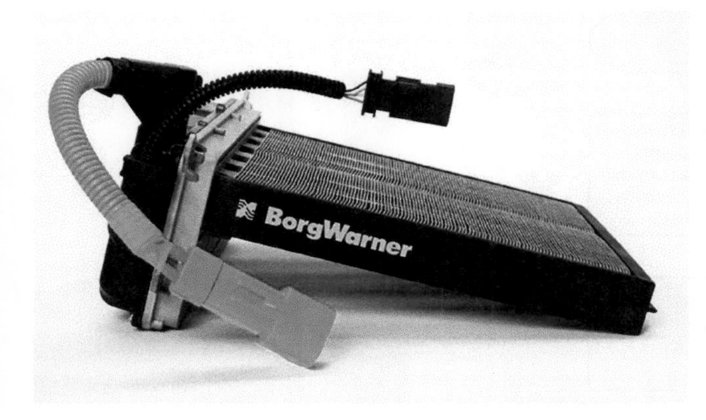

Figure 8.82 PTC heater.

Source: BorgWarner.

8.16 DIAGNOSTICS – HVAC

8.16.1 Testing procedure

The process of checking an air conditioning system is broadly as presented in Figure 8.83.

SAFETY FIRST

Warning: Do not work on the refrigerant side of air conditioning systems unless you have been trained and have access to suitable equipment.

8.16.2 Air conditioning fault diagnosis table

Symptom	*Possible fault*
After stopping the compressor, pressure falls quickly to approximately 195 kPa and then falls gradually	Air in the system, or if no bubbles are seen in the sight glass as the condenser is cooled with water, then excessive refrigerant may be the fault
Discharge pressure low	Fault with the compressor, or if bubbles are seen, low refrigerant
Discharge temperature is lower than normal	Frozen evaporator
Suction pressure too high	High pressure valve fault, excessive refrigerant or expansion valve open too long
Suction and discharge pressure too high	Excessive refrigerant in the system or condenser not working due to fan fault or clogged fins
Suction and discharge pressure too low	Clogged or kinked pipes
Refrigerant loss	Oily marks (from the lubricant in the refrigerant) near joints or seals indicate leaks

8.16.3 Heating and ventilation fault diagnosis table

Symptom	*Possible fault*
Booster fan not operating at any speed	Open circuit fuse/supply/earth Motor inoperative/seized Dropping resistor(s) open circuit Switch open circuit Electronic speed controller not working
Booster fan only works on full speed	Dropping resistor(s) open circuit Switch open circuit Electronic speed controller not working
Control flap(s) will not move	Check vacuum connections (many work by vacuum-operated actuators) Inspect cables
No hot air	Matrix blocked Blend flap stuck
No cold air	Blend flap stuck Blocked intake
Reduced temperature when set to 'Hot'	Cooling system thermostat stuck open Heater matrix partially blocked Control flap not moving correctly

8.16.4 Air conditioning receiver

A very useful guide to diagnostics is the receiver drier sight glass. Figure 8.84 shows four possible symptoms and suggestions as to the possible fault.

8.17 CRUISE CONTROL

8.17.1 Introduction

Cruise control is the ideal example of a closed loop control system as shown in Figure 8.85. The purpose of cruise control is to allow the driver to set the vehicle speed and let the system maintain it automatically. The system reacts to the measured speed of the vehicle and adjusts the throttle accordingly. The reaction time is important so that the vehicle's speed does not feel as if it is surging up and down.

DEFINITION

Negative feedback acts to reduce the input signal that caused it, is also known as a self-correcting or balancing loop. Negative feedback loops are goal seeking, for example, a temperature sensor in a system that compares actual temperature with desired temperature and acts to reduce the difference.

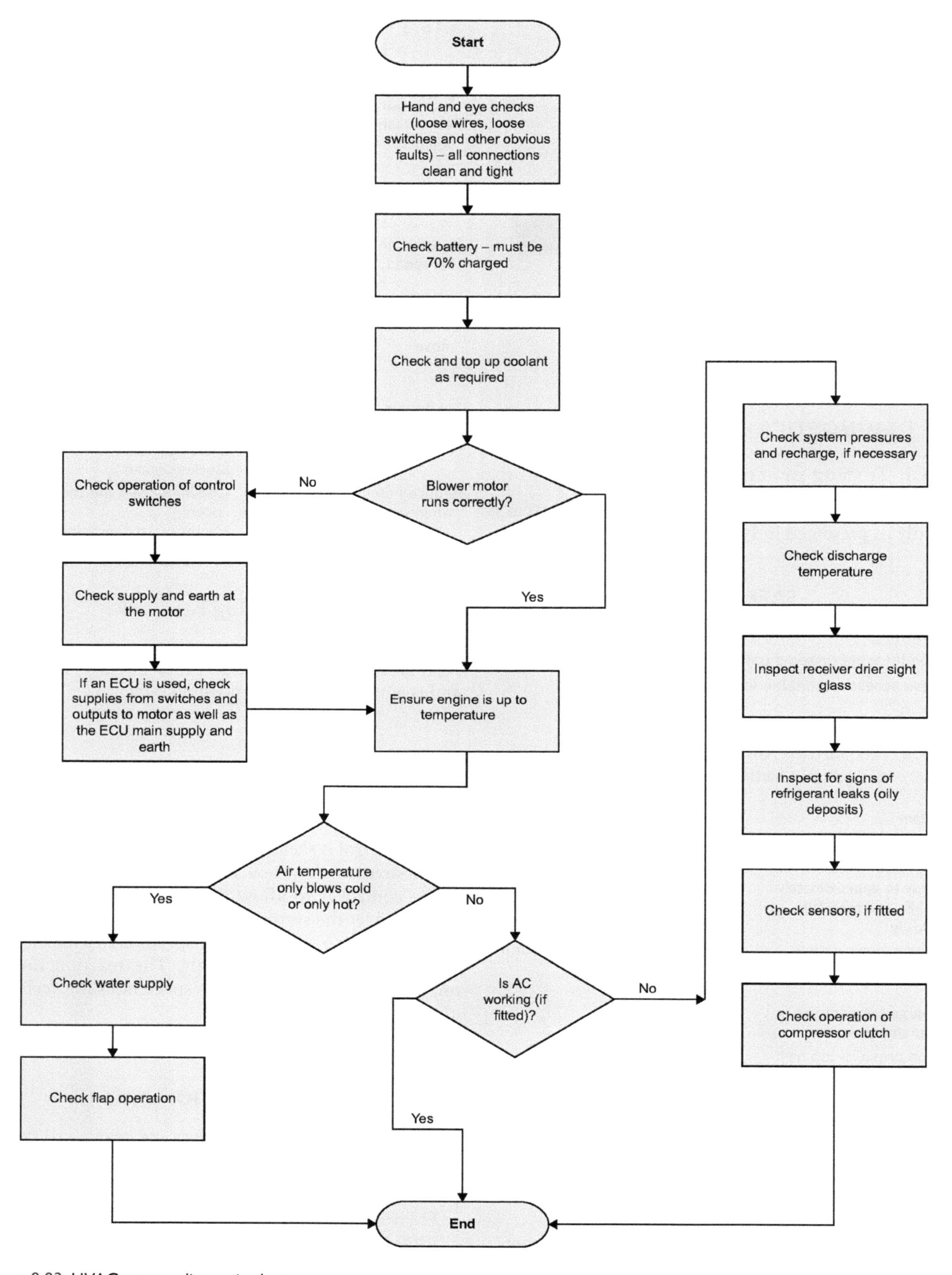

Figure 8.83 HVAC system diagnosis chart.

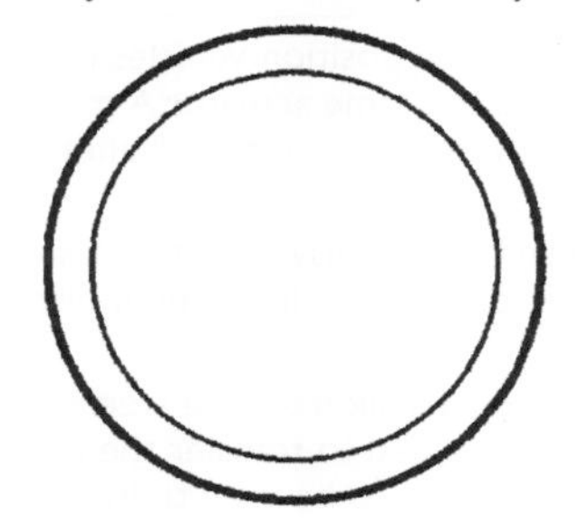

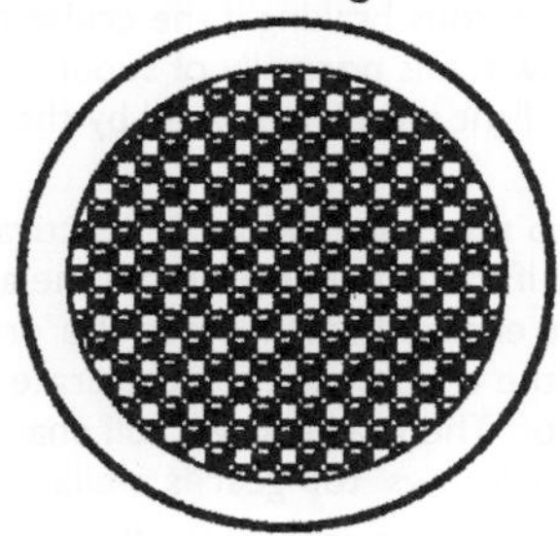

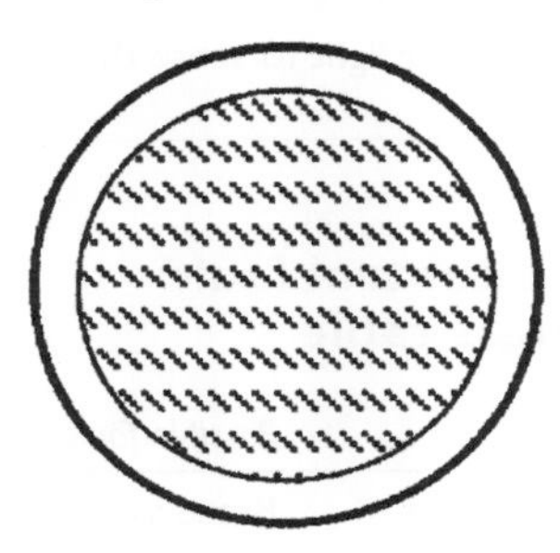

Streaky – Low on refrigerant and compressor oil circulation

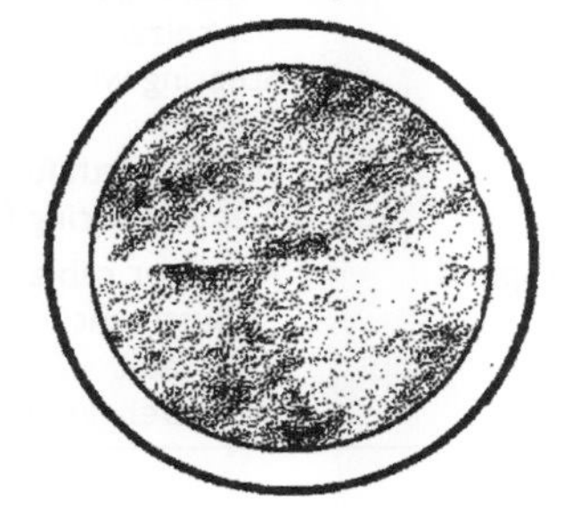

Figure 8.84 AC receiver drier sight glass.

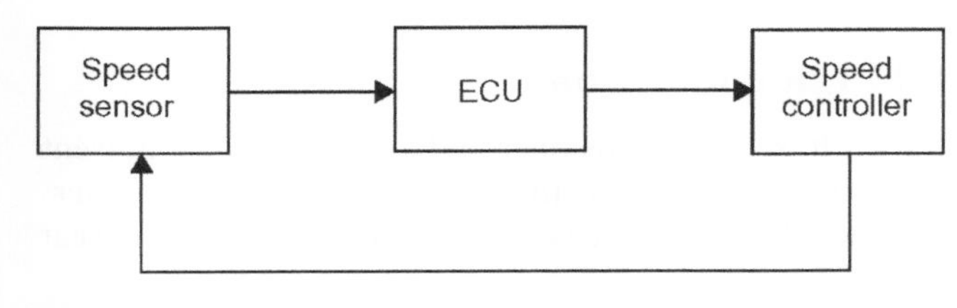

Figure 8.85 Cruise control – closed control loop negative feedback.

Other facilities are included such as allowing the speed to be gradually increased or decreased at the touch of a button. Most systems also remember the last set speed and will resume this again at the touch of a button. To summarise and to add further refinements, the following is the list of functional requirements for a good cruise control system:

- hold the vehicle speed at the selected value;
- hold the speed with minimum surging;
- allow the vehicle to change speed;
- relinquish control immediately after the brakes are applied;
- store the last set speed;
- contain built-in safety features.

8.17.2 System description

The main switch switches on the cruise control; this in turn is ignition controlled. Most systems do not retain the speed setting in memory when the main switch has been turned off. Operating the 'set' switch programs the memory, but this will normally work only if conditions similar to the following are met:

- vehicle speed is greater than 40 km/h;
- vehicle speed is less than 12 km/h;
- change of speed is less than 8 km/h/s;
- automatics must be in 'drive';
- brakes or clutch are not being operated;
- engine speed is stable.

Once the system is set, the speed is maintained to within approximately 3–4 km/h until it is deactivated by pressing the brake or clutch pedal, pressing the 'resume' switch or turning off the main control switch. The last 'set' speed is retained in memory except when the main switch is turned off.

If the cruise control system is required again, then either the 'set' button will hold the vehicle at its current speed or the 'resume' button will accelerate the vehicle to the previous 'set' speed. When cruising at a set speed, the driver can press and hold the 'set' button to accelerate the vehicle until the desired speed is reached when the button is released. If the driver accelerates from the set speed, to overtake for example, then, when the throttle is released, the vehicle will slow down until it reaches the last set position.

8.17.3 Components

The main components of a typical cruise control system are given in Table 8.3 (Figure 8.86).

Table 8.3 Cruise control components

Actuator	A number of methods are used to control the throttle position. Vehicles fitted with drive by wire systems allow the cruise control to operate the same actuator. A motor can be used to control the throttle cable or in many cases a vacuum-operated diaphragm is used which three simple valves control.
Main switch and warning lamp	This is a simple on/off switch located in easy reach of the driver on the dashboard. The warning lamp can be part of this switch or part of the main instrument display as long as it is in the driver's field of vision.
Set and resume switches	These are fitted either on the steering wheel or on a stalk from the steering column. When they are part of the steering wheel slip rings are needed to transfer the connection. The 'set' button programmes the speed into memory and can also be used to increase the vehicle and memory speed. The 'resume' button allows the vehicle to reach its last set speed or to temporarily deactivate the control.
Brake switch	This switch is very important as it would be dangerous braking if the cruise control system was trying to maintain the vehicle speed. This switch is normally of superior quality and is fitted in place or as a supplement to the brake light switch activated by the brake pedal. Adjustment of this switch is important.
Clutch or automatic gearbox switch	The clutch switch is fitted in a similar manner to the brake switch. It deactivates the cruise system to prevent the engine speed increasing if the clutch is pressed. The automatic gearbox switch will only allow the cruise to be engaged when it is in the 'drive' position. This is again to prevent the engine overspeeding if the cruise tried to accelerate to a high road speed with the gear selector in '1' or '2' position. The gearbox will still change gear if accelerating back up to a set speed as long as it 'knows' top gear is available.
Speed sensor	This will often be the same sensor that is used for the speedometer. If not, several types are available; the most common produces a pulsed signal, the frequency of which is proportional to the vehicle speed.
Headway sensor	Only used on 'active' systems, this device uses radar or light to sense the distance from the vehicle in front.

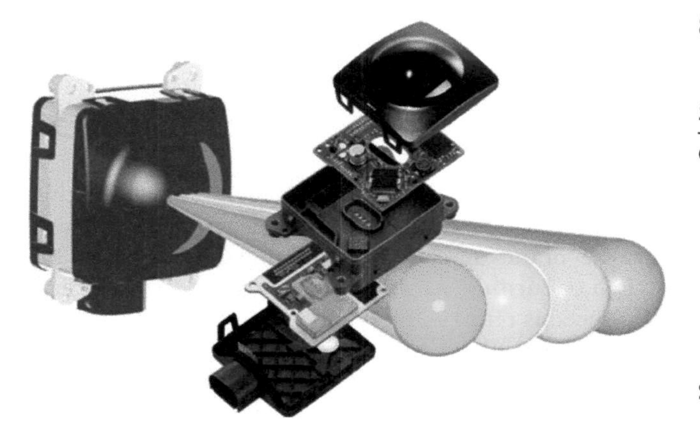

Figure 8.86 Headway sensor and control electronics.

8.18 DIAGNOSTICS – CRUISE CONTROL

8.18.1 Systematic testing

If the cruise control system will not operate then, considering the ECU as a black box, the procedure presented in Figure 8.87 should be followed.

8.18.2 Cruise control fault diagnosis table

Symptom	*Possible fault*
Cruise control will not set	Brake switch sticking on Safety valve/circuit fault Diaphragm holed Actuating motor open circuit or seized Steering wheel slip ring open circuit Supply/earth/fuse open circuit General wiring fault
Surging or uneven speed	Actuator cable out of adjustment ECU fault Engine/engine management fault

8.19 AIR BAGS AND BELT TENSIONERS

8.19.1 Introduction

Seat belt, seat belt tensioner and an airbag are at present the most effective restraint system in the event of a serious accident. At speeds in excess of 40 km/h, the seat

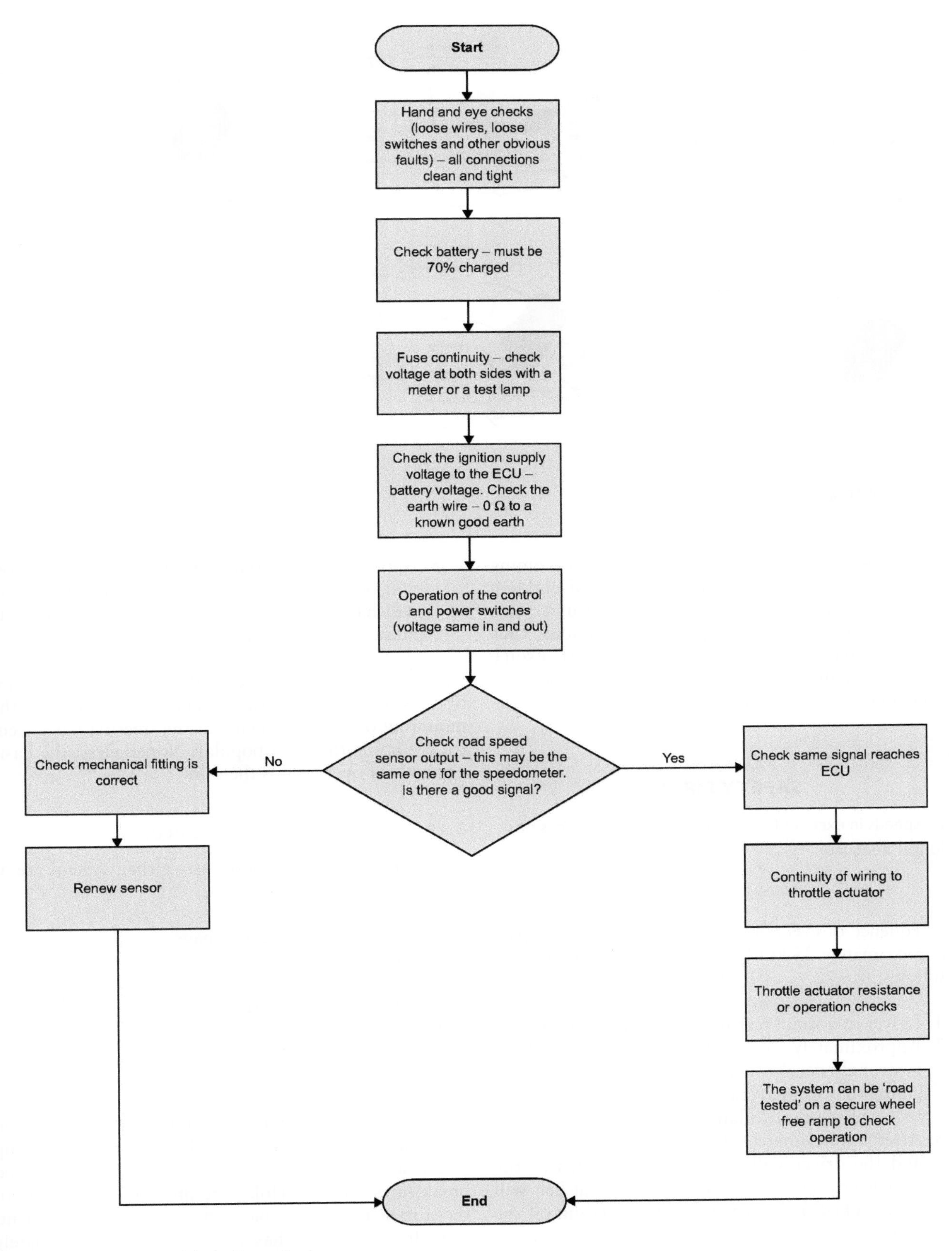

Figure 8.87 Cruise control fault diagnosis chart.

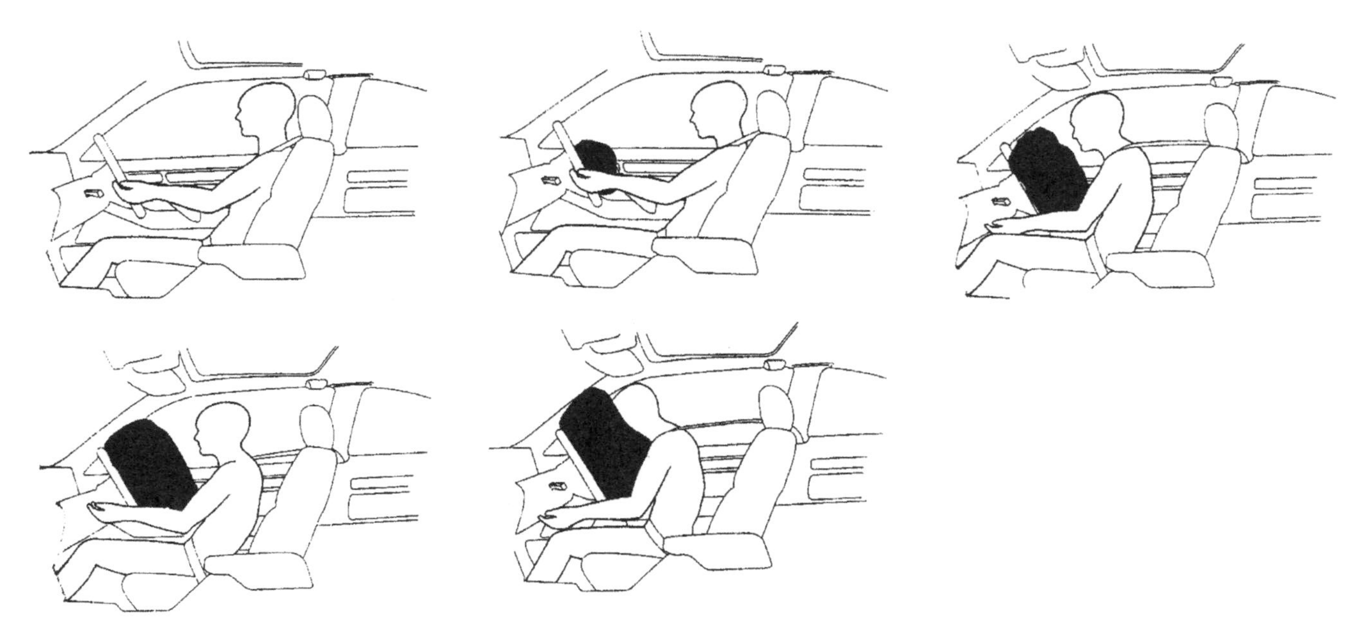

Figure 8.88 Airbag deployment.

belt alone is no longer adequate. The method becoming most popular for an airbag system is that of building most of the required components into one unit. This reduces the amount of wiring and connections, thus improving reliability. An important aspect is that some form of system monitoring must be built in, as the operation cannot be tested – it only works once.

SAFETY FIRST

At speeds in excess of 40 km/h, the seat belt alone is no longer adequate.

The sequence of events in the case of a frontal impact at approximately 35 km/h, as shown in Figure 8.88, is as follows:

1. Driver in normal seating position prior to impact.
2. Approximately 15 ms after the impact the vehicle is strongly decelerated and the threshold for triggering the airbag is reached. The igniter ignites the fuel tablets in the inflater.
3. After approximately 30 ms, the airbag unfolds and the driver will have moved forward as the vehicle crumple zones collapse. The seat belt will have locked or been tensioned depending on the system.
4. At 40 ms after the impact the airbag will be fully inflated and the driver's momentum will be absorbed by the airbag.
5. Approximately 120 ms after the impact the driver will be moved back into the seat and the airbag will have almost deflated through the side vents allowing driver visibility.

Passenger airbag events are similar to the above description. A number of arrangements are used with the mounting of all components in the steering wheel centre becoming the most popular. Nonetheless, the basic principle of operation is the same.

8.19.2 Components and circuit

The main components of a basic airbag system are as follows:

- driver and passenger airbags;
- warning light;
- passenger seat switches;
- pyrotechnic inflater;
- igniter;
- crash sensor(s);
- ECU.

The airbag is made of a nylon fabric with a coating on the inside. Prior to inflation, the airbag is folded up under suitable padding which has specially designed break lines built in. Holes are provided in the side of the airbag to allow rapid deflation after deployment. The driver's airbag has a volume of approximately 60 L and the passenger airbag approximately 160 L. Figure 8.89 shows a steering wheel with an airbag fitted in the centre.

Figure 8.89 The driver's airbag is incorporated into the steering wheel.

KEY FACT

The airbag is made of a nylon fabric with a coating on the inside.

A warning light is used as part of the system monitoring circuit. This gives an indication of a potential malfunction and is an important part of the circuit. Some manufacturers use two bulbs for added reliability.

A seat switch on the passenger side may prevent deployment when not occupied. This may be more appropriate to side-impact airbags.

The pyrotechnic inflater and the igniter can be considered together. The inflater in the case of the driver is located in the centre of the steering wheel. It contains a number of fuel tablets in a combustion chamber. The igniter consists of charged capacitors, which produce the ignition spark. The fuel tablets burn very rapidly and produce a given quantity of nitrogen gas at a given pressure. This gas is forced into the airbag through a filter and the bag inflates breaking through the padding in the wheel centre. After deployment, a small amount of sodium hydroxide will be present in the airbag and vehicle interior. Personal protection equipment must be used when removing the old system and cleaning the vehicle interior.

The crash sensor can take a number of forms; these can be described as mechanical or electronic. The mechanical system works by a spring holding a roller in a set position until an impact above a predetermined limit provides enough force to overcome the spring, and the roller moves, triggering a micro switch. The switch is normally open with a resistor in parallel to allow the system to be monitored. Two switches similar to this may be used to ensure that the bag is deployed only in the case of sufficient frontal impact. Note the airbag is not deployed in the event of a roll over. The other main type of crash sensor can be described as an

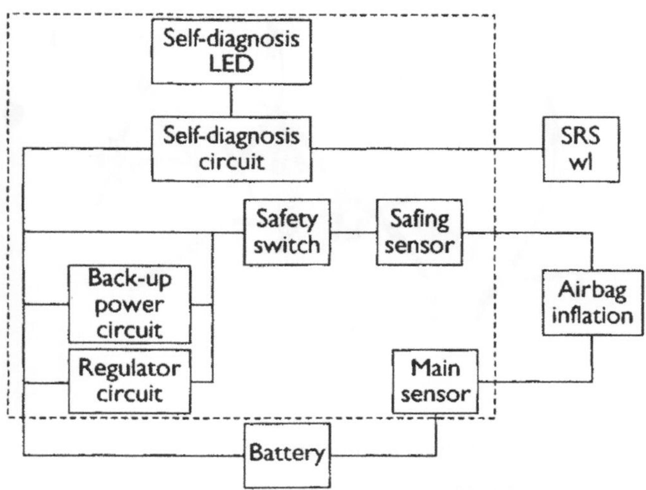

Figure 8.90 SRS block diagram.

accelerometer. This will sense deceleration, which is negative acceleration.

The final component to be considered is the ECU or diagnostic control unit. When a mechanical-type crash sensor is used, in theory no electronic unit would be required. A simple circuit could be used to deploy the airbag when the sensor switch operated. However, it is the system monitoring or diagnostic part of the ECU which is most important. If a failure is detected in any part of the circuit, then the warning light will be operated. Up to five or more faults can be stored in the ECU memory, which can be accessed by blink code or serial fault readers. Conventional testing of the system with a multimeter and jump wires is not to be recommended as it might cause the airbag to deploy.

A block diagram of an airbag circuit is shown in Figure 8.90. Note the 'safing' circuit, which is a crash sensor that prevents deployment in the event of a faulty main sensor. A digital-based system using electronic sensors has approximately 10 ms at a vehicle speed of 50 km/h to decide if the supplementary restraint systems (SRS) should be activated. In this time, approximately 10 000 computing operations are necessary. Data for the development of these algorithms is based on computer simulations but digital systems can also remember the events during a crash allowing real data to be collected.

8.19.3 Seat belt tensioners

Taking the 'slack' out of a seat belt in the event of an impact is a good contribution to vehicle passenger safety. The decision to take this action is the same as for the airbag. The two main types are

- spring tension;
- pyrotechnic.

Figure 8.91 Seat belts and tensioner.

Source: Volvo Media.

Figure 8.92 The reason for SRS!

Source: Saab Media.

The mechanism used by one type of seat belt tensioner works by explosives. When the explosive charge is fired, the cable pulls a lever on the seat belt reel, which in turn tightens the belt. The unit must be replaced once deployed. This feature is sometimes described as anti-submarining (Figures 8.91 and 8.92).

KEY FACT

A seat belt tensioner unit must be replaced once deployed.

8.20 DIAGNOSTICS – AIR BAGS AND BELT TENSIONERS

8.20.1 Systematic testing

The only reported fault for airbags should be that the warning light is staying on. If an airbag has been deployed, then all the major components should be replaced. Some basic tests that can be carried out are presented in Figure 8.93.

SAFETY FIRST

Warning: Careless or incorrect diagnostic work could deploy the airbag causing serious injury. Leave well alone if in any doubt.

8.20.2 Air bags and belt tensioners fault diagnosis table

Symptom	*Possible cause*
Warning light on	Wiring fault Fuse blown or removed ECU fault Crash sensor fault Igniter fault

SAFETY FIRST

Warning: Do not carry out any electrical tests on the airbag circuit.

8.20.3 Deactivation and activation procedures

Airbag simulators are required to carry out diagnosis and testing of the airbag system. For the frontal airbag(s), this tool may be as simple as a 2.5 Ω resistor, used to simulate an airbag module connection to the system. Do not short circuit the airbag module connections with a jumper wire. If a jumper wire is used to short circuit the airbag module connections, a lamp fault code will be displayed and a diagnostic trouble code logged by the airbag control module.

SAFETY FIRST

Warning: Do not work on airbag circuits unless fully trained.

Ford recommend the following procedure for the airbag system fitted to the Ford Focus.

8.20.3.1 Deactivation procedure

Warning: The backup power supply must be depleted before any work is carried out on the supplementary restraint system. Wait at least one minute after

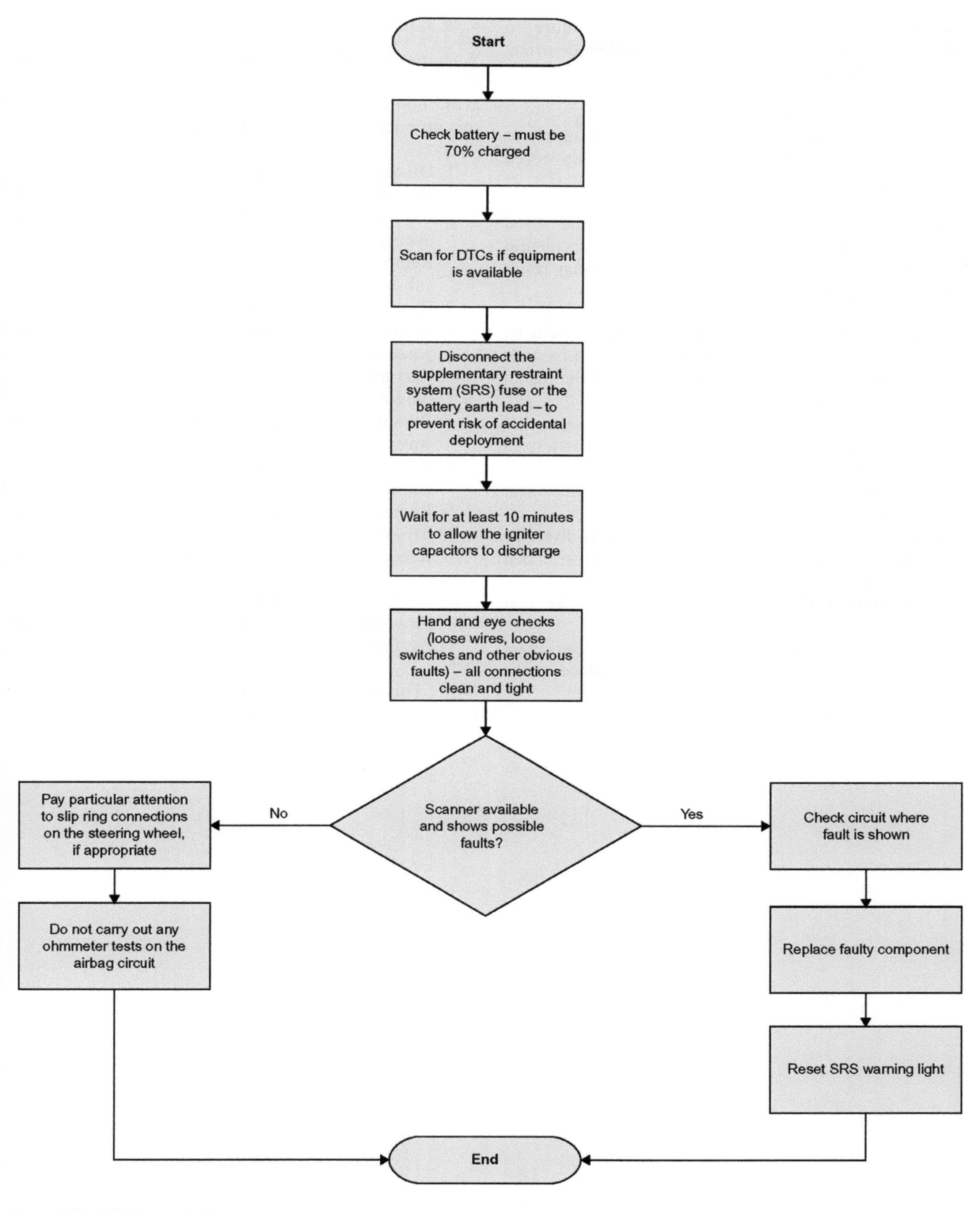

Figure 8.93 SRS diagnosis chart.

disconnecting the battery ground cable. Failure to follow this instruction could cause accidental airbag deployment and may cause personal injury.

1. Disconnect the battery ground cable.
2. Wait one minute for the backup power supply in the Airbag Control Module to deplete its stored energy.

Warning: Place the airbag module on a ground wired bench, with the trim cover facing up to avoid accidental deployment. Failure to follow this instruction may result in personal injury.

1. Remove the driver airbag module from the vehicle.
2. Connect the airbag simulator to the sub-harness in place of the driver airbag module at the top of the steering column.
3. Remove the passenger airbag module.
4. Connect the airbag simulator to the harness in place of the passenger airbag module.
5. Disconnect the driver five-way under seat connector.
6. Connect the airbag simulator to driver five-way under seat floor harness in place of the seat belt pre-tensioner and side airbag.
7. Disconnect the passenger five-way under seat connector.
8. Connect the airbag simulator to the passenger five-way under seat floor harness in place of the seat belt pre-tensioner and side airbag.
9. Reconnect the battery ground cable.

8.20.3.2 Reactivation procedure

Warning: The airbag simulators must be removed and the airbag modules reconnected when reactivated to avoid non-deployment in a collision. Failure to follow this instruction may result in personal injury.

1. Disconnect the battery ground cable.
2. Wait one minute for the backup power supply in the Airbag Control Module to deplete its stored energy.
3. Remove the driver airbag simulator from the sub-harness at the top of the steering column.
4. Reconnect and install the driver airbag module.
5. Remove the passenger airbag simulator from the passenger airbag module harness.
6. Reconnect and install the passenger airbag module.
7. Remove the airbag simulator from the driver five-way under seat connector.
8. Reconnect the driver five-way under seat connector.
9. Remove the airbag simulator from the passenger five-way under seat connector.
10. Reconnect the passenger five-way under seat connector.
11. Reconnect the battery ground cable.
12. Prove out the system, repeat the self-test and clear the fault codes.

Note: This section is included as general guidance; do not assume it is relevant to all vehicles.

Chapter 9

Transmission systems

9.1 MANUAL TRANSMISSION

9.1.1 Clutch

A clutch is a device for disconnecting and connecting rotating shafts. In a vehicle with a manual gearbox, the driver pushes down the clutch when changing gear to disconnect the engine from the gearbox. It also allows a temporary neutral position for, say, waiting at traffic lights and a gradual way of taking up drive from rest.

KEY FACT

A clutch is a device for disconnecting and connecting rotating shafts.

The clutch is made of two main parts: a pressure plate and a driven plate (Figures 9.1 and 9.2). The driven plate, often termed the clutch disc, is fitted on the shaft, which takes the drive into the gearbox. When the clutch is engaged, the pressure plate presses the driven plate against the engine flywheel. This allows drive to be passed to the gearbox. Pushing down the clutch springs the pressure plate away, which frees the driven plate. The diaphragm-type clutch replaced an earlier type with coil springs as it has a number of advantages when used on light vehicles:

- it is not affected by high speeds (coil springs can be thrown outwards);
- low pedal force makes for easy operation;
- it is light and compact;
- clamping force increases or at least remains constant as the friction lining wears.

KEY FACT

The clutch is made of two main parts: a pressure plate and a driven plate.

The method of controlling the clutch is quite simple. The mechanism consists of either a cable or hydraulic system.

9.1.2 Manual gearbox

The driver changes the gears of a manual gearbox by moving a hand-operated lever called a gear stick or shift lever. All manual gearboxes have a neutral position;

Figure 9.1 Clutch cover and pressure plate.

Figure 9.2 Clutch disc.

Figure 9.3 Manual gearbox with a cable change mechanism.

Source: Ford Media.

three, four or five forward gears; and a reverse gear. A few even have six forward gears now. The driver puts the gearbox into neutral as the engine is being started, or when a car is parked with the engine left running (Figure 9.3).

KEY FACT

Power travels in to the gearbox via the input shaft.

Power travels in to the gearbox via the input shaft. A gear at the end of this shaft drives a gear on another shaft called the countershaft or layshaft. A number of gears of various sizes are mounted on the layshaft. These gears drive other gears on a third motion shaft also known as the output shaft.

The gearbox produces various gear ratios by engaging different combinations of gears. For reverse, an extra gear called an idler operates between the countershaft and the output shaft. It turns the output shaft in the opposite direction to the input shaft.

Figure 9.4 shows the power flows through a manual box in each of the different gears. Note how in each case (with the exception of reverse) the gears do not move. This is why this type of gearbox has become known as constant mesh. In other words, the gears are running in mesh with each other at all times. Dog clutches are used to select which gears will be locked to the output shaft. These clutches which are moved by selector levers incorporate synchromesh mechanisms.

KEY FACT

A synchromesh mechanism is needed because the teeth of the dog clutches would clash if they met at different speeds.

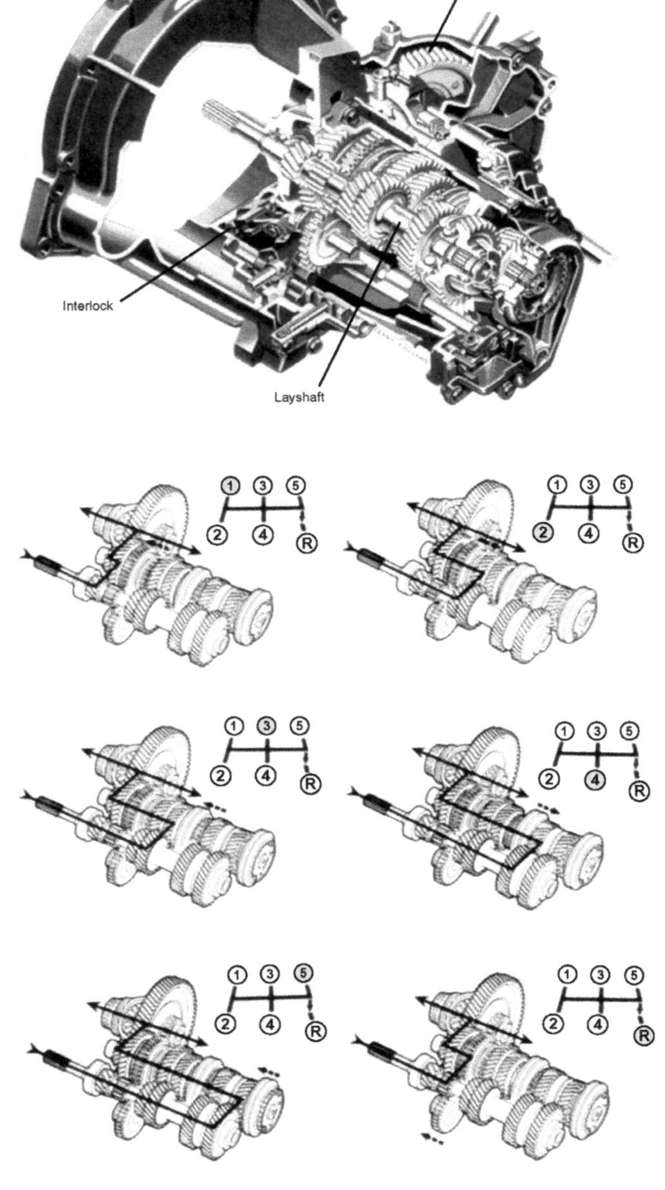

Figure 9.4 Five-speed manual gearbox and power flows.

A synchromesh mechanism is needed because the teeth of the dog clutches would clash if they met at different speeds. The system works like a friction-type cone clutch. The collar is in two parts and contains an outer toothed ring that is spring-loaded to sit centrally on the synchromesh hub. When the outer ring (synchroniser sleeve) is made to move by the action of the selector mechanism, the cone clutch is also moved because of the locking keys. The gear speeds up as the cones touch, thus allowing the dog clutches to engage smoothly. A baulking ring is fitted between the cone on the gear wheel and the synchroniser hub. This is to

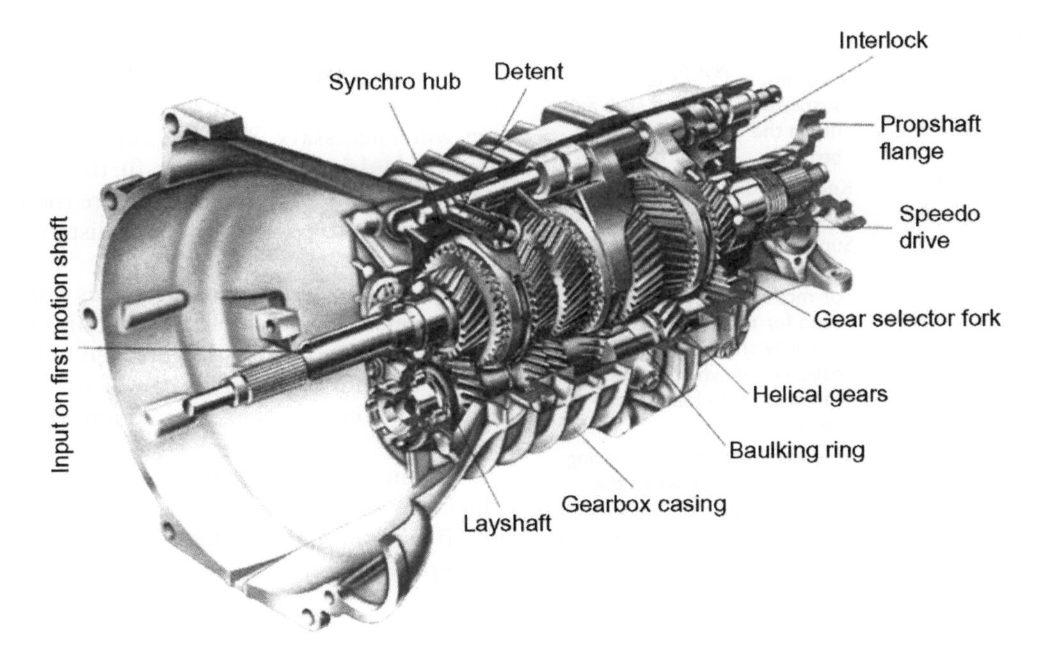

Figure 9.5 RWD manual gearbox.

prevent engagement until the speeds are synchronised (Figure 9.5).

A detent mechanism is used to hold the selected gear in mesh. In most cases, this is just a simple ball and spring acting on the selector shaft(s). Gear selection interlocks are a vital part of a gearbox. These are to prevent more than one gear from being engaged at any one time. On the single rail (one rod to change the gears) gearbox shown in the figure, the interlock mechanism is shown at the rear. As the rod is turned (side-to-side movement of the gear stick) towards first–second, third–fourth or fifth gear positions, the interlock will only engage with either the first–second, third–fourth or fifth gear selectors as appropriate. Equally when any selector clutch is in mesh the interlock will not allow the remaining selectors to change position.

KEY FACT

Gear selection interlocks prevent more than one gear from being engaged at any one time.

KEY FACT

A detent mechanism is used to hold the selected gear in mesh.

9.1.3 Drive shafts and wheel bearings

Light vehicle drive shafts now fall into one of two main categories, the first being by far the most popular.

- **Drive shafts with constant velocity joints (FWD)** – transmit drive from the output of the final drive to each front wheel. They must also allow for suspension and steering movements.
- **Propshaft with universal joints (RWD)** – transmits drive from the gearbox output to the final drive in the rear axle. Drive then continues through the final drive and differential, via two half shafts to each rear wheel. The propshaft must also allow for suspension movements.

Wheel bearings are also very important. They allow smooth rotation of the wheel but must also be able to withstand high stresses such as from load in the vehicle and when cornering (Tables 9.1 and 9.2).

Table 9.1 Front bearings

Seal	Keep out dirt and water, and keep in the grease lubrication.
Spacer	Ensures the correct positioning of the seal.
Inner bearing	Supports the weight of the vehicle at the front, when still or moving. Ball bearings are used for most vehicles with specially shaped tracks for the balls. This is why the bearings can stand side loads when cornering.
Swivel hub	Attachment for the suspension and steering as well as supporting the bearings.
Outer bearing	As for inner bearing.
Drive flange	Runs inside the centre race of the bearings. The wheel is bolted to this flange.

Table 9.2 Rear bearings

Stub axle	Solid mounted to the suspension arm, the stub axle fits in the centre of the two bearings.
Seal	Keep out dirt and water and keep in the grease lubrication.
Inner bearing	Supports the weight of the vehicle at the rear when still or moving. Ball bearings are used for most vehicles with specially shaped tracks for the balls. This is why the bearings can stand side loads when cornering.
Spacer	To ensure the correct spacing and pressure between the two bearings.
Drum	For the brakes and attachment of the wheel.
Outer bearing	As for inner bearing.
Washer	The heavy washer acts as a face for the nut to screw against.
Castle nut and split pin	Holds all parts in position securely. With this type of bearing, no adjustment is made because both bearings are clamped on to the spacer. Some older cars use tapered bearings and adjustment is very important.
Grease retainer cap	Retains grease, but should not be overpacked. Also keeps out the dirt and water.

9.1.4 Final drive and differential

Because of the speed at which an engine runs, and in order to produce enough torque at the road wheels, a fixed gear reduction is required. This is known as the final drive and consists of just two gears. These are fitted after the output of the gearbox, on front wheel drive, or in the rear axle after the propshaft on rear wheel drive vehicles. The gears also turn the drive through 90° on rear wheel drive vehicles. The ratio is normally about 4:1; in other words, when the gearbox output is turning at 4000 rpm, the wheels will turn at 1000 rpm.

KEY FACT

The final drive gears provide a fixed gear reduction.

Many cars now have a transverse engine, which drives the front wheels. The power of the engine therefore does not have to be carried through a right angle to the drive wheels. The final drive contains ordinary reducing gears rather than bevel gears.

The differential is a set of gears that divide the torque evenly between the two drive wheels. The differential also allows one wheel to rotate faster than the other does when necessary. When a car goes around a corner, the outside drive wheel travels further than the inside one. The outside wheel must therefore rotate faster than the inside one to cover the greater distance in the same time.

Some higher-performance vehicles use limited slip differentials. The clutch plates are connected to the two output shafts and hence if controlled will in turn control the amount of slip. This can be used to counteract the effect of one wheel losing traction when high power is applied.

Differential locks are used on many off-road type vehicles. A simple dog clutch or similar device prevents the differential action. This allows far better traction on slippery surfaces.

9.1.5 Four-wheel drive systems

Four-wheel drive (4 WD) provides good traction on rough or slippery surfaces. Many cars are now available with 4 WD. In some vehicles, the driver can switch between 4 WD and two-wheel drive (2 WD). A vehicle with 4 WD delivers power to all four wheels. A transfer box is used to distribute the power between the front and rear wheels:

- transfer gearbox to provide an extra drive output;
- differential on each axle to allow cornering speed variations;
- centre differential to prevent wind-up between the front and rear axles;
- extra drive shafts to supply drive to the extra axle (Figure 9.6).

Figure 9.6 4 WD transmission layout.

Source: Ford Media.

KEY FACT

The transfer gearbox on some vehicles may also contain extra reduction gears for low ratio drive.

One problem to overcome, however, with 4 WD is that if three differentials are used, then the chance of one wheel slipping actually increases. This is because the drive will always be transferred to the wheel with least traction – like running a 2 WD car with one driving wheel jacked up. To overcome this problem and take advantage of the extra traction available, a viscous coupling is combined with an epicyclic gear train to form the centre differential.

The drive can now be distributed proportionally. A typical value is approximately 35% to the front and 65% to the rear wheels. However, the viscous clutch coupling acts so that if a wheel starts to slip, the greater difference in speed across the coupling will cause more friction and hence more drive will pass through the coupling. This tends to act so that the drive is automatically distributed to the most effective driving axle. A 'Hyvo' or silent chain drive is often used to drive from the transfer box.

9.2 DIAGNOSTICS – MANUAL TRANSMISSION

9.2.1 Systematic testing

If the reported fault is a slipping clutch, proceed as follows:

1. Road test to confirm when the fault occurs.
2. Look for oil leaking from the bell housing or general area of the clutch. Check adjustment if possible.
3. If adjustment is correct, then the clutch must be examined.
4. In this example, the clutch assembly must be removed for visual examination.
5. Replace parts as necessary; this is often done as a kit comprising the clutch plate and cover as well as a bearing in some cases.
6. Road test and check operation of all the transmission.

SAFETY FIRST

Note: You should always refer to the manufacturer's instructions appropriate to the equipment you are using.

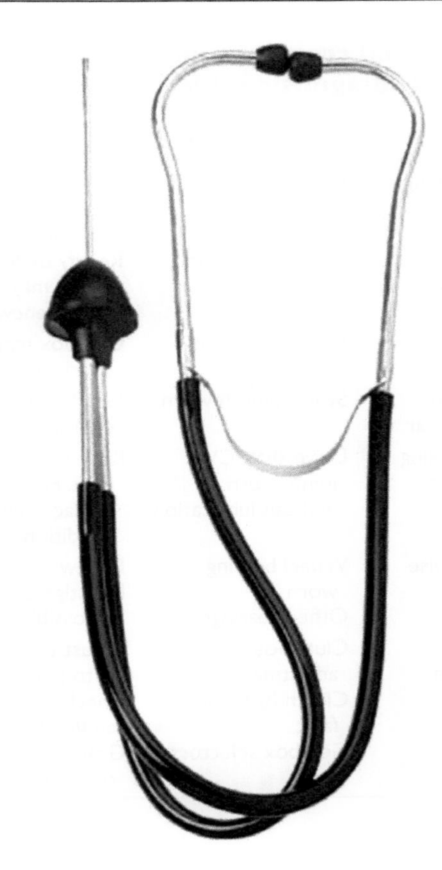

Figure 9.7 Stethoscope.

9.2.2 Test equipment

9.2.2.1 Stethoscope

This is a useful device that can be used in a number of diagnostic situations. In its basic form, it is a long screwdriver. The probe (or screwdriver blade) is placed near the suspected component such as a bearing. The ear piece (or screwdriver handle placed next to the ear) amplifies the sound. Take care though; even a good bearing can sound rough using this method. Compare a known good noise with the suspected one (Figure 9.7).

9.2.3 Test results

Some of the information you may have to get from other sources such as data books or a workshop manual is listed in Table 9.3.

Table 9.3 Tests and information required

Test carried out	*Information required*
Backlash or freeplay	Backlash data is often given as the distance component will move. The backlash between two gears, for example, should be very small.
Overdrive operation	Which gears the overdrive is meant to operate in.

9.2.4 Manual transmission fault diagnosis table 1

Symptom	*Possible causes or faults*	*Suggested action*
Clutch slipping	Clutch worn out Adjustment incorrect Oil contamination	Renew Adjust or check auto-adjuster Rectify oil leak – clutch may also need to be renewed
Jumps out of gear	Gearbox detent fault	Gearbox may require overhaul
Noisy when changing gear	Synchromesh worn	Gearbox may require overhaul
Rapid knocking noise when cornering	Drive shaft CV joints worn or without lubrication	Renew or lubricate joint. Ensure gaiter is in place and in good condition.
Whining noise	Wheel bearing worn Other bearings	Renew Investigate and renew if possible
Difficult to change gear	Clutch out of adjustment Clutch hydraulic fault Gearbox selectors worn	Adjust or check auto-adjuster Check system for air and/or leaks Gearbox may require overhaul

9.2.5 Manual gearbox fault diagnosis table 2

Symptom	*Possible cause*
Noisy in a particular gear (with engine running)	Damaged gear Worn bearing
Noisy in neutral (with engine running)	Input shaft bearings worn (goes away when clutch is pushed down?) Lack of lubricating oil Clutch release bearing worn (gets worse when clutch is pushed down?)
Difficult to engage gears	Clutch problem Gear linkage worn or not adjusted correctly Worn synchromesh units Lack of lubrication
Jumps out of gear	Gear linkage worn or not adjusted correctly Worn selector forks Detent not working Weak synchromesh units
Vibration	Lack of lubrication Worn bearings Mountings loose
Oil leaks	Gaskets leaking Worn seals

9.2.6 Clutch fault diagnosis table

Symptom	*Possible cause*
No pedal resistance	Broken cable Air in hydraulic system Hydraulic seals worn Release bearing or fork broken Diaphragm spring broken
Clutch does not disengage	As above Disc sticking in gearbox splines Disc sticking to flywheel Faulty pressure plate
Clutch slip	Incorrect adjustment Worn disc linings Contaminated linings (oil or grease) Faulty pressure plate
Judder when engaging	Contaminated linings (oil or grease) Worn disc linings Distorted or worn pressure plate Engine mountings worn, loose or broken Clutch disc hub splines worn
Noisy operation	Broken components Release bearing seized Disc cushioning springs broken
Snatching	Disc cushioning springs broken Operating mechanism sticking (lubrication may be required)

9.2.7 Drive shafts fault diagnosis table

Symptom	*Possible cause*
Vibration	Incorrect alignment of propshaft joints Worn universal or CV joints Bent shaft Mountings worn
Grease leaking	Gaiters split or clips loose
Knocking noises	Dry joints Worn CV joints (gets worse on tight corners)

9.2.8 Final drive fault diagnosis table

Symptom	*Possible cause*
Oil leaks	Gaskets split Drive shaft oil seals Final drive output bearings worn (drive shafts drop and cause leaks)
Noisy operation	Low oil level Incorrect pre-load adjustment Bearings worn
Whining noise	Low oil level Worn differential gears

9.3 AUTOMATIC TRANSMISSION

9.3.1 Introduction

An automatic gearbox contains special devices that automatically provide various gear ratios as they are needed. Most automatic gearboxes have three or four forward gears and reverse. Instead of a gear stick, the driver moves a lever called a selector. Some automatic gearboxes have selector positions for park, neutral, reverse, drive, 2 and 1 (or 3, 2 and 1 in some cases). Others just have drive, park and reverse. The engine will only start if the selector is in either the park or neutral position. In park, the drive shaft is locked so that the drive wheels cannot move. It is now quite common when the engine is running to be able to move the selector out of park only if you are pressing the brake pedal. This is a very good safety feature as it prevents sudden movement of the vehicle.

SAFETY FIRST

The selector will not move out of park unless you are pressing the brake pedal on many cars. This is a very good safety feature as it prevents sudden movement of the vehicle.

For ordinary driving, the driver moves the selector to the drive position. The transmission starts out in the lowest gear and automatically shifts into higher gears as the car picks up speed. The driver can use the lower positions of the gearbox for going up or down steep hills or driving through mud or snow. When in position 3, 2 or 1, the gearbox will not change above the lowest gear specified.

9.3.2 Torque converter operation

The torque converter is a device that almost all automatic transmissions now use. It delivers power from the engine to the gearbox like a basic fluid flywheel but also increases the torque when the car begins to move. The torque converter resembles a large doughnut sliced in half. One half, called the pump impeller, is bolted to the drive plate or flywheel. The other half, called the turbine, is connected to the gearbox input shaft. Each half is lined with vanes or blades. The pump and the turbine face each other in a case filled with oil. A bladed wheel called a stator is fitted between them.

The engine causes the pump to rotate and throw oil against the vanes of the turbine. The force of the oil

Figure 9.8 Cutaway torque converter (green, red and blue).

makes the turbine rotate and send power to the transmission. After striking the turbine vanes, the oil passes through the stator and returns to the pump. When the pump reaches a specific rate of rotation, a reaction between the oil and the stator increases the torque. In a fluid flywheel, oil returning to the impeller tends to slow it down. In a torque converter, the stator or reactor diverts the oil towards the centre of the impeller for extra thrust. Figure 9.8 shows a gearbox with a cutaway torque converter.

When the engine is running slowly, the oil may not have enough force to rotate the turbine. But when the driver presses the accelerator pedal, the engine runs faster and so does the impeller. The action of the impeller increases the force of the oil. This force gradually becomes strong enough to rotate the turbine and move the vehicle. A torque converter can double the applied torque when moving off from rest. As engine speed increases, the torque multiplication tapers off until at cruising speed when there is no increase in torque. The reactor or stator then freewheels on its one-way clutch at the same speed as the turbine.

KEY FACT

To improve efficiency, many transmissions now include a lock-up clutch.

The fluid flywheel action reduces efficiency because the pump tends to rotate faster than the turbine. In other words, some slip will occur (approximately 2%). To improve efficiency, many transmissions now include a lock-up clutch. When the pump reaches a specific rate of rotation, this clutch locks the pump and turbine together, allowing them to rotate as one.

9.3.3 Epicyclic gearbox operation

Epicyclic gears are a special set of gears that are part of most automatic gearboxes. They consist of three elements:

1. a sun gear, located in the centre;
2. the carrier that holds two, three or four planet gears, which mesh with the sun gear and revolve around it;
3. an internal gear or annulus (a ring with internal teeth), which surrounds the planet gears and meshes with them.

KEY FACT

Any part of a set of planetary gears can be held stationary or locked to one of the others.

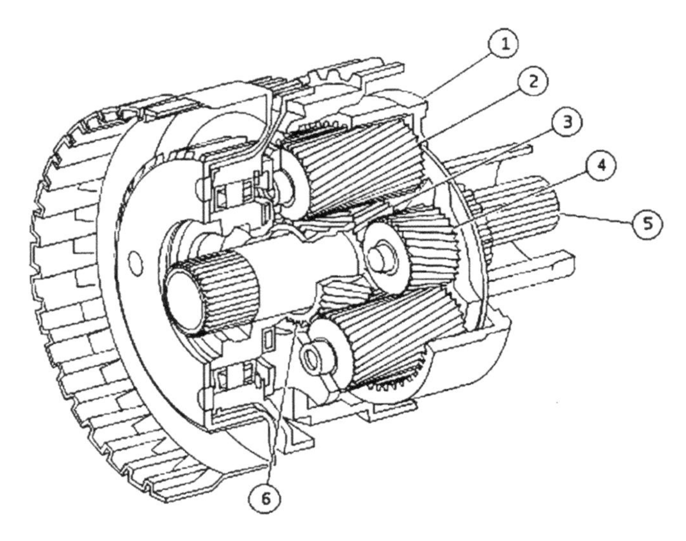

Figure 9.9 Ravigneaux gear set: 1 – ring gear; 2 – long planet gear; 3 – small planet gear; 4 – short planet gear; 5 – transmission input shaft; 6 – large sun gear.

Source: Ford Motor Company.

Any part of a set of planetary gears can be held stationary or locked to one of the others. This will produce different gear ratios. Most automatic gearboxes have two sets of planetary gears that are arranged in line. This provides the necessary number of gear ratios. The appropriate elements in the gear train are held stationary by a system of hydraulically operated brake bands and clutches. These are worked by a series of hydraulically operated valves in the lower part of the gearbox. Oil pressure to operate the clutches and brake bands is supplied by a pump. The supply for this is the oil in the sump of the gearbox (Figure 9.9).

Unless the driver moves the gear selector to operate the valves, automatic gear changes are made depending on just two factors:

1. **throttle opening** – a cable is connected from the throttle to the gearbox;
2. **road speed** – when the vehicle reaches a set speed, a governor allows pump pressure to take over from the throttle.

The cable from the throttle also allows a facility known as 'kick down'. This allows the driver to change down a gear, such as for overtaking, by pressing the throttle all the way down (Figure 9.10).

Many modern semi-automatic gearboxes now use gears in the same way as in manual boxes. The changing

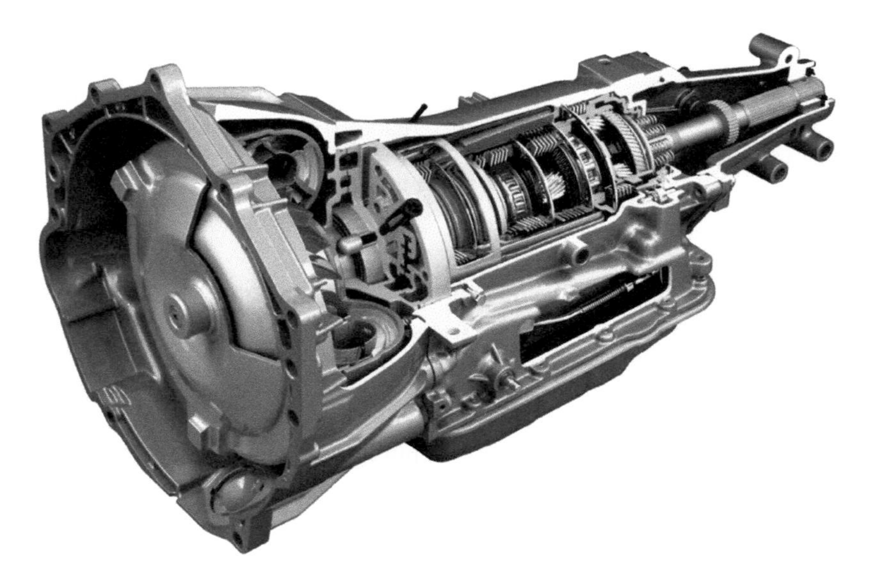

Figure 9.10 RWD automatic gearbox.

Source: GM Media.

of ratios is similar to the manual operation except that hydraulic clutches and valves are used.

KEY FACT

Modern semi-automatic gearboxes have paddle change but also work in a fully automatic mode.

9.3.4 Constantly variable transmission

Figure 9.11 shows a constantly variable transmission (CVT). This kind of automatic transmission uses two pairs of cone-shaped pulleys connected by a metal belt. The key to this system is the high friction drive belt.

DEFINITION

CVT: Constantly variable transmission.

The belt is made from high-performance steel and transmits drive by thrust rather than tension. The ratio of the rotations, often called the gear ratio, is determined by how far the belt rides from the centres of the pulleys. The transmission can produce an unlimited number of ratios. As the car changes speed, the ratio is continuously adjusted. Cars with this system are said to use fuel more efficiently than cars with set gear ratios. Within the gearbox hydraulic control is used to move the pulleys and hence change the drive ratio. An epicyclic gear set is used to provide a reverse gear as well as a fixed ratio.

Figure 9.11 Constantly variable transmission.

Source: Ford Media.

KEY FACT

A CVT transmission can produce an unlimited number of ratios.

9.3.5 Electronic control of transmission

The main aim of electronically controlled automatic transmission (ECAT) is to improve on conventional automatic transmission in the following ways:

- smoother and quieter gear changes;
- improved performance;
- reduced fuel consumption;
- reduction of characteristic changes over system life;
- increased reliability.

The important points to remember are that gear changes and lock-up of the torque converter are controlled by hydraulic pressure. In an ECAT system, electrically controlled solenoid valves can influence this hydraulic pressure. Most ECAT systems now have a transmission ECU that is in communication with the engine control ECU (Figure 9.12).

With an ECAT system, the actual point of gear-shift is determined from pre-programmed memory within the ECU. Data from other sensors is also taken into consideration. Actual gearshifts are initiated by changes in hydraulic pressure, which is controlled by solenoid valves.

The two main control functions of this system are hydraulic pressure and engine torque. A temporary reduction in engine torque during gear shifting allows smooth operation. This is because the peak of gearbox output torque which causes the characteristic surge during gear changes on conventional automatics is suppressed. Because of these control functions smooth gearshifts are possible and, due to the learning ability of some ECUs, the characteristics remain throughout the life of the system (Figure 9.13).

The ability to lock up the torque converter has been used for some time even on vehicles with more conventional automatic transmission. This gives better fuel economy, quietness and improved driveability. Lock-up is carried out using a hydraulic valve, which can be operated gradually to produce a smooth transition. The timing of lock-up is determined from ECU memory in terms of the vehicle speed and acceleration.

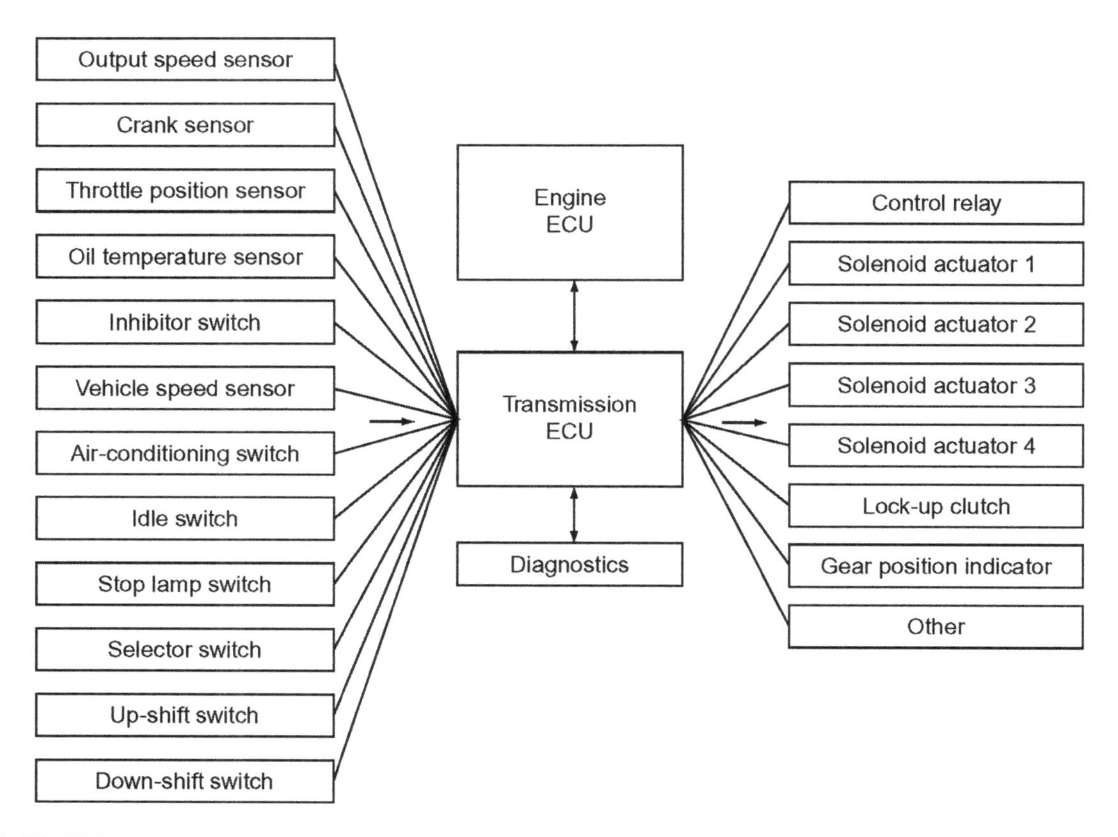

Figure 9.12 ECAT block diagram.

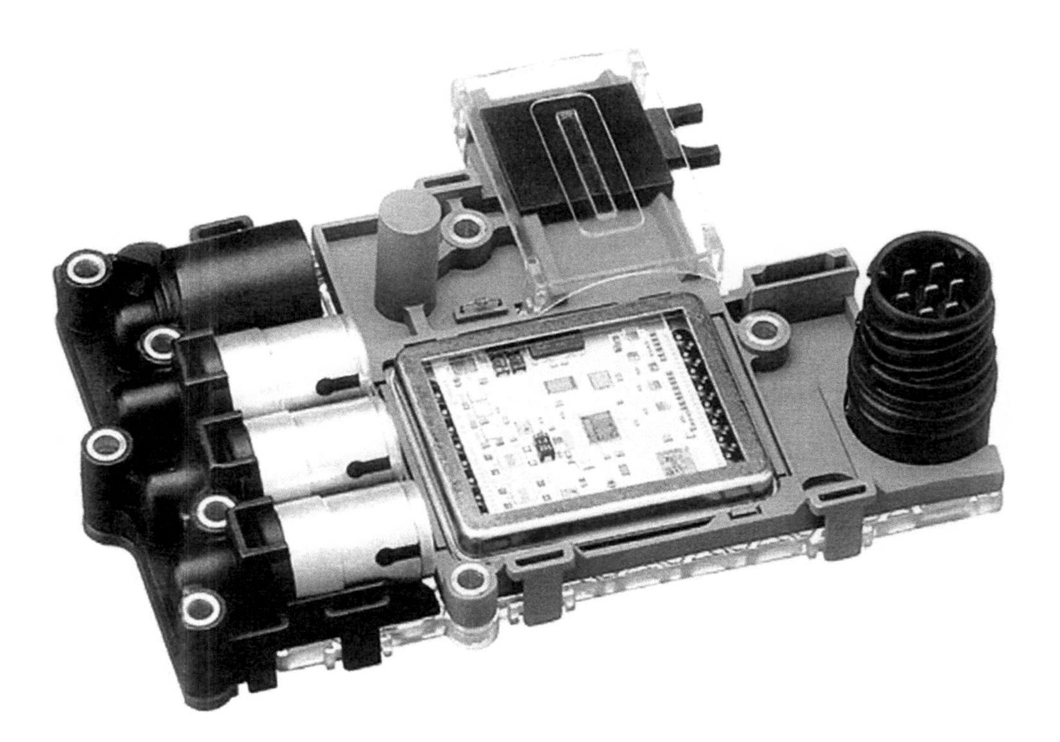

Figure 9.13 Electrohydraulic valve block.

9.3.6 Direct shift gearbox

The direct shift gearbox (DSG) is an interesting development as it could be described as a manual gearbox that can change gear automatically. It can be operated by 'paddles' behind the steering wheel, a lever in the centre console or in a fully automatic mode. The gear train and synchronising components are similar to a normal manual change gearbox (Figure 9.14).

KEY FACT

A DSG can be operated by 'paddles' behind the steering wheel.

The direct shift gearbox is made of two transmission units that are independent of each other. Each transmission unit is constructed in the same way as a manual gearbox and is connected by a multiplate clutch. They are regulated, opened and closed by a mechatronics system. On the system outlined in this section:

- 1st, 3rd, 5th and reverse gears are selected via multiplate clutch 1.
- 2nd, 4th and 6th gears are selected via multiplate clutch 2.

DEFINITION

DSG: Direct shift gearbox.

One transmission unit is always in gear and the other transmission unit has the next gear selected ready for the next change, but with its clutch still in the open position.

KEY FACT

One transmission unit is always in gear and the other transmission unit has the next gear selected ready for the next change.

Torque is transmitted from the crankshaft to a dual-mass flywheel. The splines of the flywheel, on the input hub of the double clutch, transmit the torque to the drive plate of the multiplate clutch. This is joined to the outer plate carrier of clutch 1 with the main hub of the multiplate clutch. The outer plate carrier of clutch 2 is also positively joined to the main hub.

Torque is transmitted into the relevant clutch through the outer plate carrier. When the clutch closes, the torque is transmitted further into the inner plate carrier and then into the relevant gearbox input shaft. One multiplate clutch is always engaged (Figures 9.15 and 9.16).

Figure 9.14 DSG.

Source: Volkswagen Media.

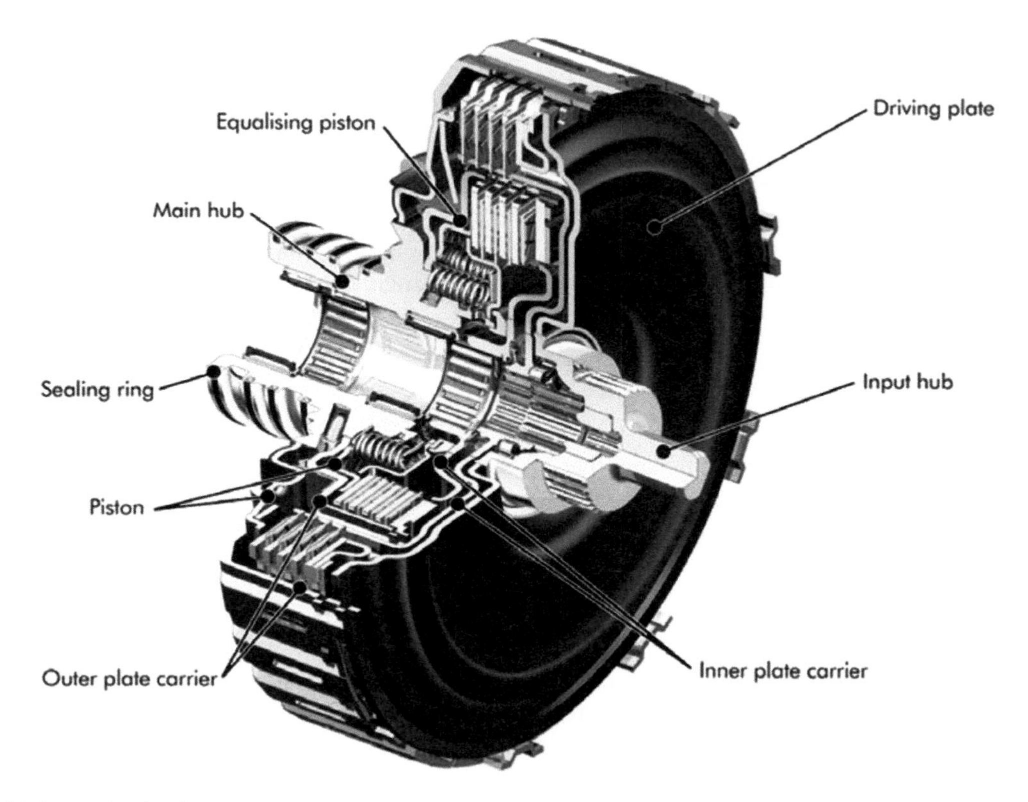

Figure 9.15 Multiplate twin clutch.

Source: Volkswagen Media.

Figure 9.16 Ford twin-clutch transmission components.

Source: Ford Media.

9.4 DIAGNOSTICS – AUTOMATIC TRANSMISSION

9.4.1 Systematic testing

If the reported fault is that the kick down does not operate, proceed as follows:

1. Road test to confirm the problem.
2. Is the problem worse when the engine is hot? Check the transmission fluid level. Has work been done to the engine?
3. If fluid level is correct, then you must investigate further. Work on the engine may have disturbed the kick down cable.
4. Check the adjustment/fitting of the kick down cable.
5. Adjust if incorrect.
6. Run and repeat road test.

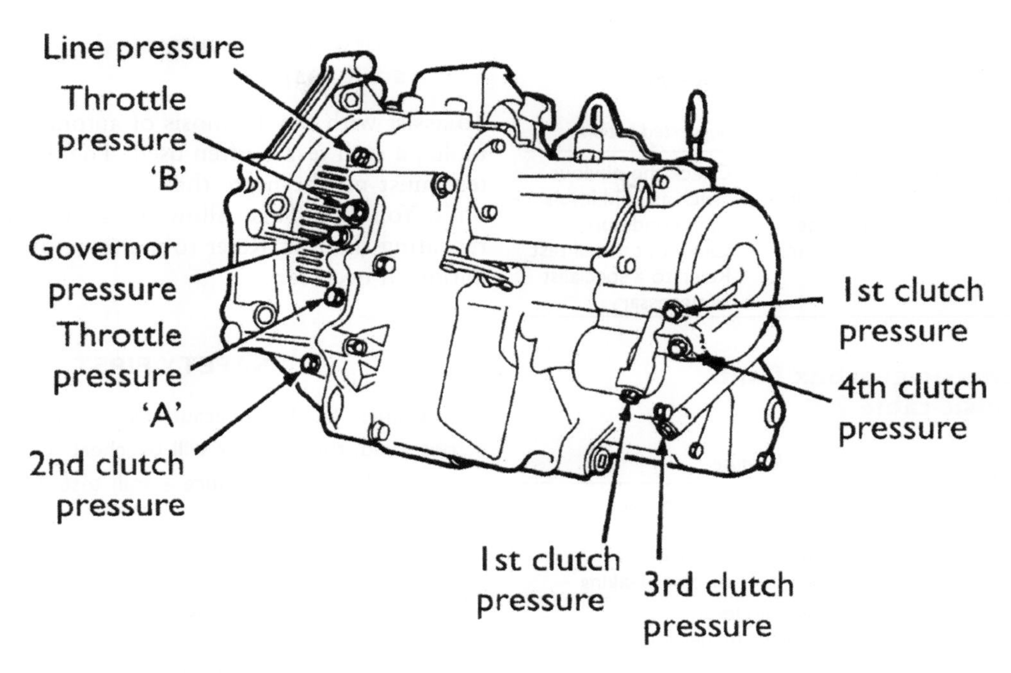

Figure 9.17 Transmission pressure testing points.

SAFETY FIRST

Note: You should always refer to the manufacturer's instructions appropriate to the equipment you are using.

Table 9.4 Tests and information required

Test carried out	*Information required*
Stall test	Highest revs expected and the recommended duration of the test.
Kick down test	Rpm range in which the kick down should operate. For example, above a certain engine rpm, it may not be allowed to work.

9.4.2 Test equipment

9.4.2.1 Revcounter

A revcounter may be used during a stall test to check the operation of the torque converter and the automatic gearbox.

9.4.2.2 Pressure gauge

This is a standard type of gauge but with suitable adapters for connection to a gearbox. Figure 9.17 shows where various tests can be carried out on an automatic gearbox (Figure 9.18).

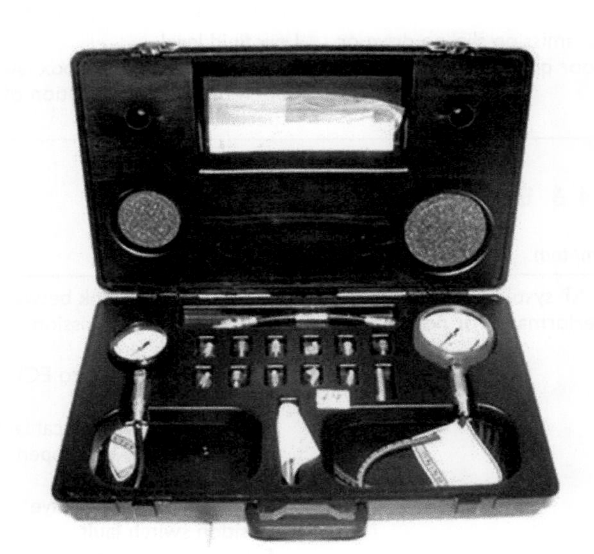

Figure 9.18 Transmission system pressure test kit.

Source: Snap-on.

9.4.3 Test results

Some of the information you may have to get from other sources such as a data book or a workshop manual is listed in Table 9.4:

9.4.4 Automatic gearbox fault diagnosis table 1

Symptom	*Possible faults*	*Suggested action*
Slip, rough shifts, noisy operation or no drive	There are numerous faults that can cause these symptoms	Check the obvious such as fluid levels and condition Carry out a stall test Refer to a specialist if necessary

9.4.5 Automatic gearbox fault diagnosis table 2

Symptom	*Possible cause*
Fluid leaks	Gaskets or seals broken or worn Dip stick tube seal Oil cooler or pipes leaking
Discoloured and/or burnt smell to fluid	Low fluid level Slipping clutches and/or brake bands in the gearbox Fluid requires changing
Gear selection fault	Incorrect selector adjustment Low fluid level Incorrect kick down cable adjustment Load sensor fault (maybe vacuum pipe, etc.)
No kick down	Incorrect kick down cable adjustment Kick down cable broken Low fluid level
Engine will not start or starts in gear	Inhibitor switch adjustment incorrect Faulty inhibitor switch Incorrect selector adjustment
Transmission slip, no drive or poor quality shifts	Low fluid level Internal automatic gearbox faults often require the attention of a specialist

9.4.6 ECAT fault diagnosis table

Symptom	*Possible fault*
ECAT system reduced performance or not working	Communication link between engine and transmission ECUs open circuit Power supply/earth to ECU low or not present Transmission mechanical fault Gear selector switch open/short circuit Speed sensor inoperative Position switch fault Selection switch fault

9.4.7 Automatic transmission stall test

To assist with the diagnosis of automatic transmission faults, a stall test is often used. The duration of a stall test must not be more than approximately seven seconds. You should also allow at least two minutes before repeating the test. Refer to manufacturer's recommendations if necessary.

SAFETY FIRST

Warning: If the precautions mentioned are not observed, the gearbox will overheat. Check manufacturer's data to make sure a stall test is an acceptable procedure.

The function of this test is to determine the correct operation of the torque converter and that there is no transmission clutch slip. Proceed as follows:

1. Run engine up to normal operating temperature by road test if possible.
2. Check transmission fluid level and adjust if necessary.
3. Connect a revcounter to the engine.
4. Apply handbrake and chock the wheels.
5. Apply foot brake, select 'D' and fully press down the throttle for approximately seven seconds.
6. Note the highest rpm obtained (2500–2750 is a typically acceptable range).
7. Allow two minutes for cooling and then repeat the test in '2' and 'R'.

Chapter 10

Practical diagnostics

10.1 REAL CASE STUDIES

10.1.1 Introduction

The case studies present in this chapter are all real jobs that were written up and sent to me by the technician who did the fault-finding. I asked people in several of my Facebook groups to do this. My thanks to everyone who took the time to respond. Unfortunately, I had to limit the number that I could reproduce for space reasons. All the case studies sent to me were really interesting, so thank you again, even if you didn't make the cut!

I would recommend joining some Facebook groups as it is a really good way to share experiences and learn from others. I'm not going to list the groups here, but if you search Facebook for words or combinations of words like 'diagnostic', 'diagnosis', 'diag', 'technician', 'automotive' and so on, there are plenty of groups to choose from.

For each of the following case studies, I have added a sentence or two to comment on specific techniques or methods or say why I have chosen it. They are not presented in any particular order, except for the final two.

Some of the images are lower quality than the others in this book. This is because they were taken on a camera phone, in most cases as a job was being done. I edited and cropped a few of them, but many are as presented, and the important details are clear. Likewise, I have done some light editing of the text, but most of the content is as presented in the words of the technicians. As already mentioned, these are real case studies from real people.

Some of the technicians who contributed, and I fully agree with them, wanted to say that a well-respected instructor, Brandon Steckler, should be acknowledged here. Thank you, Brandon, great job. Brandon has, among many other things, been instrumental in developing and sharing techniques associated with pressure pulse measurement.

10.1.2 Fiat Coupe 5-cylinder turbo with intermittent miss on number 3

Stephen Kasapis, Northside Master Auto Tech

COMMENT:

Here is a good example of how Stephen carried out a number of different tests to confirm the diagnosis.

I had this very rare car into my workshop, a Fiat Coupe five-cylinder turbo (Figure 10.1) with a miss on number 3, but only sometimes! This car has had new coils, new plugs, new injectors and the motor (engine) has just been rebuilt and has compressions of 200 PSI. The workshop that gave me the car said it must be the ECU and asked if I could have a look.

I tested the ignition system and the injectors and did a compression test and all check OK. How can this be if we have a misfire on number 3 cylinder but not all the time. The spark plug is very wet with what looks like fuel. Is somehow the number 3 over-fuelling? I did a few more tests and it was not over-fuelling, so why is

Figure 10.1 Fiat Coupe.

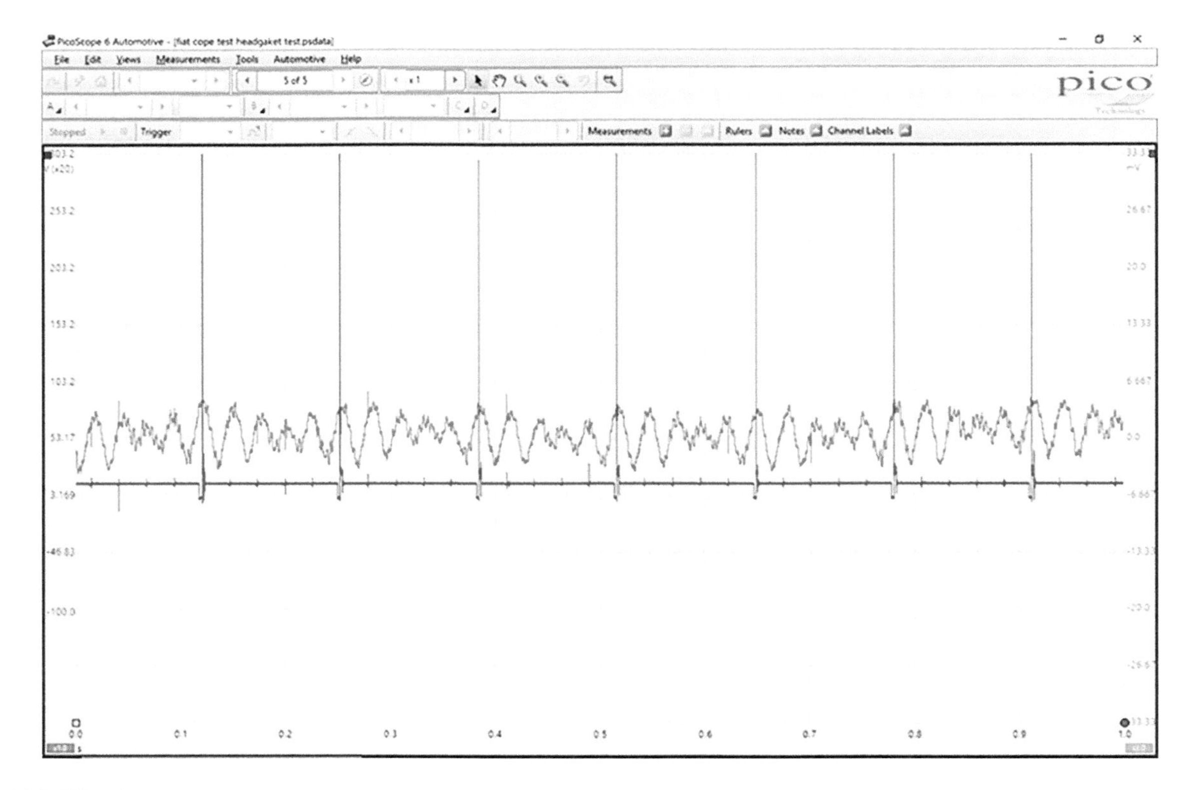

Figure 10.2 Waveforms.

it wet? If you clean the spark plug, it will run OK for about 10 minutes.

I next did a test to see what's going on with the number 3 cylinder. I set up the PicoScope with the first-look sensor in the cooling system and used channel two with the number 3 cylinder primary ignition system (Figure 10.2). We had pressure in the cooling system that lines up with number 3 cylinder. We now know that there is something not right with the cooling system, something like a head gasket probably. Now we know why it misses sometimes.

The last test I did was to put a gas analyser in the top of the radiator, and wow, we had more than 2000 PPM hydrocarbons in the cooling system. We had enough evidence to make the call that this car needed the head to come off.

10.1.3 2015 BMW 530d – unfixable!

Ryan Colley, Elite Automotive Diagnostics www.eliteautomotivediagnostics.co.uk

COMMENT:

Intermittent faults are always a challenge. I really like here how Ryan followed a logical process from the stored codes, through checking a wiring diagram, to then using a scope to monitor the CAN bus.

We recently had in a 2015 plate BMW 530d, which has been to our local dealer where it spent 3 weeks without a diagnosis and to many local BMW specialists again, without any diagnosis. But labour time was billed in excess of £1000. We were told it was unfixable BMW; we will see!

We started with a full system scan and found fault codes stored for the Powertrain CAN-Bus 2 and multiple communication error fault codes.

We were told by the customer this vehicle fails very intermittently and it can take months before it will act up. We analysed a wiring diagram and found only three control modules shared this network, the engine, gearbox and park-neutral switch assembly. Armed with this information, we removed the centre console and connected directly to the CAN bus that was reporting the error. At first, everything seemed normal, so we road-tested the vehicle for three hours before it started to act up. There were no warning lights and none of the customer's symptom to act on; however, we could see clearly the signal was being corrupted.

Next, we removed each component, in turn, monitoring the signal. We found when the transmission control module was disconnected the signal would return correct, indicating our issue (Figures 10.3 and 10.4). Powers and grounds were verified proving our fault was internal to the mechatronics unit in this vehicle. Yet again, another vehicle beating the dealer and many specialists, but not us!

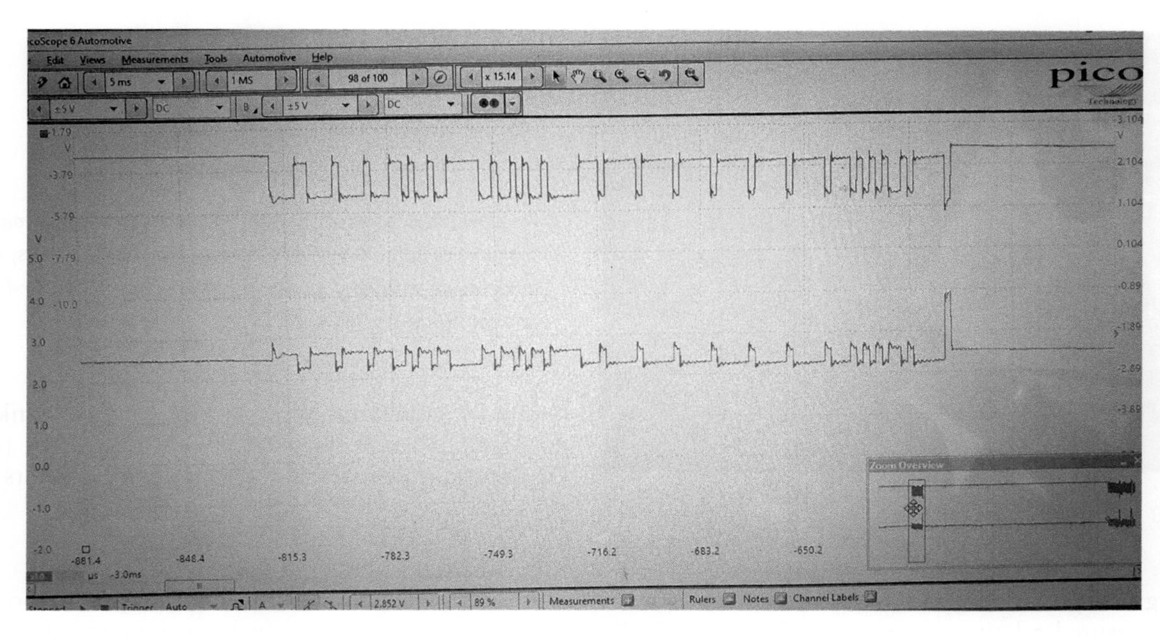

Figure 10.3 Waveform 1.

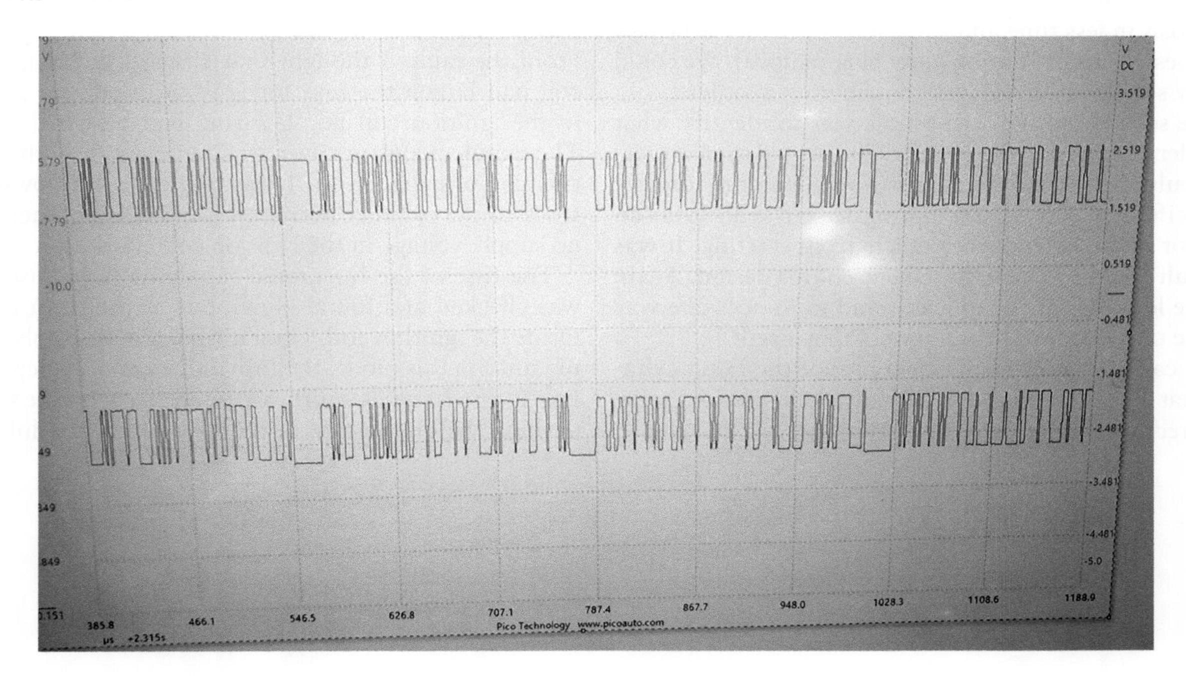

Figure 10.4 Waveform 2.

10.1.4 Misfire after going through a flood

Ryan Colley, Elite Automotive Diagnostics www.eliteautomotivediagnostics.co.uk

COMMENT:

A simple diagnosis here but it illustrates the value of non-intrusive testing, in particular the exhaust pressure pulse measurement.

Unfortunately, due to dramatic weather change we have encountered over the last few days, we had this vehicle towed to us after going through a flood. After the event, the customer reported the vehicle had a lack of power and the engine was shaking (Figure 10.5).

Our inspection began simply by checking the cranking cadence of the engine; it was clear to us that a cylinder was not contributing correctly. In order to carry out a compression test on this vehicle, the glow plugs would need to be removed, which can usually end with breakages due to seizure and this is also very time-consuming.

Figure 10.5 Engine.

Using a scope and current measurement, we nonintrusively measured relative compression (Figure 10.6). By taking measurements this way, we still have an accurate diagnosis in less time and substantially reduced the risk of issues occurring during glow plug removal. We could clearly see one cylinder was contributing a lot less.

The second piece of the puzzle was to identify what cylinder is contributing less. To do this we used a pressure pulse sensor in the exhaust pipe in order to identify a specific cylinder as attempting to get sync from an injector and disabling the vehicle from starting. It was difficult to take our measurement on this design. Again we are looking for the quickest and most accurate way to give our customers the answers they need!

We can now inform the customer of the exact cylinder that is contributing less and advise on the next steps required for repair.

10.1.5 BMW with melted ignition coil

Tony Henry

> **COMMENT:**
>
> *I suspect the smell of burning plastic was also an aid to diagnostics here! What I like is how Tony checked all aspects in detail to ensure a proper diagnosis was made and the correct repair carried out.*

Customer complaint was verified (engine cranks but no start). Vehicle was tested with ISTA-D (BMW Diagnostic software) and the following faults were stored:

- Misfire, Cylinder 1: detected
- Ignition, cylinder 1: combustion period too short
- Relay, ignition and fuel injectors, supply voltage, ignition: short circuit to ground

From the faults I thought that either a faulty ignition coil had blown the fuse for ignition supply or a short in the ignition coil no. 1 wiring had blown the fuse. The engine acoustic cover was removed, and the ignition coil of cylinder no. 1 was found melted down and stuck to the cylinder head cover. In addition, there was no supply voltage in the ignition coil connector.

The fuse in the Integrated supply module (fuse box) was checked and found blown due to the short circuit inside the ignition coil. I quickly got a new supply module since in this model the fuse only cannot be replaced. I performed a quick short to ground check on the supply terminal before putting in the new supply module and

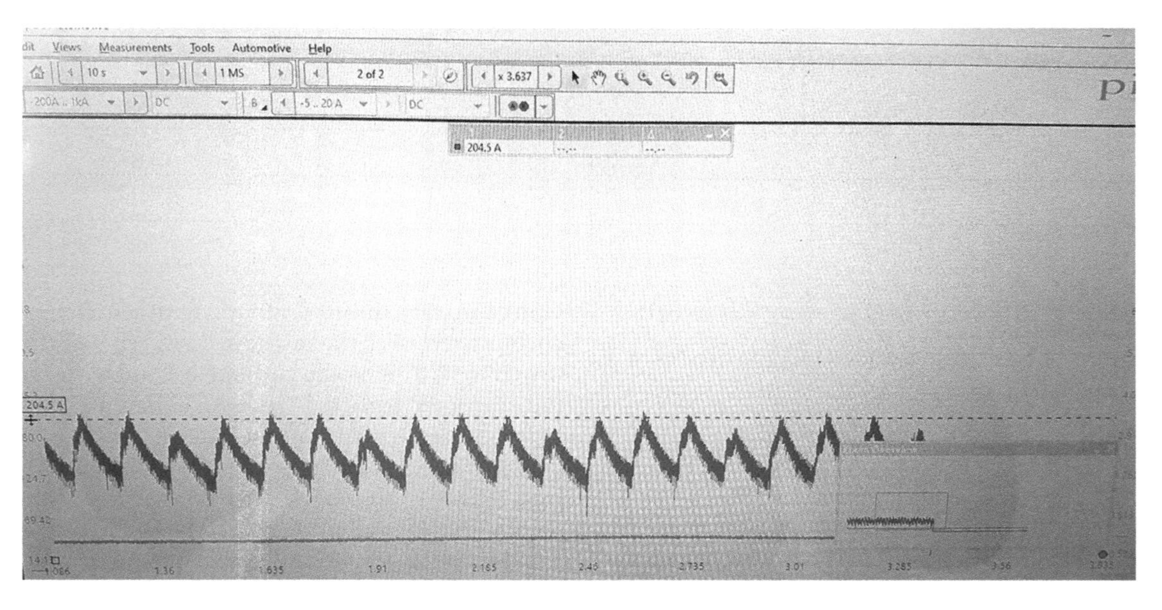

Figure 10.6 Current waveform.

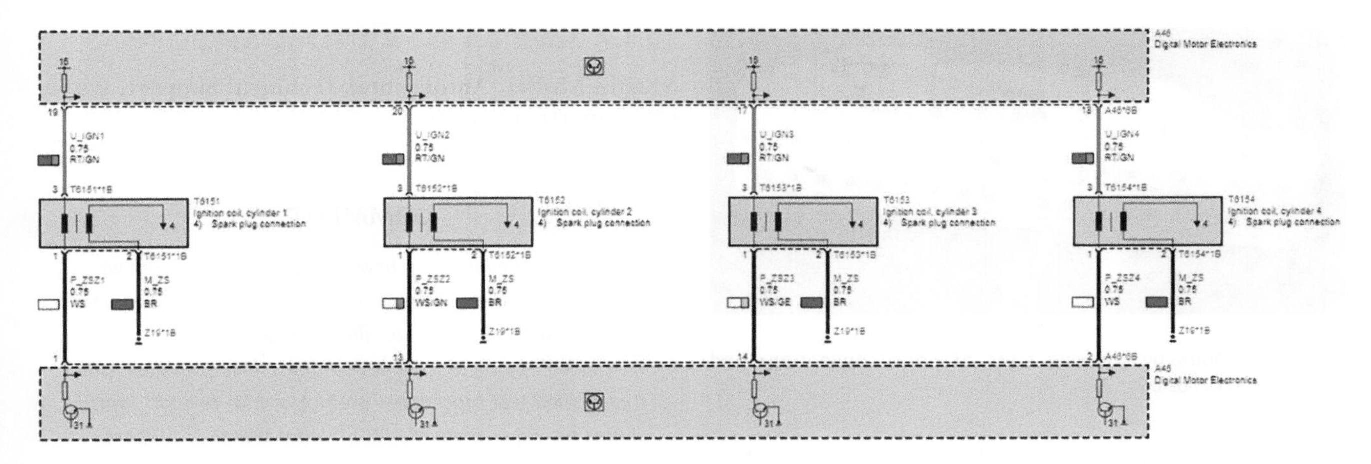

Figure 10.7 Circuit diagram.

removed the plug connection of ignition coil no. 1. The vehicle can be started now with the other three cylinders. For now we will think that it's over and these parts need to be changed:

- cylinder head cover with gasket;
- integrated supply module (fuse box);
- ignition coil;
- ignition coil sleeve.

It will be a big problem if all of the diagnostic procedure isn't completed till the end (that's what I was about to miss) because you will give your customer a quote for these parts and he may agree. But then you will send to get these parts (if they aren't available at your shop) and after replacing the cylinder head cover, ignition coil . . . the vehicle will not start again.

Checking the wiring diagram (Figure 10.7), there is a big component that we didn't check. The engine control unit may have a damaged output stage for this cylinder.

When the plug connection of ignition coil no. 1 is plugged into a new ignition coil, the signals from the DME are omitted in all four cylinders; there is a short inside the control unit as a result of the short resulted from the ignition coil melting (Figure 10.8). But when the plug connection of cylinder no. 1 is removed, the engine control unit triggers the ignition in the other three cylinders.

Figure 10.8 Burnt out coil.

Also, before replacing any ignition coil, the spark plugs should be checked. In this case after removing the spark plugs to check them, I performed a compression test for further check.

So, these parts need to be replaced:

- cylinder head cover with gasket;
- integrated supply module (fuse box);
- engine control unit;
- ignition coil;
- ignition coil sleeve;
- spark plugs.

Someone else may prefer to replace all ignition coils in such a case.

10.1.6 PHEV's won't charge when plugged in

Mike Turner, Drayton Motors Kia (Louth)

COMMENT:

It would have been very easy here to start thinking that there was a serious issue with the hybrid system. This study shows how apparently unrelated systems often communicate.

Customer complaint was their PHEV won't charge when plugged in. On testing I found that when plugged in it would work briefly (green light on dash) but then after 15 seconds it would change to non-charge state red.

The important point to know is that this vehicle will not charge unless the charge-door is locked. A 5 V reference signal is used for this information. When

Figure 10.9 Multiplug showing signs of poor connections and corrosion.

I checked the powers and grounds to the lock solenoid on the charge-door, they were fine, exactly what should be there when locking/unlocking. However, the 5 V reference wire spiked +/- only when locking and very intermittently it would drop to 0 V, then battery volts, then 5 V.

I traced the circuit back to the junction box and the plug that distributes power to central locking. I wasn't able to test the 5 V reference wire for voltage deviation while plugged in as I had no access. But I assumed the voltage was fluctuating when it was plugged in, and the junction box therefore assumed the lock had been opened. The power junction box inside the cab is shown in the images below (Figures 10.9 and 10.10). As the system requires the lock pin in charge door to be in the lock position with the door switch open, it wasn't allowing it into a charge state.

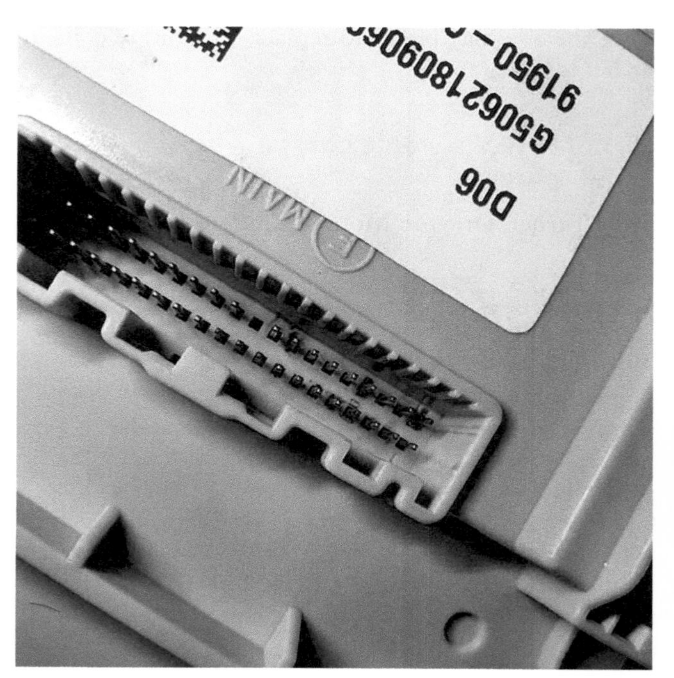

Figure 10.10 Power distribution to the central locking is from this unit.

10.1.7 Audi A6 4G 2.0 TDI Ultra, DPF

Martin Møller, AutoFrontal Technical Support, www.elektropartner.com/en

COMMENT:

This is a great example of how difficult it is to diagnose when the information you need is not available. It also illustrates Martin's determination to find the problem even when things don't go right the first few times. It further illustrates that guided test procedures are not always correct. Martin shares his experiences via AutoFrontal technical support.

My first meeting with Audi's SCR generation 2 – I had this case in 2016, when I was working for Audi in one of the largest garages of the country. The NOx scandal was upon us, and suddenly we started seeing car models called Ultra in the garage.

I clearly remember my thoughts back then: they seem to take the NOx problem seriously now by making the new SCR system, which I call SCR generation 2 at Audi. We had seen the SCR system earlier, as Audi introduced the first SCR system in their 3.0 V6 TDI in the Audi Q5 back in 2008. However, there had been no major challenges on this generation 1 SCR.

Then one day in 2016, I had a waiting customer who needed the engine indicator turned off in his new Audi A6 4G 2.0 TDi Ultra (CNHA), which had run 75,000 km (46,600 miles).

- Fault code P200200 was stored – Diesel Particulate Filter Efficiency Below Threshold, Bank 1.

As the matter was dealt with as a warranty case, I ran a guided troubleshooting with my diagnostic tool ODIS. At first, I wanted to check pressure sensor 1 (G505) to SCR (DPF). To my great astonishment, there were two identical pressure sensors right next to each other (Figure 10.11).

Figure 10.11 G450 and G505 components.

Figure 10.12 Space crunch!

So, which one of the sensors was triggering this fault? In Erwin, I checked the location of the G505 pressure sensor 1. I could not see where the hoses led, as the engine compartment in this model was very compact (Figure 10.12). I applied pressure to G505, and the sensor did not react in measuring values. I then removed the connector from the sensor to see which fault code was stored. A fault was stored in G450. It turned out that there was a fault in the OE garage literature. This is also something that can happen when you work in an OE garage with brand-new systems.

I found the correct pressure sensor and applied various pressures to the sensor and then compared this with the measured values in the diagnostic tool. All pressure measurements showed correct values. I came across a strange observation. When I applied pressure to the G450, it also slightly changed the pressure display of G505. They seemed to be connected in terms of software. However, I chose to stop the guided troubleshooting now to check the soot and ash mass in measuring values. There was nothing unusual to see. Both figures were low.

I chose to go for a drive to measure the backpressure with a pressure gauge but also via the diagnostic tool values. I noticed that the backpressure was normal at idle (below 5 mbars) but at full acceleration, it increased to 291 mbars. Until then, I had had no measurements above 250 mbars in DPF systems with low soot and ash mass, so I wanted to compare with a new flawless reference car. This showed the same backpressure. I also observed all sensors to the exhaust temperature, and there were no faults here. They all showed a realistic temperature in the entire work area of the sensor.

Back to guided troubleshooting. Now I had to complete the regeneration (despite low soot quantity) and afterwards delete the fault code. This was done and the impatient taxi driver was sent off again. It only took a few days until he called again, now seriously fed up with his new Audi as MIL was on again. I got the car in for repair, and this time I had more time available. The same fault code was stored. I analysed freeze frame data:

- The fault code was stored at 2000 rpm;
- The engine load was 30%;
- Hot engine was 99°C.

However, there was one freeze frame data that stood out. SCR efficiency control −41.7%. The limit for SCR efficiency control is −20%. However, we did not have any data for this. The fault code was intermittent and stored in the part load area each time.

I started to study the low-pressure EGR function, but at that time information was limited. I wondered whether the flap in the exhaust pipe clogs up when the low-pressure EGR is active. This flap is monitored by a potentiometer, so it should trigger a fault code (Figure 10.13).

After that, I spent several hours comparing various measuring values to a flawless reference car but I could not find any differences in the measurements. Once again, guided troubleshooting was carried out and as a regeneration had not remedied the problem, it was now concluded that the SCR catalytic converter (DPF) was to be replaced, according to the guided troubleshooting. However, as this fault was intermittent, I chose to replace both G505 and G450 as well as the pins to these.

These connectors have always been a weakness in the differential pressure sensors because it does not take much to create a bad connection to a sensor that operates in a low voltage range (0.5–4.5 V). After a road test of 100 km (62 miles) without any faults, the car was sent off again. A couple of days went by. Then the customer called again about the same fault. This time he talked about his right to a repurchase, if a fault was not corrected within three attempts.

Figure 10.13 Actuator with a potentiometer sensor.

Figure 10.14 Soot in the EGR system.

This time, my boss decided that we should do as guided troubleshooting suggested and replace the SCR catalytic converter. I started the task of disassembling; digging down to the SCR catalytic converter was a huge effort. Fortunately, I found some alarming signs of soot during this disassembly. It turned out that the flap to low-pressure EGR as well as the cooler to low-pressure EGR was full of white/grey soot. Finally, there was a visible fault (Figure 10.14).

New low-pressure EGR flaps and low-pressure EGR cooler were purchased. They were replaced along with the SCR catalytic converter, as I could not take any risks that would cause a repurchase. Fortunately, it turned out that the fault was now remedied.

From then on, increasingly more cases like this one occurred, and since I was now the expert in this field, I could try to only replace the low-pressure EGR cooler and clean the flap to low-pressure EGR. This always remedied the fault, and subsequently an OE TSB (bulletin) was made about this problem.

However, I had difficulty understanding how a clogged EGR cooler could trigger fault code: P200200 - Diesel Particulate Filter Efficiency Below Threshold, Bank 1. However, it makes perfect sense when you see a system drawing like the one in Figure 10.15.

Here you can see that when the engine is running with low-pressure EGR, the flap clogs a bit in the exhaust pipe (max 73°). If the cooler to low-pressure EGR is clogged now, a backpressure is created in DPF which can be registered by pressure sensor 1 (G505). I had expected that pressure sensor 2 would help getting a better diagnosis for this, but that was not the case in the cars I worked on back then.

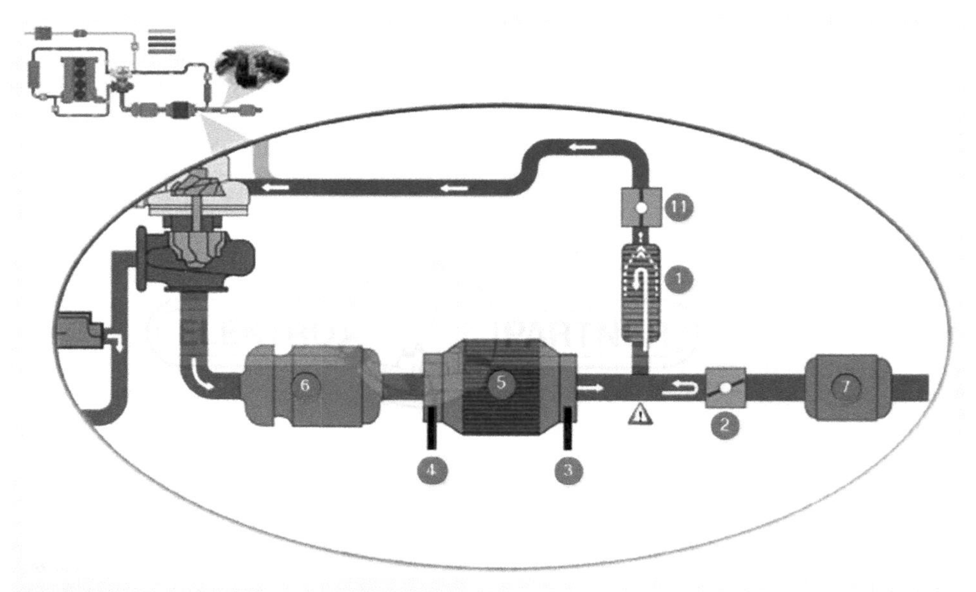

1. **Low-pressure EGR cooler**
2. **Closing valve**
3. **Pressure sensor 2**
4. **Pressure sensor 1**
5. **DPF/SCR**
6. **Oxidation catalytic converter**
7. **Ammonia blocking catalytic converter (model dependent)**
8. **Flap to low-pressure EGR**

Figure 10.15 System schematic.

10.1.8 2014 Skoda Yeti, cruise control not working

Kristopher Williams, Fords of Winsford

COMMENT:

This is a good example where the results of a series of logical tests indicated that there was nothing wrong. It is a good reminder how important it is to keep software updated.

Stalks faulty? Steering column electrics control module faulty? Wiring? Or software? Testing components goes a long way in saving a lot of money.

A car was sent in for cruise control not working and there being a list of fault codes (Figure 10.16). There was no history for the car as it was a sale car. I checked and found the cruise control inoperative. First step was live data for cruise control switches on the indicator stalk, no changes. Next I tested the stalk and checked if there was indeed an open or short circuit as fault code suggests. As seen in Figure 10.17 the switch/stalk is working perfectly and tested at the clock spring. This proves the wiring is OK as well and that the module is incorrectly listing faults.

So, at this stage I know for sure there is no need to replace stalks, so I am left with the module or software.

VEHICLE INFORMATION

MAKE	Skoda
YEAR	2014
MODEL	Yeti
ENGINE	2.0L TDI-CR (CFJA)
SYSTEM	16-Steering Wheel Electronics

VEHICLE CODES

CodeType	
U112100	Data-BUS Missing Message
B114913	Windshield Wiper Intermittent Mode Switch, Open Circuit
B115411	Cruise Control System Switch, Short Circuit To Ground
B115413	Cruise Control System Switch, Open Circuit
U000200	High Speed Can Communication Bus Performance
B116854	Steering Angle Sensor No Basic Setting

Figure 10.16 Vehicle information and codes.

I tested the module wiring inputs and outputs and all were OK. Looking at the behaviour of the stalks (wipers working correctly despite fault code) made me believe there was some sort of software/coding issue. Asked OSCA (Ford system) to check coding and they found it to be incorrect, with that sorted out, our faults are gone and cruise control is now back working with no replacement parts.

This is about my 10th car over the last year where faults are fixed with software. It shows that it's worth testing and ruling out – if your tests prove everything else is OK.

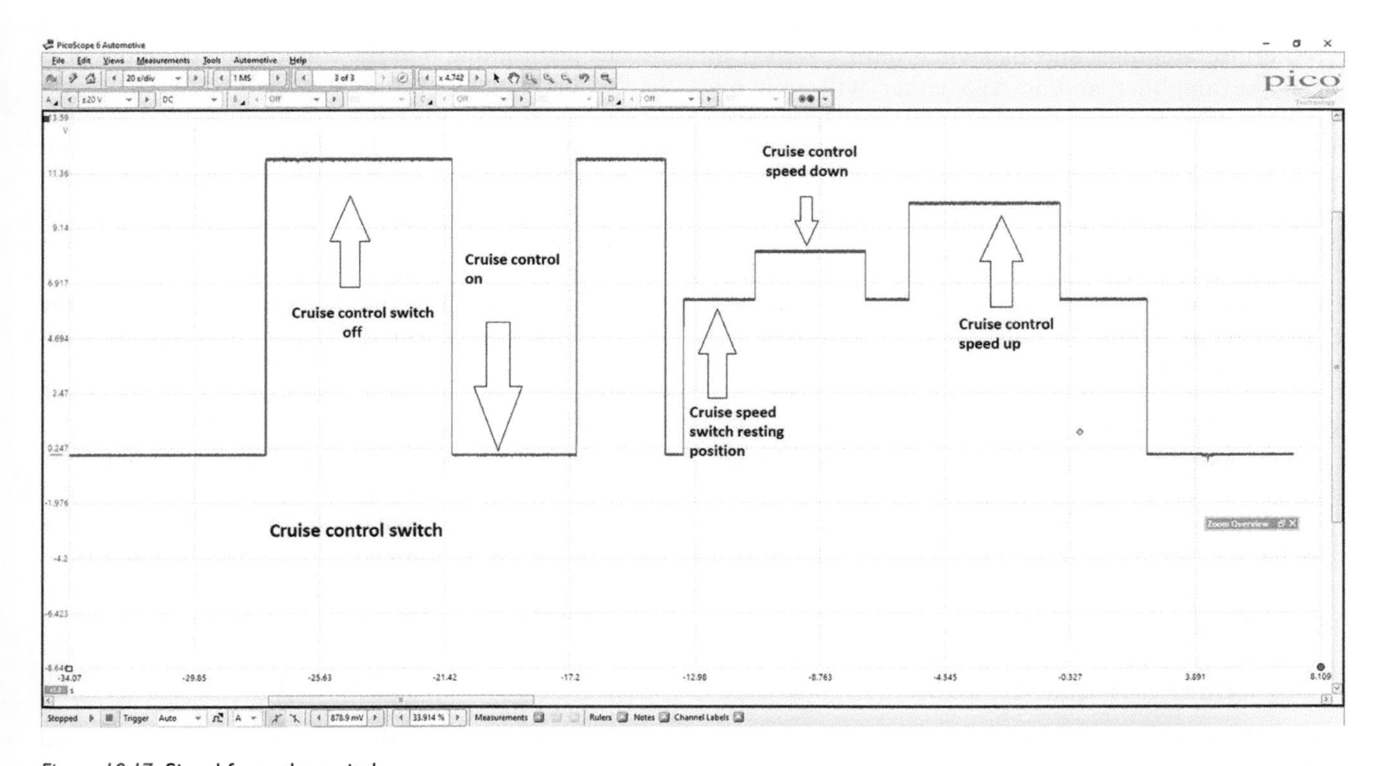

Figure 10.17 Signal from the switch.

10.1.9 1989 F250 service truck, cruise becomes inoperative with tail lights on

Johnny Crotts

COMMENT:

This example from Johnny illustrates how a change in one system can affect another even when not expected. It also shows the value of being able to load a circuit to cause a current to flow.

The problem was that cruise becomes completely inoperative with tail lights on (works fine with headlight switch off). The additional issue is the LED tail lights are dimly lit (maybe half the LEDs had a slight glow) any time key is on.

The diagnosis steps: the dimly lit tails make me think that voltage is present on tail or brake circuit, leading me to consider it could be related to the cruise issue. First turn the key on and put electronic LED test light on both tail and brake light side of lights – nothing. Plug tail in and it is still dimly lit. This tells me there's voltage but no current. I checked ground to tail and brake using a DMM and find 4.9 V on brake light wire. Now I may be getting somewhere on cruise. Cruise amplifier is seeing brake voltage – when it shouldn't be.

One by one now, I unplugged ECM, front ABS, rear ABS, cruise amplifier and t/s switch until I zeroed in on cruise amplifier and/or ABS either will emit my 4.9 V. So now to see if that's normal or if both modules are fried. Now to the issue to the tail lights killing the cruise, and sure enough if I turn the tail lights on the voltage becomes 5.3 V (telling cruise that brake is depressed).

I stick an incandescent test light in the wire where I've been testing voltage and verify that if I apply the brake it lights. Dimly lit tails go out; I test-drive and cruise works OK with the tails on.

Switch one tail light for an incandescent and all issues are resolved. The problem from the beginning was LED tail lights not pulling down brake light connected modules and letting tail lights back feed 1 V.

I will be installing LED flasher resistors for permanent fix.

10.1.10 2002 Dodge Caravan SE (Flex Fuel) hard or no-start, poor idle quality, stall

James Morton matsauto@verizon.net

The vehicle was a 3.3-liter, six-cylinder, multi-port fuel-injected, EI ignition system four-speed auto transaxle, with 71,245 miles. It was a flex fuel vehicle.

COMMENT:

I was pleased to include this case study not just because Jim is a great technician and well-respected instructor, but also because it illustrates the basics and the power of using the PIDs. The parts cannon was loaded up three different times before this vehicle reached the experts!

This vehicle was towed into a shop for a no-start and sometimes a hard start. Customer stated van has been running just fine and yesterday morning would not start. (First really cold morning we've had here, about 40°F [4.4°C] overnight). Van arrived and of course it did not start. The tech checked for spark and fuel and said it had both, also stated the fuel pressure was within specs.

The tech pulled the spark plugs for compression test and they were soaked with fuel. Checked for codes and none found current or history. Replaced plugs. Van started, but only if pedal held to floor, and then it still runs very poorly and loaded up like it was flooded. Van is not making vacuum, only 5–10 hg when we did get it to stay running on part throttle, and the vacuum gauge needle was bouncing up and down and all over.

This is a perfect example of missed information. After listening to this tech's information, it is easy to see that this engine is receiving way too much fuel, but why? The question that needed to be answered was, is the fuel just leaking into the combustion chamber or is the PCM adding all of this extra fuel?

One of my first procedures is to hook up a scanner and do exactly what the PCM does, look at the input PIDs (on-board diagnostics parameter IDs) to the PCM, one of the important inputs for the correct amount of fuel is the engine temperature. That input PID was within specs. As this was a flex fuel vehicle, most of these vehicles will show the alcohol content – bingo, the scanner PID was showing 67% alcohol. The technician said he knows that there is E10 (10% ethanol) gasoline/petrol in the vehicle because he had to add fuel while working on it.

In order to prove this was the problem, I disconnected the flex fuel sensor and using my sensor simulator, I drove a 60 hz digital signal into the PCM and verified it by looking at the scanner. The vehicle started up and after clearing out some of the extra fuel, it started running the way it should.

What is wrong here is the PCM 'thinks' it has to add more fuel because alcohol has approximately two and a half times more oxygen in it. This is why a flex fuel vehicle's mpg is about a third less when E85 is installed vs E10. The flex fuel sensor has a three-wire electrical harness connector (Figure 10.18). The ECM provides an internal pull up to 5 V on the signal circuit, and

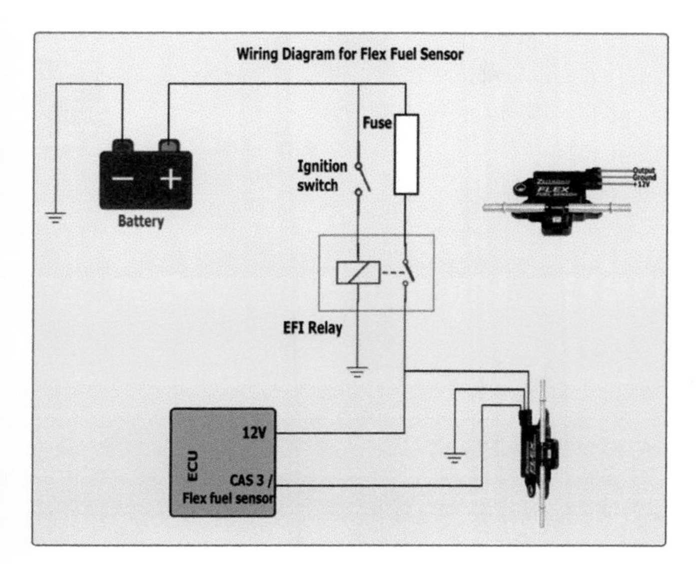

Figure 10.18 Sensor circuit.

the fuel composition sensor pulls the 5 V to ground in pulses. The normal range of operating frequency is between 50 and 150 Hz.

- 50 Hz = 0% alcohol
- 55 Hz = 5% alcohol
- 60 Hz = 10% alcohol
- 65 Hz = 15% alcohol
- 70 Hz = 20% alcohol
- 80 Hz = 30% alcohol
- 90 Hz = 40% alcohol
- 100 Hz = 50% alcohol
- 125 Hz = 75% alcohol
- 150 Hz = 100% alcohol

Every 5 Hz = 5% alcohol

10.1.11 2011 Volkswagen CC came in with a bulb out warning indicator

Hans Jorgensen

COMMENT:

I like Hans' use of the VCDS in this case study and how he fixed one aspect before moving on to the next. We all take longer on some faults especially when finding the right information (circuit diagram in this case) is difficult. Nonetheless, he stuck with the case and fixed the fault even though the information given by the customer and dealer was wrong. It is always important to collect all this information but sometimes we need to ignore it.

The other day I had a 2011 Volkswagen CC come in with a bulb out warning indicator on in the dashboard. I pulled the fault codes and it has several, both pending and current, for both rear fog lights, both reverse lights, and both tail lights. Dealer says it needs a J519 Central Electric Module.

First thing I noticed is there is no switch for even the front fog lights, let alone rear fog lights. I check the soft coding with the VCDS long coding helper, and I discover there are a couple of boxes checked for the fog light options the car doesn't have. I uncheck them and the fog light faults go away, but there are still electrical faults in circuit codes for the right reverse light and the right tail light.

My next step is to check the bulbs. They check out good so it's time to pull up the wiring diagram. When I had opened the trunk, I noticed the lid bolts to the hinges had been off (Figure 10.19). It also looked like someone hadn't even used the correct tools to undo them. I unplugged the connector out of the bulb holder to check what the wires are doing. I get nothing out of pin 2 and pin 4 on the right side. No power, no ground, just dead. When I see this, I know there is an open somewhere. I compare to the left side, and it had a signal in every pin. I am able to run an output test to the bulbs, confirming there isn't anything getting to the right-side bulbs that are out.

Right here is where I took the long way around what could have been a much faster diagnosis. The extra time was partly due to not being able to find the correct wiring diagram for the vehicle. Each one I checked had a different pin arrangement at the J519. I ended up going back to a 2010 early production (vehicle is 9/2010) and that is when I found the correct diagram (Figure 10.20). The main reason for taking so long is I had overlooked the trunk harness connectors behind the left trunk interior panel. Knowing the trunk had been off, I should

Figure 10.19 Trunk has clearly been moved around (see the mounting bolts).

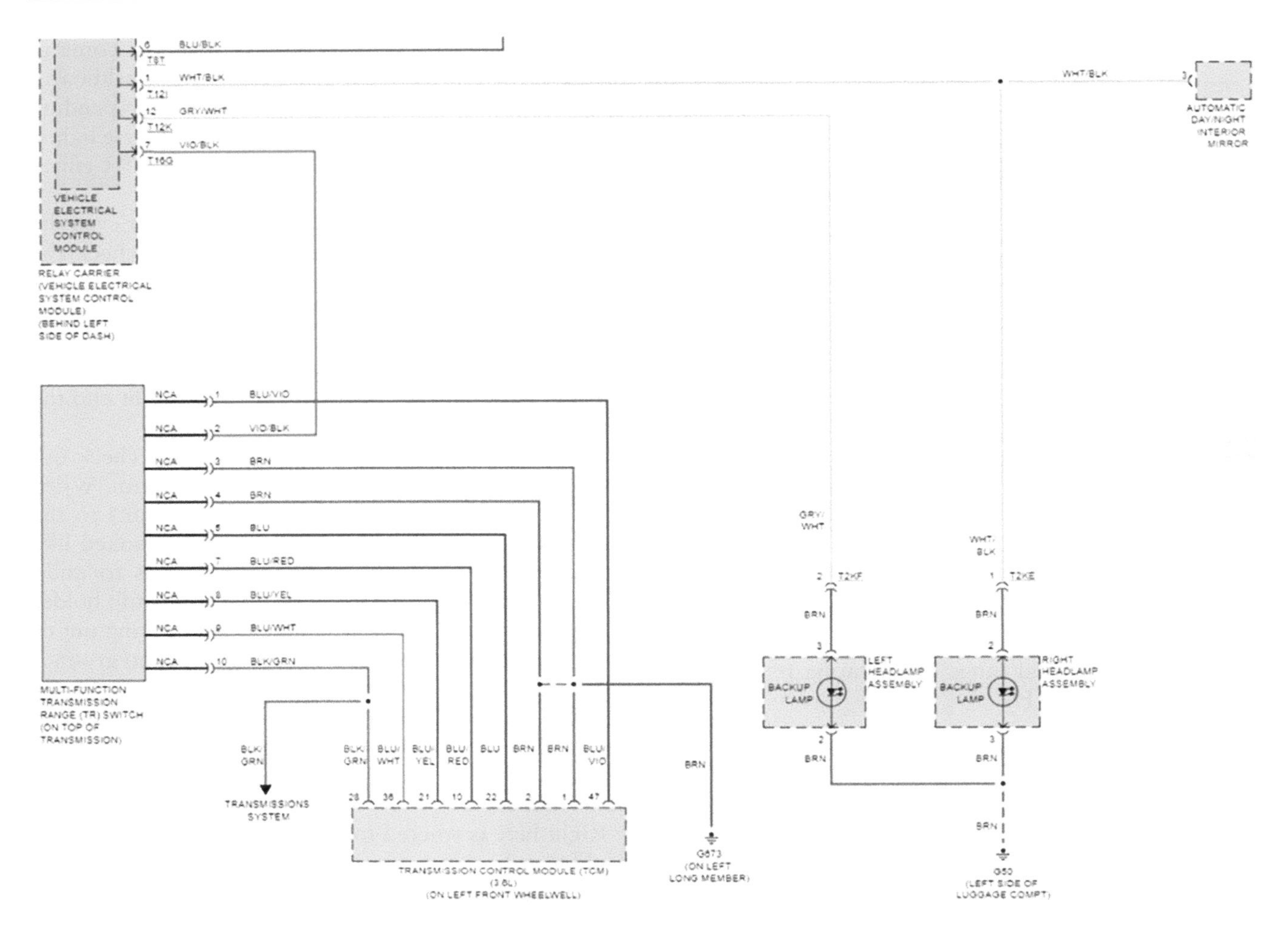

Figure 10.20 The correct wiring diagram, but it does label the reverse light assembly as the headlight assembly!

have started at the next closest connector to the problem, and checked for power and ground.

But I didn't do that, I went straight for checking the wiring coming out of the module. Now that I had the correct wiring diagram, it went a lot faster. First, I compared what my power probe had told me (Figure 10.21).

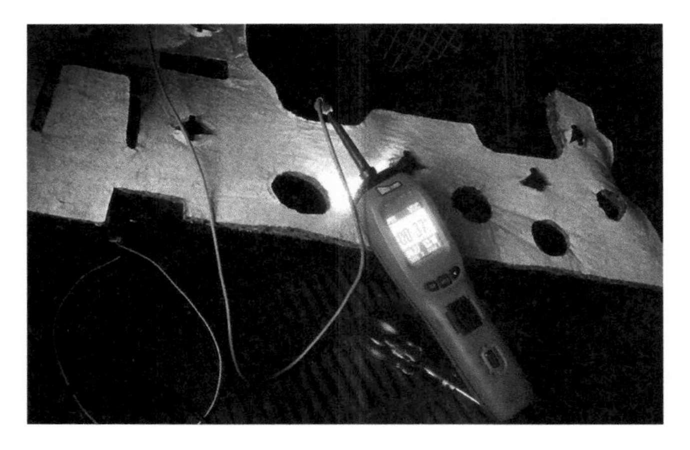

Figure 10.21 Back-probing the connector T12K, pin 12 for left reverse light, matches signal at the bulb.

Leaving the connectors plugged in, the left reverse light was the same as at the bulb holder connector. But the right reverse light was shorted to ground, not what I saw at the bulb holder. I unplugged the connector, still short to ground, even with the bulb holder unplugged. Then I used my pin drag tester straight out of the module and did an output test on the reverse lights (Figure 10.22). It commands both reverse lights to turn on. Now I know for sure it's not a module.

I go back and find the trunk wiring harness connectors and unplug them for their respective lights and run the output tests again. I'm getting the correct signal to turn on the bulbs there. Now I know the wiring between the trunk harness and the J519 is good, and I can fail the trunk wiring harness. I also use my old incandescent test light to give the wires a little bit of a load over the power probe.

This really should have gone faster than I did my tests, but the service information threw me for a loop. Also knowing the trunk had been off, I wasn't sure what I would find. This on top of knowing the dealer said it needed a module had me second-guessing my work all along. I mean, they are the dealer, right? While I like to

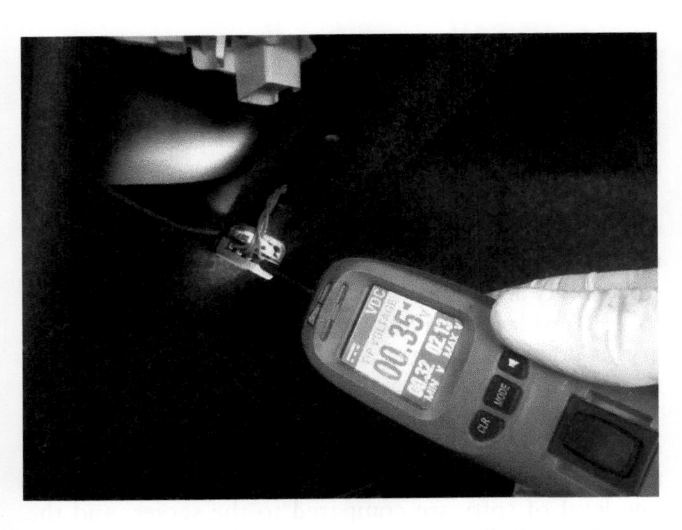

Figure 10.22 Pin 3 for the left reverse light inside the trunk.

know more about the history of a vehicle, sometimes I would just rather hear 'Bulb out warning indicator on in dashboard, fix it'.

> **COMMENT:**
>
> *I always like to hear the phrase 'root cause' as Gareth mentions in this case study. I also think the logical process used here is really good and shows an excellent understanding of the diagnostic process. Some good examples of double-checking too (glow plug resistances, but also plugged into the known good circuits). Interesting how the 'can you just...' job turned out to be as complex as the first job – often the case!*

10.1.12 BMW 123d Case Study – Engine management issues

Gareth Davies CAE AMIMI, Director, www.euro-performance.co.uk

A BMW 1 series coupe was booked in for engine management issues, with an added 'Can you take a look at this light' twist! EML active, with a reported intermittent lack of power. Initially the mandate was only to answer the engine issues, no other work required.

An initial scan after customer consultation reported multiple faults in Engine (DDE). Figure 10.23 shows how before any real testing commences, we need to consider the relevance of the reported symptom, and the likelihood of the faults stored being relevant. In other words, would we consider a glow plug issue to give the customer a lack of power or limp home mode?

After a pause to consider the faults present, and their probable root causes, a test plan was created. There were four main areas of concern:

1. Charge air pressure and EGR – A relationship between both, mechanical EGR, mechanical testing required. Electronic vacuum control testing required.
2. Glow Plug control and operation – Wiring diagrams and electrical testing required for control and output.

Code	Description	Faults
9CBF	Communication with step motor controller on the left disrupted	
9CC2	Stepping-motor controller (SMC), right, signalling fault	
9CC0	Communication with step motor controller on the right disrupted	
9CB4	Ride-height sensor, rear, faulty	
A8AB	Dipped beam, right, faulty	
E59C	No message (steering angle), receiver FRM, transmitter SZL/LWS/DSC	
A8C1	Reversing light, left, faulty	
INSTRUMENT CLUSTER (INSTR) (komb87)		**2 Faults**
A559	Instrument cluster: Power supply has been switched off (closed-circuit current cutoff relay)	
931B	Outside temperature sensor	
DYNAMIC STABILITY CONTROL (DSC) (dsc_87)		**1 fault**
5DE1	Brake-pad wear: plausibility, rear axle	
ELECTRONIC POWER STEERING (EPS) (eps_90)		**1 fault**
63F5	Reduction, servo power, due to low supply voltage	
DIGITAL MOTOR/DIESEL ELECTRONICS (DME/DDE) (d71n47a0)		**11 Faults**
4A6E	Glow plug, cylinder 1, activation	
4B1C	Fuel-filter heating, activation	
4C2F	DDE control unit internal	
4C2E	DDE control unit internal	
4C29	DDE control unit internal	
4507	Exhaust-gas recirculation, control deviation	
3FF0	Air-mass flow sensor	
4530	Charge-air pressure control, control deviation	
4A5E	Glow plug, cylinder 2, activation	
4A54	Glow plug, cylinder 2, activation	
483D	Throttle actuator	
CAR ACCESS SYSTEM (CAS)		**2 Faults**
A125	incorrect or implausible signal from engine management	
A0B3	Starter motor, terminal 50	

At the early stage of data acquisition we ascertain likely groups in fault cluster analysis within a control unit. we can then begin to think about despite 1 EML illuminated to the customer, how many faults are we dealing 1, with responsible for all, or multiple issues, all with their own cause and rectification.

We can take some time to answer the customer reported symtoms, and the relevance to the fault entry.

Figure 10.23 Scan results.

3. Fuel filter heating – Wiring diagrams and electrical testing required for control and output.
4. DDE Internal control unit fault – This could be a terminal EEPROM fault, GFF dealer test plan required.

Testing began and it was discovered the mechanical operation of the EGR (vacuum operated) was compromised. The control element was tested by vacuum measurement to the EGR, when duty cycled in final control diagnosis (the taking over control by laptop of the function on demand) and no control fault was found, only the output (the EGR itself).

A charge air pressure test was carried out. By using a pressurised smoke test machine, we were able to prove a leak-free charge air tract. Although more complex on a twin turbo assembly as fitted to this vehicle, the testing was conclusive, and no leaks were present.

Phase 1 complete. The likelihood of the lack of performance being experienced by the customer almost certainly lies with the sticking EGR, which because it shares a relationship with charge air pressure (compressed exhaust gas) is the reason a charge air pressure anomaly has been reported in DTCs.

Phase 2, Glow plug operation. This system uses a 12 V supply feed to a controller; the controller, instead of a traditional relay, is a pulse width modulation (PWM) which offers greater control over the output.

Some simple testing revealed supply and ground to the module were correct, the module was activated by final control diagnosis, and it was found the output to cylinders 1 and 2 glow plug was not present, but the output for 3 and 4 was. Finally, to ensure accurate diagnosis, each glow plug was removed, and resistance checked. Then they were connected to the supplies for either 3 or 4 to conclude if simply a controller was required, or also glow plugs. It was found that glow plugs 1 and 2 were indeed faulty, and this additional test just made sure all items affected were captured in the initial report back to the customer.

Phase 3, fuel heating. Some simple checks establishing power and ground at the heater unit (attached to rear of inline fuel filter) validated wiring. These heaters are known to be troublesome as a brand specialist; happy the unit itself is faulty and a replacement required.

Phase 4, the internal control unit fault. When the GFF test plan was run through on ISTA (BMW dealer tool), this offers some checks to be carried out regarding engine start ability and performance, and software comparisons to the latest level available on the BMW server. Critically at this point an important observation was made about the engine operation, particularly when cold.

On start-up the exhaust note is noticeably louder and has a pronounced whistle compared to other models of the same era. In conjunction with the fault stored, the low level of software compared to the server, and the exhaust behaviour, a point was raised to talk to the customer regarding previous ECU tuning/DPF solutions. Depending on this additional information, flashing the control unit to try to fix the fault code, could create a bigger problem if no internals of the DPF are present in terms of safety. It could also result in a drop in performance compared to before the repair performance when not in a faulted state.

With this stage of diagnosis reached, the customer was consulted with the repair strategy, the costs and how he would like to proceed.

All repairs were authorized and affected, with the exception of the control unit software, as it was confirmed the DDE had been tuned, and a DPF 'solution' was in place. After issuing the comprehensive diagnosis, the customer questioned some faults stored in the light module (FRM) and asked were these responsible for the amber warning light that comes up in relation to the headlights. He requested a further diagnosis to confirm the issues present, and the repair required. This is where the real fun begins.

After engine repairs were completed and validated, we then moved to the FRM faults. We had multiple fault entries stored (Figure 10.24) and started by assessing

VIN	**WBAUR52020VF30507**
Manufacturer	**BMW**
Vehicle	**E82 123d N47 50 door Car**
Year	**12/2007**
Country / Area	**Europe**
Battery voltage	**14.9V**
Autologic information	**A020735, BMW v3.187.6, Main v6.10.6**
Time / Date	**15:07 13/12/2019**

Modules with faults:

FOOTWELL MODULE (FRM) (frm_70)		**2 Faults**
A8C1	Reversing light, left, faulty	
9CC0	Communication with step motor controller on the right disrupted	
JUNCTION BOX ELECTRONICS (JBE) (jbbf70)		**1 fault**
A6CF	AUC Sensor	

Figure 10.24 Second scan.

whether any bulbs were inoperative. We found that one normal filament reverse bulb was inoperative (blown filament). This was replaced and faults cleared. We found the only faults that returned were:

- 9CBF communication with step motor right disrupted;
- 9CB4 rear ride height sensor (eventually returned).

For its age, this vehicle has a fairly rare option of Bidirectional adaptive Xenon headlights (AHL). Upon checking the vehicle on the ramp, it was noted a new rear ride height sensor had already been fitted. A function test of lights at this stage showed no level change possible by loading the car but was by laptop final control diagnosis.

Upon checking live data, the voltage for the rear ride height sensor had an 'actual' and a 'desired' pair of readings. The desired change with loading, likely referencing front change, but the actual never changed.

Checking the wiring diagram (Figure 10.25), the sensor shows three wires, a 5 V feed, a ground and a signal wire. The power and grounds were validated, 5 V OK using sensor ground, signal wire open circuit back to FRM module.

A physical inspection of the harness soon found our problem, a green corroded wire. The section was repaired, and initial thoughts suggested the stepper motor may have been a resulting fault of the level sensor entry. A quick recheck of live data now shows voltage change for the sensor and no return of the DTC.

However, when the headlights were switched on to confirm the directional self-check and change of beam pattern, the N/S dipped beam came on, and went straight out. A quick check of the bulb and it appeared to be blown. Coincidence maybe? A new replacement ordered and fitted, and a recheck of lighting function, not good as now no dipped beams on either side. In addition to this, the directional yellow headlight warning is still present on the dashboard, and now we have two entries for stepper motors NS and OS, and in addition AHL active headlight corning function faults for NS and OS. Time for a cup of tea.

A re-evaluation of what we had and where we were. We discussed the possibility of an FRM check control fault count being the issue. There is a reset for this, but these units can be very temperamental when bulbs blow for shutting circuit driving down. They also suffer with firmware corruption easily when programming or battery voltage events occur. Was this our issue?

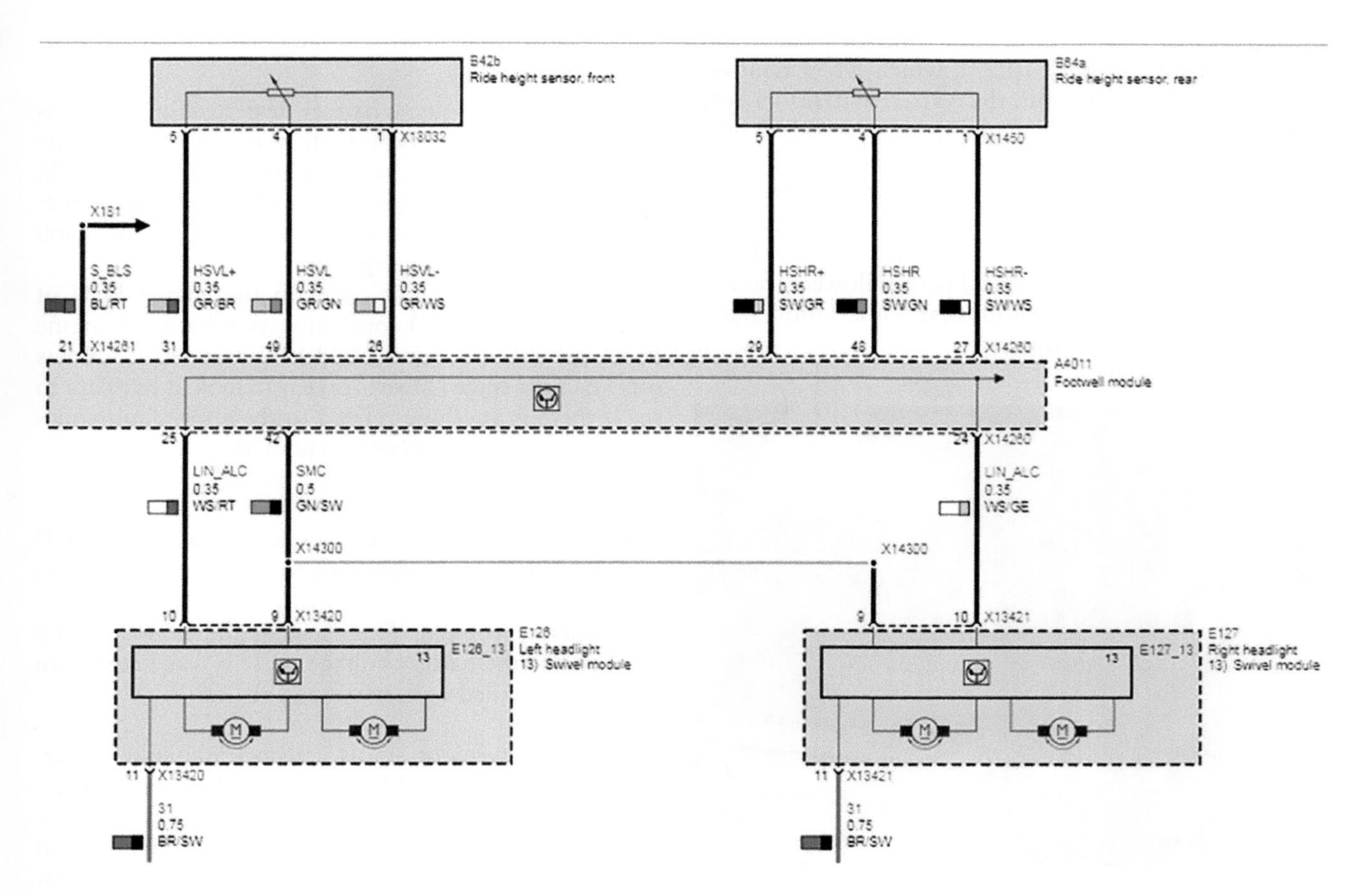

Figure 10.25 Lighting circuit.

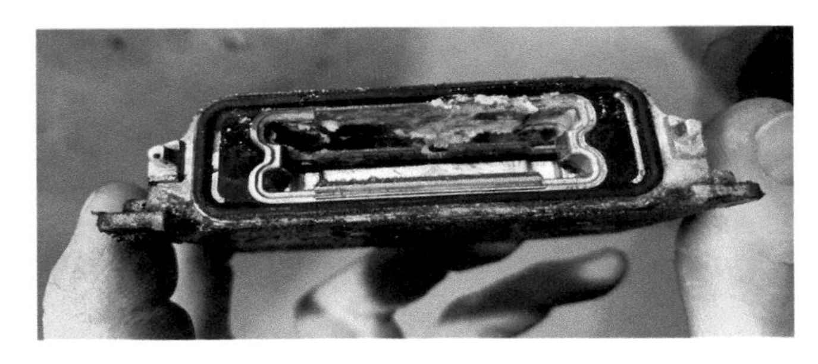

Figure 10.26 Corrosion in the connection plug.

Independent technical support suggested it could be, but also suggested to check some pin out data to the lamps themselves.

Wiring checks were carried out, and with the aid of an assistant some readings were taken when headlights were actuated by switch and final control, and all appeared correct. One reading wasn't though.

Although activity on the LIN message wire between AHL units, and the FRM, this was not surprising as they were not working/responding to any form of instruction. So, the question now was whether this is a control or output fault. Some further investigation led to a significant discovery shown in the photos (Figures 10.26 and 10.27).

The first picture (Figure 10.26) is the driver for the Xenon gas discharge headlight, and the second the receiving saddle (Figure 10.27), which when removed houses a small control unit, the AHL. All items on both NS and OS were sat submerged in water and had been for some time. It was merely pure coincidence that the output of the headlights was compromised while in our possession.

After a long conversation with the customer detailing our findings, and the logical path followed, the customer thoroughly understood, that while unfortunate, he had to bear a responsibility.

Figure 10.27 Receiving saddle for the connection.

The replacement parts for this were not cheap by any stretch, but sourced fitted, units sealed from water ingress and the appropriate coding measures complete. We now had a vehicle with no FRM or DDE faults, no warning messages present on the dash and the satisfaction that we had overcome some reasonably complex and challenging faults.

10.1.13 2016 Iveco Daily battery light illuminated on the dashboard

Edward Grigg

COMMENT:

Doing a full scan of a vehicle whatever the stated symptoms is to be recommended, the result can be saved and is evidence of the state of the vehicle when it first arrived. I also like and have quoted the excellent advice offered by Edward in this case study:

"A technique that I have learnt is if I put an alternator on this van and it didn't fix the fault, what would be my next test? That's the test I need to do now before I fit the alternator!"

He also highlights the value of training; I couldn't agree more.

Firstly, I carried out a battery test to eliminate this essential component. The battery tested OK and furthermore the alternator was charging at just under 14 V. I then carried out a global scan of the vehicle, something I'm in the habit of doing on every diagnostic job in order to see the bigger picture.

There were two fault codes in the engine ECU of concern for me. C120 communication between engine ECU and alternator and C028 LIN data bus CAN bus off. The next stage I always take is to look at live data to see if there is anything related to the above fault codes and give me some direction to go with.

The live data gave me:

- Battery specified voltage which is what the ECU is wanting or expecting to see. This value was 14.8 V.
- Battery voltage. This is what the ECU actually thinks the battery voltage is. This was lower than the specified battery voltage at 13.8 V.

Lastly, the alternator status categorically stated alternator defective. Should I trust this and order the alternator? A technique that I have learned is if I put an alternator on this van and it doesn't fix the fault, what would be my next test? That's the test I need to do now before I fit the alternator.

Using my oscilloscope, I back probed the LIN control wire at the alternator. From recent training courses I knew what I was expecting to see. I had a stable 12 V showing on the oscilloscope screen, which I instantly knew was wrong. At this stage I thought it was time I gathered some technical information. But first I would carry out a technique that had been fiercely drummed into my head over the various training courses and by the various top trainers I've had the pleasure of being taught by – a simple visual inspection.

I carried out a visual inspection of the engine wiring loom from the alternator back to the engine ECU. After following it almost back to the ECU I found there was a large multi plug. I could see straight away there was a red cable broken, the same colour as the signal wire to the alternator (Figure 10.28). I repaired the cable and cleared the fault codes (Figures 10.29 and 10.30). Immediately the battery light went out.

Another successful diagnosis and happy customer.

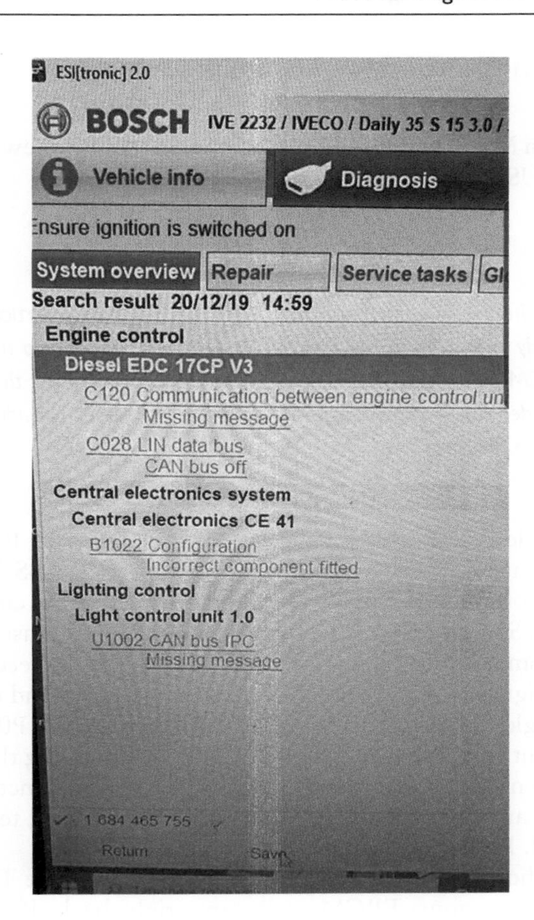

Figure 10.29 Scan results.

Figure 10.28 New alternator not needed!

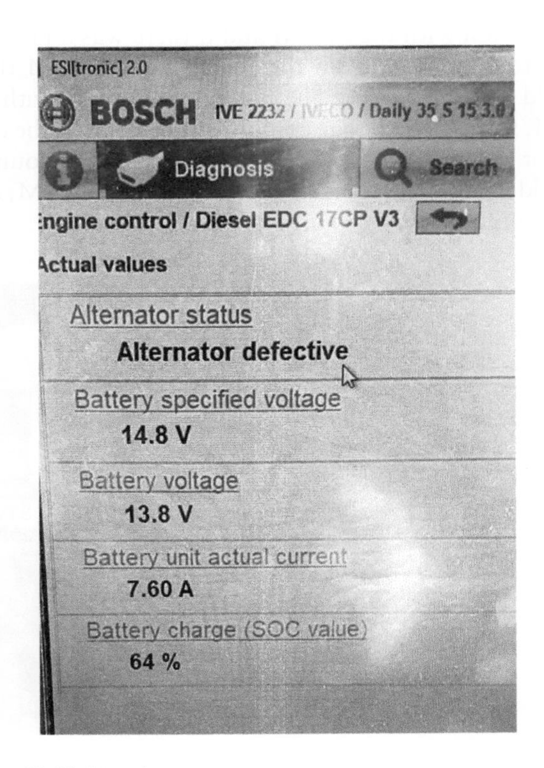

Figure 10.30 Live data.

10.1.14 Chevrolet 3500 bus with a 6.6 Duramax diesel, ABS light on

Brian Matusiewicz, Mike's Auto Service in Bridgewater, NJ, USA

COMMENT:

A short case study here from Brian, but the part I particularly liked was where he said: 'Speedo was working so the ECM was receiving the signal.' This is the sort of logic that underpins a good diagnostic routine – observe, consider, take the next step.

Vehicle was a 2007 Chevrolet 3500 bus (Figure 10.31) with a 6.6 Duramax diesel, complaint was ABS light on. Found code C0055 rear wheel speed sensor circuit fault. Speed sensor is the same as output speed sensor on transmission. Wiring diagrams show the TCM receives the signal, sends it via CAN lines to the ECM and then a single signal wire to the EBCM. I also noticed a P0722 output spend sensor circuit in the TCM. Starting there, the sensor at the TCM was scoped to verify its functionality and the wires were load tested. Everything tested good. Replaced and programmed the TCM.

I thought I was done! I cleared codes and the TCM code was gone; EBCM code came right back. If there was a problem with the CAN lines between the TCM and PCM there would be other codes. Speedo was working so the ECM was receiving the signal. I had communication with the EBCM and if there was a problem with the data line from the PCM to the EBCM there would be other codes. The problem had to lie with the PCM, so I tested voltage coming out of ECM at the connector and had 0.24 V. I did a little digging and found it should be 10 V. I replaced and programmed ECM, and then confirmed the repair.

10.1.15 Renault T DTI11 Euro 6 engine, fuel system fault

Carl Dixon, Dublin, Ireland

COMMENT:

I like where Carl said he was 'keeping something in the back of his mind' – often a good strategy even if it turns out to be not relevant. Zooming in to a waveform is also an excellent technique – as long as you understand what you are looking at!

A scan gave this result: P228F – Fuel Pressure Regulator A, Exceeded Learning Limits – Too High. The truck in question was a Renault T, DTI11 Euro 6 engine. This engine is the same as used in certain Euro 6 Volvo trucks also. It is an inline 6 configuration engine, common rail. The fault was not active at the time but stored in memory and the truck sounded smooth with no signs of misfire, etc. Freeze frame data showed it occurred reasonably frequently, about 2–3 times per day with the driver reporting a limp mode condition when it happened.

Looking for TSBs (technical service bulletins), I found one in relation to this code. It explained that this code is set after 15 seconds of continuous running where the fuel regulator is at maximum (it did not specify which regulator). It explained that fuel pressure is still within spec so does not set a 'fuel pressure too low' code. This injection system has three regulators.

After a little bit of research, I confirmed the operation of the system. While being a common rail system this system utilises three standard solenoid injectors and three pumping injectors. The pumping element of the injectors, purpose is to supply the rail with pressure. These are camshaft activated and look much like the

Figure 10.31 Chevrolet 3500 bus.

Delphi E3 injectors yet internally do not operate the same. The pumping injectors also have a solenoid injector combined with them. So essentially, we have six solenoid injectors with three pumps attached to three of them. Each of the pumping injectors has a regulator and there are two pumping phases for each injector per camshaft revolution, so six in total.

The TSB suggested numerous possible causes including rail pressure sensor, air in fuel system, rotated cam lobes, electrical connection fault, PRV leaks and injector faults. Unfortunately, not having a dealer level scan tool, many tests in the TSB could not be accomplished, particularly the high pressure system check whereby we can individually actuate each pumping injector pumping element and check for pumping efficiency.

I checked and confirmed low side fuel system condition, all was in order. High side pressure commanded vs. actual was matching on live data. No codes were active. Checking the camshaft showed no visible signs of rotation although this was always in the back of my mind as I have seen them move before very slightly and become difficult to diagnose. I also checked pumping injector preload at this stage and found all in spec.

My next step was to check fuel rail pressure with PicoScope, all looked in order until I zoomed in and started to analyse the waveform details. As we can see in the waveform, it is not uniform and points towards our issue.

As you can see in the labelled picture (Figure 10.32) the problem is quite evident. We have a pumping loss in one injector. I have labelled them 1, 2 and 3 but these injectors are actually on cylinders 2, 4 and 6. While I didn't save the capture by capturing metering solenoid current I was able to confirm that it was in fact the cylinder no. 2 pumping injector that had the loss. This injector was not able to keep up pressure under load and was setting the code.

As per the customer's request, we replaced all three pumping injectors and carried out the same test. A nice even pressure across the board with no losses of pressure was the result.

10.1.16 Vauxhall, making a decision on a £2500 part

Dean Andrew, DNA Vehicle Services

> **COMMENT:**
>
> *What can I add here? I know Dean to be an excellent technician, but I'm still stressed out over the cost of this part! Excellent work, logical process, right result.*

A question often asked among diagnostic techs is: 'If I changed that part and it didn't fix it, what would I test next?' It's a way of ensuring that parts aren't replaced in a parts darts kind of procedure. This question was especially apt with a Vauxhall Antara recently.

The part required was at a cost of £2500 and a total bill close to £3500. Making a call on a £3500 bill isn't to be taken lightly. This car had already been to a well-respected garage where I know two of the techs well; they are skilled and highly trained.

The fault was relating to cylinder 1 not doing the same amount of work as the other three. The spanner light was on, but not the engine management light and

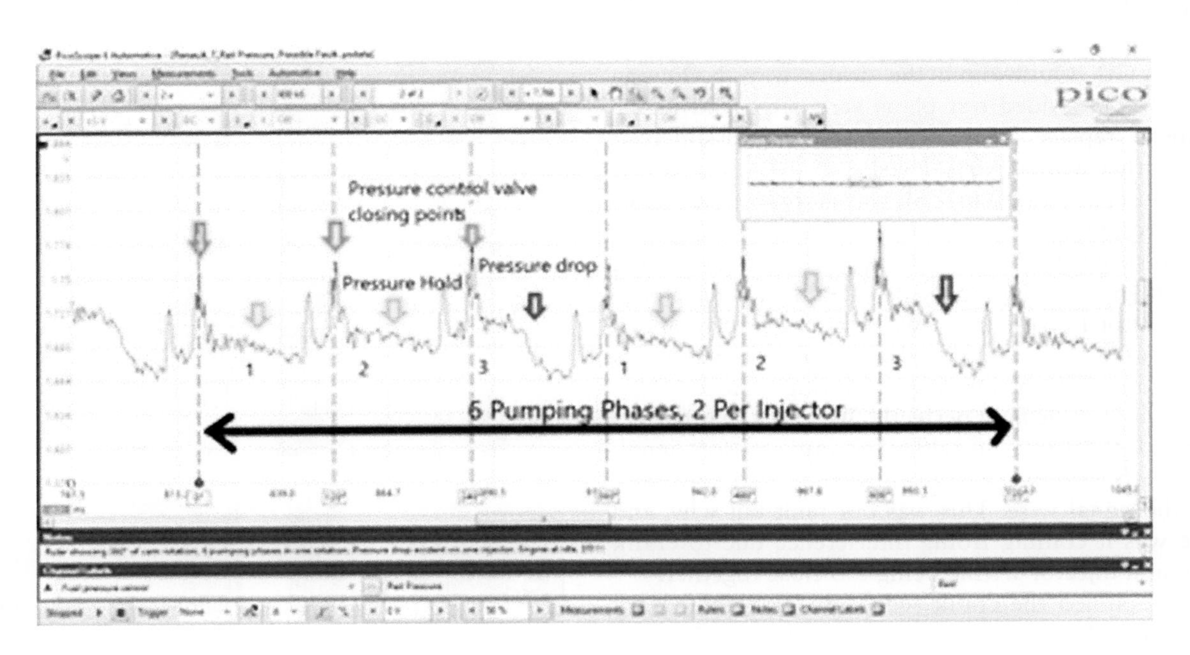

Figure 10.32 Annotated waveform.

Figure 10.33 An expensive part!

one cylinder was being shut down by the ECU as soon as the engine was started. Routine tests include compression checks, injectors and so forth. These were all fine. One injector had been replaced elsewhere with a genuine new one and swapping injectors from cylinder to cylinder didn't move the fault.

The next step was to work out how the ECU monitors the cylinder efficiency. What tells the engines computer that there is a fault with a particular cylinder?

Vauxhall's own data system was no help: test injectors, replace injectors, test wiring, replace wiring, replace ECU. Quite often the dealer way. Following Vauxhall TIS guided test plans very often end in the statement 'replace component despite successful testing'. In other words, parts darts!

The Bosch data system, ESI, was more help. It told us that the crankshaft sensor monitors changes in crank rotation speeds to detect an underperforming cylinder. We scope-tested the crank sensor and saw interference from injector 1 on the trace. We then frequency-graphed the crank sensor as well, which shows the signal voltage as a flat line with the interference sitting on top.

This showed the fault perfectly. We had a near perfect injector 1 trace mixing with the crank sensor signal. This interference wasn't on the supply or the ground. Only the signal wire. This was our fault but why, and where was it coming from? Interference due to crank sensor and injector wiring being too close together?

We de-pinned the crank sensor and ran temporary wiring to the ECU, away from the rest of the car's loom. The fault was still there.

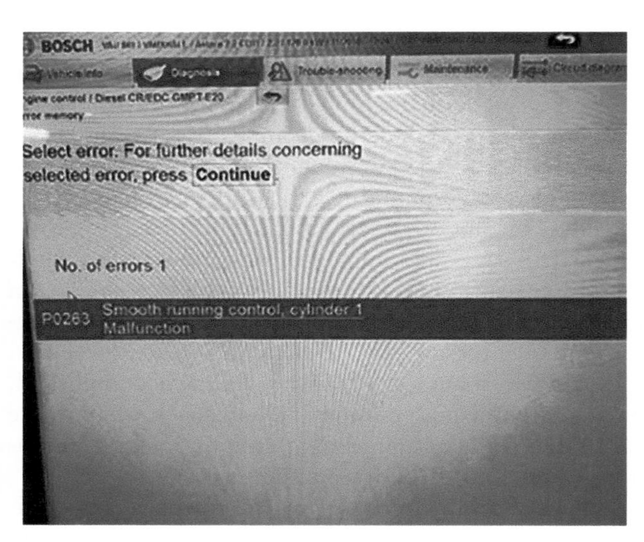

Figure 10.34 Scan results.

We then disconnected the crank sensor completely, totally removing the signal wire to the ECU at the ECU end, and also disconnected cylinder one injector. The fault was still there, coming out of the ECU and not into it with no influence from the crank sensor or injector (Figures 10.34 and 10.35).

This was an internal ECU fault (Figure 10.33). We were now happy to order the £2500 plus vat part. Not an easy call to make, after all, if we were wrong, we would have been paying for the part. This car is now fixed and gone. One of the more complex and stressful jobs this year, we learned a few things with this diagnosis, making it very worthwhile.

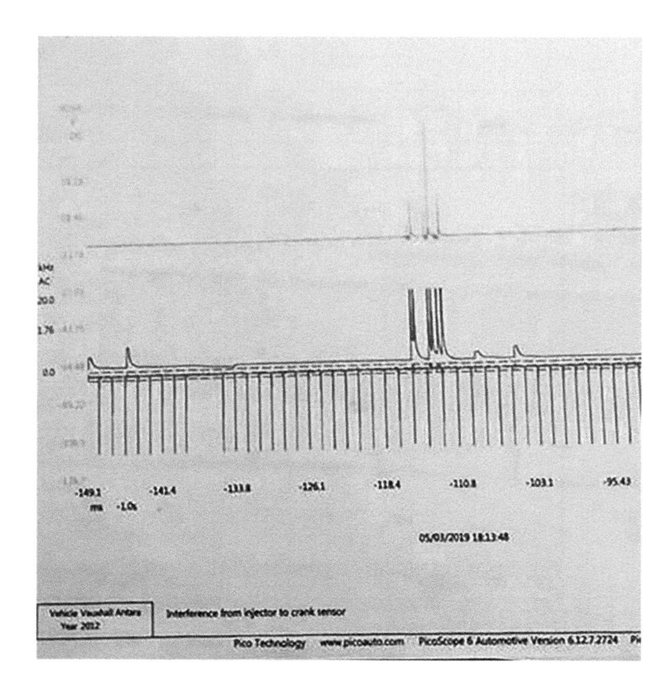

Figure 10.35 Waveform analysis.

10.1.17 Ford Fiesta with power loss

Richard Lovatt, Diagnostic Technician

COMMENT:

Great use of a pressure sensor here. As Richard says, the moral of the story is that within an hour by using the WPS500 he had enough evidence to warrant a fairly major strip.

We had a Ford Fiesta presented this morning with power loss. Test drive proved the power loss, but no codes were present and live data looked within specs. No change point was relevant as the customer had recently purchased the car with this issue.

After discussion I chose to start with the basics and a quick look over the engine bay revealed that the parts cannon had been fired. I carried out a relative compression test which looked OK, and then carried out an in-cylinder running compression test which you can see in the figures. You can clearly see the valve opening times were incorrect (Figure 10.36), which led on to a physical check of the timing, which was clearly out. The second Pico trace was after the timing was sorted out (Figure 10.37).

The valve timing was out by about 30' of crank rotation (Figure 10.38) and yet it still ran and set no trouble codes. The moral of the story is that within an hour by using the WPS500 (pressure sensor) I had enough evidence to warrant a fairly major strip.

10.1.18 Citroen DS3 1.6 Diesel – no fault codes, now what?

Glenn Norris, CAE AMIMI, Definitive Diagnosis

COMMENT:

I will use Glenn's own words as a comment here as they are spot-on: 'Even without any fault codes all that is needed is a logical thought process which usually involves proving was is working to find what is not working while gathering evidence to warrant removal of a component.'

The vehicle in this case study is a Citroen DS3 1.6 diesel. After a brief chat with the customer regarding driving style, service history and journey types, we then discussed the symptoms of excessive black smoke when the revs were increased. Driving the vehicle into the workshop confirmed the symptoms the customer described; thick black clouds of smoke were clearly visible.

I carried out some visual checks in the engine bay, oil level was perfect and also coolant level was as expected.

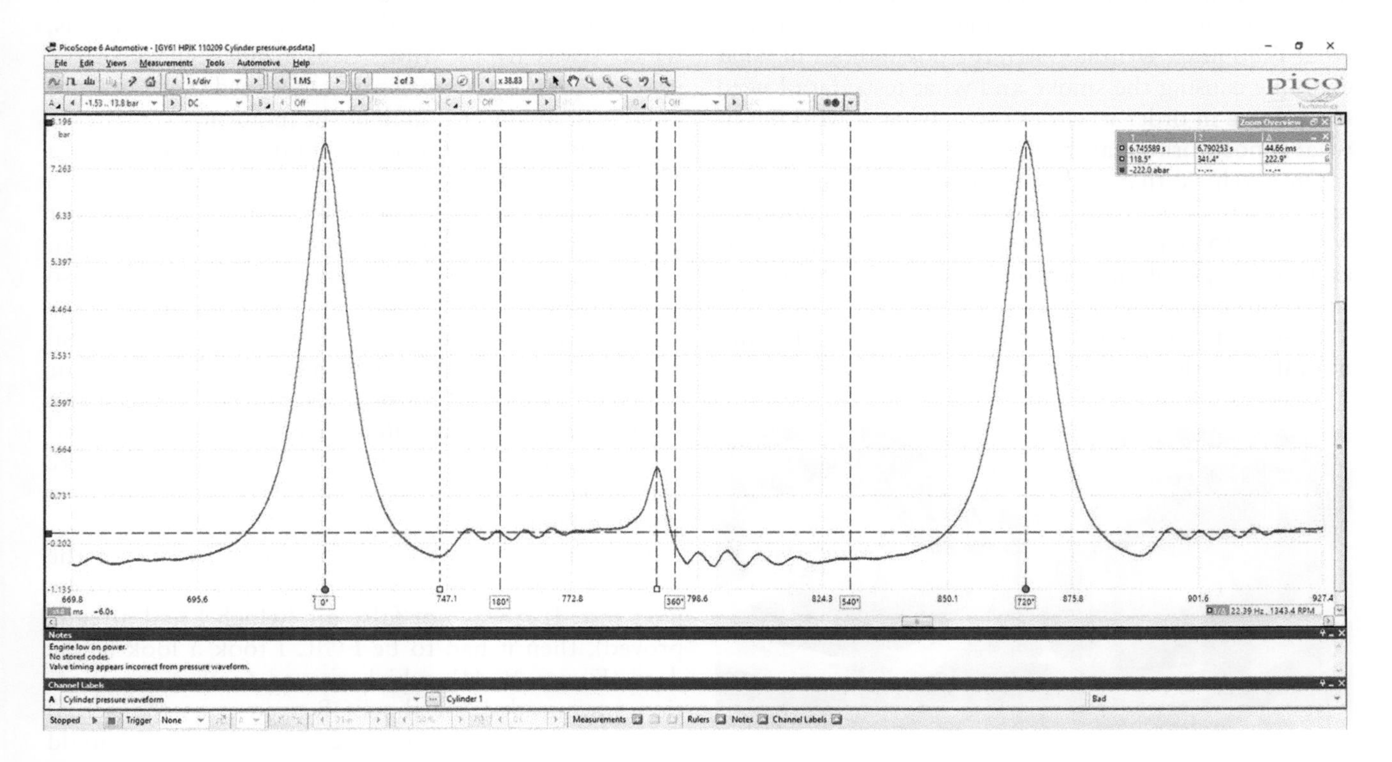

Figure 10.36 Trace showing the timing fault.

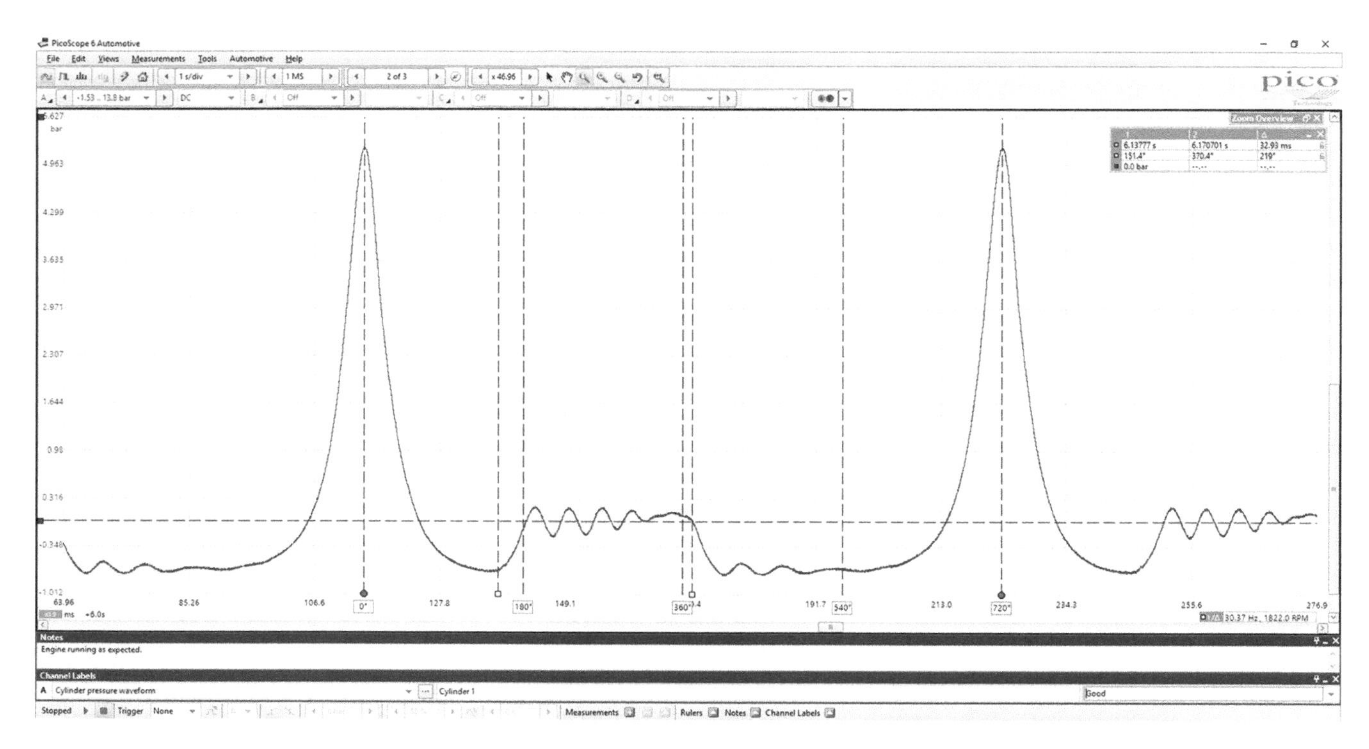

Figure 10.37 Trace after fault was corrected.

Next step was to interrogate the ECU for fault codes, but there were none (Figure 10.39).

I am a great believer in process, which often involves stepping away from the vehicle for 10 minutes and building a test plan, so I am not aimlessly testing components in the hope of stumbling across the faulty part(s). This is what I did next, writing down the possibilities of what could be causing the smoke and what tests/data I need to evaluate in order to either prove what is working or what is not working as it should.

On a vehicle that had only done 30,000 miles and had dealer servicing and mixed driving conditions, I was going to be looking at fuelling, air intake and EGR as all other possibilities I would not expect to see on a car in this condition.

First thing to check was fuel rail actual and nominal values in the data from the scan tool against engine speed, cranking and idle. They were identical with good cranking speed and under load also as expected. After raising the rpm and left to idle I switched the engine off to observe the speed at which the rail pressure dropped. This is a great indicator of injector leakage. The pressure was slowly reduced over about 8–10 seconds, which I was happy with. Also, a quick check of the injector quantity comparisons confirmed that I was not looking at an injector fault at the moment.

Figure 10.38 Tipp-Ex is not an approved timing tool!

Next up was to check the intake for leaks with a smoke test and this checked out absolutely fine. So I now know we have no false ambient air. Another parameter I checked was cam-crank synchronisation, which could point to a timing/pump problem. This also checked out OK.

I was happy that I had ruled out a few of my initial possibilities quite quickly which left EGR and mass air flow data. MAF was my next test and this was the first piece of data that indicated a clear issue. I monitored the actual value against the nominal value and there was a clear deviation between the two (plotted as a graph in Figure 10.40).

This proved that the volumetric efficiency of the engine was being compromised, so that led me into my next test. If it was not false air (which I had already proved), then it had to be EGR. I took a look at the data (Figure 10.41), which proved further investigation was needed for the EGR system as the actual value for EGR position was 255% and at idle should be 0–2%.

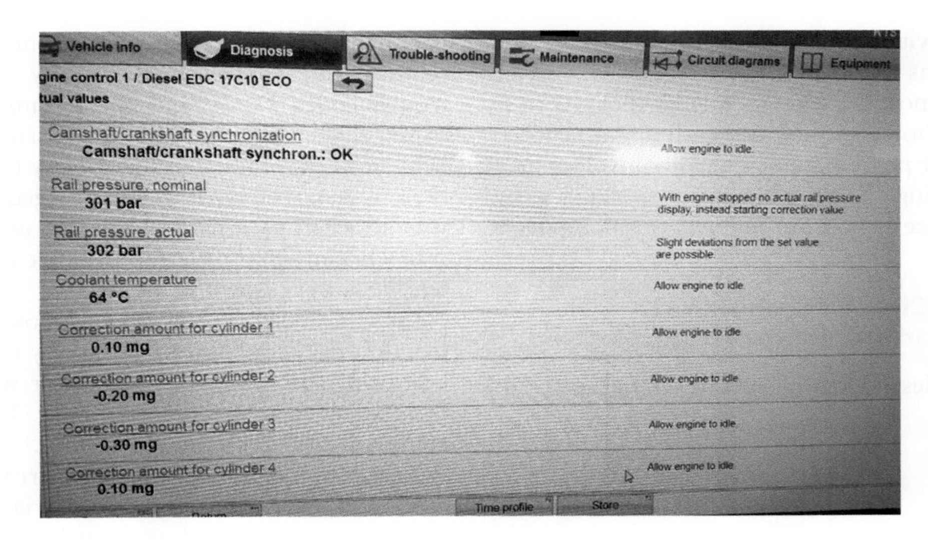

Figure 10.39 Live data.

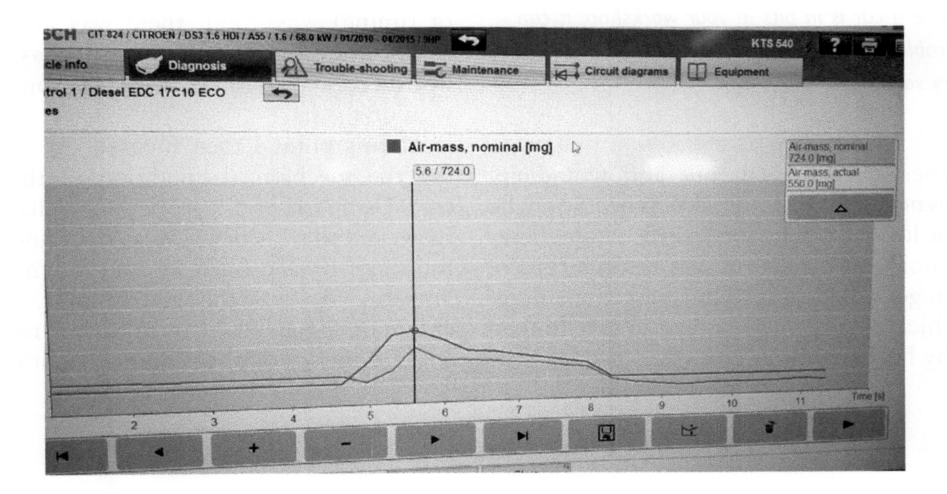

Figure 10.40 Comparing actual value against nominal value.

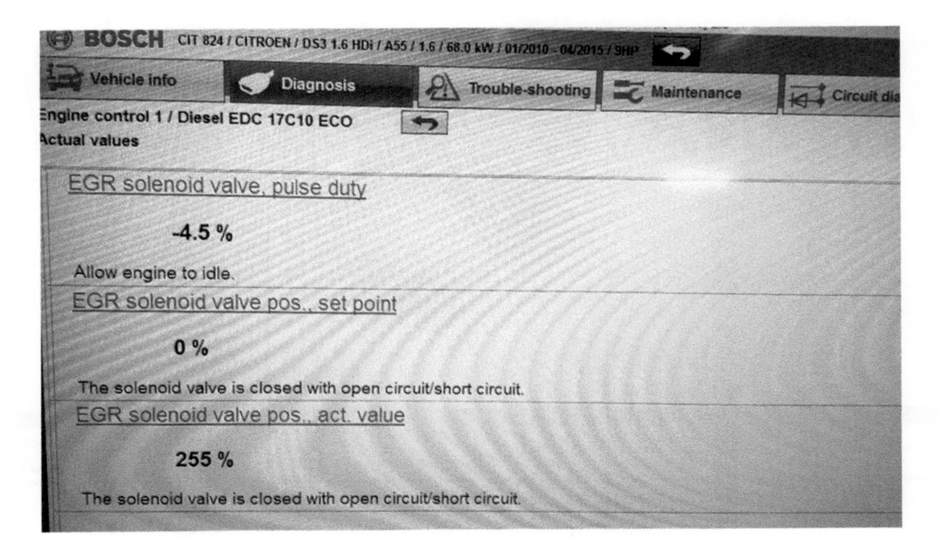

Figure 10.41 EGR data.

After a chat with the customer explaining my findings, the EGR was removed, inspected and renewed and the vehicle was now back to its former glory with no smoke. Even without any fault codes all that is needed is a logical thought process, which usually involves proving was is working to find what is not working whilst gathering evidence to warrant removal of a component.

10.1.19 2010 Citroen Dispatch cut out and would not re start

Gary Wood, Diesel Doctor Mobile Diesel and DPF Specialist

COMMENT:

The key aspect of this case study, as well as the clear logical process, was the testing was all done without any major disassembly. Once a car is in bits in your workshop, it can become your problem, so a diagnosis while everything is in one piece is very valuable.

This 2010 Citroen Dispatch cut out and would not re-start. The owner called his breakdown assist who recovered it to a local branch of a fast-fit chain. They took an initial look and decided it was not the type of job they'd like to get involved in. They passed my number on to the vehicle owner and I agreed to give him an answer as to why his van wouldn't start.

Using a compatible diagnostic interface, I initially scanned the whole vehicle for any stored diagnostic trouble codes (DTCs) and found communication faults stored across most modules. This turned out to be a red herring as the van had a flat battery that had triggered these issues. The battery was charged, and the vehicle was attached to a battery support unit (BSU) while I was working on it to ensure I didn't inadvertently flatten the battery again.

I pulled up some data streams from the engine control unit (ECU) to ascertain initially that the ECU was unlocking, proving the immobiliser was working and the controller area network (CAN). The vehicle's own communication network was intact.

An attempt was then made to start the engine, while monitoring data from vital sensors via the ECU to identify if anything looked untoward and could cause the non-start. It had adequate fuel rail pressure, the synchronisation of cam and crank (electronic monitoring of timing) was good, there was air flowing into the engine and the manifold pressure was also OK. To sum up, there appeared to be no reason why this engine wouldn't start.

At this point I then focused my attention on proving the mechanical condition of the electronic control. Using an oscilloscope and inductive amps clamp I checked that the ECU was triggering the injectors. At the same time using a hi amps clamp I monitored to amperage draw by the starter motor. This gives an indication of engine health as uneven current draw can be a good indicator of poor engine condition (Figure 10.42).

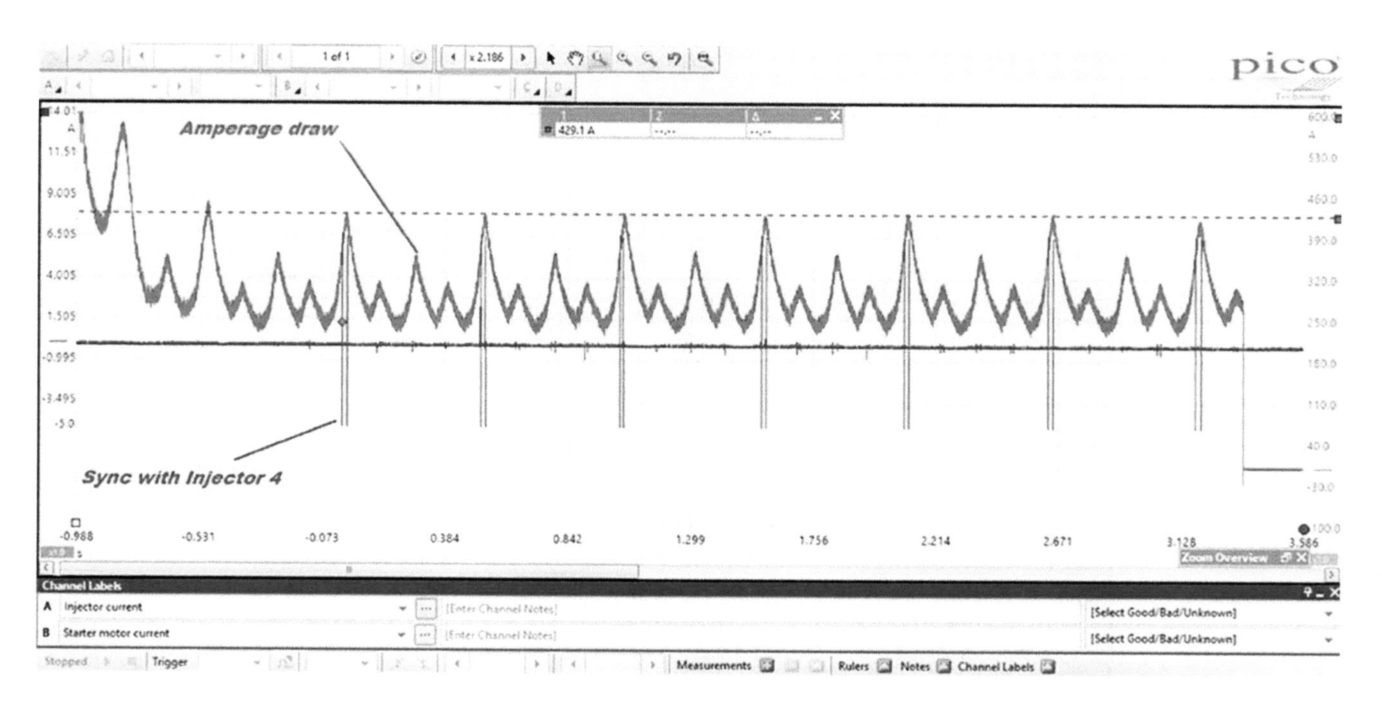

Figure 10.42 Red is the measurement of amperage drawn through the starter circuit, as you can see this is not uniformed proving relative compression across the 4 cylinders to be poor. In blue we see the switching event of the monitored injector.

Figure 10.43 Pico pressure sensor connected to measure cylinder pressure.

It was verified that injectors were being switched proving start permission was active within the ECU.

This proved that there was an imbalance across the four engine cylinders. I then removed the injector from cylinder 4 and placed a pressure transducer in cylinder to measure the pressure waveform of that cylinder and this is where everything started making sense (Figure 10.43).

I found a second compression event occurring halfway through the four cycles of the engine. Using the same method, I then performed the same test on cylinder 3 and found very low compression but again a second compression event occurring mid cycle.

The engine used in this van is a twin overhead camshaft setup and the high-pressure injection pump is driven off the end of the exhaust camshaft. The crankshaft and inlet cam are connected by a timed belt drive; the two camshafts are connected by a short timed chain drive internally within the cylinder head.

Using the information gathered so far, I could now make my diagnosis.

The relationship (timing) between the crank and inlet camshaft was good as this is monitored by position sensors, as mentioned earlier this was verified with a data stream of cam crank sync, which is being achieved so the timing belt has to be intact. Also, the timing chain between the two camshafts had to be good as the fuel pump is generating pressure; this is driven off the exhaust cam.

After analysis of the in-cylinder pressures the only conclusion can be that the exhaust camshaft must have spun and moved as the second compression event proves the exhaust valves are not opening at the correct time. A conclusive diagnosis was given to the owner without the need to strip the engine so he could then make an informed decision on how to proceed with the repair to his van.

10.1.20 2014 Mitsubishi Outlander hybrid no ready mode

Dan Versluis, Diagnose Dan, https://www.youtube.com/channel/UCR73yN7vT7-P541g2qlfRrg/

COMMENT:

Dan presents some excellent case studies on his YouTube channel. I recommend watching this one as well as others. I like this example because as well as Dan's methodical approach, it shows how a series of tests were carried out before removing the HV battery. It is the equivalent, for example, of cranking current relative compression tests and pressure pulse tests being carried out, before stripping an engine.

Please note Dan's warnings about only working on high voltage if you are qualified.

When you push the start button on a petrol version, the engine will start to crank and hopefully the engine will start. On a hybrid or an electric vehicle when you push the start button the car will tell you it's 'Ready'.

The problem on this car is when you push the start button the ready sign doesn't come up and there is a message on the dash telling us there's something wrong with the high voltage system. The high voltage battery was totally empty, and when we try to charge it, the battery won't accept a charge.

I hooked up the scan tool and I did a full system scan of this car. There is a code stored in the BMU (battery management unit), which is responsible for everything involving the high voltage battery. The stored fault code (Figure 10.44) is P1AAD: Battery current sensor fail (CAN). This is an active DTC so we can now investigate further why this particular code is showing.

The battery current sensor is responsible for measuring the current going into the battery, for example, while it's being charged, and current going out of the battery while it's powering the high voltage systems. Without the information from the current sensor, the BMU doesn't know how much current is going into

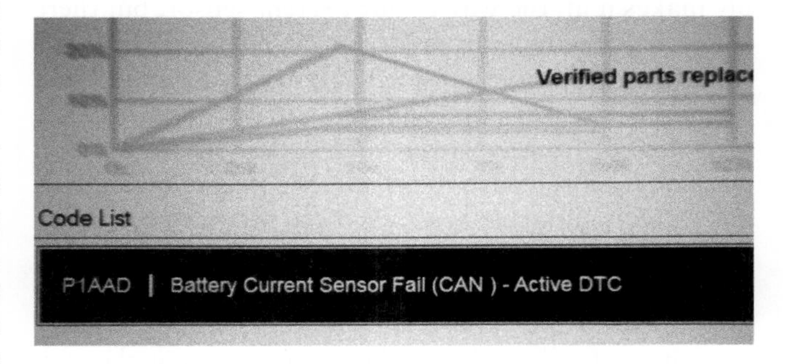

Figure 10.44 Fault code PIAAD.

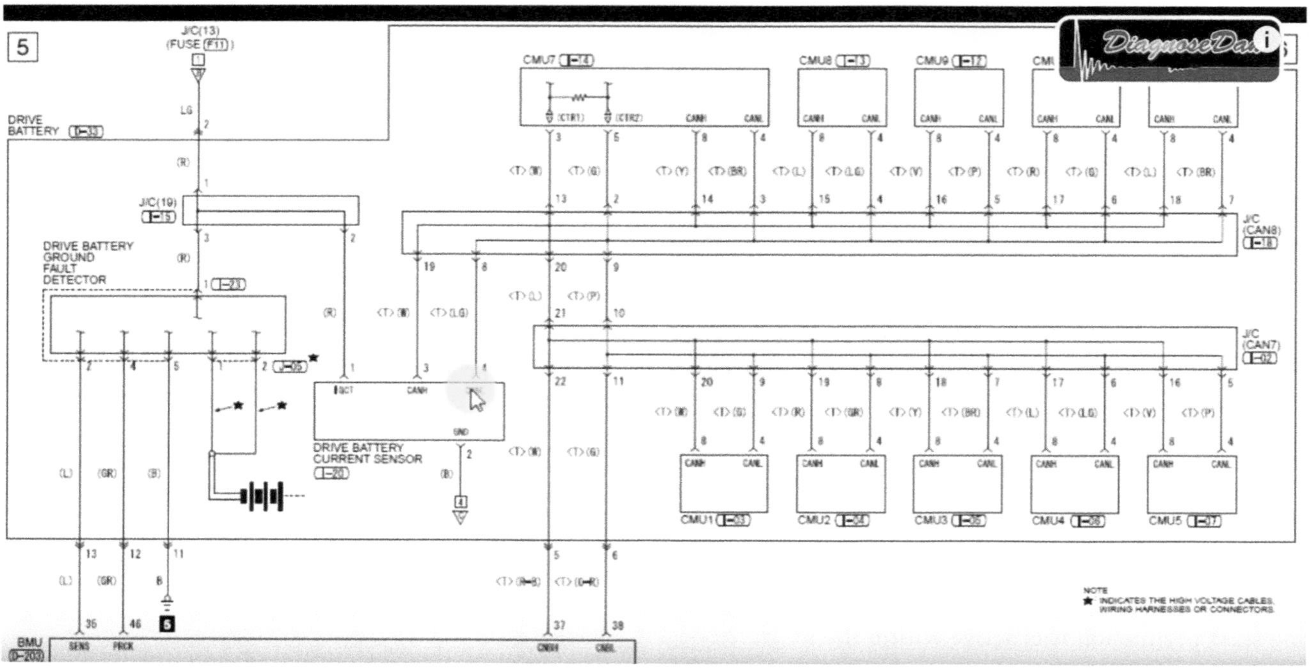

Figure 10.45 Wiring diagram.

or out of the battery so it can't operate properly. This explains why our high voltage battery is no longer accepting a charge. I also noted that the word 'CAN' is in the fault code description.

Now let's take a look at a wiring diagram (Figure 10.45) to get a better understanding why 'CAN' is in the fault code. The current sensor is a four-wire sensor. We've got a red power feed, a white and light green communication wire and a black ground wire. With the red power supply as you can see the power supply is being shared internally between the ground fault detector and the current sensor. The BMU is not flagging a code for the ground fault detector, so we can assume this sensor is working just fine.

If this sensor is working, it means the power supply is making it into the battery. Now just for peace of mind, I check the light green wire at pin two just outside the battery with the test light and sure enough we had a good voltage supply. This doesn't mean the voltage supply makes it all the way to our current sensor, but there is no way of telling without opening the battery.

There are two communication wires that run between the BMU and the high voltage battery. The BMU is receiving information from each of the other modules, just not from the current sensor. We know the BMU is able to communicate with all these modules so there's nothing wrong with the network inside the battery.

However, we don't know if the communication is making it all the way to the current sensor but there is no way of telling this without measuring at the current sensor itself. To do that the battery needs to be opened. The fourth and final wire is this black ground wire going to the current sensor. You can see it's being shared internally with all the other modules, so we know the ground going into the battery is just fine. Since the current sensor is located inside the high voltage battery, we've got no choice but to remove the high voltage battery from underneath the vehicle and open it up to see what's going on (Figure 10.46).

Working on high voltage systems can be dangerous so when doing so make sure you've got the right paperwork. Make sure you're an EV qualified technician. Also, always follow the rules and regulations in your country.

The whole process of removing and opening up the battery pack is quite a lengthy one so I will not outline every step. Now that we've got the battery pack out from underneath the car and have opened it up, we can take a look inside and try to find out what's wrong.

Figure 10.46 High voltage battery pack removed from the vehicle.

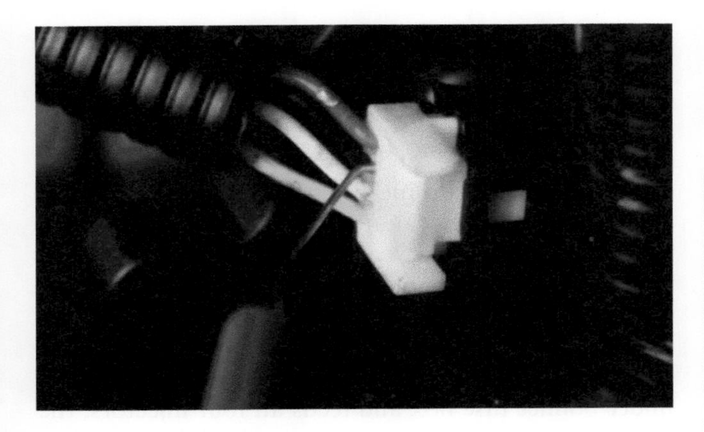

Figure 10.47 Sensor connector.

What I did first was to reconnect the low-voltage connectors from the car to the battery pack so right now we've got a live working high voltage battery pack.

As mentioned earlier, the current sensor inside this battery is a four-wire sensor. In the next step I want to check the power the ground and both communication wires. The red and the black wires are our power and our ground. When I touch the red wire with one side and the black with the other side of the test light, if the power and ground are good the test lights. Note how close we are working to the high voltage system – hence the safety equipment and gloves.

Next, I hooked up the scope to those two communication wires (Figure 10.47) to see what the CAN high and CAN low wires of our current sensor produce. There seems to be absolutely no issue with the communication on our network we've got good ground, good power and good communication. Everything is there for the sensor to work properly but it doesn't. So my final diagnosis is we've got a bad current sensor.

After fitting the new sensor, which works the same as an amp clamp, we cleared the codes, repeated the scan and no fault codes were present. The vehicle was rebuilt, tested and charged. Another successful job done (Figure 10.48)!

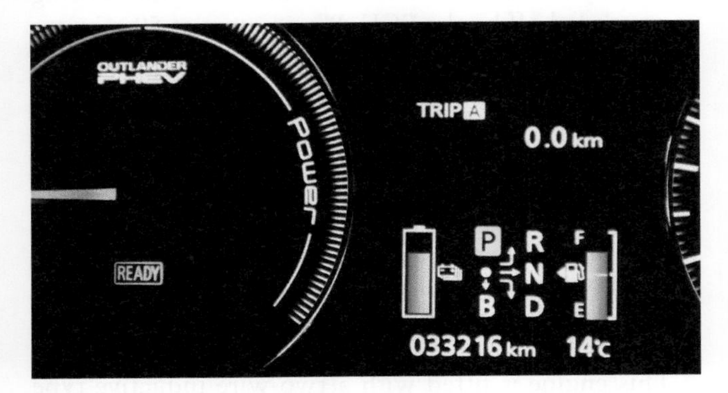

Figure 10.48 What should happen when the start button is pressed!

10.1.21 Fiat Panda 2010 1.1 petrol, stalling

Jason Sprake, Jay's Autocentre, Nottingham, England

COMMENT:

It was difficult because there were lots of excellent examples, but this is the winning case study!

What is really impressive here is how Jason built a diagnosis plan based on what they knew to be good or correct, NOT on information they did not have. As each system was proved good, they moved on to check the next. This is an excellent case study as it also shows the benefit of careful communication with the customer.

Vehicle starts from cold fine. It runs well till it's warm and at idle then it just cuts out. Then it's hard to start sometimes; other times it starts straight back up but cuts out again after a few seconds or minutes. When started if you put a few hundred revs on it runs fine. No fault codes and all live data looks fine.

Our first steps are to ask the customer for background on the fault. How long has it been happening? Does it happen all the time or is it intermittent? Has any work been done to the vehicle just before or after the fault?

The customer informed us he first noticed the issue about two week ago. He drove to his local shop and parked without issue, but on his way back he pulled up at some lights and the car just died. The engine would not restart but sounded like it was trying. He got a local garage to come and tow the car in for repair.

The garage replaced the crank sensor, plugs and coil packs; sent the ECU away for testing; and cut into the engine loom to repair some poor wiring they had noticed. After nearly two weeks and a several hundred-pound bill, they told the customer he needed to take it to a dealer as they could not find the issue.

The vehicle arrived to us on the back of a recovery truck. It started fine and drove off the truck and into the workshop. We authorised our normal one-hour labour initial inspection fee, which in turn gave us time to carry out some basic testing to indicate the best direction to follow in diagnosing the cause of the fault.

Our first hour is used as follows: Confirm the customer's complaint of cutting out when warmed up so we know the fault is present. We did not want to do this on road test so we carried out a quick drive-test around the industrial estate just to confirm the vehicle drives fine (not in limp mode, no engine noises, any warning lights on when driven like ABS/ESP that are not always present when engine is running but vehicle is not driving).

The vehicle drives fine so we bring it in to the workshop and leave it running to warm up. Our next step is to carry out a full system scan on the vehicle to see if any fault codes are stored. A full scan shows no fault codes in any systems at all. This could be due to the last garage clearing the faults, so we turned to live data. A good look down live data shows all values present and correct. We now monitor live data and keep a close eye on the engine temperature.

By this time, we are 20 minutes into the hour; we have no fault codes, all live data looks good and the engine is running fine, so we monitor. The engine is now warming up and the revs are starting to settle down to base idle, then it cuts out. First, we try to restart it. It started straight up and ran for a few seconds as the revs settle and cut out again. Customer's complaint is confirmed; so we carried out another system-wide scan to check for fault codes.

After the scan reveals no fault codes stored, we try to start it again. It starts straight up and runs fine for two minutes before shutting off once again. We are now about 45 minutes in to the agreed one hour so it's time to build a test plan based on what we know, calculate what time we need to carry out the plan so we can explain to the customer what the next steps are and for the customer to know the most important thing, how much it's going to cost.

Our test plan was based on the facts we know at the moment even though it's not much. From the symptoms it could be any one of the fundamentals (air, fuel and spark) causing the fault. With no fault codes to indicate a direction, we have to go back to basics, carry out tests and eliminate the basics first. So, we call the customer, explain what we have found, explain the next steps and request four hours labour in order to work through them. The customer agrees and we start our test plan.

Now when I said above, we know very little, the truth is we know more than we think. Because the engine runs fine for 20 minutes, we can rule a few things out. It's safe to say that we don't have a mechanical timing issue as the engine runs fine and starts back up running fine. We know it's not a coil or spark plug as there are no misfires. It's also safe to say it's not an injector fault with no misfires. The fault is something that when it happens effects all four cylinders at the same time. Some form of control like injector pulse or power feed, fuel pressure or even ECU fault.

Our first option is a visual check of the engine. We can see two new coil packs bolted to the side of the engine but nothing else stands out. A quick wiggle test on the battery leads confirms they are tight, terminals are clean, the cables look in good condition and the ground connection is good. It's time to test. We noticed a quick connector on the fuel pipe to the rail, so we removed the pipe and plumbed in our fuel pressure gauge. Checking for leaks we turned the ignition on. We could hear the tank pump start and the gauge goes straight up to 3.5 bar; we start the engine and monitor the gauge. It sits stable at 3.5 bar. After a few minutes the engine cuts out, but the fuel pressure is still at 3.5 bar after the engine stalls. We can now rule out fuel pressure, so we move to fuel delivery.

We look up the engine wiring diagram on Autodata (other data sources are available) and print it. Looking at the injector circuit tells us several things. First, we can see a relay switched 12 V feed comes from the under bonnet fuse box and feeds all four injectors, the same feed is used for the coil packs. Each injector then has an independent wire going direct to the ECU which is a switched earth controlled by the ECU. If we think about this, we can use our PicoScope to carry out a number of tests at the same time using just one probe.

As you can see from Figure 10.49, we back probed number 1 injector, from this single probe we can monitor the injector feed, the injector earth pulse signal and the engine RPM. The capture shows the engine stall event at the right-hand side. I'll explain what we are looking at. First the power feed is clear. It's the thick blue line running from left to right; if you look at the blue scale on the left it's sitting at just over 14 V, normal alternator idle voltage. We can also see the injector trigger events, which are the evenly spaced vertical lines. These show the feed dropping to zero every time the ECU turns on the earth and the consumer (the injector) uses up the voltage (voltage drop). Using vertical rulers over the last few injector pulses you can clearly see when the next injector pulse should have happened but did not. However, we can see from the feed that voltage is still present and has not yet started to fall, which means both the engine and the alternator are still at idle speed.

After this point you see the engine speed starting to slow as indicated by the voltage starting to drop. Then we see the relay switch off and cut the feed. We know that both the relay and the injector triggers are controlled by the engine ECU and that the engine is being commanded off by the ECU. We are now at 50 minutes into our four hours and have some good direction to follow.

We now know why the last garage has gone down the path it did, replacing coil packs and sending the ECU away for test. Thinking about where the ECU gets its information from in order to control cylinder events, like fuel injection and spark, points us to the crank sensor, but we know this has been changed. Is the new sensor faulty? There is only one why to find out and that's to test it.

This engine is fitted with a two-wire inductive type crank sensor. It's different from the Hall type three-wire

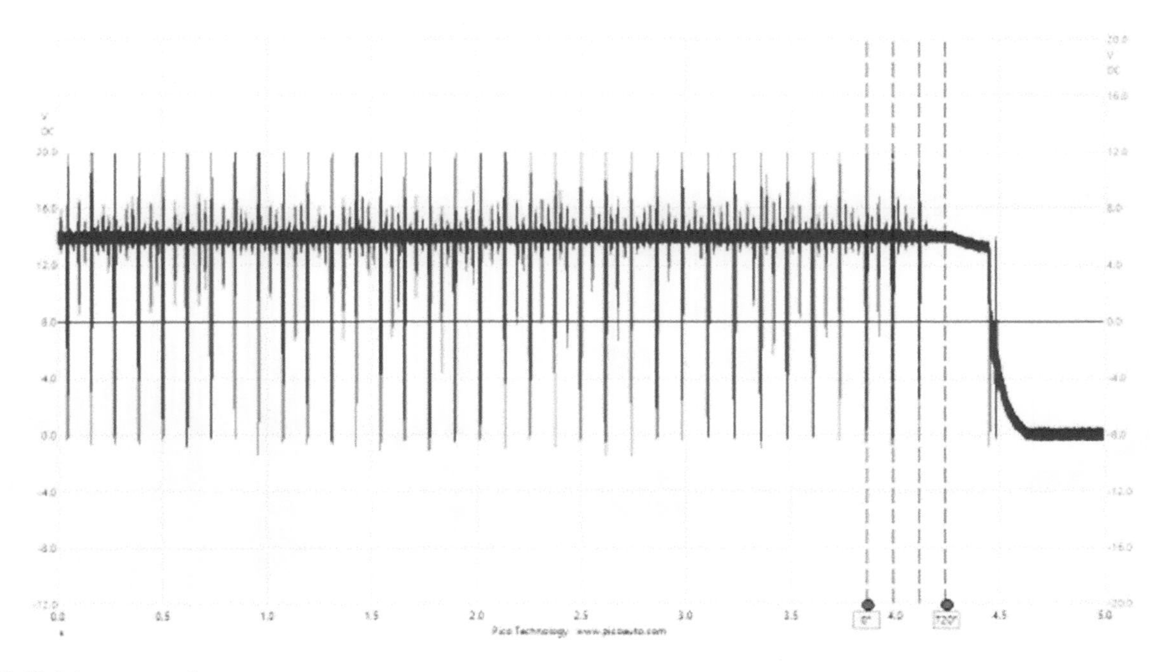

Figure 10.49 Injector waveform.

sensor, which has a feed, an earth and a signal wire going to the ECU. The inductive sensor has only two wires going direct to the ECU. It has no feed or earth because it generates its own voltage by metal teeth on the timing plate passing a fixed magnet, which in turn induces a voltage that is sent direct to the ECU. The ECU can use this information to calculate both engine speed and position, so it knows when to inject fuel and when to create a spark. The voltage inducted through the sensor is alternating current and not direct current which the Hall type three-wire sensor uses. The voltage induced normally has a 3 V sweep going from plus 1.5 V to −1.5 V. The ECU needs to see at least a 2 V sweep in order to recognise the signal; the only way to test it is to use the PicoScope again and look at the output signal.

Below are three images (Figures 10.50–10.52), the first shows crank sensor and cam sensor on start up. The pink track is the cam sensor and the red and blue show the three signals from the crank sensor. The second is engine at idle, and the third image is the same but shows the stall event.

Using rulers, we can see the cause of the stall in Figure 10.52. You will notice the two rulers going from left to right; one is at −1 V and the other is at +1 V on the blue scale for the blue trace. This is to show the minimum voltage needed in order for the ECU to recognise the signal. We also have vertical rulers as a crude way of measuring the engine RPM. You will notice the first three vertical rulers line up perfectly with the leading edge on the cam sensor indicating stable idle speed at this point. Just after the fourth vertical ruler you can see that the crank sensor voltage drops just below the 2 V indicator ruler and is no longer feeding the ECU information. We also see at this point the engine speed is still the same and the leading edge on the cam sensor lines up perfectly.

After this point we see both the cam and crank speed getting longer on the timescale indicating the engine is slowing down due to no longer being powered by the cylinders firing. Then ECU can no longer confirm the engine position and so shuts down the engine. Because this is an inductive type crank sensor, the speed the teeth pass by the magnet increases the amount of voltage being induced into the sensor. When the engine is cold and the idle speed is higher it's generating enough voltage signal for the ECU to recognise it. However, as the engine warms up and the revs drop to the signal voltage also dropped to the point it could net be recognised and shuts the engine down.

Inspecting the crank sensor, which is positioned by the bottom pulley, we could see impact damage to the phonic wheel. When running we could see it had a slight buckle causing the crank sensor air gap to be 0.4 mm too big, so, when it got up to temperature and revs settled at idle, the inductive voltage drops below a 2 V spread (AC) by just 0.02 V, and the injectors shut down. Looking at the waveform the signal voltage goes up and down slightly like the phonic ring has a slight buckle in it.

A new phonic wheel was fitted and tested. The fault was now cured. Three hours in total to diagnose; we used the remaining two hours to fit the phonic wheel and confirm the fix.

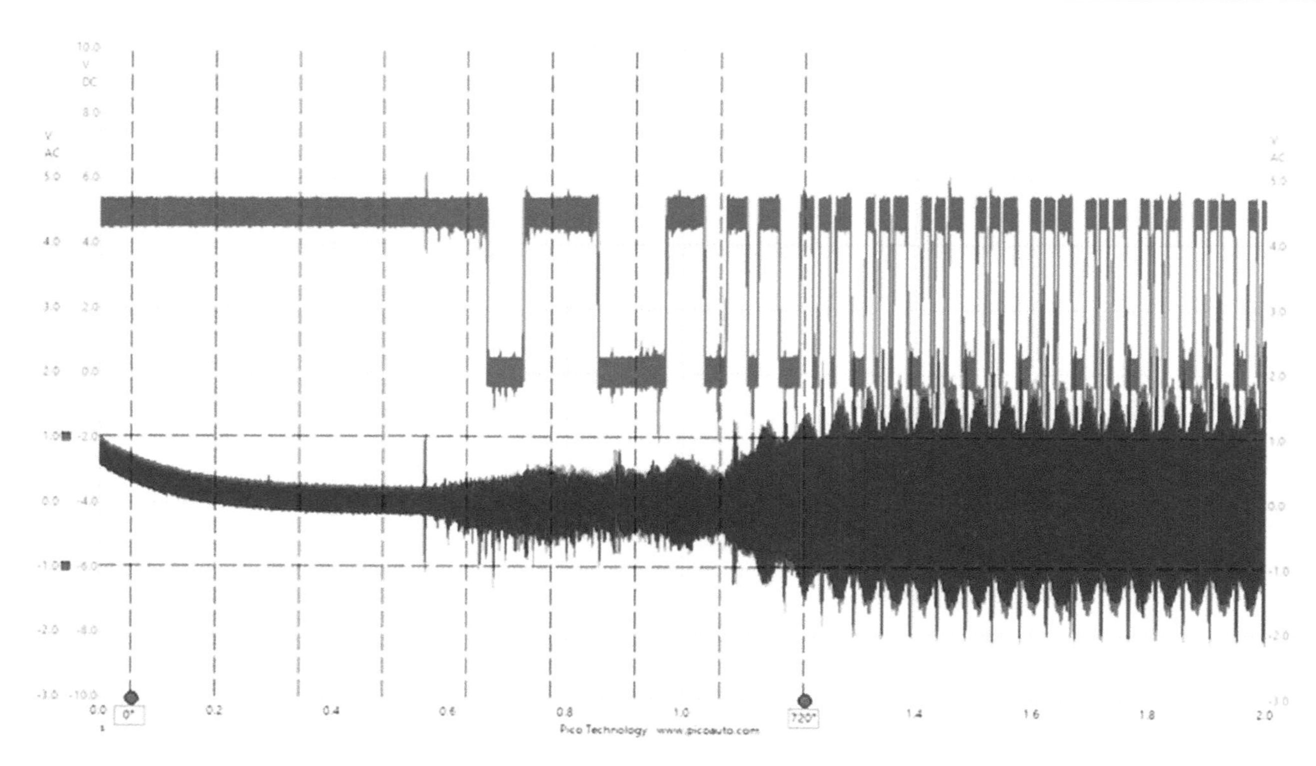

Figure 10.50 Crank and cam sensor on start up.

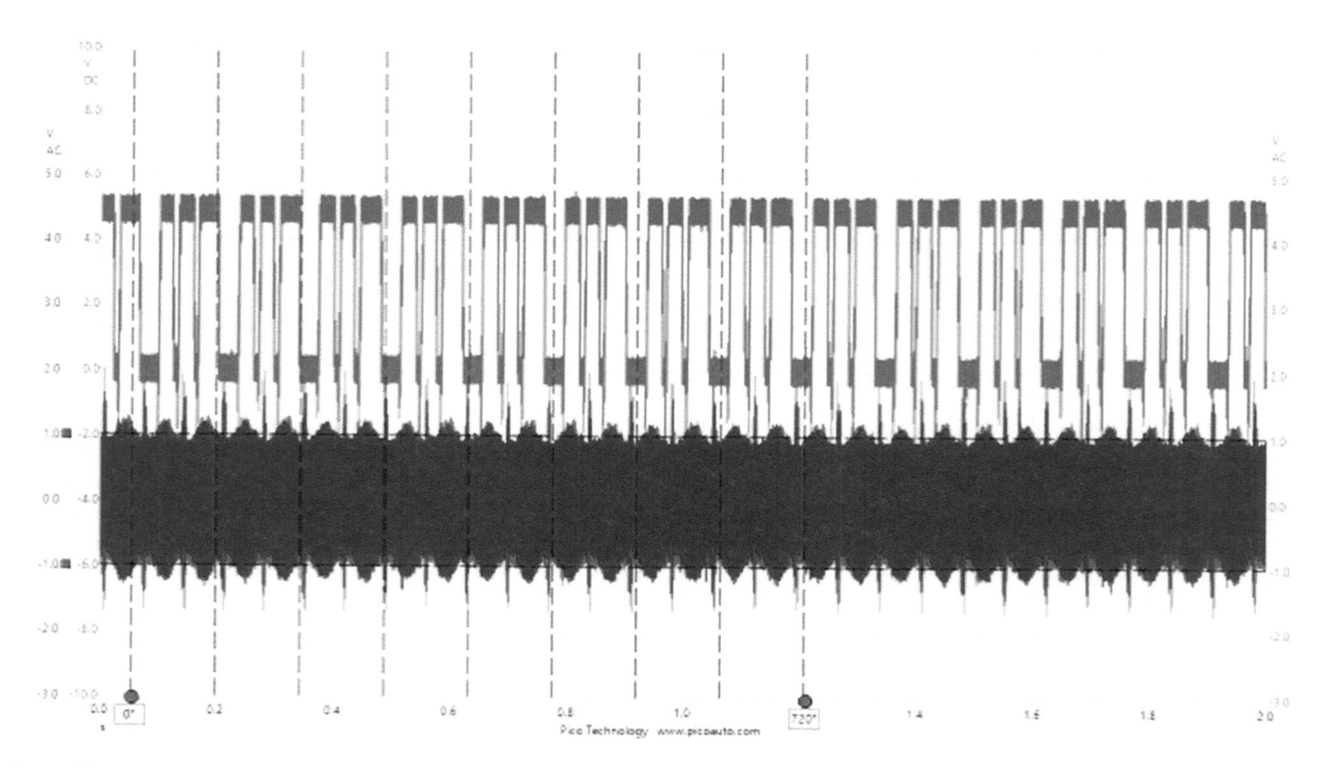

Figure 10.51 Engine at idle.

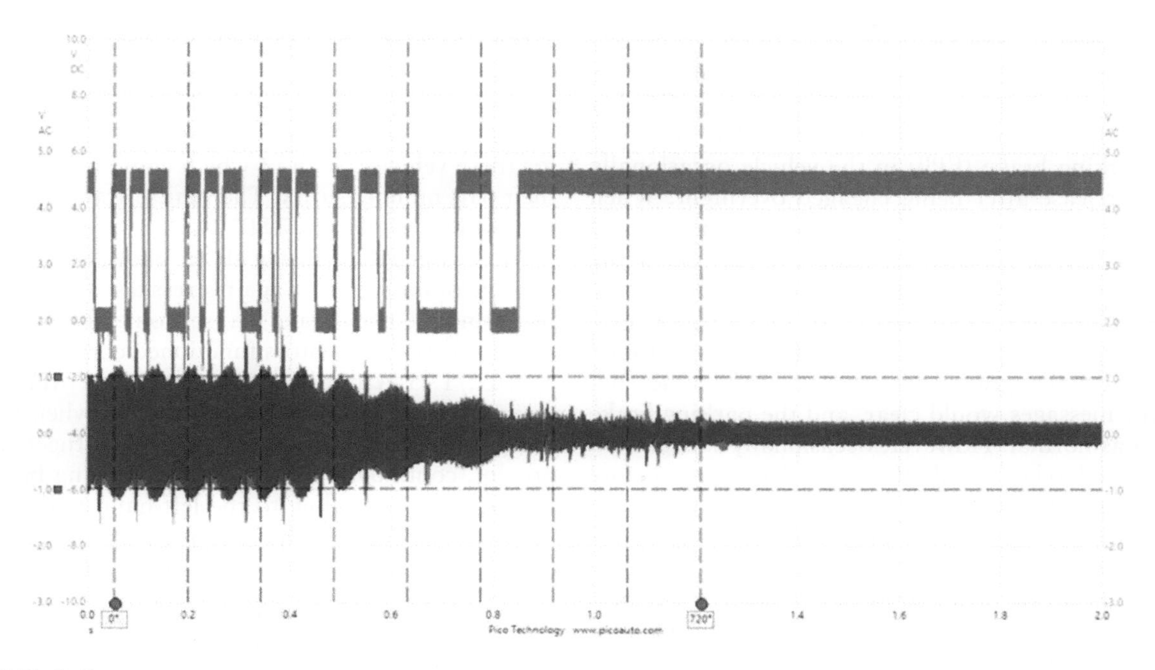

Figure 10.52 Stall event.

10.1.22 And finally . . .

Liam Peers

> **COMMENT:**
>
> *A reminder from Liam to always check for the obvious things first!*

Spend years learning how to diagnose faults quickly, training courses, time in the evenings reading books, reading wiring diagrams. And you get given 'seat won't go all the way back' for a warranty job (Figure 10.53).

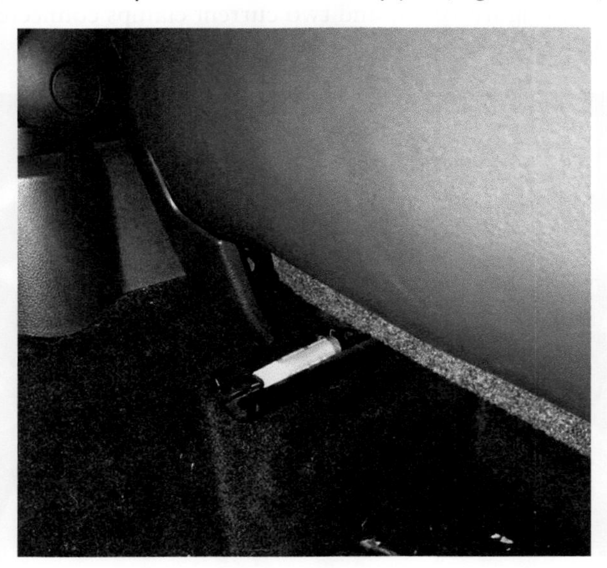

Figure 10.53 Lip gloss apparently!

10.2 COMPANY CASE STUDIES

10.2.1 Introduction

I have included a few case studies here presented by companies. It will be clear to readers that I am a big fan of the PicoScope, but alternatives are available. I have also included three examples from a technician support organisation in Australia.

10.2.2 PicoScope: Audi A6 Quattro S-Line electric parking brake (EPB) failure

Steve Smith, PicoScope
www.picoauto.com/library/case-studies

> **COMMENT:**
>
> *I would have expected nothing less than an excellent case study from the PicoScope team. I edited this case study down a little for space reasons, so I strongly recommend following the above link to read the full details.*
>
> *I particularly like how not only was the PicoScope used to great effect, a very logical and detailed process was followed. This included looking at bulletin information and similar data. Also, the use of simple tests like listening to each actuator – good to remember the basics!*

Our customer reports the following messages appear within the instrument panel: Parking Brake Malfunction, Start/Stop system deactivated. Further questioning of the customer revealed that the electronic parking brake (EPB) in the vehicle occasionally failed to release after being engaged overnight. When the fault occurs, the vehicle can be driven without any brake binding or drag, suggesting the parking brake has released. The symptom had been very intermittent for over 12 months regardless of temperature or driving conditions. Additional feedback from the customer added that if the ignition was turned off and back on, the error messages would clear, and the parking brake released as normal. However, occasionally the ignition was required to be turned off for over 20 minutes before the parking brake function was restored.

A road test of the vehicle proved fruitless, given the parking brake function and Start/Stop feature performed normally. However once stationary, with repeated operation of the parking brake (On/Off) it became apparent when listening to their operation that the left-hand (LH) rear calliper parking brake mechanism appeared to lag compared to the right-hand (RH) rear calliper. During this stress test a chime was emitted from the instrument panel and the Parking Brake Malfunction message appeared. With the error message displayed, further switching of the parking brake confirmed no operation of the left-hand parking brake actuator whereas the right-hand functioned normally. Cycling the ignition restored both parking brake actuators to normal operation. The customer complaint was verified, and the vehicle ID and specification were confirmed.

The customer interview highlighted a full-service history, numerous accessories installed and the recent replacement of the LH parking brake actuator – but not the complete calliper assembly.

An initial inspection confirmed correct installation of a new left-hand rear parking brake actuator, the rear brake linings to be serviceable while freely operating and all visible electrical connections and harness routing under the vehicle to both rear brake calliper to be secure.

A vehicle scan of all on-board control units revealed a number of fault codes listed below:

- ENGINE- 17403 - Implausible Data Received from Parking Brake Control Module U0417 00
- ABS- 0486 - Function Restriction due to Faults in Other Modules U1113 00
- ABS- 0692 - Function Restriction due to Faults in Other Modules U1113 00
- EPB- 67840 - Left Parking Brake Motor voltage supply C100D 01 [008] - Electrical Failure
- EPB- 71424 - Left Parking Brake Motor voltage supply C100D 12 [008] - Short to Plus

Based on the evidence acquired so far, pursuing the parking brake fault codes became the priority.

Prior to diving in here, taking a step back and checking for technical bulletins, software updates, recalls and campaigns, etc., is paramount. This action revealed that a number of parking brake control units were utilised for these vehicles, each of which could be identified by their part number suffix. The vehicle scan identified the control unit as 4H0 907 801 G.

After saving and erasing all the above fault codes, the parking brake was once again stress tested by repeatedly operating the park brake whilst monitoring live data.

Initially both actuators functioned correctly drawing approx. 17.5 A until the Parking Brake Malfunction error message appeared. This occurred when releasing the parking brake, not applying. With the fault now present, continual operation of the parking brake confirmed no current flow to the left-hand parking brake actuator, right-hand operation was normal. A brief check of the fresh fault codes confirmed the return of:

- EPB- 67840 - Left Parking Brake Motor voltage supply C100D 01 [008] - Electrical Failure
- EPB- 71424 - Left Parking Brake Motor voltage supply C100D 12 [008] - Short to Plus

The PCM and ABS modules also reported U codes, indicating communication errors with the EPB module. At this stage we can attribute these codes to missing messages from the EPB module as the parking brake malfunction results in communication anomalies generating multiple warnings for the driver. The action plan is predominately governed by accessibility and probability. Based on the fact the left-hand parking brake actuator is repeatedly indicated as the offender, we need to qualify the actuator circuit.

With the fault present on the vehicle, the current flow and voltage supply to both right-hand and left-hand parking brake actuators was captured using breakout leads (Figure 10.54) and two current clamps connected to PicoScope.

Figure 10.54 Breakout leads.

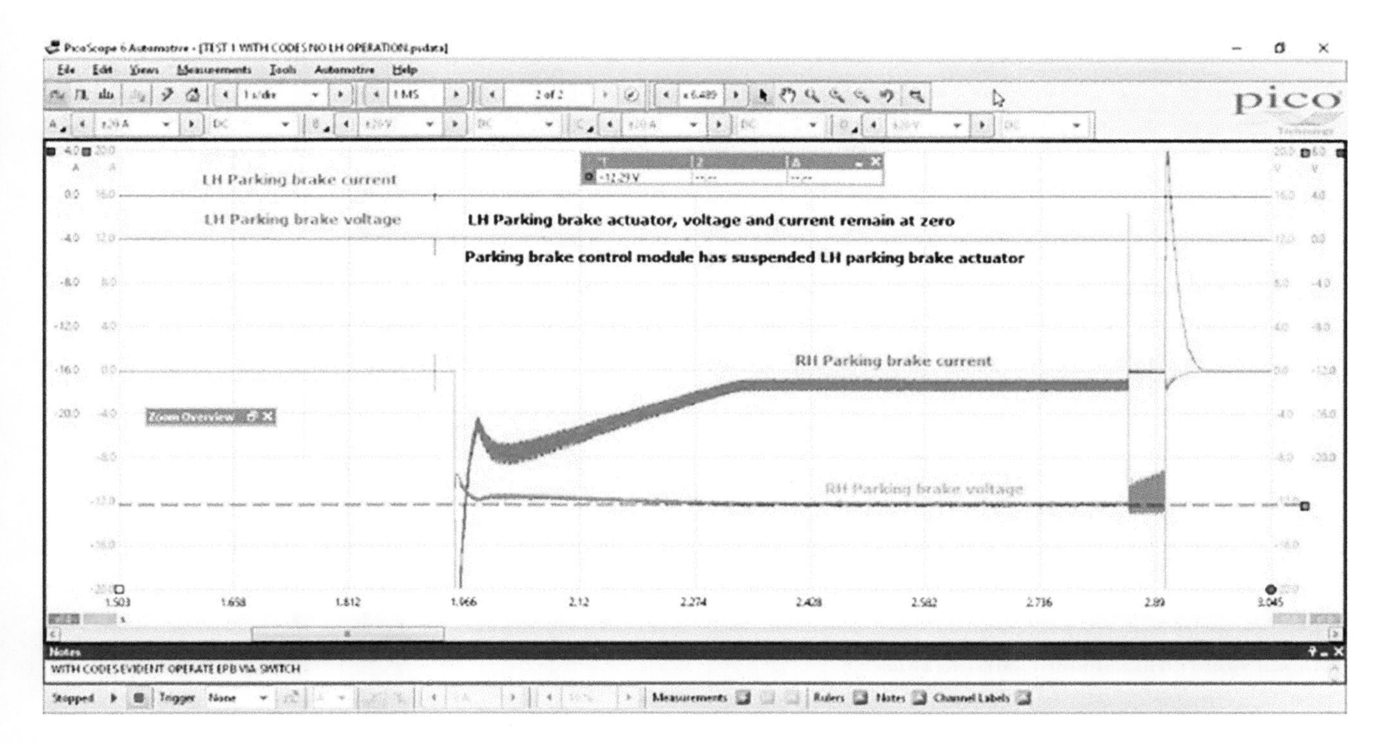

Figure 10.55 Current measurement on the PicoScope.

Figure 10.55 confirms there is no voltage supply or current flow to the left-hand parking brake actuator. However peak inrush current to the RH actuator exceeds 20 A.

Given the fault was becoming easier to reproduce, the fault codes were once again erased, and the parking brake operated while monitoring voltage and current to both actuators. The theory here was to catch the exact moment the fault appeared and identify what events were taking place in the parking brake circuit (Figure 10.56). Below I have revised all the input ranges and test lead orientation at both brake actuators to provide a clearer view of voltage and current during EPB operation.

Our parking brake control module suggests the left-hand parking brake actuator has a short to battery positive. This can be seen in the time taken for the voltage on Channel B red to fall to zero volts after the parking brake release function has completed. The left-hand actuator voltage decay time measures approx. 350 ms

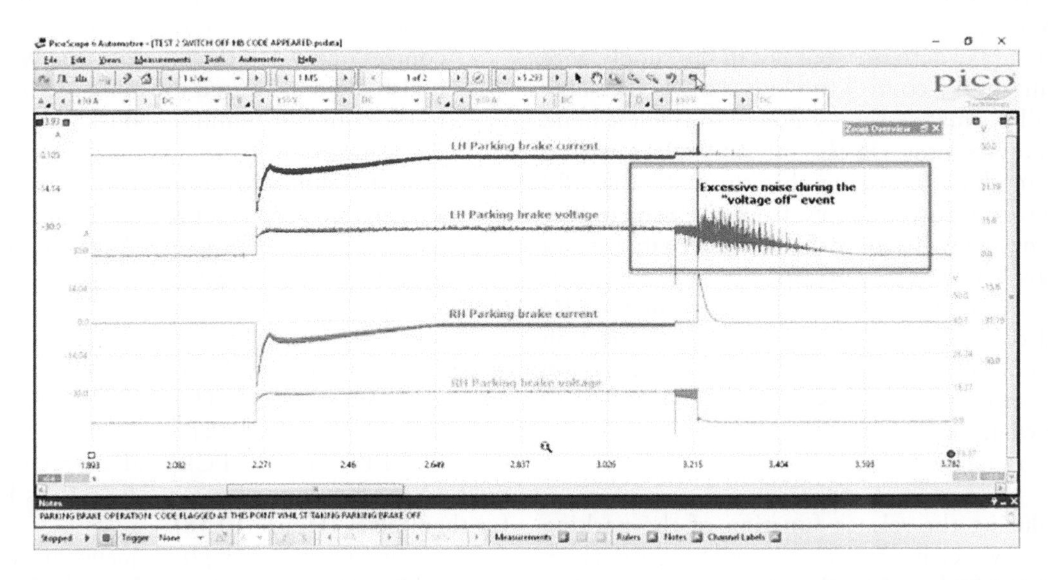

Figure 10.56 We have captured one of those rare intermittent glitches that is directly responsible for our EPB fault codes. Closer analysis reveals a number of concerns with the voltage supply and current flow through the left-hand parking brake actuator.

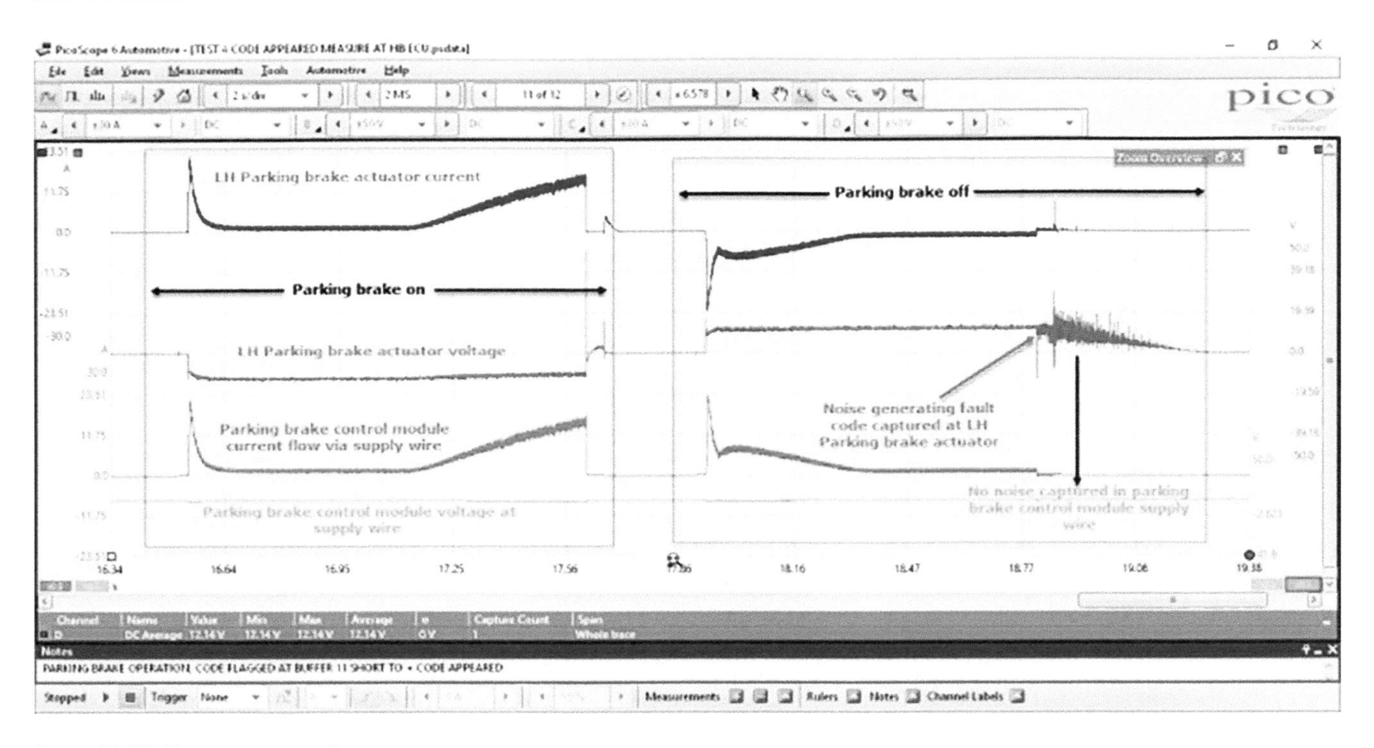

Figure 10.57 Comparing waveforms.

compared to the right-hand actuator at approximately 15 ms, therefore interpreted as fault code 71424 - Left Parking Brake Motor voltage supply - Short to Plus.

I guess the question raised from the above capture (Figure 10.56) is where is the electrical noise being generated, causing the parking brake fault code and suspension of the left-hand parking brake actuator? In Figure 10.57 I have changed Channel C green and Channel D yellow to capture the voltage and current flow through the power supply wire to the EPB module.

We can see how the noise generated during the release stage of the parking brake is evident at the left-hand parking brake actuator, but not in the supply wire to the parking brake control module. This is the same wire that will ultimately supply power to the left-hand parking brake actuator via the internal relays within the control module. Thinking this through, the EPB module receives a power supply from the vehicle fuse box, which in turn will supply the brake actuator via an internal relay. The earth return for the brake actuator once again will pass via an internal relay inside the EPB module and onto chassis ground. A visual inspection of the EPB module chassis ground point confirmed no areas of concern and so at this point a decision has to be made on what we know.

Summary of diagnosis:

- Only during the release function of the parking brake do we see intermittent excessive noise which generates a fault code within the EPB system referring to the left-hand parking brake actuator.
- The application of the parking brake does not generate noise or induce a fault code.
- The left-hand parking brake actuator has been replaced. Had the actuator been responsible for the noise, e.g. arcing motor brushes, the noise would have been present during parking brake on and off.
- The circuit between the EPB module and left-hand actuator is good based on the values compared at both the module and actuator.
- The power and ground to the EPB module are good with zero noise on the power supply to the control unit when the fault occurs. With hindsight I wish I had measured the noise on the ground line to the controller simultaneously.

Based on the summary above the EPB control module was replaced as the source of our noise during the parking brake release function. With the EPB module replaced, stress testing was once again carried out to confirm a fix.

Happy in the knowledge we have a fix, the temptation to dismantle the original parking brake control module was too much to refuse. Thanks to the hardware team here at Pico the control relays were carefully removed from the EPB module PCB and inspected under the microscope for signs of wear. The results below speak for themselves (Figure 10.58).

The ability to capture high frequency noise prior to fix Channel B red highlights the need for a high-performance scope where such anomalies occur. Without sufficient

Figure 10.58 Burnt relay contacts.

bandwidth and sample rates such high frequency noise would not have been displayed so leading to a prolonged and ill-informed diagnosis. Once the repair had been completed, PicoScope also revealed finite detail surrounding the operational characteristics of the parking brake control system that we are not gifted when it comes to technical data. The post fix analysis becomes a form of reverse engineering that can only be concluded when measuring component operation in real time with infinite detail at lightening speeds, all of which require PicoScope.

10.2.3 Autoscope: Diagnosing an engine using the pressure waveforms

By Vasyl Postolovskyi and Olle Gladso, Riverland Technical and Community College, Albert Lea, MN, USA
https://usbautoscope.eu/en

COMMENT:

I like the discussion of cylinder pressure in this case study. The software for the Autoscope also has some amazing features.

In this article, we will similarly discuss an alternative approach to diagnosing an engine using the pressure waveforms from an in-cylinder pressure transducer.

We are going to diagnose engines based on the pressure waveform obtained from the cylinder(s) of a running engine. We will need to safely disable the ignition for the chosen cylinder. One method is to connect the coil or plug wire to a spark tester. To avoid any problems from ignited fuel, from perhaps a hot surface in the cylinder, we should disable fuel delivery to the cylinder being tested as well. To display the waveform, we use a transducer that converts pressure into voltage. The voltage output can then be displayed as a trace on an oscilloscope screen.

The transducer is replacing the spark plug in the cylinder being tested (See Figure 10.59). Due to heating of the transducer from the running engine, the test should not run more than approximately three minutes. If the spark plug is recessed, it may be necessary to use an extension. A flexible extension, such as from a compression gauge set, should not be used. This is because the flexing may cause a loss of detail in the waveform.

Although the transducer shown here is incorporated in the oscilloscope kit used, other transducers can be used as well. The pressure transducers used should have low inertia so they react quickly to pressure changes. They should also be accurate, retain accuracy as they warm up and accept as well as recover quickly from pressure overload situations.

After setting up the transducer and the oscilloscope, a waveform (Figures 10.60 and 10.61 or similar) will be obtained once the engine is started. It's best to save and/or record the waveform and then analyse it. This way, there are no time limitations present with regards to running the engine.

It may take some time to become comfortable with using these waveforms to analyse engines. The particular software shown here has provisions for performing automatic analysis, which can be very helpful, especially when you're starting out performing this type of diagnosis.

Figure 10.59 The pressure transducer replaces the spark plug.

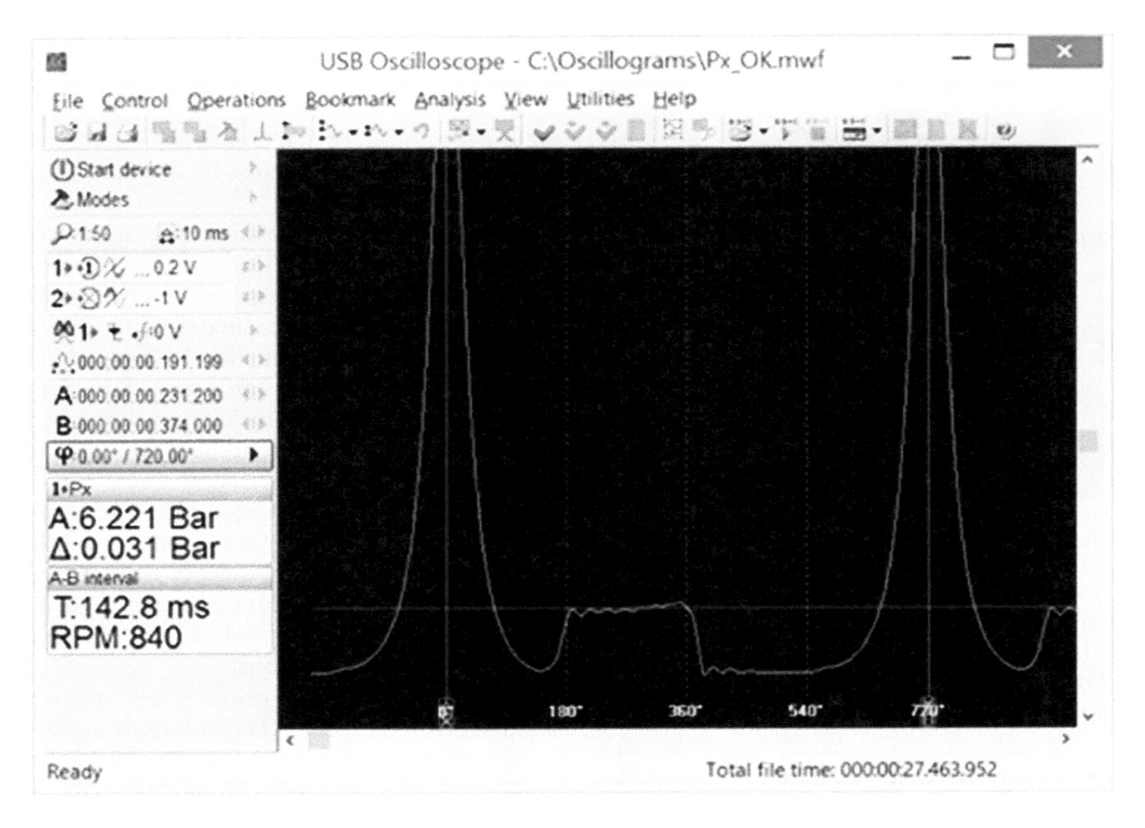

Figure 10.60 Recorded pressure waveform from a cylinder of an engine in good condition.

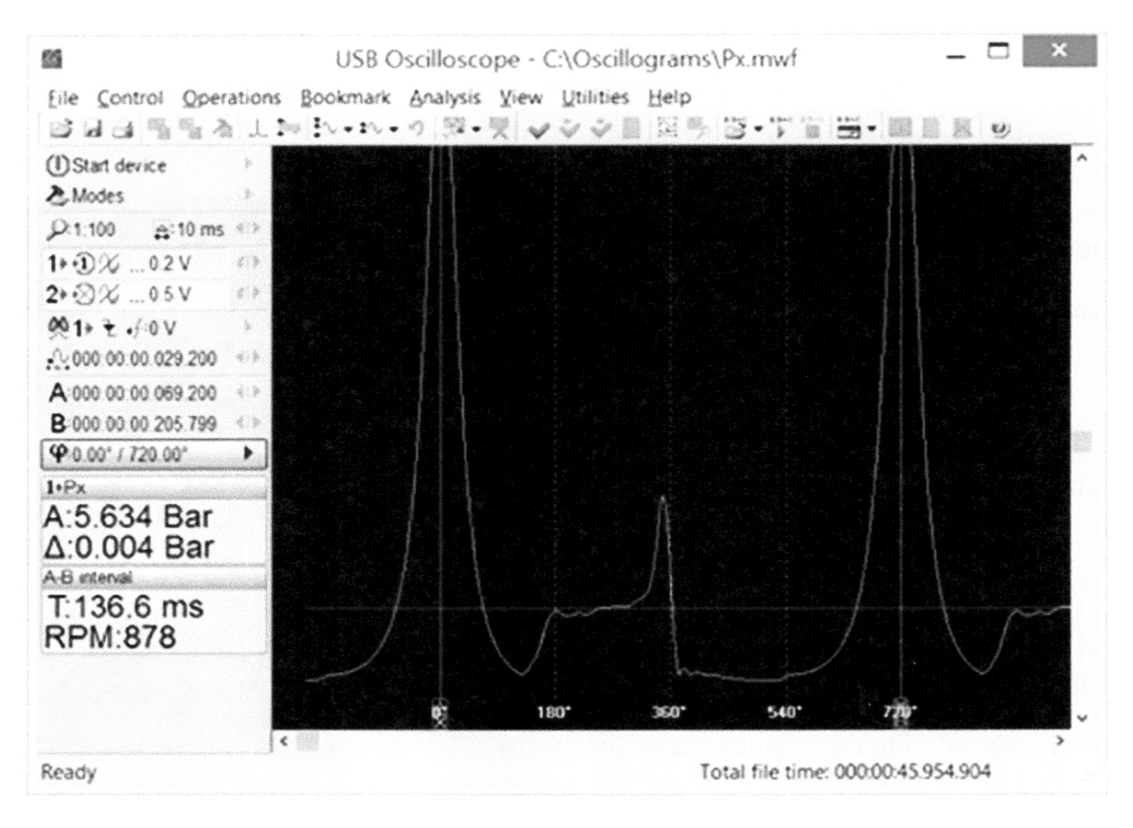

Figure 10.61 Recorded pressure waveform from a cylinder of an engine with a malfunction.

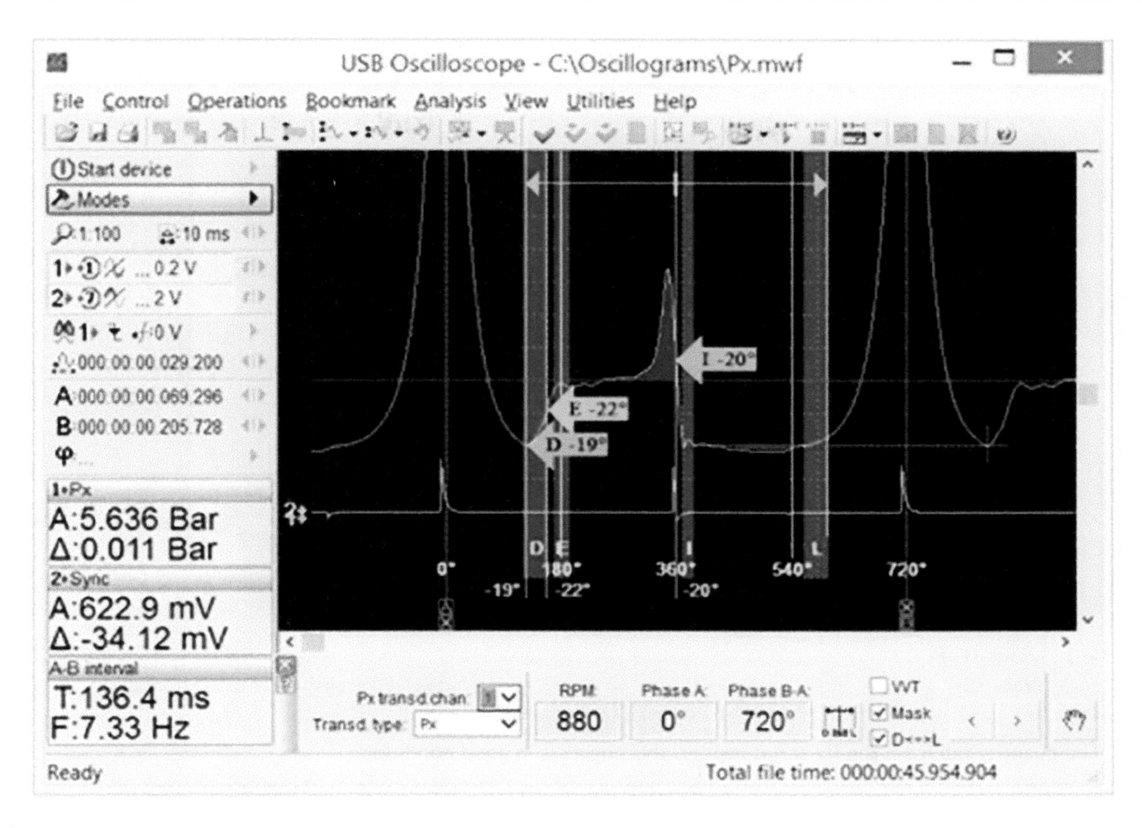

Figure 10.62 Deviations in the cylinder pressure waveform automatically detected and displayed by the oscilloscope program.

In Figure 10.62, the software has automatically detected deviations from the positions of characteristic points in the graph as well as shape distortions. Due to this automatic capability, the probability of detecting a mechanical defect in the engine increases, and the time spent on waveform research is significantly reduced.

This particular software automatically analyses the cylinder pressure waveform and generates a printout or report with a number of additional parameters and characteristics of the engine and the associated control unit. The calculated values are pneumatic and geometric characteristics of the cylinder; the list of found deviations is displayed in the form of text messages. To improve speed and accuracy of valve timing research, the cylinder pressure waveform is converted into a diagram of the gas amount in the cylinder and is displayed in two different ways, using a script.

A detailed diagram of the cyclic filling of the cylinder during the intake stroke, which characterizes the properties of the entire intake manifold of the engine is also provided. A diagram showing the energy consumption for scavenging exhaust gases from the cylinder is provided as well. Using these diagrams and the ignition timing signal, the ignition timing diagram is built and can be displayed.

Naturally, the script cannot replace an expert diagnostician, it can only convert cylinder pressure waveforms into a more readable form, but it does allow for a decrease in the time a diagnostician spends on analysis, and also to limit missed implicit/hidden deviations/faults. For more detailed information on the possibilities of the script, we will review each of its report tabs (Figure 10.63).

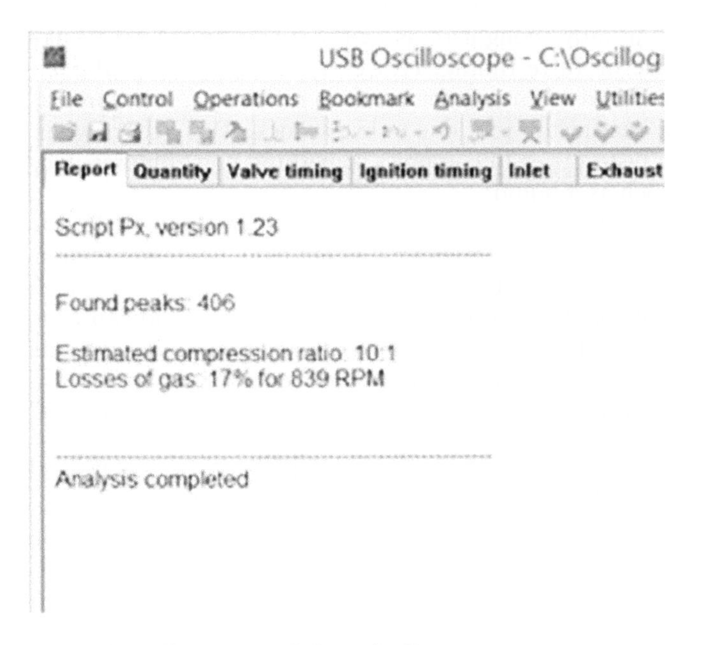

Figure 10.63 The report tab from the Px script.

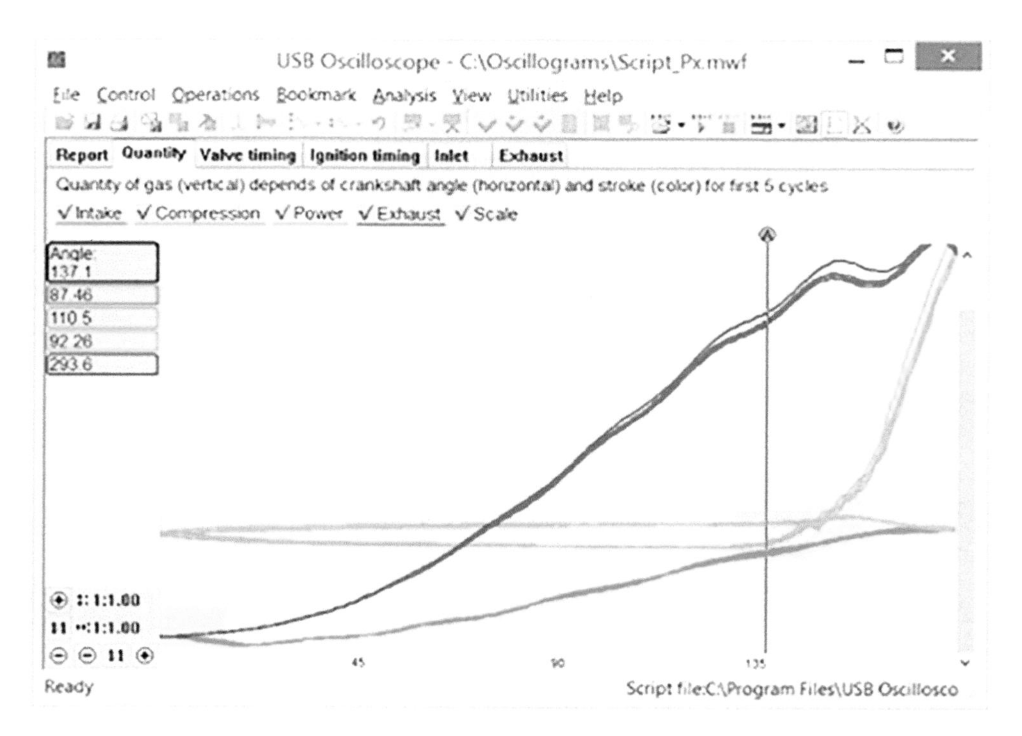

Figure 10.64 The quantity tab from the Px script report, this engine is in good condition.

The conventional or classic tool for assessing the state of an engine cylinder and piston is a compression gauge. It is designed to measure the compression or peak pressure in the cylinder obtained while cranking the engine. The measurement is a complex value and depends on losses through cylinder leakage, the compression ratio, the valve timing, the cranking speed and the state of the intake and exhaust ports or manifold.

A reduction of compression pressure in a cylinder is usually thought of as being caused by cylinder leakage or valve timing. However, the reason can also be reduced geometric compression ratio from, for example, a bent piston rod, due to hydro-lock. This occurs when a piston tries to compress something non-compressible, such as a liquid.

The Px script, as shown in Figures 10.63 and 10.64, can distinguish cylinder leakage from low compression ratio because it independently calculates gas losses and the compression value. The compression ratio can usually be found in the service information, under general engine data, and depends on the engine's design.

Normal pressure or gas loss for an engine in good condition is in the 10–18% range. A loss of more than 20% could indicate excessive leakage in a cylinder. The algorithm for calculating cylinder losses is complex, with some variables that are difficult to account for. A typical problem is the heat loss of the gas in the cylinder. The heat loss arises from the fact that the gas temperature in the cylinder during compression, even without ignition, is rising above the temperature of the cylinder walls. Consequently, part of the heat energy of the gas in the cylinder is transferred to the piston, cylinder and cylinder head. The loss of heat causes a loss of pressure. In practice, the calculated cylinder pressure loss of an engine in good condition is about 10%.

The graph indicates the amount of gas in the cylinder relative to the position of the piston in the cylinder and the stroke. Marker A is set to coincide with the end of the intake valve closing and the opening of the exhaust valve beginning.

10.2.4 The Automotive Technician (TaT), Australia

Jeff Taylor
www.tat.net.au

COMMENT:

I have included these case studies (Figures 10.65–67) as I particularly like the way they are presented. TaT is an Australian organisation, but it is not expensive to join and may provide useful information no matter where you are based. They have some excellent resources for technicians.

AUDQ709118
AUDI Q7
2009
Six-cylinder

Customer complaint

The vehicle, fitted with the TDI V6 diesel engine, was slow to fire up and the glow-plug light was flashing on dash.

Problem summary

The vehicle required excessive cranking on a dead-cold start. It would run OK when it did start and there were no power-loss issues while driving.

The vehicle would restart fine when it was hot. But the longer it was left, the more cranking needed to restart it.

Diagnostic sequence

The diagnostic equipment was hooked up to allow system interrogation lice-data monitoring. The following code came up:

Electronics section **P0087** – Fuel-rail pressure too low, sporadic fault.

This code had logged three drive-session times in two days. The logged pressure reading at times varied from 106 bar to 695 bar.

The fact that the code related to the fuel-rail pressure did not necessarily mean it was the main pump not doing the right thing.

Fault description

The first step was checking how the low-pressure feed was getting to the pump. The engine's top cover was removed and that was when a very fine stream of fuel shooting towards the windscreen was spotted.

The engine top-cover underside insulating material was also soaked in diesel fuel. It had been pressing and rubbing on the clear low-pressure fuel line (pic 1) that goes into the main pump feed.

The hose had developed a pin-hole leak that allowed the fuel to bleed out and air to get in when the vehicle was turned off (pic 2), hence the longer cranking time.

Fault solution

Two options were available. We could replace the complete hose set with one provided by Audi or fit a new 9.5mm fuel line and reroute it so it would not suffer the same issue again.

The second option was authorised by the owner (pic 3). Each end of the line had hard plastic barb fittings so it would seal up well.

Next came clearing the code, then a road test. The problem was solved.

Note: the vehicle had only done 62,500km.

Recommended time

Diagnostic time was 1.5 hours, taking into account preparation and research.

Repair time was 2.5 hours, taking into account location of parts and carrying out the repair to a tested outcome.

To access the entire Repair Solutions database
www.tat.net.au/tats-a-fact

The Automotive Technician 28

Figure 10.65 Audi Q7.

CHRVO98450
CHRYSLER VOYAGER
1998
Six-cylinder

Customer complaint

The vehicle was towed in with a dead-flat battery. It would not crank over.

Problem summary

The vehicle was dead on key even when it was jump-started. The battery was removed and put on charge.

Diagnostic sequence

Fitted a fully charged battery in order to test the vehicle systems. There was no action from either of the two keys.

Located the starter relay and bridged it out. The engine did crank over but it would not fire up.

There was no injection pulse, indicating the engine management system had shut down.

Removed the steering-column shrouds and checked the circuit to the ignition switch. All fuses and relays appeared OK.

The only fault code related to a circuit-relay malfunction.

There were no fault codes relating to an immobiliser function error.

Fault description

The warning lights on the top section of the dash were illuminated but the instrument cluster's gauges were not working, nor were the LCD displays for the odometer or gear-selection indicator (pic 1).

Removed the instrument-cluster surround. As the instrument cluster was unbolted and manipulated (pic 2), the dash lit up.

In this state the vehicle could also be started.

Unplugged the cluster and removed it (pic 3).

The cluster is the gateway that the immobiliser must go through; with the cluster out, the vehicle will not start.

Fault solution

Pulled the cluster apart (pic 4), then removed the circuit board off the back of the gauges set and housing (pic 5).

Close inspection revealed some of the pin to circuit-board connections had cold solder joints that had cracked, causing an open circuit (pic 6).

The connections were all treated, resoldered and the circuits were repaired.

The cluster was reassembled and refitted to the vehicle. The cluster and warning lights lit up as required (pic 7).

The vehicle now started and ran fine. Problem solved.

Recommended time

Diagnostic time was 3.5 hours, taking into account preparation and research.

Repair time was 1.5 hours, taking into account location of parts and carrying out the repair to a tested outcome.

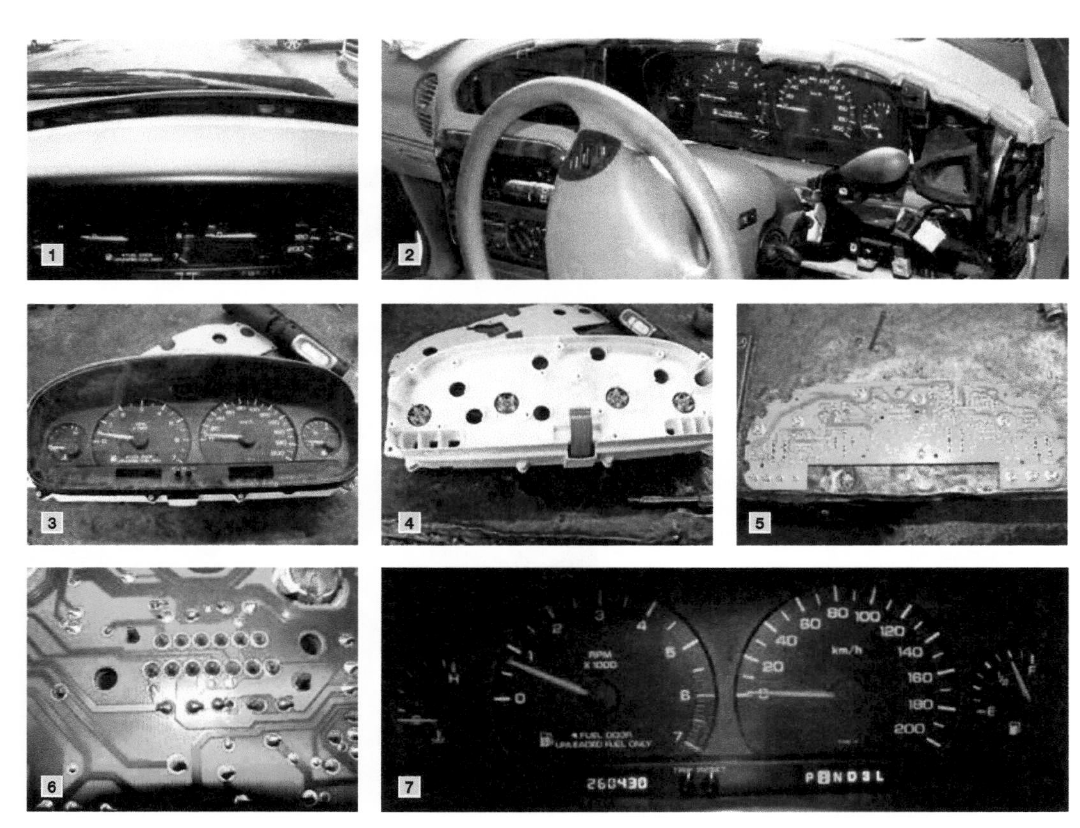

The Automotive Technician 25

Figure 10.66 Chrysler Voyager.

MERML07460
MERCEDES ML280
CDI W164 2007
Six-cylinder

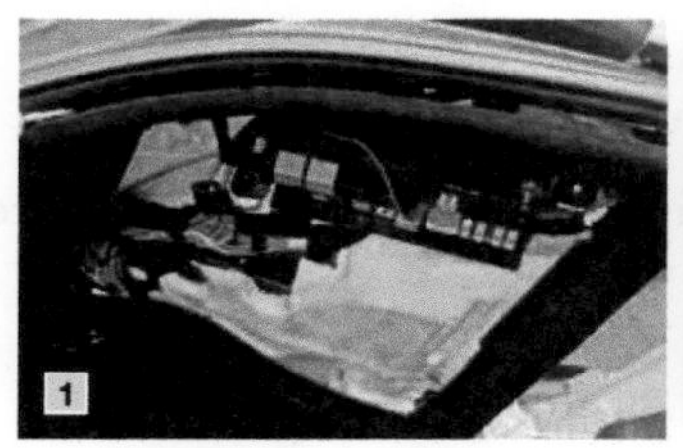

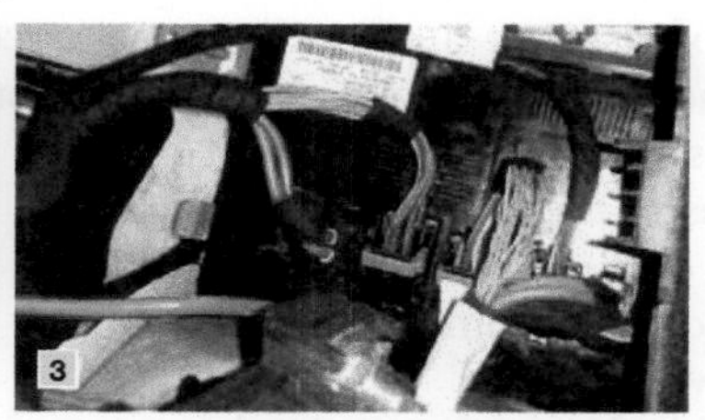

Customer complaint

The vehicle had been referred from another workshop with multiple issues.

Problem summary

The wipers, electric seats, fuel gauge, tailgate, indicators and key remote were not working.

Diagnostic sequence

Used an Autologic system to scan the vehicle. The printed report showed 48 fault codes from various systems and a large list of modules that would not communicate with the scan tool.

There were too many to write down but the most obvious involved the engine (ME-SFI 2.8 AMG), transmission (TCM164), electric power steering, Keyless Go (key remote) and rear signal acquisition and actuation module (SAM).

Fault description

Concentrated on the lack of communication from the rear SAM, as well as the the code: 9007 (B1007) – Airbag circuit hard fault [current and stored], undervoltage at terminal 15R.

Acquired a wiring diagram, then checked the main fuses in the battery area. They were OK.

Most of the issues related to the rear SAM controlling these circuits (pic 1).

Checked the relay and fuse. There was no power supplied to these circuits in the rear SAM (pic 2).

Fault solution

Removed the rear SAM (pic 3) and dismantled it. Traces of corrosion were found all over both sides of the main board (pic 4 to 8).

Removed as much corrosion as possible with a fine brush and electronic cleaner spray. Then treated the circuit board with protection spray before reassembling it and cancelling the codes.

All issues listed by the customer were now resolved. A code for a glow-plug issue was not cleared and the owner was advised.

Told the customer that a new SAM would cost more than $1500, possibly involve programming and we would not be able to warrant the job but that it was recommended if the problem returned.

The owner was also advised to have the cause of the water investigated because it was not obvious where it had come from.

Recommended time

Diagnostic time was five hours, taking into account preparation and research.

Repair time was three hours, taking into account location of parts and carrying out the repair to a tested outcome.

Tips for *TaT*

TaT thanks Wayne Broady from Barrie Automotive Service in Hornsby, NSW for this report and images.

To access the entire Repair Solutions database
www.tat.net.au/tats-a-fact

The Automotive Technician 23

Figure 10.67 Mercedes ML280.

10.3 WEBSITE

10.3.1 Introduction

The online Automotive Technology Academy has been created by the author (Tom Denton), who has over 40 years of relevant automotive experience, and over 30 published textbooks that are used by students and technicians worldwide.

The aims of the online academy are to:

- improve automotive technology **skills** and **knowledge;**
- provide **free access** to study resources to support the textbooks;
- create a worldwide **community** of automotive learners;
- freely **share** automotive related information and ideas;
- **reach** out to learners who are not able to attend school or college;
- improve automotive training **standards** and **quality**;
- provide online access to **certification** for a range of automotive subjects.

To access the academy visit: www.automotive-technology.org and create an account for yourself. To access the free courses that work in conjunction with the textbooks, you will need to enter an enrolment key. This will be described something like this:

'The third word on the last line of page ## of the associated textbook' You will therefore need to own the book!

All you need to do is enter the word in a box and you will have full unrestricted access to the course and its associated resources.

WEBSITE

www.automotive-technology.org

10.3.2 Resources

The following is a list of some of the resources that will be available to you:

- images;
- videos;
- activities;
- 3d models;
- hyperlinks;
- assignments;
- quizzes;
- forums;
- chat features;
- social media;
- interactive features, games and much more.

A progress bar is used in the course, so you can see at a glance how you are getting on, if you are working to cover all the content. Alternatively, you can just dip in and out to find what you need.

More formal assessments will also be available for those who are not able to attend traditional training centres. It will be possible to obtain certification relating to theory and practical work. A charge will apply to this aspect, but all other resources are free.

Updates and interesting new articles will also be available, so what are you waiting for? Come and visit and join in!

Glossary of abbreviations and acronyms

OBD2/SAE TERMINOLOGY

ABS	antilock brake system
A/C	air conditioning
AC	air cleaner
AIR	secondary air injection
A/T	automatic transmission or transaxle
B+	battery positive voltage
BARO	barometric pressure
CAC	charge air cooler
CFI	continuous fuel injection
CKP	crankshaft position sensor
CKP REF	crankshaft reference
CL	closed loop
CMP	camshaft position sensor
CMP REF	camshaft reference
CO	carbon monoxide
CO_2	carbon dioxide
CPP	clutch pedal position
CTOX	continuous trap oxidiser
CTP	closed throttle position
DEPS	digital engine position sensor
DFCO	decel fuel cut-off mode
DFI	direct fuel injection
DLC	data link connector
DPF	diesel particulate filter
DTC	diagnostic trouble code
DTM	diagnostic test mode
EBCM	electronic brake control module
EBTCM	electronic brake traction control module
EC	engine control
ECL	engine coolant level
ECM	engine control module
ECT	engine coolant temperature
EEPROM	electrically erasable programmable read only memory
EFE	early fuel evaporation
EGR	exhaust gas recirculation
EGRT	EGR temperature
EI	electronic ignition
EM	engine modification
EPROM	erasable programmable read only memory
ESC	electronic stability control
EVAP	evaporative emission system
FC	fan control
FEEPROM	flash electrically erasable programmable read only memory
FF	flexible fuel
FP	fuel pump
FPROM	flash erasable programmable read only memory
FT	fuel trim
FTP	federal test procedure
GCM	governor control module
GEN	generator
GND	ground
HC	hydrocarbon
H_2O	water
HO2S	heated oxygen sensor
HO2S1	upstream heated oxygen sensor
HO2S2	up or downstream heated oxygen sensor
HO2S3	downstream heated oxygen sensor
HVAC	heating, ventilation and air conditioning system
HVS	high-voltage switch
IA	intake air
IAC	idle air control
IAT	intake air temperature
IC	ignition control circuit
ICM	ignition control module
IFI	indirect fuel injection
IFS	inertia fuel shutoff
I/M	inspection/maintenance
IPC	instrument panel cluster
ISC	idle speed control
KOEC	key on, engine cranking
KOEO	key on, engine off
KOER	key on, engine running
KS	knock sensor
KSM	knock sensor module
LTFT	long-term fuel trim
MAF	mass airflow sensor
MAP	manifold absolute pressure sensor
MC	mixture control
MDP	manifold differential pressure

MFI	multiport fuel injection
MIL	malfunction indicator lamp
MPH	miles per hour
MST	manifold surface temperature
MVZ	manifold vacuum zone
NOx	oxides of nitrogen
NVRAM	non-volatile random access memory
OBD	on-board diagnostics
OBD I	on-board diagnostics generation one
OBD II	on-board diagnostics, second generation
OC	oxidation catalyst
ODM	output device monitor
OL	open loop
O2S	oxygen sensor
OSC	oxygen sensor storage
PAIR	pulsed secondary air injection
PCM	powertrain control module
PCV	positive crankcase ventilation
PNP	park/neutral switch
PROM	program read only memory
PSA	pressure switch assembly
PSP	power steering pressure
PTOX	periodic trap oxidiser
RAM	random access memory
RM	relay module
ROM	read only memory
RPM	revolutions per minute
SAP	accelerator pedal
SC	supercharger
SCB	supercharger bypass
SDM	sensing diagnostic mode
SFI	sequential fuel injection
SRI	service reminder indicator
SRT	system readiness test
STFT	short-term fuel trim
TB	throttle body
TBI	throttle body injection
TC	turbocharger
TCC	torque converter clutch
TCM	transmission or transaxle control module
TFP	throttle fluid pressure
TP	throttle position
TPS	throttle position sensor
TVV	thermal vacuum valve
TWC	three-way catalyst
TWC+OC	three-way+oxidation catalytic converter
VAF	volume airflow
VCM	vehicle control module
VR	voltage regulator
VS	vehicle sensor
VSS	vehicle speed sensor
WOT	wide open throttle
WU-TWC	warm up three-way catalytic converter

OEM AND OTHER TERMINOLOGY

A	amps
AAV	anti-afterburn valve (Mazda)
ABS	antilock brake system
ABSV	air bypass solenoid valve (Mazda)
A/C	air conditioning
AC	alternating current
ACTS	air charge temperature sensor (Ford)
AERA	Automotive Engine Rebuilders Association
A/F	air/fuel ratio
AFM	airflow meter
AFR	air/fuel ratio
AFS	airflow sensor (Mitsubishi)
AIR	Air Injection Reaction (GM)
AIS	Air Injection System (Chrysler)
AIS	automatic idle speed motor (Chrysler)
ALCL	assembly line communications link (GM)
ALDL	assembly line data link (GM)
API	American Petroleum Institute
APS	absolute pressure sensor (GM)
APS	atmospheric pressure sensor (Mazda)
ASD	automatic shutdown relay (Chrysler)
ASDM	airbag system diagnostic module (Chrysler)
ASE	Automotive Service Excellence
A/T	automatic transmission
ATC	after top centre
ATDC	after top dead centre
ATF	automatic transmission fluid
ATMC	Automotive Training Managers Council
ATS	air temperature sensor (Chrysler)
AWD	all-wheel drive
BARO	barometric pressure sensor (GM)
BAT	battery
BCM	body control module (GM)
BHP	brake horsepower
BID	Breakerless Inductive Discharge (AMC)
BMAP	barometric/manifold absolute pressure sensor (Ford)
BP	backpressure sensor (Ford)
BPS	barometric pressure sensor (Ford & Nissan)
BPT	backpressure transducer
BTC	before top centre
BTDC	before top dead centre
Btu	British thermal units
C	Celsius
C3	Computer Command Control system (GM)
C3I	Computer Controlled Coil Ignition (GM)
C4	Computer Controlled Catalytic Converter system (GM)

CAAT Council of Advanced Automotive Trainers
CAFE corporate average fuel economy
CALPAK calibration pack
CANP canister purge solenoid valve (Ford)
CARB California Air Resources Board
CAS Clean Air System (Chrysler)
CAS crank angle sensor
CC catalytic converter
CC cubic centimetres
CCC Computer Command Control system (GM)
CCD computer controlled dwell (Ford)
CCIE Coolant Controlled Idle Enrichment (Chrysler)
CCEV Coolant Controlled Engine Vacuum Switch (Chrysler)
CCOT clutch cycling orifice tube
CCP controlled canister purge (GM)
CCV canister control valve
CDI Capacitor Discharge Ignition (AMC)
CEAB cold engine air bleed
CEC Crankcase Emission Control System (Honda)
CECU central electronic control unit (Nissan)
CER cold enrichment rod (Ford)
CESS cold engine sensor switch
CFC chlorofluorocarbon
CFI Cross Fire Injection (Chevrolet)
cfm cubic feet per minute
CID cubic inch displacement
CID cylinder identification sensor (Ford)
CIS Continuous Injection System (Bosch)
CMP camshaft position sensor (GM)
COP coil on plug ignition
CP canister purge (GM)
CP crankshaft position sensor (Ford)
CPI Central Port Injection (GM)
CPU central processing unit
CSC Coolant Spark Control (Ford)
CSSA Cold Start Spark Advance (Ford)
CSSH Cold Start Spark Hold (Ford)
CTAV Cold Temperature Actuated Vacuum (Ford)
CTO Coolant Temperature Override Switch (AMC)
CTS charge temperature switch (Chrysler)
CTS coolant temperature sensor (GM)
CTVS choke thermal vacuum switch
CVCC Compound Vortex Controlled Combustion system (Honda)
CVR control vacuum regulator (Ford)
dB decibels
DC direct current
DEFI Digital Electronic Fuel Injection (Cadillac)
DERM diagnostic energy reserve module (GM)
DFS deceleration fuel shutoff (Ford)
DIS Direct Ignition System (GM)
DIS Distributorless Ignition System (Ford)
DLC data link connector (GM)
DOHC dual overhead cams
DOT Department of Transportation
DPF diesel particulate filter
DRBII Diagnostic Readout Box (Chrysler)
DRCV distributor retard control valve
DSSA Dual Signal Spark Advance (Ford)
DVDSV differential vacuum delay and separator valve
DVDV distributor vacuum delay valve
DVOM digital volt ohm meter
EACV electronic air control valve (Honda)
EBCM electronic brake control module (GM)
EBM electronic body module (GM)
ECA electronic control assembly
ECCS Electronic Concentrated Control System (Nissan)
ECM electronic control module (GM)
ECS Evaporation Control System (Chrysler)
ECT engine coolant temperature (Ford & GM)
ECU electronic control unit (Ford, Honda & Toyota)
EDIS Electronic Distributorless Ignition System (Ford)
EEC Electronic Engine Control (Ford)
EEC Evaporative Emission Controls (Ford)
EECS Evaporative Emissions Control system (GM)
EEPROM electronically erasable programmable read only memory
EFC electronic feedback carburettor (Chrysler)
EFC electronic fuel control
EFCA electronic fuel control assembly (Ford)
EFE Early Fuel Evaporation system (GM)
EFI electronic fuel injection
EGO exhaust gas oxygen sensor (Ford)
EGRPS EGR valve position sensor (Mazda)
EGR-SV EGR solenoid valve (Mazda)
EGRTV EGR thermo valve (Chrysler)
EI electronic ignition (GM)
ELB Electronic Lean Burn (Chrysler)
EMI electromagnetic interference
EOS exhaust oxygen sensor
EPA Environmental Protection Agency
EPOS EGR valve position sensor (Ford)
EPROM erasable programmable read only memory
ESA Electronic Spark Advance (Chrysler)
ESC Electronic Spark Control (GM)
ESS Electronic Spark Selection (Cadillac)

EST Electronic Spark Timing (GM)
EVP EGR valve position sensor (Ford)
EVRV electronic vacuum regulator valve for EGR (GM)
F Fahrenheit
FBC feedback carburettor system (Ford & Mitsubishi)
FBCA feedback carburettor actuator (Ford)
FCA fuel control assembly (Chrysler)
FCS fuel control solenoid (Ford)
FDC fuel deceleration valve (Ford)
FI fuel injection
FLS fluid level sensor (GM)
FMVSS Federal Motor Vehicle Safety Standards
ft. lb. foot pound
FUBAR Fracked Up Beyond All Repair
FWD front-wheel drive
gal gallon
GND ground
GPM grams per mile
HAIS Heated Air Intake System (Chrysler)
HEGO heated exhaust gas oxygen sensor
HEI High Energy Ignition (GM)
Hg mercury
hp horsepower
IAC idle air control (GM)
IAT inlet air temperature sensor (Ford)
IATS intake air temperature sensor (Mazda)
IC integrated circuit
ICS idle control solenoid (GM)
ID inside diameter
IGN ignition
IIIBDFI If it isn't broke don't fix it
IM240 inspection/maintenance 240 program
IMI Institute of the Motor Industry
I/P instrument panel
ISC idle speed control (GM)
ISO International Standards Organisation
ITCS Ignition Timing Control System (Honda)
ITS idle tracking switch (Ford)
JAS Jet Air System (Mitsubishi)
kHz kilohertz
KISS Keep It Simple Stupid!
km kilometres
kPa kilopascals
KS knock sensor
kV kilovolts
L litres
lb. ft. pound feet
LCD liquid crystal display
LED light-emitting diode
MACS Mobile Air Conditioning Society
MAF mass airflow sensor
MAMA Midwest Automotive Media Association
MAP manifold absolute pressure
MAP Motorist Assurance Program
MAT manifold air temperature
MCS mixture control solenoid (GM)
MCT manifold charge temperature (Ford)
MCU Microprocessor Controlled Unit (Ford)
MFI multiport fuel injection
MIL malfunction indicator lamp
MISAR Microprocessed Sensing and Automatic Regulation (GM)
mm millimetres
MPFI multi point fuel injection
MPG miles per gallon
MPH miles per hour
MPI multi-port injection
ms millisecond
MSDS material safety data sheet
mV millivolts
NACAT National Association of College Automotive Teachers
NATEF National Automotive Technician's Education Foundation
NHTSA National Highway Traffic Safety Administration
Nm Newton metres
OBD on-board diagnostics
OC oxidation converter (GM)
OD outside diameter
OE original equipment
OEM original equipment manufacture
OHC overhead cam
ORC oxidation reduction catalyst (GM)
OS oxygen sensor
OSAC Orifice Spark Advance Control (Chrysler)
P/B power brakes
P/N part number
PA pressure air (Honda)
PAFS Pulse Air Feeder System (Chrysler)
PAIR Pulsed Secondary Air Injection system (GM)
PCM powertrain control module (supersedes ECM)
PECV power enrichment control valve
PERA Production Engine Rebuilders Association
PFI port fuel injection (GM)
PGM-FI Programmed Gas Management Fuel Injection (Honda)
PIP profile ignition pickup (Ford)
PPM parts per million
PROM program read only memory computer chip
PS power steering
PSI pounds per square inch
pt. pint
PVA ported vacuum advance

PVS	ported vacuum switch
QS9000	Quality assurance standard for OEM part suppliers
Qt.	quart
RABS	Rear wheel Antilock Brake System (Ford)
RFI	radio frequency interference
RPM	revolutions per minute
RPO	regular production option
RWAL	Rear Wheel Antilock brake system (GM)
RWD	rear-wheel drive
SAE	Society of Automotive Engineers
SAVM	spark advance vacuum modulator
SCC	Spark Control Computer (Chrysler)
SDI	Saab Direct Ignition
SES	service engine soon indicator (GM)
SFI	Sequential Fuel Injection (GM)
SIR	Supplemental Inflatable Restraint (airbag)
SMPI	Sequential Multiport Fuel Injection (Chrysler)
SOHC	single overhead cam
SPOUT	Spark Output signal (Ford)
SRDV	spark retard delay valve
SRS	Supplemental Restraint System (airbag)
SS	speed sensor (Honda)
SSI	Solid State Ignition (Ford)
STS	Service Technicians Society
TA	temperature air (Honda)
TABPV	throttle air bypass valve (Ford)
TAC	thermostatic air cleaner (GM)
TACH	tachometer
TAD	Thermactor air diverter valve (Ford)
TAV	temperature actuated vacuum
TBI	throttle body injection
TCC	torque converter clutch (GM)
TCCS	Toyota Computer Controlled System
TCS	Transmission Controlled Spark (GM)
TDC	top dead centre
TIC	thermal ignition control (Chrysler)
TIV	Thermactor idle vacuum valve (Ford)
TKS	throttle kicker solenoid (Ford)
TP	throttle position sensor (Ford)
TPI	Tuned Port Injection (Chevrolet)
TPMS	Tyre Pressure Monitor System
TPP	throttle position potentiometer
TPS	throttle position sensor
TPT	throttle position transducer (Chrysler)
TRS	Transmission Regulated Spark (Ford)
TSP	throttle solenoid positioner (Ford)
TV	throttle valve
TVS	thermal vacuum switch
TVV	thermal vacuum valve (GM)
V	volts
VAC	volts alternating current
VAF	vane airflow sensor
VCC	viscous converter clutch (GM)
VDC	volts direct current
VDV	vacuum delay valve
VIN	vehicle identification number
VSM	vehicle security module
VSS	vehicle speed sensor
WOT	wide open throttle
WSS	wheel speed sensor

Index

Note: Page numbers in *italics* indicate figures and **bold** indicate tables in the text.